Resources for the Future Library Collection
Energy Policy

Volume 9

Unitization of Oil and Gas Fields in Texas

A Study of Legislative, Administrative, and Judicial Policies

Full list of titles in the set
Energy Policy

Volume 1: Analyzing Demand Behavior
Volume 2: Limiting Oil Imports
Volume 3: Discounting for Time and Risk in Energy Policy
Volume 4: Economic Aspects of Oil Conservation Regulation
Volume 5: Petroleum Conservation in the United States
Volume 6: The Leasing of Federal Lands for Fossil Fuels Production
Volume 7: Energy, Economic Growth, and the Environment
Volume 8: Energy in America's Future
Volume 9: Unitization of Oil and Gas Fields in Texas

Unitization of Oil and Gas Fields in Texas

A Study of Legislative, Administrative, and Judicial Policies

Jacqueline Lang Weaver

Washington, DC • London

First published in 1986 by Resources for the Future

This edition first published in 2011 by RFF Press, an imprint of Earthscan

Earthscan LLC, 1616 P Street, NW, Washington, DC 20036, USA
Earthscan Ltd, Dunstan House, 14a St Cross Street, London EC1N 8XA, UK
Earthscan publishes in association with the International Institute for Environment and Development

For more information on RFF Press and Earthscan publications, see www. rffpress.org and www.earthscan.co.uk or write to earthinfo@earthscan.co.uk

ISBN: 978-1-61726-024-7 (Volume 9)
ISBN: 978-1-61726-002-5 (Energy Policy set)
ISBN: 978-1-61726-000-1 (Resources for the Future Library Collection)

A catalogue record for this book is available from the British Library

Publisher's note

The publisher has made every effort to ensure the quality of this reprint, but points out that some imperfections in the original copies may be apparent.

At Earthscan we strive to minimize our environmental impacts and carbon footprint through reducing waste, recycling and offsetting our CO_2 emissions, including those created through publication of this book. For more details of our environmental policy, see www.earthscan.co.uk.

Unitization of Oil and Gas Fields in Texas

Unitization of Oil and Gas Fields in Texas

A Study of Legislative, Administrative, and Judicial Policies

Jacqueline Lang Weaver

PUBLISHED BY RESOURCES FOR THE FUTURE, INC.
WASHINGTON, D.C.

Published by Resources for the Future, Inc.
1616 P Street, N.W., Washington, D.C. 20036
Resources for the Future books are distributed worldwide by
The Johns Hopkins University Press.

Library of Congress Cataloging-in-Publication Data
Weaver, Jacqueline.
Unitization of oil and gas fields in Texas.

Includes index.
1. Unit operation of oil fields—Texas.
I. Title.
KFT1458.A1W43 1985 343.76404'682313 85-24475
ISBN 0-915707-20-9 347.64034682313

1616 P Street, N.W., Washington, D.C. 20036

Resources for the Future is a nonprofit organization for research and education in the development, conservation, and use of natural resources, including the quality of the environment. It was established in 1952 with the cooperation of the Ford Foundation. Grants for research are accepted from government and private sources only on the condition that RFF shall be solely responsible for the conduct of the research and free to make its results available to the public. Most of the work of Resources for the Future is carried out by its resident staff; part is supported by grants to universities and other nonprofit organizations. Unless otherwise stated, interpretations and conclusions in RFF publications are those of the authors; the organization takes responsibility for the selection of significant subjects for study, the competence of the researchers, and their freedom of inquiry.

This book is the product of RFF's Small Grants Program. Jacqueline Lang Weaver is Associate Professor of Law at the University of Houston Law Center, specializing in oil and gas law, environmental law, water law, and energy policy.

This book was edited by Jo Hinkel and designed by Elsa Williams. The index was prepared by Florence Robinson.

Contents

Foreword xi

Preface xiii

1 Introduction and Overview 1

A Note on the Frame of Reference 5

2 The Setting for Unitization 9

The Engineering Background of Unitization 9
The Benefits of and Need for Unitization 20
The Problems of Voluntary Unitization 29
The Need for Unitized Operations in Texas 33
Summary 34

3 The Texas Legislature and Unitization: The Early Years 37

The 1931 Anti-market Demand Prorationing Act 38
The 1932 Market Demand Prorationing Act 60
The 1932–33 Prorationing Orders and the Marginal Well Act 63
The 1935 Gas Conservation Act 68
Summary 74

4 The Texas Legislature and Unitization: The Later Years 77

Passage of the 1949 Voluntary Unitization Act 77
Analysis of the 1949 Voluntary Unitization Act 82
The Failure of Compulsory Unitization Legislation in Texas 97
Compulsory Pooling in Texas 124
Summary of Texas Legislation Affecting Unitization 131

5 The Railroad Commission and Unitization: Field Rules and Statewide Orders 137

Field Rules Affecting Unitization 138
Statewide Orders and Unitization 166

6 The 1949 Voluntary Unitization Statute and the Railroad Commission 173

Requiring Railroad Commission Approval 174
The Railroad Commission's Definition of Secondary Recovery Operations 175
The Required Findings 177
Correlative Rights and the Required Findings 183
The Effect of Railroad Commission Approval 194
Summary of the Commission's Administration of the 1949 Voluntary Unitization Act 195
Summary of the Effect of the Railroad Commission's Decision Making on Unitization 196

7 Unitization and the Courts: The Common Law 201

No Equitable Pooling in Texas 202
The Pooling of Nonexecutive Interests 206
The Cross-Conveyancing Theory 207
Cotenancy 209
Judicial Interpretation of Pooling and Unitization Clauses in Leases 215
Implied Covenants 219
Tort Liability of the Unit 234
Liability to the Surface Estate 248
Summary of the Effect of the Common Law on the Progress of Pooling and Unitization in Texas 253

8 Judicial Review of Railroad Commission Orders 261

Judicial Review of the Commission's Statutory Authority to Prevent Waste 262
Judicial Review of the Commission's Statutory Authority to Protect Correlative Rights 272
Conservation Versus Correlative Rights 282
The Commission's Unused Authority 283
Summary of Judicial Review of the Commission's Statutory Authority 286
Judicial Review of the Reasonableness of the Railroad Commission Orders 288
Summary of Judicial Review of the Reasonableness of Commission Orders 301
The Antitrust Laws 302
Summary of the Role of the Courts 310

9 Statistical Assessment of the Degree of Unitization Achieved in Texas 315

The Extent of Unitization 315
Characteristics of the Unitized Projects 317

10 Conclusions and Proposals for Reform 323

Does Texas Need a Compulsory Unitization Law? 323
Proposed Reforms in the Absence of Compulsory Unitization Legislation 349
Epilogue 357

Appendix I Texas Natural Resources Code, Section 85.046 (Vernon Supp. 1982) 359

Appendix II Texas Natural Resources Code, Sections 101.001–.018 361

Appendix III Rule 37 Applications, 1940–81 367

Appendix IV Number of Unitization Agreements Approved by the Railroad Commission 369

Notes by Chapter 371

Table of Cases 533

Index 541

Foreword

Between 1980 and 1984 I served as director of Resources for the Future's Small Grants Program. Through that program, RFF is able to fund each year a handful of researchers in universities or other nonprofit organizations around the world doing work related to energy, natural resources, or the environment. I came quickly to realize that such a position carried with it many frustrations. These included the considerable time required to read and farm out for independent reviews the many hundreds of proposals we received, requesting several orders of magnitude more money than we could hope to give out; the unpleasant task of notifying many talented authors of interesting and thoughtful proposals that we could not fund them; and occasionally, being disappointed by those we were able to fund.

Such a job was not without its rewards, however. First, it provided a rare opportunity to learn by reading proposals from researchers in all branches of the social and natural sciences. Next, it enabled me to pass on good news once a year to a small number of talented and hopeful scholars. And finally, it occasionally resulted in a project which delivered *more* than was promised—a real rarity in the grant-making world, I have come to realize.

This book is such a gem. Jacqueline Weaver was awarded a small grant to prepare a law review article on an important though obscure subject—the unitization of oil and natural gas fields in Texas.

But, through a process she describes, she produced this book, one which in addition to its original intent also sheds light on the complex relationships between a state legislature, a powerful regulatory agency, and the court system. It will fascinate anyone interested in the complexities of public policymaking, including lawyers, economists, political scientists, and the lay reader as well. I can speak for Resources for the Future when I say that all our small grants should turn out this well!

October 1985

Paul R. Portney
Resources for the Future

Preface

In 1981 I received a research grant from Resources for the Future under their Small Grants Program to write a law review article on the attitude of the Texas courts toward unitization issues. The article was to assess whether the courts had attempted to fill the void in Texas' oil and gas statutory scheme, resulting from the lack of a compulsory unitization statute, by creating common law doctrines which directly or indirectly coerced Texas oil and gas producers into unitizing. My proposal to Resources for the Future had been couched in terms of traditional legal research: I would read and analyze all of the cases decided by Texas courts that involved litigation over the unitization or pooling of oil and gas fields in the state.

When I applied to RFF for a grant, I was the faculty advisor to the University of Houston Law Review's annual symposium volume on oil, gas, and energy topics. I suggested to the Review that they ask renowned practitioners and law professors in Texas to write on the same topic that had been proposed to RFF. Not one of the authors solicited wanted to do so. A well-known law professor in Texas declined to write on the issue of compulsory unitization in the state because, in his words, "There was nothing to write about."

Thus, when I received notification of my grant award, I was not elated. I faced the task of doing the appointed research with grave doubts and even despair about its worth. It seemed that I was being paid to do something which was insignificant and unneeded in the world of legal scholarship. To earn my grant and produce an article, I would have to spend my summer making mountains out of molehills.

This book is the law review article—the molehill—which consumed my research interest for the next three years. My enthusiasm for this lengthy project was sustained largely because the research ultimately led me to many sources of information not commonly used in legal scholarship: questionnaires to oil companies about the unitization process; interviews with the staff of the Texas Railroad Commission and with former commissioners; books by economists and political scientists; empirical data from obscure industry and government sources; and real-world examples of unitization from sources such as *Drill Bit* magazine and American Petroleum Institute studies of actual secondary recovery projects. In my determination to find something worth writing about, I discovered hundreds of pages of transcripts covering the legislative debate that occurred in Texas in 1931 and 1932 over the enactment of market-demand prorationing and unitization statutes. These transcripts, seemingly forgotten as primary source material by scholars of all disciplines, were enormously enlightening on the issue of why Texas is the only oil- and gas-producing state that has failed to pass a compulsory unitization statute. By the time the first year of my research had ended, I had written more than two hundred pages of legislative history and analysis explaining the politics of Texas lawmakers' anti-unitization stance. That research now forms chapters 3 and 4 of this book.

The second year of my research involved discovering and documenting the hidden law of unitization in Texas—the Railroad Commission's use of its administrative power to coerce and encourage oil and gas producers to unitize. A few court opinions from the late 1940s hinted at some of the commission's arm-twisting techniques, but case analysis gave a very sketchy picture of exactly how the commission operated behind the scenes. Nontraditional sources served to solve the mystery of how unitization was occurring in Texas without legislative support—indeed, with sustained legislative antipathy. Chapters 5 and 6 cover this story. Finally, in the third year, I turned to my original goal of analyzing the role of the courts in inhibiting or encouraging unitization through judicial decision making. This traditional form of legal scholarship constitutes chapters 7 and 8 of this book.

By the time I had completed the second year's research, I had gained an insatiable curiosity about the extent to which Texas oil and gas fields are actually unitized. My burgeoning collection of reference sources provided raw data that could be manipulated to measure the effect of the three sources of law—legislative acts, administrative orders, and judicial decisions—on the actual progress of unitization in Texas. Thus was born chapter 9, an empirical study of the extent and type of unitization occurring in Texas. Finally, chapter 10 became more than a simple conclusion. The "article" had become a full-fledged study of an administrative agency, the Railroad Commission, working with its partner, the Texas Supreme Court, to supplant and override the failure of the legislative branch to enact statutes that would facilitate unitization and enable Texas to maximize the efficient recovery of its oil and gas resources. Chapter 10 thus became an analysis of administrative law principles and the proper functioning of our democratic system of government by checks and balances.

During the long years of research and writing this book, I accumulated many debts of gratitude for help and support. The Resources for the Future grant forced me to pursue a project from which I otherwise might have been dissuaded. The Energy Lab of the University of Houston provided funding for secretarial assistance, and this enabled me to hire Joni Morgan who performed both word processing and statistical tasks with equal skill. My colleague, Professor Irene M. Rosenberg, who specializes in constitutional and juvenile law, made her way through the entire manuscript without any prior background in oil and gas law—a Herculean task performed in friendship to make sure no "fatal flaws" appeared in the analysis. Professor Stephen F. Williams at the University of Colorado Law School also read early drafts of the manuscript and sent thought-provoking comments. I was most fortunate to have Sandra McFarland, a first-year law student, as a research assistant in the final stages of preparing the manuscript for publication.

Of greatest importance to my completion of this work was the support and encouragement of my husband Kirk. It was he who cared for our two sons and kept the household running almost every weekend for three years so that I could scale the mountain of research that I had discovered. This book is dedicated to Kirk and to our two sons, Kyle and Kenyon. They climbed with me all the way.

Houston, Texas
July 1985

Jacqueline Lang Weaver
University of Houston Law Center

1

Introduction and Overview

> If the public some day in the near future awakens to the fact that we have become a bankrupt nation as far as oil is concerned, and that it is then too late to protect our supply by conservation measures, I am sure they will blame both the men of the oil industry and the men who held public offices at the time conservation measures should have been adopted.[1]
>
> —*Letter From Henry L. Doherty to President Coolidge, August 11, 1924*

In the decade of the 1970s, our nation's bankruptcy in oil and gas was announced to the public through widespread energy shortages, tremendous price increases for energy products, blackouts, factory and school closings, lengthy lines of motorists at gasoline pumps, and instances of violence and death.[2] Predictably, the roused public blamed the oil industry for this state of bankruptcy, and elected officials were called to account for the shortages in our national oil and gas deposits. The political response was twofold: an outpouring of legislation, both state and federal, to mitigate the effects of the oil and gas deficits and to prevent their recurrence; and massive antitrust litigation and legislative proposals designed to break up the oil industry giants.[3]

Yet, while this multitudinous legislation and litigation fulfilled Henry Doherty's prophecy, his policy message went unheeded. The federal government did not pass a statute compelling the unitization of oil and gas fields in the United States. It was this conservation measure that Doherty propounded in his letter to President Coolidge in 1924.[4] *Unitization* is the joint, coordinated operation of all or large parts of an oil or gas reservoir by the owners of the separate tracts overlying the reservoir. Only through such cooperative agreements and efforts by the many owners of interests in an oil and gas field can efficient, low-cost operations be realized and the optimal recovery of oil and gas achieved.

Because it is often difficult—if not impossible—to obtain unanimous approval of unitization through voluntary agreements among all the producers and landowners in a field, many states passed compulsory unitization statutes during the half-century after Doherty's call for universal legislation in this area. These acts allow the regulatory agency administering the state's oil- and gas-conservation laws to compel an unwilling minority of owners to join a unitization plan—already endorsed by the majority of owners—which the agency finds is necessary to increase the recovery of oil and gas. The energy crisis of the 1970s spurred even "diehard" states to pass such legislation.[5] Yet Texas, the largest of all oil- and gas-producing states, still lacks such a statute.[6] Despite the energy crisis, Texas oilmen and public officials do not seem to be held accountable for deficiencies in oil and gas production resulting from wasteful and inefficient methods of extraction.

This book seeks to assess the need for a compulsory unitization statute in Texas. In so doing, it traces the economic, legal, and political forces that explain the state's lack of a compulsory unitization law, and it focuses on the role that legal institutions have played in either promoting or retarding the actual unitization of oil and gas fields in Texas. Perhaps the failure of state and federal lawmakers to pass a compulsory unitization statute during the past decade—in the face of what would appear to be an overwhelming need for increased oil and gas recovery—tolls the final requiem for such a law in Texas. However, three factors underscore the importance of this topic and justify our taking another look at the issue before its interment. First, the United States is still an oil and gas debtor in that it is a net importer of oil; and although no longer threatened by immediate bankruptcy, the nation is by no means liquid in its energy accounts.[7] Domestic oil and gas are valuable, strategically critical commodities, and it is of more than passing local interest to inquire whether Texas is maximizing its current and potential oil and gas production. The federal government is clearly interested in promoting domestic oil and gas production, both to decrease dependence on insecure foreign sources and to increase supplies and availability, thereby reducing any inflationary pressure from rising oil and gas prices. If Texas cannot ensure the maximum efficient recovery of its oil and gas resources to the nation, the federal government could, and probably should, pass legislation forcing Texas to adopt unitization.

Second, the tenfold increase in oil prices in the last decade has made secondary and tertiary recovery of oil more profitable.[8] These recovery technologies often require cooperative action by all the lessees in an oil field in order to operate successfully and efficiently. While increased profitability surely encourages more voluntary unitization by self-interested lessees, voluntary processes alone may not be adequate to achieve maximum efficiency in the development of such fields. The fields most easily unitized—that is, those having the greatest potential increases in oil from secondary recovery or the fewest legal problems in implementing voluntary action—will be unitized first, leaving a legacy of lost oil in some fields and of unrecovered, but still recoverable, oil in other fields.

Third, the Texas Supreme Court recently has stated that lessees may have a duty to unitize oil and gas fields in Texas.[9] While the statement is dictum, it is deliberate and should give oil and gas producers pause to consider whether their leasehold duties might require more active attention to unitization.

In light of these three factors, it seems important to reassess the need for compulsory unitization legislation in Texas. The question is not whether unitization of oil and gas fields is a good thing; it is universally acknowledged as the best method of producing oil and gas efficiently.[10] The issue is whether compulsory process is required to attain the fullest implementation of unitized production practices. The failure of Texas to enact a compulsory unitization statute may reflect the state's unique ability to achieve unitization voluntarily. Texas does have a statute governing voluntary unitization agreements, and it may be that this singular act encourages a remarkable degree of cooperation among lessees. Texas also has many statutes prohibiting waste in oil and gas production. These statutes are enforced through rules and orders of the Texas Railroad Commission, the state agency with primary jurisdiction over oil and gas conservation in Texas.[11] If the Texas Railroad Commission administers its conservation rules and regulations in such a manner as to promote unitization, and if the courts construe the applicable statutes, regulations, and common law on unitization to similar effect, the lack of compulsory process may not retard the efficient recovery of oil and gas in Texas. On its face, this would seem a remarkable finding to be supported by the evidence. No other significant oil- and gas-producing state has achieved such a high degree of voluntary behavior. Yet, perhaps it would be no less remarkable a finding than the fact that Texas has survived a decade of disastrous oil and gas shortages, as well as frequent federal intervention into the oil and gas industry, without being compelled to pass a unitization statute that would demonstrably increase the ultimate recovery of precious domestic oil and gas resources.

This book amasses and analyzes the evidence required to ascertain which of these findings, remarkable as each is, best mirrors the current state of affairs. Chapter 2 presents the scientific and engineering background of oil and gas production; the benefits of unitized operations in terms of conservation, protection of correlative rights, and economic efficiency; the difficulties of achieving voluntary unitization; and the potential significance of unitized projects to oil and gas production, particularly in Texas. Chapters 3 through 8 examine the three sources of legal authority that define the ease of securing unitization in Texas: (1) the relevant statutes enacted by the legislative branch governing unitization and the conservation of oil and gas; (2) the administrative rule making and orders of the Railroad Commission promulgated under this statutory authority; and (3) the courts' review of the legality of these statutes and the commission's decision making thereunder, and the courts' own common law approach to unitization issues. Chapter 9 presents a statistical analysis of the amount and type of unitization that nas been achieved in Texas to date. Here, we will see that the pattern of unitization in Texas is directly traceable to the legal environment that has been described and analyzed in the earlier chapters.

In light of this statistical evidence, the final chapter of the book summarizes the respective roles of the legislative, administrative, and judicial authorities in Texas in achieving unitization without the aid of a compulsory unitization statute, and then assesses the need for such a statute. The book concludes that Texas has achieved a fairly high degree of unitization in its oil and gas fields despite legislation that is most often inimical to the very concept of unitization. The Railroad Commission and the judiciary have actively, and often successfully, thwarted the anti-unitization stance of the legislative branch. Ironically, in so doing, the commission and the courts have reduced the need for a compulsory unitization statute and have thus increased the political strength of those who have triumphantly opposed compulsory unitization legislation in Texas for more than fifty years. There is little doubt that unitization of many fields in Texas would have been achieved more quickly and at lower costs had compulsory process been available. The commission and the courts—no matter how dedicated to the pursuit of efficient recovery of oil and gas—are fettered by statutes and case law that necessarily restrict the fullest implementation of unitization and limit our ability to produce this nation's heritage of valuable oil and gas in Texas.

But in this realm, the good is the enemy of the best. The fact that unitization in Texas could have been, and could still be, achieved more easily with compulsory process is in many respects less important than the fact that the partnership between the Railroad Commission and the courts often has succeeded in promoting unitized operations and in preventing the large-scale waste of oil and gas that otherwise would have occurred. Passage of a strong compulsory unitization law would enable Texas to move from a good framework of conservation law and practices to an excellent one, thereby achieving an additional recovery of oil and gas measured by the margin between the good and the best. Passage of such a law is recommended. However, a half-century of history teaches that enactment of a compulsory unitization statute is politically unrealistic in Texas. Thus, the book concludes by advancing proposals for legislative reform that recognize and strengthen the unique methods by which Texas has achieved unitization without compulsory process. Several proposals to facilitate voluntary unitization in Texas are discussed in the final section. While the good is second to the best, it is nonetheless a monumental achievement by a state administrative agency and by the Texas judiciary to have secured as much unitization as now exists without legislative support—indeed, with sustained legislative antipathy. This achievement is uniquely Texan and, in this regard, it is a first.

A Note on the Frame of Reference

This book focuses on the legal institutions which have affected oil and gas conservation in Texas by their positive or negative impact on the unitization of oil and gas fields in the state. In so doing, the book often uses the economist's standard of economic efficiency as an analytical framework for evaluating the success of the conservation results achieved by these legal institutions. However, it is not written only—or even primarily—from the economist's perspective but is written from the perspective of a scholar of legal institutions. Thus, the research seeks to explain why economic efficiency has not always been the goal of the conservation policies actually enacted and implemented by Texas legislators, administrators, and judges.

Economic efficiency is only one of many normative goals that constitute the public interest. When legislators or judges choose a goal other than efficiency, their decisions are not necessarily "wrong" or "bad." The decision may serve goals of equity, income redistribution, or national security, all of which are perceived to be superior to the goal of efficiency. However, economic analysis still can be applied usefully to measure the costs of choosing goals other than efficiency. An economist's perspective is also useful in assessing the validity of politicians' or judges' actions which profess to serve the public interest of conservation, but which economic analysis reveals to have counterproductive effects on conservation. A close examination of legislative history and case precedent can then show what other motivations may have influenced these decision makers.

Because this book often draws on the economist's perspective as a useful analytical framework for assessing the conservation policies implemented by legal institutions, a brief explanation of the economist's definition of conservation is in order.[12] Economic efficiency is achieved when resources are allocated in such a way as to maximize society's total income or welfare. In the context of natural resource use, *conservation* is defined as action that, from the viewpoint of society as a whole, achieves the optimal distribution of resource use over time; that is, as action that maximizes the present value of the resource. Thus, the most efficient production rate for a given reservoir is not necessarily that which maximizes the total amount of oil and gas physically recovered from the reservoir.

A single example suffices to show the difference between maximizing the value of the resource and maximizing its total physical recovery. To date, only about one-third of the oil discovered in the nation's reservoirs has been recovered; the other two-thirds, or 300 billion barrels, remains underground.[13] If the cost of recovering a barrel of this remaining oil is $40, and society currently values this oil at a market price of $30 a barrel, then it is not efficient to produce these remaining reserves, even though it is technically possible to do so. Government policies, such as supporting the domestic price of oil above the market price of $30 a barrel, may be enacted to accomplish greater physical recovery of this oil, perhaps for national security reasons. This policy may be in the public interest, but it is probably not efficient.[14] The converse of conservation is waste, and the prevention of waste is an important objective of oil and gas regulation. Economists would not prevent all waste however—only that which causes a preventable loss of resources whose value exceeds the cost of eliminating the waste. These economic concepts of efficiency will be used throughout the book to assess the costs and benefits of the decisions made by the triumvirate of legal institutions influencing oil and gas conservation in Texas. Because the lack of unitization generally results in less physical recovery of oil and gas in the long term than is economically optimal, the ensuing discussion often speaks in terms of increasing the quantity of oil and gas recovered as the primary goal that conservation policy should seek to achieve. This shorthand terminology is not meant to ignore or in any way diminish the importance of the goal of maximizing the present value of these resources.

Regarding terminology, this book carefully distinguishes the term *unitization* from that of *pooling,* even though the two words are often used interchangeably by judges and scholars. *Pooling* is the process of combining small tracts into an area of sufficient size to merit a well permit under the field's applicable spacing rule. For example, if the regulatory agency has ruled that only one well can be drilled on a 40-acre unit, and four different landowners each own adjacent 10-acre tracts in the field, these four landowners may pool their acreage into a 40-acre drilling unit and share the costs and proceeds of the drilling and production operation equally. Pooling generally occurs while the field is in the primary stage of recovery. *Unitization,* on the other hand, is the process of combining all or a large part of the acreage of an entire field into a unit, and may involve the joint operation of hundreds of wells on hundreds of individual drilling units covering several thousand acres. Unitization often occurs during secondary recovery after the natural pressure in the field has been dissipated in primary recovery.

The history of pooling in Texas illuminates many of the same legal, political, and economic issues that are involved in unitization, but there are distinct differences also. Texas does have a compulsory pooling act, which can be used in certain circumstances to force unwilling owners to join a drilling unit, but Texas does not have a compulsory unitization law.

2

The Setting for Unitization

No understanding of state policy, administrative regulation, or judicial decision making on unitization can be had without an understanding of the physical nature of oil and gas reservoirs and the engineering requirements for efficient production. Also, because the necessity for and feasibility of unitization depend on the particular physical characteristics of a reservoir, one possible explanation for the lack of a compulsory unitization statute in Texas is that the state's reservoirs are not physically conducive to the types of production operations that generally require unitization.[1] This chapter examines the engineering requirements for efficient production, the role of unitization in meeting these requirements, the difficulty of securing voluntary unitization agreements, and the importance of unitized operations to Texas in particular.

The Engineering Background of Unitization

Oil Reservoirs

The efficient recovery of crude oil is technically rather complicated.[2] Crude oil is trapped in the pore spaces of the reservoir rock under great pressure. When a well bore penetrates an oil reservoir, resulting in an area of low pressure, the pressure differential causes oil to flow into the well. Yet oil has very little compressibility, so if this factor alone were relied on to produce oil, very little would be produced. Crude oil simply cannot expel itself from the reservoir rock. It is the compressed gas or water within the reservoir, expanding in response to the lower pressure, that expels the oil from the rock. In essence, the production of oil is a displacement process by which gas or water in the reservoir expands to fill the space voided by the withdrawn oil. The expanding gas or water provides the continuing pressure essential for producing oil. Three types of oil displacement mechanisms exist either alone or in combination. They are the dissolved gas drive, the gas-cap drive, and the water drive.

The dissolved gas drive. In this type, the oil in the reservoir is saturated with dissolved gas. When oil flows into the well bore as a result of the pressure gradient, the dissolved gas expands and escapes from solution. This free gas fills the void once occupied by the oil. As oil production proceeds, pressure is reduced throughout the reservoir, and more and more gas is freed from solution everywhere. Eventually, the free gas begins to flow into the well along with the oil. The gas–oil ratio of the well increases and reservoir pressure declines continuously, until all of the gas originally contained in the reservoir has been withdrawn. At this point, there is no more reservoir pressure to expel the oil, and oil production ceases.

In many reservoirs, the agency of this dissolved gas is the primary mechanism of oil expulsion. But, unfortunately, only 10 to 30 percent of the oil in the reservoir is recovered by this method.[3] The gas in the reservoir is exhausted before all the oil can be produced, and the oil cannot expel itself.

The gas-cap drive. In this type, the upper part of the reservoir is filled with gas, and saturated oil lies beneath this gas cap (see figure 2-1). As a well is drilled into the lower oil zone, the pressure reduction originating at the well bore causes the compressed gas in the gas cap to expand. Since the gas cannot expand upward through the impervious rock or shale that traps the oil and gas in the reservoir, it expands into the lower oil zone, driving the oil

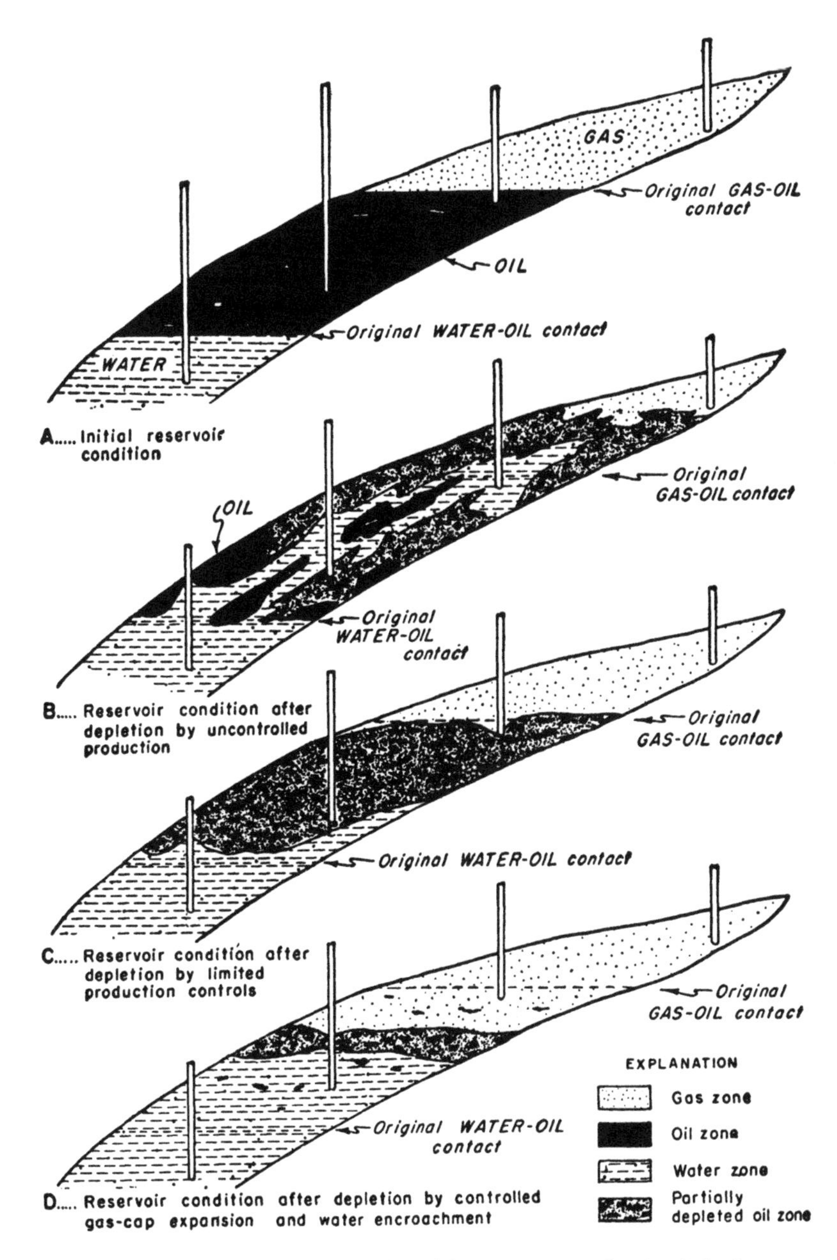

Figure 2-1. Distribution of fluids without a reservoir. At right is a magnified view of sand grains, with a film of water around each grain and oil in remainder of pore space. *Source:* H. Williams, R. Maxwell, and C. Meyers, *Oil and Gas: Cases and Materials* (4 ed., Mineola, N.Y., Foundation Press, Inc., 1979) p.8; reprinted by permission of the publisher.

toward the low-pressure area at the well bore. This oil migration continues as the gas cap expands to fill the space voided by withdrawn oil, until the gas invades the lower part of the formation and enters the producing oil wells. Then, as with a dissolved gas drive, the gas–oil ratio rises sharply, reservoir pressure declines rapidly, and when the gas completely fills the original oil zone, oil production ceases, except for that produced by gravity drainage.[4] However, a properly controlled gas-cap drive can result in a much greater oil-recovery rate than does a dissolved gas drive. If gas production is prohibited from the gas cap, and oil wells are located to produce only from the bottom of the reservoir (called "downdip," in the industry vernacular), then the gas in the gas cap can be retained within the formation for a longer period of time. This maintains the pressure in the reservoir and allows more oil to be flushed out. The same effects can be had if any gas produced from the gas cap is returned continuously to the formation. A properly maintained gas-cap drive (using gravity drainage as well) can yield oil-recovery rates of 25 to 50 percent.[5]

The water drive. In this type of pool, the oil lies atop a layer of water. The compressed water held in the pore spaces of the formation expands slightly when the pressure reduction from the well bore spreads outward. Water is much less compressible than gas, and therefore much less expandable. But its disadvantage as a displacement agent is overcome by its greater weight and viscosity, which allow water to "scrub" the reservoir rock more thoroughly as it migrates through the pore spaces. The expanding water moves upward into the oil zone, driving the oil in front of it toward the well bore.

Eventually, as the water invades the oil zone, more water and less oil are produced from the wells. Oil production ceases when the cost of operating the uppermost wells exceeds the value of the oil withdrawn from them. A water drive is thus similar to a gas-cap drive, except that in the former the oil is displaced in an upward rather than a downward direction. However, it is not loss of pressure that causes abandonment of the reservoir (as in a gas-cap field), but the fact that the oil content of the rock has been depleted. Reservoir pressure is still relatively high when a water-drive field is abandoned. Because water is a better "pushing" agent than gas, oil recovery rates of 75 to 80 percent are possible if the reservoir is operated properly to preserve the flushing action of the water.[6] This requires that the oil be produced largely from wells having low water–oil ratios that are located in the upper part of the reservoir. Downdip wells producing large amounts of water dissipate the reservoir's source of pressure and should be shut in, or the withdrawn water should be reinjected into the formation to maintain adequate pressure.

From these descriptions it is easily understood why a dissolved gas drive is less efficient than a gas-cap or water drive. In a dissolved gas drive, gas is released from solution throughout the reservoir. It is not segregated into a zone separate from the oil-producing zone. The gas must be produced simultaneously with the oil. Thus the reservoir pressure dissipates with every barrel of oil produced. When the gas is exhausted, no other expelling agent exists to drive the remaining oil into the wells. On the other hand, with a gas-cap or water drive, wells can be selectively completed only into the oil zone, preserving the gas or water as a pressure agent over an extended period of time. The progressive displacement and migration of the oil by the invading gas or water results in more efficient recoveries.

Factors affecting efficient production. With this knowledge of reservoir mechanics, it is now possible to discuss the two major factors that must be controlled in order to produce efficiently: (1) the rate of production; and (2) the location of wells. Control of these factors enables operators to substitute an efficient gas-cap or water drive for an inefficient dissolved gas drive.[7]

A controlled rate of production maintains reservoir pressure, and this is important for several reasons. First, it prevents the rapid release of dissolved gas from solution, which could create an inefficient dissolved gas drive as the dominant production mechanism. Second, control of production rates assures that the boundary between the migrating oil and the advancing gas or water front is fairly uniform so that the entire reservoir is effectively flushed without bypassing areas of oil-saturated rock. And third, pressure maintenance increases recovery efficiency by its effects on the viscosity, shrinkage, density, and interfacial tension of the reservoir fluids.[8] Figure 2-2 illustrates these effects.

The proper location of wells is the second important factor in increasing recovery rates. In a gas-cap drive, the oil-producing wells should be drilled into the lower part of the reservoir; and no wells should be allowed to produce gas from the gas cap, since this would dissipate the reservoir's pressure source. In a water drive, the oil-producing wells should be located high on the structure; downdip wells would produce excessive amounts of water and dissipate the reservoir's driving source.

Thus, for each reservoir there is generally a dominant displacement mechanism, an optimal pattern of well locations, and a maximum efficient rate of production (MER), which, if exceeded, would lead to an avoidable loss of ultimate oil recovery. The MER of a particular reservoir is determined technically through engineering studies. The use of the MER to control production rates is probably the single most important conservation measure for assuring efficient oil and gas recovery.[9] Even a dissolved gas reservoir sometimes can be operated to achieve significant increases in oil recovery, which would not occur under uncontrolled production practices. For example, if this type of reservoir is produced slowly, gravity will drain oil to the bottom of the reservoir and the released gas can be segregated at the top of the reservoir. If oil wells are completed only into the bottom zones, gas dissipation can be minimized and the reservoir pressure maintained longer. The resultant increase in oil recovery is directly attributable to the slow rate of production and the selected well pattern.[10] Similarly, but on a larger scale, the recovery efficiencies of gas-cap, water-drive, and combined gas-cap–water-drive fields are very sensitive to the rate of oil production and well placement. Excessive rates of withdrawal and poorly placed wells would lead to a rapid fall in reservoir pressure, the release of dissolved gas, loss of a uniform front between the gas or water and the oil, bypassing of the oil rock, and—in extreme cases—dominance of the entire recovery process by an inefficient dissolved gas drive.

Further improvements in production efficiency can result from artificial pressure maintenance activities such as injecting water or gas into the reservoir, either to supplement a natural water or gas-cap drive or to create such a drive.[11] The injected gas or water

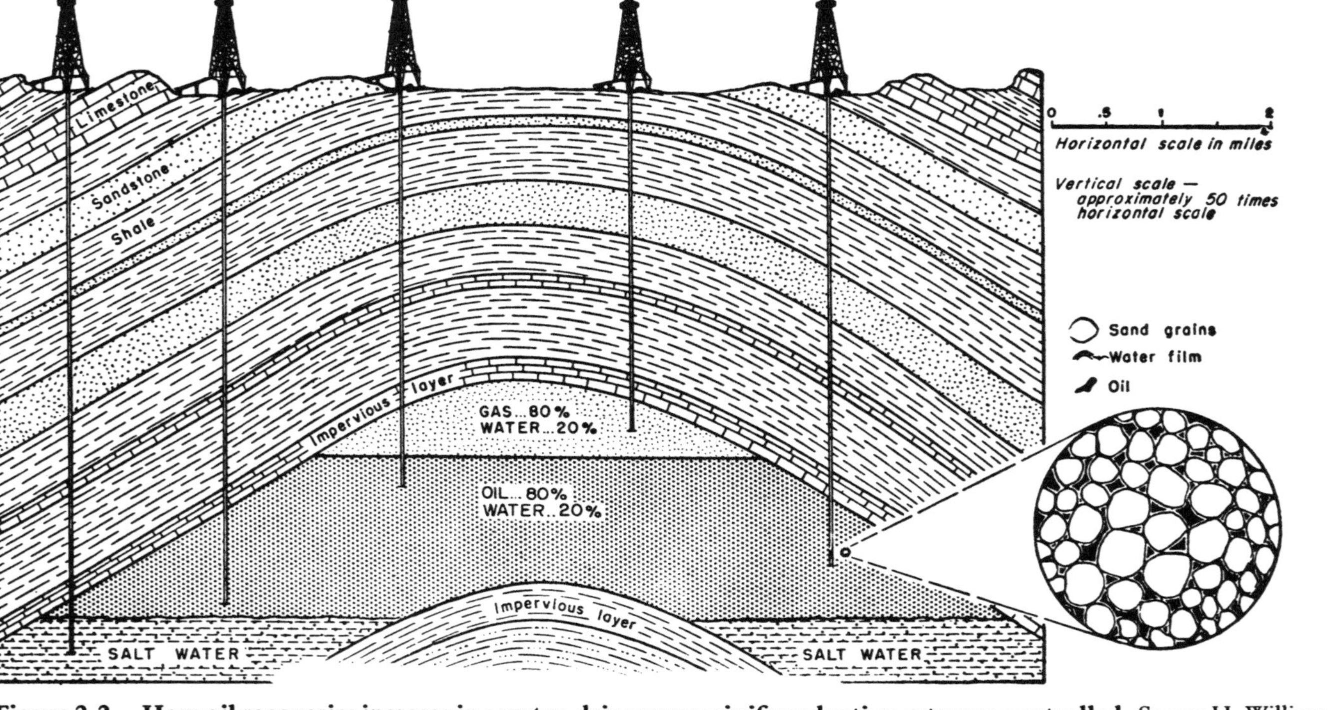

Figure 2-2. How oil recoveries increase in a water-drive reservoir if production rates are controlled. *Source:* H. Williams, R. Maxwell, and C. Meyers, *Oil and Gas: Cases and Materials* (4 ed., Mineola, N.Y., Foundation Press, Inc., 1979) p. 10; reprinted by permission of the publisher.

may be recycled from the reservoir itself or brought in from an extraneous source. The placement of injection wells is as important to recovery efficiency as is the placement of producing wells. Gas should be injected high in the reservoir, and water should be injected around the bottom edges of the pool. While pressure maintenance operations cannot be used in all reservoirs, because of adverse geological conditions, such operations can be successfully employed in many instances.

The distinction between pressure maintenance and secondary recovery operations—both of which involve injecting gas or water into the reservoir—is that the former is done early in the life of the field, when the original pressure is still high, and the latter is done after the pressure in the field has declined to a low level. Secondary recovery is generally not practical in fields depleted by an efficient primary production method that leaves little residual oil behind.
after the pressure in the field has declined to a low level. Secondary recovery is generally not practical in fields depleted by an efficient primary production method that leaves little residual oil behind.
discovered long before the science of pressure maintenance was understood, the only way to secure additional oil recovery is by secondary recovery operations. Because older fields were typically operated without any restrictions on their rates of production, an inefficient, dissolved gas drive often left substantial amounts of oil in the reservoir. Large quantities of oil were produced in a short period of flush production, but this was followed by many years of "stripper" operations—that is, production from wells at low pressure, averaging only a few barrels a day.[12] Thus after many years of stripper production, much recoverable oil may remain in the reservoir.

In a typical secondary recovery operation, injected gas or water is used to drive oil away from the injection well (a high-pressure area) toward the producing stripper well (a low-pressure area). A water-flood operation commonly uses a five-spot pattern which places injection wells at the corners of a square and a producing well at the center.[13] The injected water pushes a bank of oil toward the producing well and, after some delay (sometimes several months), the output of oil increases significantly at the producing well as the oil bank arrives. After this, the well produces increasing amounts of water and is abandoned when the oil recovery no longer pays for the cost of operating the well. In many water-flooding operations, injection wells are located only 200 to 400 feet away from the producing wells. This requires significantly more drilling than does production by efficient primary methods in which one oil well can often drain 40 to 80 acres.

Secondary recovery by gas injection was once more widely used than water flooding.[14] Until the 1960s, gas injection was relatively inexpensive because natural gas was in excess supply and very low priced. Natural gas is extremely soluble in oil, thus increasing the oil's volume and reducing its viscosity and surface tension, making it easier to produce. However, gas injection achieves less oil recovery because gas is a less effective pushing agent than water. Gas is very mobile and migrates rapidly through the reservoir, seldom producing an oil bank in front of it. During the secondary recovery operation, the gas–oil ratio of the producing well rises continuously until it is so high that the value of the oil recovered no longer justifies the costs of injection.

While large quantities of oil still remain in depleted fields in the United States, secondary recovery is only feasible in certain reservoirs. Improper production techniques in the early stages may have caused irreparable injury to the formation, or the remaining oil may not be sufficient to support the high costs of recovering it.[15] Also, successful secondary recovery requires certain geological conditions, such as a continuous and uniform formation without fractures, fissures, and streaks of high permeability, which cause injected gas or water to flow only in certain channels. Many secondary recovery attempts have failed because the physical characteristics of the reservoir were not adequately analyzed in advance and because of the substantial costs involved in reworking or plugging old wells, drilling new injection and production wells, securing a supply of gas or water, and constructing the necessary surface facilities.

Despite the success of some secondary recovery, it is generally a poor substitute for producing oil efficiently from the start. The early selection and use of an efficient displacement mechanism allow wells to be placed optimally from the start and reduce the need to drill and rework wells for the secondary recovery phase. Total development and operating expenses are also minimized because efficient primary techniques can recover the oil over a shorter period of time than a two-stage production process. Also, the physical changes in the reservoir induced by inefficient primary recovery methods often reduce (or, indeed, preclude) the recoverability of the remaining oil, especially in water-drive fields.

Enhanced oil recovery. In recent years, a new term has entered the oil industry's parlance—*enhanced oil recovery* (EOR). EOR is an advanced form of oil recovery over and above that which can be economically recovered by conventional primary and secondary methods.[16] It includes (1) thermal processes; (2) carbon dioxide miscible flooding; and (3) chemical flooding.[17]

Thermal processes add heat and pressure to the reservoir to reduce oil viscosity or vaporize the oil, thus making it more mobile. *Steam injection,* commercially used to produce heavy oil in California for the last fifteen years, operates much like a patterned water-flooding operation except that, in this instance, steam is injected into the well. *In-situ combustion* is another thermal process which generates heat in the reservoir by injecting air and burning part of the crude oil in the ground. This partially vaporizes the oil and drives it to the producing wells by a combined steam, hot-water, and gas drive.

Carbon dioxide miscible flooding is an advanced form of gas injection. Carbon dioxide can miscibly displace oil from the pore spaces of the reservoir rock; that is, it causes oil to become soluble in the carbon dioxide provided that the reservoir pressure is kept fairly high. To achieve higher sweep efficiency, water and carbon dioxide may be injected alternately.

Chemical flooding is the newest and least tested of the EOR processes. Surfactants, polymers, or alkaline chemicals are added to the water-flooding process to "scrub" the reservoir rock more thoroughly. Surfactant flooding is expected to follow conventional water flooding in many fields; polymer and alkaline flooding will probably be initiated early in the life of the field.

EOR processes could have a significant impact on oil production from existing reservoirs. Currently, only about 4 percent of U.S. production comes from EOR,[18] but if the technology continues to advance and the price of oil remains high, EOR may add 24 billion barrels of oil to domestic production through the year 2000.[19] However, these new ventures are risky and often require large capital investments and long lead times before results are achieved.

The success of any particular EOR project may be at least as uncertain as exploratory drilling.

Gas Reservoirs

Recovery of natural gas is much less complicated than recovery of crude oil because gas is highly expansible. It can be recovered from porous rock simply by allowing it to escape by expansion into the low-pressure area around the well bore. This process commonly can recover 90 percent or more of the original gas in place.[20] Recovery rates less than this occur only when the permeability of the reservoir is so low that gas cannot flow freely into the well bore. In some instances, a naturally existing water drive may be used to displace the gas rather than relying solely on gas expansion. The water drive maintains reservoir pressure, which in turn may reduce the number of wells required to deplete the reservoir and also save compression costs for delivering the gas into the pipeline. The gas reservoir is then operated just as an oil reservoir under water drive, with well location and controlled production rates having as much importance.

One type of gas reservoir requires more complicated recovery methods. In some deeper gas pools, liquid hydrocarbons are under such high pressure that they turn into a gaseous state in the reservoir. When this "wet" gas is produced, the liquid hydrocarbons can be separated from the gas at the surface by cooling and pressure reduction. This recovered fluid is called condensate. In the past, condensate was often more valuable than the natural gas from which it was separated. If the gas is produced by simple expansion in such a reservoir, the pressure in the reservoir falls continuously and the condensate may turn to a liquid within the formation. Once in liquid form, there is no method of recovering the condensate from the rock. Proper operation of a condensate reservoir therefore requires a high degree of pressure maintenance, either by use of a water drive or by gas cycling. In a cycling operation, the wet gas is produced, condensate is processed from it at the surface, and the stripped dry gas is compressed and returned to the reservoir to maintain pressure. The injected dry gas drives the remaining wet gas toward the producing wells. Obviously, maximum recovery of condensate requires that the dry gas injection wells be placed as far as possible from the producing wells so that the dry gas can sweep the entire reservoir before being produced. Eventually, of course, cycling stops when the reservoir produces only marketable dry gas.[21]

Secondary recovery has not been applied to natural gas production in the past, but research now suggests that certain depleted gas fields, which previously had been operated with strong water drives, might profitably be reworked.[22] Generally, however, the closest counterpart to EOR oil in the industry's vernacular is UGR gas—*unconventional gas recovery*. This term does not usually refer to the reworking of depleted fields. It applies to the production of gas from new and difficult formations that heretofore have been uneconomic to produce at the low gas prices of the pre-1970 period. UGR is primarily applicable to low-permeability reservoirs, called "tight sands" reservoirs, which trap the natural gas in such small pore spaces that it does not flow easily. Most current research and field testing involve massive hydraulic fracturing of the formation in order to create cracks in the rock extending for several thousand feet through which the gas can flow.[23] Once fractured, the gas is produced by simple expansion. The potential for UGR is difficult to ascertain, but some estimates are huge.[24]

The Benefits of and Need for Unitization

This summary of efficient production techniques for oil and gas fields provides the necessary background for understanding the importance of unitization. As seen, the most efficient production methods are those that segregate the fluids in the reservoir and force oil or gas to migrate from one part of the reservoir to another. Since few reservoirs are owned by a single person, this type of production method inevitably has a large and unavoidable effect on the individual property interests of the many owners of the common oil and gas pool. Oil is transferred from beneath the land of one owner to the land of another. Because of this, the conservation essentials of controlled production rates, pressure maintenance, and optimal well location patterns are often irreconcilable with individual property rights, as defined by the common law and by the limits placed on the state's power to regulate private property. The property relations governing the ownership of oil and gas produced from a common pool are called correlative rights.

The Rule of Capture

The common law governing the correlative rights of the owners of a common source of supply is the rule of capture. This rule allows an owner to drill a well on his tract and drain oil and gas from his neighbors without liability. The neighbors cannot enjoin his production or share in the proceeds from his well. They can only "go and do likewise,"[25] namely, drill their own wells and drain the minerals back. The incompatibility of this rule with sound conservation practices is immediately obvious. It forces each and every owner of the reservoir to drill and pump quickly lest his neighbors drain the oil away first. The rule of capture resulted in the wasteful expenditure of enormous sums of capital and labor on unnecessary wells to recover the oil and gas in the reservoir. When all these wells flowed wide open, the energy drive in the reservoir was quickly dissipated and the fields were depleted within a few years of their discovery, leaving vast quantities of oil in the ground.

In the 1920s and 1930s most oil and gas states—recognizing that the pell-mell scramble for advantage under the rule of capture was antithetical to conservation—passed statutes authorizing agencies to regulate oil and gas production in order to prevent waste. Initially, these agencies performed this function primarily by passing regulations to control the spacing and location of wells and to control production rates through a prorationing system.[26] Statewide spacing rules today typically allow no more than one oil well on 40 acres and one gas well on 640 acres. In fields with particularly low or high porosities, these ratios will, of course, differ.

After well-spacing laws were passed, the need for compulsory pooling legislation to protect correlative rights became evident. For example, when four landowners each own 10-acre tracts and only one owner will be granted a drilling permit under a 40-acre well-spacing rule, the oil and gas belonging to the other three will be drained away without liability under the rule of capture. Because the owner with the well will not voluntarily share his bounty with the others, many states passed compulsory pooling laws in the 1940s and 1950s. These laws allow the other owners who do not have a well on their land to force the well owner to share the production and expenses on an equitable basis.[27]

While well-spacing, pooling, and prorationing laws prevent much waste as compared with the unbridled rule of capture, these laws cannot always maximize the recovery of oil and gas and, at the same time, afford each owner the opportunity to recover a fair share of the reservoir's bounty. This can be demonstrated as follows: assume a field in which the western part of the reservoir is underlaid by water, and a water drive displaces the oil upward and to the east. The owners of wells on the western edge quickly must withdraw the oil in place underneath their tracts before it migrates eastward and their wells "water out." But if the western wells are allowed to produce at their "wide open" potential, the water drive is unlikely to advance uniformly through the formation. Oil-saturated rock would be bypassed, and reservoir pressure would be reduced excessively. These effects might be so severe that even the owners of western tracts might recover less oil under unrestricted production than under controlled production. To prevent the permanent loss of oil, the state regulatory agency will want to control the rate of production of all wells in the field by MER prorationing. But at the slower, controlled rate, the owners of wells on the western edge of the field may not recover all of the oil originally underneath their tracts. Under prorationing, more oil will be swept eastward, and the eastern wells will produce many times the amount of oil originally lying beneath them.[28] Much of this oil will have migrated from the western tracts, but, under the rule of capture, the oil belongs to the owners of the eastern wells.

Herein lies the dilemma. A prorationing order that prevents waste but greatly magnifies the natural migration of oil away from the western tract owners may violate the constitutional constraint against taking private property without just compensation.[29] Moreover, the state regulatory agencies are often charged by statute with the duty to protect correlative rights, that is, to assure fairness in the allocation of the field's oil and gas. A waste-prevention order, which results in large, uncompensated drainage from one landowner to another, may be invalidated as unreasonable and discriminatory by the reviewing courts. Even if the courts allow such orders to survive judicial challenge, on the basis that waste prevention is a more important duty than the protection of correlative rights, the agency order, while legal, can still be criticized as fundamentally unjust.

Optimal well placement creates an even more obvious conflict between correlative rights and conservation. For maximum cost efficiency, some, if not many, of the western tracts should not have wells drilled on them at all. The wells are not needed to withdraw the oil from the reservoir; they will water out relatively quickly, and then must be plugged and abandoned. Yet, under the common law, the owner of a tract who does not drill and produce has no right to the oil or gas that is drained away from his tract onto another's land. Each owner of the common pool must drill or suffer drainage. If the regulatory agency prohibits an owner of a western tract from drilling, his oil and gas property has effectively been taken by the state.

Well-spacing and compulsory pooling orders do not by themselves result in optimal well placement and the most efficient number of wells. Such orders can ensure that only one well is drilled on a 40-acre unit and that all owners share equitably in its proceeds, but efficiency in production might require that no wells at all be drilled on some of the 40-acre units along the western edge.

Further, to maximize production efficiency, pressure maintenance operations should probably be instituted in this field to maintain the water drive. Water should be injected through the western wells in order to push the oil to the east. Yet, the owners of the western tracts are unlikely to forgo oil production and place water-injection wells on their land; this only will have the effect of moving the oil away from their tracts more quickly. The state's authority to regulate well locations does not extend to forcing an owner to drill and operate injection wells on his land for the benefit of other tracts.

The same type of impasse is created in a gas-cap drive except that the downdip tracts now have the structural advantage of being on the receiving end of the oil migration, and the updip tracts are disadvantaged. Indeed, updip tracts overlying the gas cap should not be allowed to produce at all lest the gas-cap energy be dissipated. However, a complete prohibition against producing gas from wells in the gas cap will allow the owners of wells in the oil stratum to maximize their recovery of oil and then to produce all the migrated gas that has been retained in the reservoir and displaced from the owners of tracts overlying the gas cap. Such an order is clearly confiscatory so far as the latter owners are concerned. The tension between the state's interest in conservation and its obligation to protect private property rights often results in a compromise rule: the owners of gas-cap wells can produce gas, but only in carefully controlled and very limited amounts.[30] Such a rule only partly serves the cause of conservation, and only partly protects the correlative rights of the gas-cap owners.

Likewise, gas cycling in condensate gas fields requires that dry-gas injection wells be located at the opposite end of the pool from the producing wells and that few wells be drilled and produced from the middle of the reservoir. Inevitably, the property interests of some tract owners will be seriously injured by this arrangement because of the rule of capture. Again, compulsory pooling legislation, which enforces regularly spaced wells, does not solve the problem of optimal well placement on a fieldwide basis.

Secondary recovery operations also create this type of conflict. If gas injection is used, the reservoir pressure increases locally around the injection well and forces oil and gas away from the well. Gas is very mobile and travels rapidly, so that the injected gas may benefit tracts some distance away. In water-flooding operations, the injection wells and producing wells are generally located fairly close to each other so that the oil migrating away from the injection wells often can be captured by the nearby producing wells on the same tract. Nonetheless, a large water-flooding operation in one part of the reservoir may induce a general migration of oil across property boundaries, thereby benefiting tract owners who paid no part of the expenses of the project. Also, the water-flooding operation may prematurely drown out producing wells on nearby tracts. It is one thing for owners to accept the regional migration of fluids that is inherent in any natural gas-cap or water-drive field. It is quite another matter to accept the deliberate, manmade migrations of fluids resulting from artificial pressure maintenance or secondary recovery operations, when these migrations destroy existing oil wells and radically alter the relative amounts of valuable oil and gas and worthless brine underlying the separately owned tracts. Long before the courts announced the rule of capture, injurious invasions of one person's property by another were subject to the common law of trespass. Any owner who initiates pressure maintenance or secondary recovery operations risks liability for trespass if the operations injure an adjoining owner's property (see full discussion of this point in chapter 7).

For these reasons, many oil and gas reservoirs cannot be operated efficiently under the common law. Well-spacing and prorationing regulations, which modify the common law in order to prevent waste, often result in unfair allocation of the increased recovery made possible by such rules. These regulations may have such a severe impact on the property rights of some owners of the common reservoir that the regulatory orders issued thereunder cannot survive legal challenge. States have struggled to balance the irreconcilable dictates of efficiency and equity in the regulations, but in many instances every gain in equity costs a loss of recovery.

Unitization offers a solution to this impasse between conservation and correlative rights. With unitization, all the tracts in the field are combined so that the entire reservoir can be treated as a single production unit. This allows the implementation of the most efficient method of operating the pool without regard to individual property lines. The owners of the separate tracts receive a share of the unit's total recovery according to a formula that reasonably approximates each tract's contribution to the total oil and gas recovered by the unit. Such a formula might allocate production on the basis of the number of productive surface acres of each tract or the amount of recoverable oil and gas under each tract.

The unit members usually select one producer to be the unit operator in charge of the day-to-day management of the unit. A unit operating committee composed of all the working interests usually retains supervisory authority over the unit operator. Thus, if, in the water-drive example used above, the pool is unitized, the owners of western tracts will secure a share of the total oil produced from the field, even though no wells are placed on their tracts. The water drive can be effectively operated without regard to the rule of capture and the law of trespass; development and operating expenses are minimized by avoiding the drilling of nonessential wells; and injection wells for pressure maintenance can be placed optimally, further enhancing the recovery efficiency. With unitized development, conservation and correlative rights no longer conflict. Especially in gas-cap drive, water drive, pressure maintenance, and condensate cycling operations, unitization is the only satisfactory solution to development that is both fair and efficient. These operations are generally conducted when the pressure in the reservoir is still high, and the migration of fluids from tract to tract is especially difficult to prevent and control. In low-pressure water floods and enhanced oil recovery processes, unitization is not as essential because the migration of fluids is easier to control. Nonetheless, these processes still depend on gas, water, steam, or chemical injections that are most efficiently conducted on a large scale, often transcending individual tract boundaries. Even the massive fracturing of tight sands in unconventional gas recovery may be facilitated by unitization. The large fractures created by one producer may cross lease-boundary lines and cause drainage from one tract to another. The traditional pooled gas unit of 640 acres may be too small to contain the fractures. Unitization would avoid any possible liability for trespass damages caused by the extension of fractures into another owner's mineral estate.

Thus, there are very few types of reservoirs that would not be served by unitization. The only fields in which controlled production rates, pressure maintenance, and careful well placement are not essential to increased recovery rates are oil reservoirs under certain dissolved gas drives and dry-gas reservoirs containing little condensate. In these reservoirs, unitization is not required to prevent the physical waste of oil and gas.[31] Regular well spacing, pooling, and prorationing formulas that allocate production fairly on the basis of surface acreage or productive acre-feet of sand, generally suffice to prevent wasteful drilling and to protect correlative rights. However, even in dry-gas fields, unitization can perform the valuable function of simplifying the state's administration of its gas-prorationing system. Indeed, unitization of dry-gas fields may be the only method to achieve fairness in the allocation of gas production from fields producing at 100 percent of their potential as a result of the great demand for natural gas. Notwithstanding the undisputed fairness of prorationing gas according to a formula based on each tract's share of the productive acreage in the field, the gas-prorationing system that exists today in Texas often cannot ensure the protection of correlative rights in this new era of high demand except by drilling wells that may be unnecessary from the standpoint of ultimate recovery (see chapter 8, at notes 32 to 36). If these gas fields were unitized, correlative rights could be protected without additional drilling.

This discussion of unitization has stressed its virtues in terms of eliminating the conflict between correlative rights and the ultimate physical recovery of oil and gas. Unitization also offers large benefits from the perspective of economic efficiency. In economists' terms, the common law rule of capture is inefficient because it results in a too rapid rate of production, which does not maximize the present value of the oil and gas resource to society.[32] Ordinarily, when resources are privately owned, a private producer's desire to maximize profits results in efficient behavior that maximizes social welfare as well. However, this is not the case with oil and gas in a commonly owned reservoir. Each producer's optimal rate of production over time is influenced by the fact that the oil or gas which he does not produce immediately may not remain in the reservoir for his later production. Other owner-producers in the field may drain this oil or gas away. This drainage is a very real cost to the individual producer and skews the decision to produce now rather than later in order to maximize profits. From society's viewpoint, however, the drainage is simply a transfer of income between producers. The optimal rate of production from society's point of view is not affected by this income transfer between private parties. Thus, the rule of capture drives a wedge between the socially optimal rate of production of oil and gas and the private operator's profit-maximizing behavior. Private actions become socially inefficient. Economists have demonstrated that the rule of capture leads to excessive drilling and production rates and to the physical loss of oil and gas. Economists also have analyzed the effects of the rule of capture on the price of oil. The short-term effect of the rush to produce and avoid drainage is that the price of oil falls below that which would obtain were the field unitized and produced at the socially efficient rate.[33] However, in the intermediate and long time frames, the rule of capture drives the price of oil above that which would prevail with unitization. The too-high price results because, under the rule of capture, more wells are drilled and less oil is obtained ultimately from each reservoir.

Higher costs and lower revenues reduce the expected profitability of developing newly discovered reservoirs, so that fewer available discoveries are developed at any given projected price. The quantity of oil supplied is less than optimal from society's vantage point, causing the price to be above the socially optimal level. In the long term, producers react to the anticipated higher costs and lower recovery rates induced by the rule of capture by decreasing the amount of exploration performed. Each exploratory prospect is economically less attractive, and so fewer are explored. With fewer than the optimal number of prospects explored from society's perspective, the long-run supply is smaller and the long-run price is higher than would exist with unitization.

With unitized development and operation of reservoirs, no difference exists between the actual amount of oil and gas privately supplied and the socially optimal amount; hence no difference exists between the actual and the socially optimal price. Under the rule of capture, private and social costs and benefits diverge. This divergence reflects the cost of the excessive number of wells and the excessive loss of ultimate recovery induced by the rule of capture. Well-spacing, pooling, and prorationing rules can reduce, but not eliminate, the difference between the unrestrained rule of capture and society's optimal rate of production.[34] Fieldwide unitization can completely eliminate the differential.[35] A fieldwide unit embracing all owners in the common pool cannot be drained by anyone else. The unit operator can maximize the profits of the unit as a whole by maximizing the present value of the resources in the field. The unit will produce at the optimal rate of production with the optimal degree of ultimate recovery and the optimal number of wells. The unit will use the most efficient natural drive available in the field and enhance it, if necessary, with the most efficient type of artificial drive. The unit's rate of output will vary as expected prices and costs change with market conditions. The unit will adjust its output rate flexibly as it maximizes present value continuously. The unit will place surface equipment in the best locations and invest the optimal amount of capital in such equipment. Wells producing large quantities of wasteful by-products will not be drilled or will be shut in at the optimal time, thus reducing the costs of disposing of brine and wastes and reducing oil field pollution. The state administrative agency need no longer intervene in the field's production operations through well-spacing rules, gas–oil and water–oil ratio rules, and prorationing orders. The unit operator will select the most efficient wells and rate of production in his own self-interest. The cost to the state and to private operators of administering many conservation regulations can be eliminated. When producers expect unitization to occur as a matter of course in the fields they discover, the margin of profitable exploration is extended because producers will anticipate greater recovery rates and reduced costs. Bonuses and royalties to landowners will be higher because the present value of their oil and gas resource is greater when producers can expect unitization to occur. In sum, unitization not only protects correlative rights and enhances the physical recovery of oil and gas, but also provides significant benefits in economic efficiency.

The Problems of Voluntary Unitization

Tremendous increases in recovery rates have been achieved in unitized fields.[36] This fact alone provides a great incentive for voluntary unitization agreements among all the owners of a reservoir.[37] Unitized operations create a much larger pie to divide among the unit members. The previous discussion of reservoir mechanics also indicates, however, why voluntary unitization may often be difficult to achieve: the allocation formula for sharing the pie will be highly controversial. Those owners of tracts with a natural structural advantage will want to retain the value of this advantage in the unitization formula. Such owners are unlikely to agree to a unitization agreement that does not give them at least as much oil or gas as they would have received by "going it alone." Even if the increase in ultimate recovery from unitization is so great that these owners will receive more from unit operations under almost any reasonable allocation formula than from individual development, the owners have a much stronger bargaining position in the negotiations than less-favored tract owners. They can hold out for the most favorable allocation formula, secure in the knowledge that the regional migration of oil will continue toward their tracts during any delay in negotiations. Indeed, holding out may increase the value of a structurally advantageous location. If the others in the reservoir unitize without the participation of the owners of better located tracts, the pressure maintenance operations of the unit may well increase the amount of oil migration toward the unsigned tracts. The holdouts then benefit from the unit without incurring any costs of the pressure maintenance activity. This disincentive to unitize voluntarily is called "profitable obstructionism."[38]

Other obstacles to voluntary unitization are the lack of reservoir data and the riskiness of some pressure maintenance and secondary recovery operations. It is essential that the physical characteristics of a reservoir be adequately tested and analyzed so that the production techniques most suitable to its geology are utilized. Improper analysis may result in an unprofitable operation. It takes time and money to obtain these reservoir data.[39] Yet without it, property owners may not be able to find a common ground for agreement on relative tract-participation values.[40] Even if the data are reasonably available, many secondary recovery and EOR operations remain risky, both technologically and economically. Some owners may refuse to join the unit because they assess the risks and rewards differently than do the proponents of the unit. They refuse, not because of profitable obstructionism, but because they have an honest difference of opinion about the profitability of the undertaking.

Another related reason that voluntary unitization may be difficult to achieve also results from the effects of reservoir performance under controlled operations. The time pattern of oil and gas production is often changed by unitization, and payments to those with royalty interests and working interests may be reduced in the near term,[41] even though the total returns from the pool are expected to increase substantially in the long run. For example, this could happen if the unit instituted a cycling operation in a condensate field, and dry gas—rather than being marketed and sold immediately after production and processing—was reinjected into the reservoir. The effect of reduced royalty payments in the near term is often cited as the major reason for opposition to unitization on the part of the royalty owner.[42] The fear of reduced, short-term returns from unitization is not as serious now as it was in the 1930s, when unitization was advocated as a method of controlling rates of production from wells operated at wide open rates. Today, state prorationing regulations control production rates in all fields, so that unitizing a field seldom results in a large reduction in short-term income flows.

Many other obstacles to voluntary unitization exist which are not related to the reservoir's structure or performance. The incentive to "go it alone" rather than to join a unit is sometimes created by state regulatory orders that favor small-tract owners by granting liberal exceptions to the well-spacing rules and by prorationing fields on the basis of formulas that allocate production according to the number of wells drilled on each tract, irrespective of the tract's acreage. In this case, the state regulatory agency—not mother nature—has created the "structural advantage" that is the source of the disincentive to unitize. Unitization aims at minimizing the number of wells drilled. A prorationing system that allocates allowables on the basis of the number of wells drilled on each tract works counter to this aim. Those owners who have a regulatory advantage under the per-well allowable system will refuse to join a voluntary unitization agreement that credits them only with the amount of oil and gas reserves underlying their tracts. These owners will prefer to drill on their own and drain oil and gas from others under the rule of capture and the field's existing prorationing rules.

Other obstacles to voluntary unitization include:[43]

Pride of ownership and operational control. Unitized operations do not appeal to individuals who prize independence and complete control of their own operations over joint decision making by a unit committee and supervised directives from the unit operator. Efficient unitized operations often reduce the number of wells needed to develop a field and can cause unemployment problems for certain individuals.[44] In some instances, unitization may eliminate a small operator's *raison d'être* by shutting in and abandoning the marginal wells he once so carefully nurtured.

Mistrust. Instances of unfair treatment of minority participants by the majority of unit members have caused some royalty interest owners and small operators to distrust unitization in principle.[45] Small independent producers especially fear that the major oil companies, which are likely to be the unit operators, will not treat them fairly. History records a long struggle for competitive advantage between the majors and the independents, and unitization revives the battle. The mistrust of a single dominant operator can be guarded against by a strong committee of operating interests in which the small producers are represented and have voting power on all important issues. But, then, the individual producer is subsumed in a committee environment with all its attendant bureaucracy.

Multiple parties. The very existence of hundreds, sometimes thousands, of different royalty and working interest owners in some large fields discourages voluntary unitization. The costs of negotiating with so many different interests is high.[46] Very often, some owners cannot be located at all. Title defects in the unit area must be cleared or litigated. All of this is expensive.

Fear of violating antitrust laws. In some states, especially those with strong antitrust laws, operators have been unwilling to unitize for fear that cooperative agreements to control a field's development would violate the antitrust laws. Because of this, the American Petroleum Institute (API) at an early date called on the state and federal governments to pass legislation exempting unitized operations from the antitrust laws.[47] Many states enacted such legislation.[48] It is generally considered today that unitized operations do not violate any antitrust laws, and that this fear is most often used as a bargaining chip by those seeking a better allocation formula.[49]

Fear of increased legal problems and uncertainties. Few oil and gas leases expressly allow lessees to unitize. Indeed, some leases prohibit unitization or recycling of natural gas. When unitization was a relatively new phenomenon, uncertainties about its effects on maintaining leases, the taxable status of the unit revenues, and other issues discouraged some individuals from joining the unit. While many such issues have now been litigated and resolved, some still remain as possible deterrents. (Some of these legal deterrents are discussed in chapter 7.) To a great extent, this fear of increased legal uncertainty now seems to reflect a general attitude in favor of the tried-and-true staus quo with all its customary rules and vested interests rather than any fear of a specific legal problem. Customs are comfortable, especially for those whose correlative rights are favored under the status quo.

Thus, despite the awesome potential for increased recovery and profit that unitization promises, many factors militate against its voluntary adoption. Even if a field is ultimately unitized by voluntary agreement, lengthy delays in the negotiating process may result in significantly higher costs of operation. During the delay, some existing wells that would be useful as injection or output wells may be plugged and abandoned. The subsequent project will then incur higher costs of drilling and reworking wells.[50] Thus, it is not surprising that states have enacted compulsory unitization legislation to overcome some of the obstacles to voluntary agreements. The only surprise is that Texas has not.

The Need for Unitized Operations in Texas

If Texas geology consisted largely of gas reservoirs with little condensate content and oil fields producible only with dissolved gas drives, there would be little need for unitization. This is not the case, however. The Texas Gulf Coast contains many of the nation's largest condensate gas fields. The East Texas oil field—the largest in the continental United States—is a water-drive field. Pressure maintenance operations in fields such as these are essential for efficient recovery. More than half of Texas' current production comes from fields discovered more than forty years ago, and almost 75 percent is from fields that have been in existence more than thirty years.[51] Many old fields were produced inefficiently with dissolved gas drives. The only way of increasing their recovery rates is by secondary recovery or EOR operations. Fieldwide unitization is not always as essential to these types of operations as it is to pressure maintenance activities; but, in many instances, unitization is required to obtain economies of scale, to minimize the drilling of injection and output wells, and to avoid possible trespass liability. Oil fields discovered during the spurt of drilling in the 1970s supply less than 5 percent of Texas' current production,[52] but proper operation is even more important in these fields during primary recovery so as to avoid repeating the mistakes made in the older fields.

Much secondary recovery and pressure maintenance is being done in Texas fields already. In 1978–79, 3,298 injection projects were active in Texas.[53] Most of these projects (2,767, or 84 percent) were low-pressure water floods, but pressure maintenance operations were conducted in 518 projects (about 16 percent of the total). These 3,298 active injection projects involved 1,070 different fields, 4,876 different leases or units, and 943 different operators.

The magnitude of these operations is gauged by the following: the projects used 24,897 injection wells, 52,420 producing wells, more than 3 billion barrels of injected water, and more than 425 trillion cubic feet of injected gas to produce 584 million barrels of oil and 869 billion cubic feet of gas. The 584 million barrels of oil produced is 63 percent of all the oil produced in Texas in 1980. Some of this oil would have been recoverable by primary production methods, but about 30 percent of Texas' output of crude oil is estimated to result from the effects of these injection operations. The estimated ultimate oil recovery from the injection operations is more than 6.3 billion barrels.

At present, enhanced oil production in Texas amounts to about 25 million barrels per year, less than 3 percent of Texas' total production, but Texas fields hold much potential for EOR. The Texas 2000 Commission Report conservatively projects an increase in EOR oil to 150 million barrels a year by the year 2000.[54] Another study, by the Texas Governor's Energy Advisory Council, made a detailed analysis of 242 Texas oil fields, representing about 70 percent of the state's recoverable oil reserves, and concluded that 4.7 billion barrels of oil could definitely be recovered by EOR methods and that double that amount could be reasonably expected.[55]

Clearly, secondary recovery, EOR, and pressure maintenance projects have major roles to play in many Texas oil fields. The large migrations of oil and gas artificially induced by such operations inevitably affect the property rights of the co-owners of the reservoirs. Texas geology does not explain any lack of need for unitization. (See chapter 9 for a discussion of the extent to which Texas oil and gas fields are actually unitized.)

Summary

The efficient recovery of oil and gas often requires the careful control of production rates and well placement and often results in the large-scale displacement and migration of fluids from one person's tract to another. The common law of the rule of capture and trespass conflicts with these conservation essentials. Legislation, enacted to conserve the state's oil and gas resources through prorationing and well-spacing orders, often cannot maximize the efficient recovery of oil and gas without severely harming the property interests of some of the owners of the common reservoir. In some cases, state conservation orders that intensify drainage between tracts may amount to an unconstitutional taking or an abdication of the agency's statutory duty to protect correlative rights. Unitization allows sound conservation practices to co-exist with fairness in the allocation of the reservoir's increased bounty. For many reasons, such as profitable obstructionism and differences in opinion about tract values, voluntary unitization agreements are difficult—and sometimes impossible—to achieve. Compulsory unitization legislation allows the state conservation agency to force an unwilling minority of owners of a reservoir to join a unit that has been found to promise additional ultimate recovery of oil and gas. Texas stands in the unique position of not having a compulsory unitization statute, despite a subsurface topography that warrants controlled production rates, optimal fieldwide well patterns, and pressure maintenance and secondary recovery. The question remains whether the absence of such a statute has retarded unitization in Texas oil and gas fields. Legislative, administrative, and judicial forces may have combined to create a unique legal framework that has managed to achieve unitization without compulsory process. The rest of this book seeks to answer this question.

3

The Texas Legislature and Unitization: The Early Years

Henry Doherty's words in 1924 were but the beginning of a stream of academic, industry, and government writings on the issue of unitization. At first, almost all members of the oil industry denounced Doherty's call for a federal unitization law, and treated him as a pariah. However, by 1929, the oversupply of crude oil that had developed worldwide, coupled with the growth of knowledge in the reservoir sciences, led to a flow of endorsements of unitization from such well-established organizations as the American Institute of Mining and Metallurgical Engineers, the Federal Oil Conservation Board, the American Petroleum Institute, the American Bar Association, and the Midcontinent Oil and Gas Association.[1] The early writers often felt that voluntary unitization could be achieved and was preferable in all respects to laws that might coerce unwilling owners into a unit. However, as the years passed and the many difficulties of securing voluntary agreements became evident, the literature almost unanimously favored the passage of compulsory unitization statutes (although these statutes still required a large percentage of voluntary agreement before compulsory process could be had). In 1940 Louisiana passed the first compulsory unitization statute, applicable to cycling in gas condensate fields. In 1945 Oklahoma passed the first compulsory unitization statute applicable to both oil and gas fields. From then on, slowly but steadily, a number of states enacted this type of legislation.[2]

Texas swam vigorously against this tide. Its earliest enactment, in 1931, was decidedly anti-unitization. In 1949 Texas finally passed a bill authorizing voluntary unitization in oil fields. A singular piece of legislation in the annals of oil- and gas-conservation laws, it remains the major legislation governing the unitization process in Texas.

Chapters 3 and 4 analyze the legislative policies that have shaped Texas' unique approach to unitization. Chapter 3 looks at the statutes enacted in the 1930s, which have molded Texas' policies toward unitization over the subsequent half-century. These early statutes consist of the 1931 and 1932 prorationing laws and the 1935 gas conservation act which approved voluntary unitization in gas fields. Chapter 4 provides a detailed statutory analysis of the 1949 voluntary unitization act, especially in terms of its limitations in encouraging unitization; a discussion of the failure of compulsory unitization legislation in Texas from 1949 to the present; and finally, an examination of Texas' compulsory pooling act, passed in 1965, which sheds much light on the issue of unitization in Texas.

The 1931 Anti-Market Demand Prorationing Act

The first express mention of unitization in Texas oil and gas laws appeared in 1931 with the enactment of House Bill 25 (H.B. 25). In the enumeration of wasteful practices that the Railroad Commission was to prohibit, the legislature included:

> Waste incident to or resulting from the unnecessary, inefficient, excessive or improper use of the gas, gas energy or water drive in any well or pool; *however, it is not the intent of this Act to require repressuring of an oil pool, or that the separately owned properties in any pool be unitized under one management, control or ownership* (italics added).[3]

This provision still stands, more than fifty years later, as a clear denial of the Railroad Commission's power to require unitization of an oil or gas pool to prevent waste. As if Texas' lack of a compulsory unitization law were not remarkable enough, the state's conservation laws evidence a uniquely hostile rather than a merely silent attitude toward mandatory unitization. In some respects it is ironic that this anti-unitization provision appeared in H.B. 25, a bill that in many other ways strengthened the powers and duties of the Railroad Commission to prevent the physical waste of oil and gas in Texas. Yet an analysis of the legislative history of this act shows that the insertion of the above italicized phrase was deliberate. The arsenal of more powerful ammunition given to the Railroad Commission to prevent waste in 1931 was definitely *not* to include the weapon of unitization. This legislative history is important not only for an understanding of the 1931 enactment, but also because it explains much of the opposition to compulsory unitization in subsequent decades, as well as many of the limitations placed on the 1949 bill that eventually authorized voluntary unitization agreements.

The legislative background of the 1931 Act shows that it was the unfortunate, but perhaps inevitable, fate of the unitization concept to be entangled in the bitter and protracted debate over prorationing of the East Texas field. The discovery of this gigantic field on October 9, 1930, by an independent wildcatter named Dad Joiner, set off a frenzied drilling spree so that a year later more than 3,000 wells had been completed in the field.[4] The result was a glut of oil on the market and a sharp decline in its price. By May 1, 1931, East Texas was producing more than 1 million barrels of oil per day, one-third of U.S. production, at a price of 10 cents per barrel versus its 1930 price of about $1.00 per barrel.[5] On this date, the commission's first proration order for the East Texas field was implemented, but it had no effect on the depressed price of crude oil.

On July 14, 1931, Governor Sterling called a special session of the legislature to pass a conservation act which would aid the "demoralized and tottering" oil industry, the thousands of people in the industry who were going bankrupt, and the state's treasury which was losing millions of dollars in reduced production taxes.[6] Both the House and the Senate set up committees to investigate the oil and gas situation. The committee hearings lasted almost two weeks, and the roster of witnesses included the three railroad commissioners, the governor and assistant attorney general of Texas, the presidents of Humble Oil and Refining (known today as Exxon Co. U.S.A., a division of Exxon Corp., which was formerly the Standard Oil Co. of New Jersey), Gulf, and the Texas Company (known today as Texaco, Inc.), large and small independent producers, landowners, mayors, statisticians, the chairman of the University of Texas Board of Regents, geologists, and attorneys. The transcript of these hearings is probably the single most important document to an understanding of the Texas anti-unitization course.[7]

The key issue before the legislators at the special session was whether the Railroad Commission should be given the authority to prorate oil production on the basis of reasonable market demand, that is, whether the agency should be authorized to limit the total supply of oil produced from Texas fields to the amount reasonably demanded. The issue was often phrased in terms of whether the commission should be empowered to prevent "economic waste" or should be limited to preventing physical waste. Under the state's original conservation act of 1919, the commission had been granted broad powers and duties to prevent physical waste such as the escape or wasteful burning of gas and the drowning of gas strata with water.[8] Then, in 1929, the original act was amended and a proviso was added stating that waste "shall not be construed to mean economic waste."[9] To many witnesses at the 1931 hearings, economic waste meant the production of oil at an unreasonably low price so that producers could not earn a "fair profit."[10] However, some of the major oil companies—particularly Humble Oil and Refining Co.—included in "economic waste" the excessively high costs of production resulting from drilling more wells than were necessary to drain a field efficiently.[11] These excessive costs precluded producers from earning a fair profit as did prices that were too low. The prevention of both types of economic waste—depressed prices and overdrilling—was enormously controversial. Antipathy to market demand prorationing arose from its clear relationship to outright price fixing in aid of an industry already notorious for monopolistic practices.[12] Four themes, which are discussed below, recurred throughout the days and nights of testimony at the special session.

A Threatened Monopoly

The dominant theme was the threatened monopoly of Texas crude oil by a few major oil companies. In probing for the reasons that crude oil sold at 10 cents per barrel, the lawmakers heard much evidence that this price resulted from a conspiracy of the major oil companies to drive the independents out of business. Many witnesses testified that the vertically integrated majors could drive the price of crude oil down and lose profits in their producing operations, but, at the same time, maintain overall corporate profitability by keeping prices high in their pipeline and refinery operations where they possessed substantial monopoly power.[13] Also, those majors with overseas production and profits could easily weather the East Texas situation without facing bankruptcy. Indeed, by importing crude oil into Texas, they could increase the oversupply and hasten the demise of the independent producer who had no other financial base.[14] The facts showed that the majors did control more than 90 percent of the pipelines in the United States.[15] Many independents testified that this pipeline power was used to drive them out of business.[16]

Texas had passed a Common Purchaser Act in 1930 whose purpose was to assure that all producers had equal access to oil pipelines, but it admittedly was not enforced by the Railroad Commission.[17] Independent producers testified to their inability to secure pipeline connections and to sudden cutoffs in existing connections. Other witnesses testified that the majors conspired to keep the price at 10 cents per barrel in order to force small companies, who faced bankruptcy at this price, to sell their leases to the majors at cut-rate bargains.[18] The majors had been late in recognizing the potential of Dad Joiner's initial strike in East Texas, and hundreds of independents had moved in quickly and secured acreage in the field before the majors woke up.[19]

Once awakened, the majors entered the field in force, fiercely determined to purchase acreage. Humble's own data showed the success of this movement. By July 8, 1931, the nineteen largest operators had secured 57 percent (or 69,730 acres) of the field's total productive acreage, which measured 122,229 acres. Humble alone had secured 15.75 percent (or 19,248 acres) of the total; Gulf had 8.75 percent (or 10,692 acres); and Shell had 5.53 percent (or 6,760 acres).[20] If market demand prorationing were imposed on top of the depressed price situation, the independents' production would be severely curtailed. Then both the price and quantity of oil sold would be low, and the independents' bankruptcy hastened. Many independents had little faith that prorationing would raise prices. Independent refiners also testified that they were doomed by the majors' pipeline monopoly, their exclusive Patent Club, and their control of prorationing.[21]

The advocates of market demand prorationing were primarily the major oil companies and the larger, established independents.[22] They strove to convince the lawmakers that the 10-cents-per-barrel price was due to free market forces, not monopoly.[23] They argued that the country's deepening recession had lowered demand, and the East Texas field had flooded the market with supply. The price remained at 10 cents, even after the Railroad Commission's prorationing order, because "outlaw" and "illegitimate" producers violated the order with impunity and ran millions of barrels of "hot" oil.[24] The advocates also tried to establish that market demand prorationing was essential to prevent physical waste,[25] and that it would benefit all Texans, including independent producers, by restoring higher prices.[26]

While some of the independents' allegations of monopoly and price conspiracy were imperfectly documented, the record spread before the legislators from the majors' own testimony, or that of public officials, showed much evidence of antitrust violations and market power.[27] Also, the argument that market demand prorationing was needed to prevent physical waste rang hollow because the Railroad Commission already had been granted strong powers under the 1919 Act to accomplish this, and because the evidence showed so clearly that the prorationing orders were not based on physical waste. Commissioner Neff explained to the lawmakers how the East Texas prorationing order was made:

> We have a hearing and develop much technical knowledge as to the underground waste and the topground waste and the evaporation and other scientific information, . . . but when the hearing is adjourned, . . . the Railroad Commission then inquires of the different fields how much oil can be sold, . . . and when that is determined, then the Railroad Commission fixes its allotment to that field based upon what the field is able to sell and not based upon any scientific knowledge, . . . [or on] the conservation of either oil or gas. . . .[28]

In the middle of the hearings, a federal district court confirmed the suspicions of many lawmakers by handing down a decision in

MacMillan v. Railroad Commission,[29] which enjoined the commission from enforcing its prorationing order in East Texas. The federal judges felt no need to detail the evidence that prorationing was being used primarily as a price-raising device because this fact was "so known to every man, that this court could fairly have taken judicial cognizance of the matters disclosed by the evidence."[30] Texas' populist heritage was strongly antimonopoly. In this political atmosphere, the three elected railroad commissioners and the governor quickly sought to distance themselves from the major oil companies and any possible support of market demand prorationing. On the eleventh day of the special session, Governor Sterling sent a letter to the legislators stating that passage of a market demand prorationing act would "tend to bring a condition where the oil interests of this state might create a monopoly in this important part of the people's business."[31]

The Railroad Commission's Lack of Control

A second recurring theme of the hearing was the Railroad Commission's incompetence and inability to understand—much less regulate and enforce—the complexities of oil and gas production. The oil fields were basically regulated by industry committees dominated by the major oil companies and larger independents. The Railroad Commission's prorationing orders were enforced by "umpires," who were elected by the industry's Central Prorationing Committee (CPC). Most often, they were experienced employees of the major oil companies.[32] The umpires' salaries were paid by the Central Prorationing Committee whose source of funds derived largely from the major oil companies' contributions.[33]

The Railroad Commission's own employees, called supervisors, were concerned only with enforcing proper drilling and production practices, such as minimizing fire hazards. Even this, they seemed hopelessly incapable of doing properly.[34] The commission's staff was controlled by a personnel policy under which each commissioner hired one-third of the employees. Generally, the employees were friends or loyal political supporters of the commissioner who named them to their positions, and had no real qualifications for the job. The supervisors were clearly less well educated, less well paid, and much less powerful than the umpires who policed the prorationing orders. The Railroad Commission did not employ a petroleum engineer; and its chief supervisor of oil and gas was a civil engineer who had once handled railroad matters.[35] The only petroleum engineer to testify in support of the commission's power to prorate was Mr. Foran, a consultant to the industry's Central Prorationing Committee.[36]

A large portion of the hearings was devoted to determining who wrote the Railroad Commission's proration order for the East Texas field—the commissioners or Robert Hardwicke, the attorney employed by the CPC.[37] The commissioners who testified were unable to offer any advice to the legislators as to what new legislation was required to cope with the crisis in the oil fields, nor could they offer any opinions about the cause of the depressed market price.[38] One commissioner adamantly opposed market demand prorationing and accused his fellow commissioners of consciously refusing to enforce the existing conservation laws by taking the "lines of least resistance."[39] Governor Sterling's dedication to representing the public interest was seriously questioned when it was disclosed that he recently had received $225,000 in advance royalties from Humble Oil and Refining.[40] The governor was one of the founders of Humble and had been its first president. Nor did the Attorney General's Office seem to have any command over the oil fields.[41] In sum, no state officials, either at a high or low level, seemed to have any knowledge or control over oil field operations. The legislators could not be assured that the conservation laws were being administered with the public interest in mind rather than being manipulated by powerful private oil interests.[42]

In this situation, market demand prorationing was vilified for two seemingly contradictory reasons: that it would raise prices to farmers and consumers suffering through the Great Depression, and that it would lower prices and drive independents out of business. These conflicting effects were reconcilable as a result of the same basic force: the majors controlled the mechanics of market demand prorationing and could use it as a monopolistic device—first, to destroy the competition, and then to raise the price of crude oil.

Devising an Equitable Prorationing Plan

The third theme of the hearings testimony was the impossibility of devising a prorationing plan of any sort for Texas that would be fair and equitable to all and still prevent the physical waste of oil. The legislators from other oil-producing parts of the state were worried that East Texas, with its prodigious reserves of high-quality crude so close to the more populated areas of the state, would be allocated so large a portion of the total state allowable that other fields would be forced to shut down.[43] Intrapool prorationing formulas were just as politically sensitive as interpool allocations. Whether the East Texas field should be prorated on the basis of well potentials, surface acreage, or a flat, per-well allocation absorbed much of the solons' time.[44] The existence of hundreds of small tracts overlying the East Texas field, the hundreds of already drilled wells, and the unique nature of the East Texas water drive made any prorationing formula difficult to accept.[45] Small-tract owners and their lessees obviously desired allocation on a per-well basis. Large-tract owners preferred the "unit plan," which allocated production on the basis of 20-acre drilling units. The Railroad Commission waited for six months before enacting its first prorationing order in East Texas. During this time many wells were drilled on fewer than 20 acres, thereby making it very difficult to revert to an acreage basis. The need to apply prorationing and well-spacing rules early in the life of a field in order to achieve orderly development was well recognized by the major oil company representatives who testified. So difficult was it to devise a prorationing formula for East Texas that even the East Texas Advisory Committee of the Central Prorationing Committee had given up trying to formulate a general rule, and had left it to the umpires. The chief umpire's description of how he allocated the number of barrels to the East Texas wells only served to show how intricate and complex the job was.[46] Exceptions and "special concessions" seemed to be the rule. Certainly neither the railroad commissioners nor their staff could hope to master the system.

Unitization of Oil Pools Versus Prorationing

These three themes interwove with a fourth one to condemn the unitization concept. The clear alternative to market demand prorationing—indeed, a superior alternative, according to most of the major oil company officials who testified—was to unitize each oil pool.[47] By placing each field under the management and control of one oil company, the field could be developed without the frenzied drilling of offset wells ringing each lease line and without wide open, flush production that wastefully dissipated reservoir energy. To the majors, the "economic waste" threatening the oil industry's profitability was as much the result of excessively high costs of production from overdrilling as it was of depressed prices. Unitization would minimize the number of wells needed to produce a field efficiently. William Farish, president of the Humble Oil and Refining Co., testified that unitized pools with proper gas conservation and repressuring could produce oil at one-fourth the cost of competitive drilling operations and produce 50 percent more oil per acre in the long run.[48]

Aside from any fears that unitization would allow the majors to dominate the oil fields, the allure of unitization as a method of increasing ultimate recovery and producing more oil was decidedly weak in light of the depressed market for petroleum products in 1931 and the existing glut of supply. Even the ardent proponents of unitization admitted that its greatest benefit was to lower operating costs, not to conserve oil.[49] To many lawmakers, this professed benefit was not a virtue. It meant reduced drilling activity and increased unemployment in the oil fields at a time of national economic depression.[50]

Humble Oil's Farish was the strongest advocate of unitization at the hearings. He explained that Humble, as early as 1927, had foreseen a long-term oversupply of oil in the United States and had determined that the only way to keep producing profitably in this environment was to produce at low cost through unit operations.[51] Humble had developed a policy of only buying into oil fields in which it could secure enough acreage for efficient operations. Farish described two ideally operated fields in Texas—the Sugarland field and the Yates field.[52] A petroleum engineer testified that oil from the unitized Yates field was produced at 4 cents per barrel.[53] Thus, even at a price of 10 cents per barrel, the Yates operators could make a profit. To legislators concerned with the majors' possible monopoly of Texas oil, these two examples were not reassuring. Humble owned the entire Sugarland field, and the

Yates field was voluntarily prorated by the fifteen operators owning leases in the field, mainly because Humble had exercised its pipeline monopoly power to enforce such a system. Very quickly after the large size of the Yates field became apparent in early 1927, Farish addressed a letter to the Yates producers stating that Humble would build a pipeline to Yates provided that the producers prorated their production to the pipeline's capacity. At the time, Humble had the only pipeline system into West Texas.[54] Humble also had attempted the same sort of pipeline exercise in the East Texas field. The following letter which Humble had addressed to a large meeting of East Texas operators on January 15, 1931, was read into the record at the special session:

> To secure orderly development and lowered costs of production and to prevent the waste of oil and gas resulting from rapid development and overproduction, it is necessary that rapid and close drilling of wells be discouraged. It is easily seen that production under present conditions can be raised far beyond the demand for oil in this area. . . . It follows that slow drilling with widely spaced wells is desirable from every standpoint. . . .
>
> Considering the size of tracts in this area, we believe that a sane conservation program with equitable participation in the market outlet would be obtained by dividing the field into 20-acre units and permitting each producing unit to share in market outlet ratably on the basis of the average potential of the wells located thereon.
>
> The size and location of the Humble Company's holdings in this area justify the extension of its pipeline to serve its properties. This extension will be made. In the event a program of orderly development and production with proration along the lines above set forth is worked out by the Railroad Commission, this Company will undertake to provide a market for such quantities of crude produced in the area as it can use itself or for which it can find a market demand; and in such event it will run the oil of other producers along with its own production on the basis of the proration schedules so established. In the absence of an orderly program of development and production, it would be foolish for the Humble Company to attempt to serve the area generally. No good would result to us or to the producers generally. In the absence of such a program we

> would be compelled to seek to protect our own properties as best we could in the dog-eat-dog scramble for advantage in which the owner of each tract seeks to secure all the oil he can at once from the property owned by him whether it comes from beneath his property or is drained from his neighbor's land.
>
> It is seriously to be doubted if any substantial purchasing company will feel justified in these times in seeking to provide a market outlet for the area in the absence of some effective Commission order for orderly production and proration.[55]

Humble's reference to "any substantial purchasing company" in the letter fanned the fears of a concerted conspiracy among the major oil companies. To legislators, who had heard much testimony about the pipeline companies' abuse of independents and about the majors' conspiracy to buy up East Texas leases from independents at cut-rate prices, Humble's dedication to securing large tracts of land and to unitization was only questionably in the public interest of efficient production. Of all the witnesses, Farish was the best prepared, most knowledgeable, and most persuasive on the need to prevent the waste of oil and gas in Texas and to do so in a fair and equitable manner.[56] However, the Humble Oil and Refining Co. was owned by the Standard Oil Co. of New Jersey and, as the bearer of this standard, Farish's testimony was automatically tainted. The API's sudden advocacy of unitization at the hearings, after years of virulent opposition to Doherty's ideas, suggested that the majors' primary interest in unitization was to control production and prices, not to prevent waste.[57] The majors' change of heart concerning unitization made their quest for antitrust immunity for voluntary unitization and prorationing agreements appear to reflect an attempt to secure a "legal" monopoly of the oil fields,[58] rather than an effort to shield efficient behavior from the risk of erroneous antitrust attacks. And when the proponents of unitization recommended that a billion-dollar corporation be formed to manage the East Texas field, the shadow of corporate empire-building in the Rockefeller tradition became all too real.[59]

In this atmosphere, support for unitization was as much a vote in favor of oil monopoly as support for market demand prorationing.[60] Thus it is not surprising that when H.B. 25 was passed, the Railroad Commission was prohibited in the strongest of terms from considering economic waste,[61] and its power over physical waste expressly excluded the authority to require that "the separately owned properties in any pool be unitized under one management, control or ownership."[62]

While the transcripts of the legislative hearing vividly illuminate the atmosphere surrounding passage of the 1931 Act, it is dangerous to rely solely on this testimony as being representative of the whole truth about the events leading to the passage of the Texas anti-unitization and antirepressuring statute. In the highly charged atmosphere of the House and Senate chambers, emotions, prejudice, and political hyperbole often prevailed. One piece of information, in particular, would lead an observer to question the motives, hence the veracity, of some of the independents who characterized prorationing and unitization as monopolistic schemes of the major oil companies. The data showed as of July 10, 1931 that the nineteen largest oil companies in the East Texas field had acquired 57 percent of the acreage, but had produced only 36 percent of the field's total output. The next twenty operators in the field had 12 percent of the acreage and had produced 15 percent of the oil. The remaining 586 known operators in the field held 20 percent of the acreage, but had produced 49 percent of the oil.[63]

To most independents in East Texas, the prospect of unitization or of prorationing was a deadly threat. Either practice would necessitate establishing an equitable allocation formula in the field. A fair formula would undoubtedly include an acreage factor, and this would reduce or eliminate the independents' existing advantage. The advocates of unitization and market demand prorationing urged the adoption of a new concept of ownership of oil and gas in Texas: that each landowner was entitled to recover that portion of the oil in the pool which underlay his land.[64] This concept was diametrically opposed to the common law rule of capture that allowed a landowner to drain his neighbor and secure a disproportionate share of the oil in the reservoir without liability. Many of the major oil companies testified that the rule of capture was nothing more than a rule of piracy hurting independents and majors alike. Indeed, many independents also testified that the need to drill offset wells to prevent drainage from their tracts severely strained their limited financial resources and was sometimes used viciously to destroy them.[65] Nonetheless, given the data showing who was winning the race to drain this giant new field, many East Texas independents were not likely to embrace this new concept of correlative rights. With such enormous stakes at issue, one would expect some independents to oppose prorationing and unitization for purely self-interested reasons rather than because of any real abuses of monopoly power by the majors.

Moreover, the themes that permeated the session do not fully explain the phrasing of the statute which—in addition to opposing unitization—states that "it is not the intent of this Act to require repressuring of an oil pool." When H.B. 25 passed into law, it greatly strengthened the commission's powers over physical waste. It enlarged the categories of physical waste that were prohibited and for the first time expressly authorized the commission to prorate production in order to prevent physical waste.[66] The 1931 Act also abolished the private umpire system and levied a new production tax to provide funding for the commission to enforce its duties.[67] It is clear from the act itself and from the hearings that the lawmakers were seriously concerned with physical waste. It seems strange, then, that the legislators withheld from the commission the power to require repressuring, even though it could prevent physical waste and did not relate to the economic waste of overdrilling or of prorating production to raise prices.[68] Thus, it is necessary to look at other accounts of the events of this time, especially to assess whether the threat of monopoly was real or was simply a bugaboo raised by the independents to exploit the political process and retain a disproportionate share of the East Texas wealth.

Even the briefest review of the works of more dispassionate scholars of the era confirms the accuracy of much of the testimony heard by the lawmakers in 1931. It is indisputable that the moving force behind the proposed market demand law was to raise and stabilize oil prices;[69] that the major oil companies had formed an international cartel in crude oil;[70] that considerable pipeline monopoly existed;[71] and that the early commission was an incompetent agency.[72] One of the best documented and most insightful studies of this era appears in the official biography of the Humble Oil and Refining Co. By Humble's admission, the company's own actions were understandably, although incorrectly, interpreted in ways that fueled the antiprorationing and anti-unitization forces.[73] In Humble's view, the 1931 hearings were simply an "informal trial of Humble."[74] The prominent influence of this one company on Texas oil- and gas-conservation laws bears closer attention because the nexus between unitization and monopoly established in 1931 would pervade Texas politics for decades to come. A look at Humble's special position in the Texas oil industry also provides essential background on the Texas antitrust laws which were, until very recently, inimical to voluntary unitization agreements.

The Role Played by the Humble Oil and Refining Co.

The Humble Oil and Refining Co. was incorporated in Texas in 1917. Seven of its nine founders were independent oil producers who had come to realize that, by uniting their separate companies, they would have greater bargaining strength against the considerable power of the vertically integrated majors such as Gulf, the Texas Co., and Sun Oil who were already established in the Gulf Coast area.[75] The Texas oil industry was born in 1901 with the discovery of Spindletop, an oil field on the Gulf Coast.[76] At that time, Texas was an attractive arena for the independent producer. Standard Oil of New Jersey controlled most of the crude oil production in the older, eastern fields, but Standard Oil did not enter the Gulf Coast area as a producer, even after Spindletop presaged the great wealth lying beneath Texas' surface.[77] Standard Oil's initial decision to stay out was partly because of the state's political climate and antitrust laws, which were decidedly hostile to Standard Oil. Even before the federal government passed the Sherman Antitrust Act of 1890, Texas had enacted a very stringent state law designed to prevent the growth of trusts and monopolies. The Texas antitrust law of 1889 defined a trust as a "combination of capital, skill, or acts by two or more persons, firms, corporations, or association of persons" created for any of the following purposes:

> First: To create or carry out restrictions in trade.
> Second: To limit or reduce the production, or increase or reduce the price of merchandise or commodities
> Third: To prevent competition. . . .
> Fourth: To fix at any standard or figure, whereby its price to the public shall be in any manner controlled or established, any article or commodity of merchandise, produce, or commerce intended for sale, use, or consumption in this state.

> Fifth: To make or enter into . . . any contract . . . by which they shall agree to pool, combine, or unite any interest they may have in connection with the sale or transportation of any such article or commodity that its price might in any manner be affected.[78]

The same populist, agrarian philosophy that produced the antitrust law of 1889 led to the formation of the Railroad Commission in 1891 in order to regulate railroad practices and protect farmers, small businessmen, and the public from the rapaciousness of Eastern magnates like Jay Gould.[79]

Acting under the 1889 antitrust law, the Texas attorney general, not once, but twice, ousted Standard Oil's marketing affiliate, the Waters–Pierce Co., and perpetually enjoined it from ever doing business in Texas. In both the 1898 and the 1907 antitrust suits, Waters–Pierce's business charter was forfeited and, in 1907, a fine of $1.6 million was assessed jointly against the company and Standard Oil. The history of Standard Oil's early practices in Texas can only be described as sordid.[80] Thus, even before the federal government dissolved the Standard Oil trust in 1911, Texas had battled the Goliath in two highly publicized trials. Not surprisingly, Standard Oil was a greatly hated institution in Texas.

In 1919 the founders of Humble Oil and Refining Co. needed capital to expand their production activities and to integrate into pipelines and refineries. Standard Oil of New Jersey, eager to secure Texas production to serve its still massive network of pipelines and refineries to the east, bought 50 percent of the stock of Humble in 1919 for $17 million.[81] The Texas attorney general brought another antitrust suit in 1923 seeking to forfeit Humble's charter because of its relation with Standard Oil, but the court in 1924 ruled that a foreign corporation, though holding the majority of stock in a Texas corporation, was not doing business in the state, and Humble survived.[82]

With vast capital at its command, Humble grew quickly. By 1925, it was the largest oil producer in Texas, and by 1928, it was the largest pipeline company in the United States.[83] So profitable were its pipeline operations, even in the Depression years, that Standard Oil itself complained about the rates it had to pay on oil purchased from Humble.[84] However, despite its own large oil and gas reserves, Humble purchased four times more oil than it produced.[85] It was in its role as the largest purchaser of oil in Texas that Humble became the object of the independent producers' fury whenever the price of crude oil fell. Indeed, it was Humble's announcement of a price decrease in 1923 that led to the attorney general's filing the antitrust suit against Humble seeking forfeiture of its charter.[86]

William Farish was the towering figure of Humble and served as its president from 1922 until 1933, when he resigned to become chairman of Standard Oil of New Jersey. At first, Farish's attitude toward unitization was negative, and he scorned Doherty's advocacy of unitization as a conservation device in 1924.[87] However, by 1926, Farish had become a champion of unit operations and of the use of gas to repressure oil pools and increase recovery rates. Farish's newly found respect for unitization and repressuring derived from the growth of knowledge about the function of gas in oil recovery. His education was advanced by the experiments of Humble's own staff of scientists and engineers. By 1926, Farish was urging the Railroad Commission to require gas recycling in the Panhandle in order to increase oil recovery.[88] In 1928 Farish thanked Doherty for making the industry realize the need for conservation.

The post–World War I push to find new reserves was so successful that the price of a barrel of oil fell from $1.85 in 1926 to 99 cents in 1930.[89] From 1927 onward, Humble adopted a formal company policy to find large reserves to develop at low cost, implementing it in two ways: (1) by leasing and purchasing large blocks of land; and (2) by promoting and seeking unitization.[90] Humble's quest was to secure leases covering an entire field, so that it could develop its acreage with wide spacing and at controlled production rates, without worrying about drainage by offset operators. Humble succeeded spectacularly in its block-leasing policy. In 1930 Humble Oil experimented with pressure maintenance in its wholly owned Sugarland and Olney fields.[91] The results showed that pressure maintenance was cheaper than putting the fields on pumps to secure secondary recovery oil. Humble began paying advance royalties to landowners in exchange for the right to delay drilling and to produce at slower rates.

Farish tackled the second goal of securing unitization with equal vigor. As early as 1927, Humble sought antitrust immunity for voluntary units in Texas.[92] Under the terms of the Texas antitrust act, largely unchanged since 1889, voluntary unitization (or prorationing) agreements were virtually *per se* violations of the law. Humble's attorneys advised that the only way to avoid the strictures of the antitrust law, absent legislative immunity, was to unitize before oil and gas were discovered—that is, to form exploratory units. The legal department drew up a form for exploratory units, and even succeeded in securing such an agreement for the Van field in 1929.[93] Unitization of the Van field involved only four other major oil companies; no independents held acreage in this field. However, it was exceedingly difficult to achieve unitization voluntarily in most cases, and while Farish had never advocated compulsory unitization, the need for the state's police power to achieve unitization became more and more evident.

Humble's understanding of and push for unitization and gas repressuring during the late 1920s were far ahead of most producers, including most major oil companies, and found little support in the industry.[94] Even Teagle, the president of Standard Oil of New Jersey, thought Farish had gone too far in urging the Railroad Commission to require gas repressuring. In 1929 Humble promoted passage of House Bill 388 which—in addition to strengthening and revising the original conservation act of 1919—authorized the majority of producers in a field, with the permission of the Railroad Commission, "to make and enforce orders for orderly development of separate tracts."[95] Humble supported this bill energetically and even printed two pamphlets explaining the virtues of unitization that were distributed widely to the public. Most independents and many majors, especially Gulf Oil, strongly opposed any state regulation.

H.B. 388 was passed in 1929, but without Humble's desired provision. Instead, the bill stated that the commission's authority to prevent waste "shall not be construed to mean economic waste."[96] This proviso was clearly meant to prohibit the commission from issuing orders designed to prevent unnecessary drilling or to restrict production to market demand—two tenets of Humble's credo for effective state regulation. Thus, this 1929 proviso was really the first anti-unitization statute passed by the Texas legislature. It was enacted in direct opposition to Humble's principle that the state should prevent overdrilling and flush production.

Unable to secure support for gas repressuring and unitization, Farish stopped promoting these ideas and looked instead to state control of production rates as the solution to preventing waste.[97] By 1928–29, the worldwide surplus of oil was so large that the API and most international majors—including Standard Oil of New Jersey—sought to implement a worldwide prorationing system that divided the globe into regions and established production quotas for each. The U.S. attorney general refused to sanction this agreement under the antitrust laws, so it became even more important to secure state control of production rates. Humble campaigned for statewide prorationing in Texas and for the formation of an organization of producing states to coordinate supply with demand nationwide. Humble's crusade for prorationing was strongly based on the need to prevent the physical waste of oil and gas, not on the need for higher prices and price stability.[98] As a low-cost producer, Humble would be able to survive an era of lower-priced oil far better than other companies. Farish did not consider prorationing rational if its intent was simply to raise prices. Higher prices would lead to further drilling and a self-defeating spiral of increasing costs and reduced profits. Nonetheless, Humble did propound the benefits of higher prices and price stability that would inure to all operators in the industry if prorationing came to be.

While Humble was crusading for statewide prorationing, the oversupply of crude oil had resulted in the company's storing huge amounts of crude in expensive tanks. In an attempt to reduce its storage costs, Humble initiated a drastic, crude oil price cut in early 1930. This was its undoing. The price cut occurred only six weeks after Farish had publicly declared that no excess crude was being produced. The fury that followed this action left such a legacy of bitterness, by Humble's account, that "for at least three or four years almost any rumor or charge—no matter how baseless—as to the selfish, unscrupulous, ruthless, and arbitrary character of Humble received credence from a considerable segment of the Texas oil industry."[99]

The angry independents retaliated for the price cut by seeking legislation at both the state and national levels.[100] The Independent Petroleum Association almost secured passage of a tariff on foreign crude, a move designed to hurt Standard Oil of New Jersey which imported large amounts of crude. The Texas independents did succeed in securing passage of the Common Purchaser Act of 1930, which required any purchaser of oil in Texas affiliated with a common carrier pipeline to purchase oil without discrimination between fields or producers. This bill was intended to penalize Humble, but Farish actually welcomed the act because it granted the commission more power to curtail production in oil fields to meet the capacity of the field's pipeline outlets. Just before the Common Purchaser Act was to take effect in June 1930, Humble shocked and outraged the industry further by announcing that it would discontinue purchasing oil in seven counties in North Texas.[101] Humble made this move to spur the Railroad Commission into prorationing the production from these counties whose output exceeded Humble's pipeline capacity. When the commission started work on a prorationing program, Humble agreed to continue purchasing the North Texas oil, even though it did not need the crude. To many, Humble's action displayed the raw arrogance of a monopolist with tremendous economic and political power.

Then, on August 14, 1930, the commission issued its first statewide prorationing order. Shortly thereafter, Humble cut the price of crude oil.[102] Humble's campaign for statewide prorationing had tacitly promised that such an order would restore healthy prices to the industry. Humble explained that its price cut was a result of market forces: the commission's prorationing order was not obeyed or enforced and, as the glut of oil continued, Humble could not afford to purchase oil at prices far above the market price at which its competitors were purchasing. To many independents, especially those who had supported Humble's campaign for prorationing, the price cuts amounted to treason.

All this preceded the events in East Texas in 1931. Humble's account verifies much of the testimony at the 1931 hearings, but in one regard the witnesses were incorrect: not all the majors had overlooked East Texas as a promising geological area. Humble had purchased land overlying the East Texas field and had prepared to drill a test well in April 1930. Had it not been for title difficulties, Humble—not Dad Joiner—probably would have discovered the East Texas field.[103] Humble moved quickly to buy up acreage in the field after the discovery well came in. Within a month, the company had bought 16,000 acres, or 13 percent, of the field.

Ultimately, Humble secured 16 percent of the field and was by far the largest, single owner of reserves.[104]

As both the largest purchaser and producer in the East Texas field, as well as being an offspring of Standard Oil, Humble was in an untenable position. The Railroad Commission was not able or willing to regulate the early development of the field. Two months after Dad Joiner's well set off the drilling spree, Humble addressed the letter about pipeline access (quoted earlier) to the producers in the field in an attempt to pressure the commission into prorationing the field, just as Humble had done in North Texas. The commission took no action, and so in February 1931, Humble built its own pipeline into the field to serve its own wells.[105] When the commission finally issued a prorationing order for East Texas to be effective on May 1, 1931, Humble made pipeline connections with other producers' leases. Yet on May 26, only three and a half weeks after the prorationing order became effective, four major oil companies, including Humble, reduced the price of East Texas crude from 65 cents to 35 cents a barrel, and then to 15 cents.[106] These price cuts were a repeat of the majors' reaction to the first statewide prorationing order issued in 1930. Prorationing orders led to price decreases, not increases, by the majors. When the governor called the special legislative session in July 1931, the stage was set to enact more anti-Humble legislation, and that is precisely what the legislators did.[107]

In light of Humble's own account of the events preceding the session, the 1931 proviso that "it is not the intent of this Act to require repressuring of an oil pool or that the separately owned properties in any pool be unitized under one management, control, or ownership" can be seen as a uniquely anti-Humble statute. It was Humble that had campaigned longest and loudest for unitization; that had developed legal forms for exploratory units; that had brought producers to cooperate in the Yates field by using its influence as the only pipeline owner in the area; and that had actually implemented a unitization plan in the Van field. It was Humble that had requested the Railroad Commission as early as 1926 to require that the gas produced in the Panhandle fields be returned to the reservoirs to repressure the fields and that had actually instituted pressure maintenance in several fields. It was Humble alone that had distributed pamphlets to the public in 1929 explaining the virtues of unitization and seeking antitrust immunity for voluntary cooperative agreements.

By juxtaposing the transcripts of the 1931 special session with Humble Oil and Refining's own history, the 1931 Act can be fully understood: physical waste was to be prevented, but not by any of the three methods—market demand prorationing, unitization, or repressuring—especially promoted by Humble.

While this history explains the 1931 Act, it also shows that the independents had much to fear from Humble as a competitor, not because of any abusive use of monopoly power, but because of Humble's superior business acumen, technology, and efficient, low-cost production. The company's drive for efficiency is in many respects the very model of how a competitive firm maximizes both private and social welfare in a private enterprise system. Humble's price cuts angered the independents more than any other action. Yet these cuts better prove the lack of monopoly power rather than its existence. The evidence shows that Humble could not either singly or in concert with other majors prevent other operators from pouring crude oil onto the market, except in those fields where Humble had the only available pipeline.[108]

This pipeline control lies at the heart of the antitrust issue. Humble's pipelines did have considerable monopoly power, as witnessed by their very high, long-term profits. Humble used this power in the Van, North Texas, and Yates fields to secure "voluntary" unitization or prorationing by the producers, or to push a reluctant commission into prorationing the fields. Humble tried the same tactic in East Texas but was unsuccessful because it had no monopoly on the market outlets for this field. The regulatory vacuum created by the slow-acting commission led the major oil companies to attempt to impose orderly development on this field by using their economic muscle, both singly and in combination. Clearly, Humble was correct in attributing much of the physical and economic waste in the oil fields to the rule of capture. Humble's letter to the East Texas operators, describing the "dog-eat-dog scramble" to drain neighboring tracts, is as vivid a description of the externalities and inefficiencies induced by the rule of capture as any economist's graphs and equations. The company's efforts to counteract the common pool problem and impose orderly development were probably in the public interest. Nonetheless, the crucial point remains that if Humble and the other major oil companies had the economic power to counteract the common pool problem by themselves, then they would have the economic power to do more than simply correct inefficiency and waste; they could monopolize the oil fields. The short-term price of crude oil under the unrestrained conditions in East Texas may well have been too low in terms of economic efficiency and social welfare (see chapter 2 text at notes 32 to 35). Had the private oil companies succeeded in raising the price of oil by controlling their own production and that of others, they would have prevented waste, but would they then limit their actions only to those needed to protect the public interest in efficient production? Using the same economic muscle, they could raise the price above the socially optimal level and use their pipelines as barriers to entry by other competitors.

That the majors' efforts to control production failed in East Texas is proof that they lacked monopoly power in this field. That the effort was made is virtual proof of antitrust violations (especially when the majors discussed joint pricing behavior among themselves).[109] Even assuming that Humble's private attempts to prorate production equitably were in the public interest, no citizen, public official, or lawmaker was likely to accept this assumption as true, given the past misdeeds of Standard Oil. Nor should such a situation be accepted. Regulation of the oil fields should be done by a public agency, not by a segment of the industry itself.

To its credit, Humble knew that use of its pipeline power to pressure producers and the commission into prorationing was politically and legally untenable, and the company consistently supported legislation to give the commission stronger powers to control production and ratable takings by pipelines. Humble's biography argues rather convincingly that it was one of the few companies that tried to prorate all of its lease connections equitably so that none of the producers from whom it purchased was completely shut off from a market.[110] However, independent producers had great political clout in Texas, from the very beginning of the oil industry's growth in the state.[111] Even before East Texas was discovered, the independents, in 1929, had secured a legislative prohibition against the commission's regulation of the "economic waste" of overdrilling and flush production, and a Common Purchaser Act in 1930 to prevent discrimination by pipelines in their purchases of oil from nonaffiliated producers. This sequence of legislation shows that the independents desired equal access to and ratable takes by pipelines, but without production prorationing.[112] First, the independents wanted the opportunity to produce a disproportionate share of the reserves of a field under the rule of capture. Second, they wanted assured access to a pipeline that took ratably from all producers in the field based on the percentage shares of production that each producer offered for sale. In contrast, Humble desired equitable production prorationing in place of the rule of capture, followed by ratable pipeline takes based on the prorationing formula established for the field's production.

Thus, regardless of either the appearance or the reality of anticompetitive acts by the majors, the independents' political clout might have succeeded in killing any prorationing or unitization bill which threatened to take from them their opportunity, often already realized, to secure a disproportionate share of the oil through drainage from larger tracts via the rule of capture. The majors' use of private prorationing committees and pipeline power to bring about production control and unitization in the regulatory vacuum that existed in 1931 condemned both prorationing and unitization to a sure death.

The 1932 Market Demand Prorationing Act

H.B. 25 was passed on August 12, 1931. Its disapproval of market demand prorationing, plus the *MacMillan* court's invalidation of the East Texas prorationing order, led many producers to open up their wells. Five days later, on August 17, Governor Sterling declared martial law in East Texas and ordered the National Guard into the field to shut down all the wells. The governor's action was based on threats of violence and rioting between the flush producers and the advocates of prorationing who believed that unrestrained production would ruin the East Texas field. For more than a year, the federal and state courts and the commission struggled with the immense task of managing East Texas.[113] Their total lack of success led Governor Sterling to call another special session of the legislature in November 1932,[114] at which time the lawmakers reversed themselves and passed an act specifically authorizing the Railroad Commission to prorate production based on reasonable market demand.[115] This monumental flipflop in policy was not, however, extended to the proviso that prohibited the commission from requiring repressuring or unitization.

In light of the evidence presented in the 1931 hearings, the legislature's reversal of its position on the market demand issue seems incredible. How could market demand prorationing be any less monopolistic in 1932 than in 1931? Many factors contributed to the reversal. The effectiveness of martial law prorationing had raised the price of crude oil to 85 cents per barrel,[116] and many independents who had once opposed prorationing were no longer convinced of its evil.[117] The Railroad Commission now had a strong and forceful leader in Commissioner Ernest O. Thompson, who was so well respected that the commission was no longer seen as a pawn of the major oil companies, but as an agency having the public interest in mind.[118] Commissioner Thompson vigorously asserted that the proration orders were based on physical waste only, not on the desire to raise market prices. Moreover, a new Democratic president had been elected on a platform of active government intervention and price support in aid of all sectors of the depressed economy.[119] More facts were available about the East Texas field to support the theories of those who advocated restricted production to prevent physical waste.[120] Courts and legislators in other oil-producing states had accepted market demand prorationing.[121] Congress had passed a tariff on imported crude oil in 1932, and it was no longer as easy to blame domestic overproduction and low prices on the international majors.[122] Also, the majors, especially Humble, kept a low profile in 1932. By design, the chief lobbying organization for the market demand bill was the Texas Oil and Gas Conservation Association, a group composed mainly of independents. Humble supplied engineering data, advice, and financial aid to this organization, but otherwise did not campaign for the bill.[123] Moreover, the reversal in prorationing policy was *not* a reversal of Texas antitrust policy. The 1932 Market Demand Act added a new section to the conservation laws that was designed to assure the primacy of antitrust law over market demand prorationing. The redundant, hyperbolic style of this section shows some of the emotional tone underlying the antitrust issue in Texas:

> It is especially provided that nothing herein shall in any manner affect, alter, diminish, change or modify the anti-trust and/or monopoly statutes of this State, and that no provision of this Act shall in any manner directly or indirectly authorize a violation of such anti-trust and/or monopoly statutes, and in this connection it is hereby declared and especially provided by the Legislature of the State of Texas enacting this legislation that... it is the legislative intent that no provision of this Act shall in any manner, affect, alter, diminish or amend any provision of the anti-trust and/or monopoly statutes of this State, or in any manner authorize a violation of such anti-trust and/or monopoly statutes; and it is further especially provided that if any provision of this Act shall be so construed by any court of this State as to in any manner affect, alter, diminish, or modify any provision of the anti-trust and/or monopoly statutes of this State, then in that event any such section, subsection, sentence or clause or any provision of this Act so construed as conflicting with said monopoly and/or anti-trust statutes, it is hereby declared null and void rather than the anti-trust and/or monopoly statutes of this State. The legislative intent herein expressed is to prevail and take precedence over... any other section or sections of this Act, regardless of any statement therein to the contrary.[124]

As another safeguard to the public interest, the 1932 Act was scheduled to expire on September 1, 1935, unless reenacted.[125] Section 6-A of the 1932 Act also required that the Railroad Commission "take into consideration and protect the rights and interests of the purchasing and consuming public of crude oil and all its products."[126] This obviously was intended to restrain the commission from raising prices too much.

The 1932 Act also did not threaten a reversal of the legislature's pro-independent stance. Market demand prorationing was authorized only as a waste-prevention measure. While the definition of waste now included the production of crude oil in excess of reasonable market demand,[127] the 1932 Act did not authorize the commission to prorate oil solely to protect correlative rights. The absence of such authority was predictable and deliberate. The independents generally did not support the principle of correlative rights that would allow each owner of a common source of supply the opportunity to withdraw only his fair share from the reservoir.[128] Under the rule of capture, the independent operator could often produce more than his fair share by draining from others. Certainly, the independents in East Texas who were winning the race to drain the reservoir first would not willingly grant the state a duty to protect correlative rights.[129]

In the event that the commission prorated a field, the 1932 Act required that the allowable production be allocated among producers "on a reasonable basis."[130] This standard gave the commissioners considerable discretion. Given the enormous controversy over allocation formulas in East Texas in the 1931 hearings, it is unlikely that the legislators could have agreed on a more detailed statutory guidance for allocating allowables. A per-well formula was clearly desired by the independents, who had relatively more wells and less acreage than the majors.[131] In his testimony supporting the Market Demand Bill, Commissioner Thompson artfully dodged the issue of exactly how East Texas allowables would be distributed, but he pledged that prorationing would benefit the small producer more than the majors.[132] In light of this promise, it was unlikely that the commission would adopt a formula for East Texas that would surrender the independents' existing advantage in this field. Thus, the Market Demand Bill became more palatable to the East Texas constituency.

The 1932 Act also required that the commission prevent unreasonable discrimination in the allocation of allowables between separate pools of oil in the state.[133] This section placated producers in West Texas who feared that the East Texas and Gulf Coast fields would receive too large a share of the state's total allowable.

The 1932–33 Prorationing Orders and the Marginal Well Act

Still, the 1932 Market Demand Act did not bring peace and order to the East Texas field. The commission proceeded to vary the East Texas field's total allowable from 290,000 barrels per day to

750,000 barrels per day, vainly seeking the magic number that would be fair to all producers yet prevent physical waste. At the 290,000-barrel allowable, allocated on a flat, per-well basis, many producers violated the order in the belief that the limit was unrelated to any conservation objective and that the per-well formula was illegal. Hot oil again caused a sharp price decrease. At the 750,000-barrel level, allocated by well potentials, the price of crude fell to 10 cents and the reservoir pressure in the field dropped precipitously.[134] The commission desired to allocate the field's allowable on a per-well basis, but the federal district court repeatedly struck down this allocation method as unreasonable and discriminatory. Only after the federal court cited the commission for contempt did the agency change the allocation formula to one using well potentials to distinguish good from bad wells. Even then, well potential, which is the producing capacity of a well over a stated period such as twenty-four hours, ignores the amount of acreage drained by a well. There is no direct correlation between well potential and the quantity of oil recoverable under a tract.

The commission's seemingly irrational devotion to the per-well method of allocation was the result of two factors. First, the commission had pledged that market demand prorationing would not harm the independent producer who was favored by the per-well formula.[135] This virtually precluded prorationing based on shares of acreage or reserves. Simply put, the per-well allocation formula was the price the majors had to pay for the adoption of statewide prorationing.

Second, the Marginal Well Act of 1931 prohibited the commission from prorating the production of many wells. This act, passed on April 16, 1931, had been made effective immediately by a unanimous vote of both the House and Senate, so that it would precede the scheduled institution of prorationing in the East Texas field on May 1, 1931.[136] The stated purpose of the Marginal Well Act was to prevent waste by prohibiting market demand prorationing of wells that produced such small amounts of oil that curtailments would damage them, cause their premature abandonment, or result in the loss of recoverable oil. The act defined marginal wells as those pumping oil wells that produced only a stated number of barrels of oil daily from particular depth horizons.

As applied to the East Texas horizon, marginal wells were those pumping oil wells having a daily production of 40 barrels a day or less. Because the 40-barrel limit was so large and so many marginal wells existed in the East Texas field, these wells, if exempted from prorationing, would account for a huge share of the field's total allowable, leaving little for allocation to the more productive, high-flowing wells. After the federal court in *People's Petroleum Producers v. Smith* had invalidated as unreasonable the commission's per-well prorationing order for East Texas,[137] the commission was in a predicament. The Marginal Well Act required that marginal wells be free to produce without restriction. A prorationing formula based on well potentials was the least disruptive replacement for the outlawed, flat per-well formula, but it would necessitate that freely flowing wells be restricted to less than the 40-barrels-per-day limit given to pumping wells. Surely the federal court would not consider as reasonable this perverse situation that good wells should receive *less* than bad wells. The commission hastily wrote to the attorney general for advice: Would a prorationing formula based only on potentials be valid? Must an acreage factor be included to have a valid order? Would a valid order have to provide that no well could be restricted to an allowable less than the 40 barrels granted to marginal wells?

The attorney general replied that "[c]ertainly no order is valid . . . which does not, upon some reasonable basis, substantially give to each operator the same proportion of production to which he would be entitled if there were not proration."[138] In essence, this response gave producers a vested right in the rule of capture, the rule which would apply in the absence of prorationing. A prorationing formula based only on well potentials would closely approximate the rule of capture and so would be valid in the attorney general's opinion.

In answer to the second question, the attorney general wrote that "surface acreage is not an indispensable factor to be considered in writing a valid order, although it might be considered."[139] This answer also supported a prorationing formula based only on potentials. The attorney general warned, however, that if the commission did not see fit to include acreage in the formula, "It may result in the striking down of Rule 37 as applied to the East Texas field."[140] Rule 37 was the commission's well-spacing rule which, at that time, required 10-acre spacing. The attorney general clearly foresaw that large-tract owners—in order to protect themselves against drainage from small-tract owners who were favored by the well-potential formula and the Marginal Well Act—would assert the right to secure exceptions to the spacing rule and drill additional wells, as would have been their common law right.

As to the third issue, the opinion advised that any order that curtailed flowing wells below that of marginal wells would probably be invalid.[141] Hence, if the commission were to adopt a prorationing order based only on well potentials without raising the East Texas field's total allowable to a sum so large that it would probably cause physical waste, the Marginal Well Act of 1931 would have to be amended. Thus, on April 27, 1933, the Texas legislature revised the act and redefined marginal wells as those pumping oil wells producing fewer than 20 barrels a day from the depth of the East Texas horizon.[142]

Even so, as the number of wells drilled in East Texas increased without a proportionate increase in market demand, more and more wells received the 20-barrel daily minimum. The burden of prorationing then fell on the highest-potential wells, which had to be cut back to small percentages of their potential, resulting in an allowable to each of these wells of about 22 barrels per day.[143] So the difference between prorating on a flat per-well basis and prorating by well potentials was negligible, given the statutory constraints of the Marginal Well Act.

Of course, this was the very result desired by many independent producers. The Marginal Well Act ostensibly was passed as a conservation measure, but its legislative history—particularly its timing—shows that it was passed to protect and preserve the independent producers' existing advantage in the East Texas field.[144] There seemed to be little scientific basis for the maximum number of barrels used to define a marginal well in the 1931 Act; the numbers were halved just two years later. The legislated decrease from 40 to 20 barrels for the East Texas field does not show a turning away from the independents. The independents realized that a new prorationing order for East Texas, based only on well potentials, could not pass judicial muster unless the Marginal Well Act was revised. The well-potential formula was the next best thing to the per-well formula which the courts refused to allow.

Indeed, well-potential allowables approximated the rule of capture to a greater extent. By redefining marginal wells, the well-potential formula had a chance of surviving judicial scrutiny, thereby precluding a formula based on acreage or reserves in place. This was a chance well worth taking. On February 12, 1934, the federal district court upheld, albeit reluctantly, the commission's East Texas order dated April 22, 1933, based solely on well potentials.[145] The Marginal Well Acts of 1931 and 1933 illustrate to a greater extent than the Common Purchaser Acts of 1930 and 1931 the political power of Texas independent producers, particularly those in East Texas. This same political power well may have resulted in the 1931 proviso against unitization, even had there been no bona fide reason to fear that unitization would be used anti-competitively by Humble and the other majors.

Thus, the East Texas field left a legislative legacy that would influence oil and gas conservation in Texas for decades. It foreclosed compulsory unitization as a conservation option. It set a precedent for allocating allowables on a per-well basis, thus favoring small tracts, small producers, and small wells.[146] This precedent greatly hinders the ability to secure voluntary unitization agreements. The essence of unitized operations is to drill and produce the minimum number of wells required to recover the oil and gas, and to allocate the field's output so that each producer receives his proportionate share of the oil and gas beneath his tract. The more frequent the deviations from this pattern stemming from prorationing and well-spacing rules, the greater the difficulty in securing voluntary agreements to unitize. The East Texas regulatory scheme induced much wasteful and unnecessary drilling, resulting in enormous economic inefficiency that could have been prevented by unitization or pooling. As of January 1, 1938, 23,951 wells existed in East Texas; 16,000 of these had been drilled as exceptions to the field's spacing rule. Of these, 13,000 were unnecessary from an engineering viewpoint. Unnecessary drilling in Texas was estimated to cost more than $50 million per year.[147] Yet a half-century later, the field continues to produce huge amounts of oil at or close to its original reservoir pressure, and it will probably recover about 85 percent of the original oil in place. The water drive in the field is constantly maintained by a repressuring program, which reinjects into the reservoir the massive amounts of salt water produced with the oil. The method by which such cooperative repressuring and high ultimate recovery have been achieved, despite a prorationing system which is wholly inimical to unitization, is described in chapter 5.

The 1935 Gas Conservation Act

In 1935 the Texas legislature passed an act authorizing voluntary unitization agreements in gas fields.[148] At first blush, this seems to evidence an inconsistent, if not schizophrenic, attitude on the part of the lawmakers toward unitization. However, the legislative history and purpose of the 1935 Act show that it did not deflect significantly from Texas' anti-unitization policy. The 1935 Act, with its approval of unitized operations, was as much for the protection of independent producers as was the 1931 Act prohibiting the commission from requiring unitization.

The gigantic Panhandle gas field is to the 1935 Act what the East Texas oil field is to the 1931 and 1932 Acts. This field is an interconnected geological reservoir, 125 miles long, consisting of more than 1.5 million acres. About two-thirds of the field produces "sweet" (nonsulfurous) gas and the other one-third produces "sour" (sulfurous) gas. About fifty oil pools are scattered along the north side of the structure. By 1935, nine major pipelines held leases on 80 percent of the sweet-gas areas and produced and transported this gas for heat and fuel to large domestic and industrial customers. The remaining sweet gas was owned or leased by nonintegrated producers who had no market for their gas in the sparsely populated Panhandle. The pipelines would not purchase their sweet gas or any sour gas, nor would the pipelines purchase the casinghead gas produced from the hundreds of oil wells in the area.[149]

An 1899 statute required that gas wells be shut in unless the gas could be used for lighting, fuel, or power.[150] The nonintegrated gas producers in the Panhandle faced shut-ins and drainage of their gas to the pipeline producers unless the act was amended. In 1931 the legislature amended the 1899 statute and empowered the Railroad Commission to permit the use of gas for other purposes.[151] The commission allowed sour-gas producers to construct plants making carbon black, a substance used in the rubber industry, but the nonintegrated, sweet-gas producers were still left without a market. The commission attempted to require the pipeline companies to purchase sweet gas ratably, without discrimination, from all the producers in the pool under the Common Purchaser Act of 1931,[152] and under the agency's general authority to prevent waste. However, the federal district court twice enjoined the Railroad Commission from enforcing its orders on the basis that the agency lacked the authority to protect correlative rights when no issue of waste was involved.[153] Faced with these court rulings, the legislators, at the same 1932 special session in which the Market Demand Act was passed, also enacted a section designed to authorize market demand prorationing of gas among producers on a reasonable basis.[154] With this new grant of authority, the Railroad Commission made findings that the production of sweet gas in the Panhandle was in excess of reasonable market demand and allocated allowables so that the wells not owned or leased by the pipelines received 18 percent of the total allowable.[155] The commission hoped that this order would force the connection of pipelines to these wells. The pipeline owners challenged the order and a federal court enjoined the commission again, holding that the agency still lacked the authority to restrict production except to prevent physical waste, and that the pipeline plaintiffs were not wasting any gas.[156] The only method left to the Railroad Commission to protect the nonintegrated producers from drainage to the wells owned by the pipelines was to permit them to conduct stripping operations. Stripping plants extract the liquid hydrocarbons from natural gas and flare the residue gas, often wasting 95 percent of the heating value of the gas. Such operations would clearly exhaust the Panhandle field prematurely, but this was the only way of providing a market outlet to the nonintegrated producers.[157]

This, then, was the situation facing the legislature in early 1935: More than 1 billion cubic feet of gas per day were being blown into the air in the Panhandle as residue from forty-one stripping plants and twenty-nine carbon black plants. The field had about 200 sour-gas wells, 800 sweet-gas wells (of which 300 had no pipeline outlet), and 2,600 oil wells flaring casinghead gas.[158] The lawmakers had to devise a system that would prevent waste and protect the correlative rights of all the producers. The result was a comprehensive act prohibiting many specific kinds of gas waste and establishing a detailed system of gas prorationing.[159] For the first time, Texas conservation laws mentioned the concept of correlative rights. The declaration of policy in the 1935 Act reads as follows:

> In recognition of past, present, and imminent evils occurring in the production and use of natural gas, as a result of waste in the production and use thereof in the absence of correlative opportunities of owners of gas in a common reservoir to produce and use the same, this law is enacted for the protection of public and private interests against such evils by prohibiting waste and compelling ratable production.[160]

The 1935 Act also declared in Section 10 that it was the commission's duty to prorate and regulate the daily gas well production from each common reservoir, and that this was to be done:

> . . . for the protection of public and private interests:
> (a) In the prevention of waste . . . ;
> (b) In the adjustment of correlative rights and opportunities of each owner of gas in a common reservoir. . . .[161]

Finally, Section 21 authorized voluntary unitization agreements, as follows:

> In order that land owners and operators that have undeveloped land within a proven natural gas field may secure a market for their natural gas, and in order that the market for natural gas may be more equitably distributed among the various land owners and operators, and in the interests of the conservation and development of natural gas, it is declared to be lawful for any two or more lessors, lessees, operators, or other persons, firms or corporations owning or controlling production, leases, royalties or other interests in the separate properties of the same producing gas field, with the approval of the Attorney General of Texas, to enter into agreements for the purpose of bringing about cooperative development and/or operation of all or a part, or parts of such field, or for the purpose of fixing the time, location and manner of drilling and operating wells for the production, storage, marketing or the repressuring of gas, or for the purpose of the equitable distribution

> of royalty payments. Any such agreement shall bind only the parties thereto, and their successors and assigns of such having knowledge or notice thereof, and shall be enforceable in an action for specific performance.[162]

Thus the impetus of the 1935 Act was to protect the correlative rights of nonintegrated producers by instituting an equitable system of prorationing and end-use controls that either would encourage or force the pipeline-owners and producers to cooperate with the independents. The facts in the Panhandle field were quite the reverse of those in the East Texas field: the Panhandle independents were clearly disadvantaged by not having a prorationing system to protect their correlative rights, whereas the East Texas independents had produced more than their proportionate share of the field's oil (see this chapter text at note 63). The Panhandle operators' disadvantage in large part was the result of the inherent differences between gas and oil production. Gas is almost impossible to store aboveground or to ship by truck or rail. This fact eliminated the possibility of running "hot gas" as many East Texas producers ran "hot oil." Gas pipelines are expensive to build, especially relative to the low price of gas then prevailing, and few served the Panhandle area in an interconnected way.[163] Gas drains more quickly from a much larger area than does oil, and the ultimate maximum recovery of natural gas from gas fields is usually not sensitive to its rate of production, making it difficult to establish physical waste as a basis for prorationing.[164] For all these reasons, the lawmakers deemed it essential that the Railroad Commission be granted the express authority to protect correlative rights in gas fields independent of its authority to prevent physical waste. Compulsory market demand prorationing or voluntary unitization would accomplish this. The 1935 Act understandably permitted both.[165]

Still, it seems odd to find this legislative approval of unitization so soon after the concept of unitization had been condemned so vehemently in the House and Senate chambers.[166] However, a close analysis of Section 21 of the 1935 Act shows that the lawmakers did not veer far in the direction of encouraging unitization. First, the statute authorized voluntary agreements only. No producer could be coerced into a unitization plan. Second, the 1935 Act applied only to proven gas fields, not to oil fields.[167] Third, any voluntary agreement required the approval of the attorney general of Texas, who was bound to uphold the antitrust laws of the state.[168] The Railroad Commission was given no authority to approve these cooperative agreements. Cooperative practices that threatened monopoly were to be scrutinized and prevented by the public office having a long, proud history of fighting (and often winning) antitrust battles against the majors. Under the statute's own terms, independent oil operators in East Texas or elsewhere had little reason to fear that Section 21 signaled a new legislative respect for compulsory unitization. Indeed, by expressly authorizing cooperative development in gas fields only, the 1935 Act arguably made voluntary unitization agreements in oil fields even more legally tenuous under the Texas antitrust laws. In fact, when the legislature finally—in 1949—passed an act authorizing voluntary unitization agreements in oil fields and repealed Section 21, the new act was, in some important respects, a weaker statute than its 1935 forebear (see chapter 4 text at notes 31 to 66).

The 1935 imprimatur to unitize gas fields proved especially useful later, after the process of gas cycling was developed. Cycling made it possible to produce gas from gas wells having no immediate market outlet, to recover the liquefied products from the gas, and to return the residue to the formation. In gas fields containing large amounts of condensate, cycling is necessary to achieve maximum recovery of all the liquid hydrocarbons. As noted, efficient cycling operations absolutely require operating the entire reservoir as a unit (see chapter 2 text at notes 20 to 21). In 1934 the Railroad Commission had started to issue orders prohibiting the flaring of gas from gas wells in condensate fields. These orders, fully discussed in chapter 5, had the effect of forcing operators to cycle their gas, and this required voluntary unitization agreements. Thus, Section 21 of the 1935 Act, originally passed to protect correlative rights, came to serve a conservation function also.

Armed with the 1935 Act, the Railroad Commission issued new prorationing orders for the Panhandle field. Incredibly, the federal courts again enjoined the order, still asserting, despite the 1935 Act, that the commission had no statutory authority to prorate gas except to prevent physical waste.[169] In addition, the courts found that the orders were an arbitrary and discriminatory taking of private property, because their intent was to force the pipeline owners to buy from and share their private marketing contracts and pipeline facilities with the owners of wells who had not contributed money, services, negotiations, skill, or foresight to the development of these markets.[170] On appeal, the U.S. Supreme Court upheld the lower court's injunction on the basis that the commission order was an unreasonable and discriminatory taking of the pipeline producers' property.[171] While the Supreme Court stated that the Texas legislature and the Railroad Commission probably had the authority to pass acts and issue orders with the sole objective of protecting correlative rights,[172] this was a meager victory in light of the Court's finding that the specific order at issue was invalid in that it was contrary to the protection of private property rights. The effect of this decision was to leave natural gas prorationing in a state of considerable confusion and uncertainty for many years.[173]

The Railroad Commission's ability to force equitable takings from gas fields, using the Common Purchaser Act, was also of questionable legality after this series of federal court rulings. Consequently, the Railroad Commission largely withdrew from active pursuit of gas prorationing for many years.[174] As a result, a private system of pipeline prorationing evolved which was understandably disliked by nonintegrated producers. When a pipeline company could not market all of the gas offered for sale by the producers connected to the pipeline, the pipeline company would itself proration the production from the wells. In a field served by several different pipelines, one company might have a market for 75 percent of all the gas from its wells, while another company might have a market for only 40 percent of its gas. Thus, even though each pipeline company took ratably from all of its wells, the producers whose allowables were at the 40 percent level suffered drainage to those producers whose pipelines had more market demand.[175] In recent years, Texas has often experienced natural gas surpluses. In response, the Railroad Commission has adopted a number of "maddeningly complex" written and unwritten rules to prevent discriminatory prorationing by gas-pipeline purchasers.[176] One way of avoiding the legal entanglements of intrapool prorationing is to unitize gas fields, and the current regulatory morass therefore provides incentive to do so. Ironically, the Railroad Commission no longer has the statutory authority to approve unitization agreements which are made solely to protect correlative rights. Section 21 of the 1935 Act was repealed in 1949 when Texas passed the now existing statute authorizing voluntary unitization agreements in oil and gas fields.

Summary

In the early years of the Railroad Commission's regulation of Texas oil and gas fields, two important statutes were enacted which have significantly affected the course of unitization in Texas. In 1931 the Texas legislature declared that the Railroad Commission was never to require repressuring or unitization of oil and gas fields in its efforts to prevent waste. In 1935 Texas lawmakers authorized the attorney general to approve voluntary unitization agreements made in gas fields. Both the 1931 anti-unitization statute and the 1935 pro-unitization statute were passed to aid independent producers against the vertically integrated majors in the race to drain oil and gas from commonly owned pools. The independents were winning the oil race in East Texas and thus opposed unitization. The independents, losing the gas race in the Panhandle, were anxious to establish unitization agreements.

These two acts show that the independent producer in Texas possessed considerable political clout. The timing and substance of the Marginal Well Act of 1931, and its 1933 amendment, vividly illustrate this clout. Thus, when the Texas legislators had to choose among competing methods of preventing the large-scale waste of oil and gas wrought by the rule of capture, it is not surprising that they picked market demand prorationing rather than unitization. Prorationing placed control of the oil fields in the hands of a state agency rather than in the hands of the majors, who, of necessity, would be the unit operators. The private strategies of major oil companies—especially Humble—to control production rates in the oil and gas fields by dint of their pipeline power, umpire committees, and combined purchasing power indelibly linked unitization with monopolization. Not all of the independents' charges alleging anticompetitive acts of the majors were credible. Nonetheless, there was enough evidence so that legislators solely concerned with preventing waste and monopoly were understandably as much opposed to unitization as those lawmakers who may have supported the independent producers solely to protect the selfish interests of this politically popular group of businessmen.

In lieu of unitization, market demand prorationing became the device for preventing waste. Through the subsequent years, this prorationing counteracted the inefficient distortions between private and socially optimal production rates caused by the rule of capture. Market demand prorationing prevented the physical waste of much oil and gas and, in this regard, has served the public interest. However, this same prorationing also has been a source of waste and inefficiency. Market demand prorationing does not coincide with MER prorationing, or with the production rate which maximizes the present value of the resource to society over time. Market demand prorationing is a form of cartelization providing producers the opportunity to secure higher prices at the expense of the general public. The fact that market demand prorationing raised oil prices in the 1930s is not a *per se* sign of its inefficiency; short-term crude oil prices are suboptimal under the rule of capture. The problem inherent in state-controlled production rates is that the long-term price of crude oil no longer is determined by market forces. If fields were unitized and unit operators competed against one another—each producing at the rate which maximized the present value of the resource to themselves (and, coincidently, to society)—efficient operators and low-cost fields would survive at the expense of high-cost fields, and the market price of oil would fluctuate according to its value to society. The idea of liberating the oil and gas fields from state control and officially turning them over to the umpires and prorationing committees formed by the majors was politically untenable at a time when it was impossible to distinguish monopolistic behavior of the majors from their strategic behavior designed to prevent waste and promote efficiency. More state control rather than less seemed to be the answer. Market demand prorationing was finally acceptable; unitization was not.

4

The Texas Legislature and Unitization: The Later Years

In 1949 Texas passed an act providing for the cooperative development of oil and gas fields. In 1973 Texas almost passed a compulsory unitization statute. This chapter examines the legislative history and the substantive content of both the act that passed and the one that got away. The chapter analyzes the reasons for Texas' continuing failure to enact a compulsory unitization statute, especially in contrast to the successful passage of a compulsory pooling act in 1965.

Passage of the 1949 Voluntary Unitization Act

In 1949 the Texas lawmakers authorized the Railroad Commission to approve private agreements for the cooperative development of oil and gas fields.[1] This act is the foundation of Texas' current unitization policy. It was passed to provide antitrust immunity to operators who joined in voluntary unitization agreements. This immunity long had been sought by producers, especially by the major operators who had the most cause to fear possible antitrust actions against them. Nonetheless, even so small a change in Texas' policy toward unitization engendered much opposition. A voluntary unitization bill was proposed and widely debated in

1947, but it died when the House and Senate passed different versions of it and the joint conference committee failed to agree on a compromise bill before the legislative session ended.[2]

Yet the march of events clearly ordained the passage of such a bill. By 1949, independent producers had survived many years of market demand prorationing with beneficial results, and unitization was no longer tainted by its association with the major oil companies' efforts to control production rates. World War II had inured many producers to the concept of unitization. The Petroleum Administration for War (PAW) allocated scarce oilfield supplies under a system of priorities which encouraged wide well spacing, unitized pressure maintenance, and secondary recovery operations.[3] Unitization was wrapped in a patriotic flag and was perceived less readily as a monopolistic tool of the major oil companies to drive out independents.

Also, the technology of production had advanced. New, deep gas fields were discovered which required cycling to prevent retrograde condensation and to produce the maximum amount of valuable liquid condensate.[4] By 1948, forty cooperative agreements in gas fields had been approved by the attorney general under Section 21 of the 1935 Act.[5] The older oil fields in Texas, discovered in the 1920s and 1930s, now required secondary recovery if production declines were to be reversed.[6] As a result of the enormous demand for wartime petroleum supplies and the pent-up, postwar consumer demand, most fields in Texas were operating at their maximum efficient rates (MERs), and the economic incentive was strong to find new, deeper pools and to increase the ultimate recovery from existing pools. In addition, World War II had increased the prices of crude oil and natural gas so that the expensive facilities required to process and recycle casinghead gas and gas from condensate reservoirs were now economically justified.

When operators did not move quickly enough to conserve casinghead gas or repressure gas cap reservoirs, a forceful Railroad Commission issued no-flare orders that shut down oil fields until wasteful practices were corrected. In 1947 the commission had prohibited oil production in the Seeligson field unless the casinghead gas was put to a beneficial use, and in 1948 the commission shut down every oil well in sixteen oil fields in which casinghead gas was being flared (see chapter 5). If no immediate market for the flared gas existed, the alternative was to reinject the gas. This provided considerable incentive to unitize oil fields.

At the same time that economics, technology, and the Railroad Commission's orders were pushing producers into the types of operations requiring unitization, the threat of antitrust liability for entering into cooperative agreements seemed to increase. In 1947 the U.S. Department of Justice filed suit against the operators of the Cotton Valley unit in Louisiana, alleging violations of the Sherman Antitrust Act. The Cotton Valley field was a condensate field which had been voluntarily unitized by most of the producers in order to conduct cycling operations. While the suit did not attack the joint activity of the producers in the cycling operation itself, but only the joint processing, refining, and marketing of the products removed from the wet gas, it caused considerable stir in the oil industry.[7] Then, in 1948, Yale law professor Eugene Rostow published a book advocating the passage of a federal compulsory unitization law to supplant state regulation of oil fields based on prorationing, which he contended was wasteful, inefficient, and a price-fixing device.[8] This, too, caused a stir and provoked states to improve their conservation laws to encourage unitization and avoid federal intervention. Professor Rostow also renewed fears of antitrust liability because he recommended compulsory unitization under federal law only after the major oil companies were dissolved and disintegrated. Otherwise, he argued, unitization would heighten, rather than diminish, the major oil companies' dominance of the oil fields.

All these factors led major oil company spokesmen to assail the lack of a statute authorizing cooperative development in Texas oil fields.[9] The Texas antitrust laws were still largely unchanged since 1889. Cooperative agreements by producers were virtually *per se* violations of the law, which defined an illegal trust as a combination of capital, skill, or acts by one or more persons to regulate, fix, or limit the output of any product.[10] A state statute expressly granting immunity from state antitrust law to commission-approved voluntary unitization agreements might also provide federal antitrust immunity.[11] These factors ordained successful passage of the voluntary unitization bill in 1949. Indeed, in many other states, these circumstances presaged compulsory unitization legislation, especially as the difficulties of securing voluntary agreements were becoming more and more evident.[12] Texas was clearly not ready to go this far, however.

Robert Hardwicke, a leading authority on oil and gas conservation law in Texas at the time, both as a scholar and a practitioner, opposed compulsory unitization legislation. His experience with state regulation in the East Texas field in the 1930s had convinced him that government regulation inevitably brought "arbitrary and unwise action by legislative, executive, and judicial agencies" and enormous "hazards, expense, and losses."[13] He urged the industry to lessen the need for regulation by making unusual efforts to solve problems voluntarily. Hardwicke lauded the Railroad Commission's encouragement of voluntary efforts through its practice of adopting as field rules any unanimous recommendations of operators that were consistent with good conservation practices.[14] In testimony before the Cole Committee in 1940,[15] Hardwicke stated that the problem of establishing fair participation formulas in compulsory unitization orders was "insurmountable," and that courts would undoubtedly strike down such orders as unreasonable.[16] Further:

> This much is certain: the merits of compulsory unit operations, as against reasonably efficient and effective control under existing conservation statutes, are not well enough established to overcome in legislative bodies the reluctance to give administrative agencies far greater powers than are now being exercised. I think it can be said that no compulsory unit operations law would have a chance to pass either in Congress or in any state legislature.[17]

Even as late as 1951, Hardwicke saw little need for compulsory unitization because voluntary agreements were becoming more and more common.[18] Other Texas scholars shared his views.[19] Voluntary unitization was considered the only clear course of action open to Texas.[20]

Moreover, the Texas legislature's longstanding antipathy for unitization, rooted in the belief that unitization would tend to monopolize the oil fields under the major oil companies' control, virtually precluded compulsory legislation.[21] Indeed, it would significantly affect the degree of antitrust immunity and the scope of voluntary agreements authorized by the 1949 bill which Texas ultimately enacted.

When Texas finally did pass a statute authorizing voluntary unitization agreements in oil fields, it was not a simple one-paragraph act, as recommended by the Interstate Oil Compact Commission (IOCC) and adopted in other states, or as used in Texas for gas fields under Section 21 of the 1935 Act.[22] Rather, it was a much longer bill detailing the many conditions under which voluntary unitization could *and could not* be approved. Much of the wording in the 1949 Act can be directly attributed to the attempt in 1947 to pass House Bill 67, a one-paragraph bill modeled after Section 21 of the 1935 Act. The Texas attorney general's opinion had been sought on this 1947 proposal, and it stated that:

> H.B. 67 authorizing unitization agreements for cooperative exploration, development, and operation of oil and gas properties, and the marketing of gas, without the requirement that the same be necessary for the prevention of waste and conservation of natural resources, if enacted as written, would constitute a serious threat to the antitrust laws of this State.[23]

In particular, the attorney general disapproved of the proposal because: (1) it allowed cooperative agreements for exploration and for marketing, two activities which seemed unrelated to conservation objectives; (2) it allowed cooperative agreements which would affect or control prices, the cost of production, and the rate of output of oil and gas, all of which were inconsistent with the antitrust laws; and (3) by exempting the oil and gas industry from the antitrust laws, the latter would be held void as denying the equal protection of the law to all citizens.[24] While Texas statutes exempted labor unions and agricultural producers from the antitrust laws under certain conditions, the attorney general could not justify the broad exemption in H.B. 67 for an industry like oil which was "highly organized, closely integrated, and of tremendous power and wealth."[25] The attorney general recognized that unitization was highly desirable for conservation and the protection of correlative rights, but in his opinion this would justify an exception from the antitrust laws *only if* the voluntary agreements were necessary to achieve these goals.[26] The proposed bill authorized voluntary agreements without any finding that the agreements were in the public interest of conservation rather than for the mere convenience and profit of oil and gas producers desirous of "monetary savings,"[27] that is, of preventing the economic waste of overdrilling.

Faced with this opinion, the proponents of voluntary unitization drafted a revised version of H.B. 67 and secured a favorable opinion on it from the attorney general.[28] The new, lengthy H.B. 67 passed both the House and Senate, but died on the calendar while in a joint subcommittee established to resolve differences between the House and Senate bills.[29] The 1949 bill closely tracked this revised 1947 proposal, but was even more restrictive in its authorization of voluntary unitization agreements. The 1949 bill finally passed the Senate by a 21 to 7 vote and the House by a 75 to 54 vote.[30] Even such a restricted bill did not receive the whole-hearted support of the lawmakers.

Analysis of the 1949 Voluntary Unitization Act

The 1949 Act authorizes voluntary agreements by persons owning interests in separate property in the same oil or gas field for either of the following purposes, as stated in Section 101.011:

> (1) to establish pooled units, necessary to effect secondary recovery operations for oil or gas, including those known as cycling, recycling, repressuring, water flooding, and pressure maintenance and to establish and operate cooperative facilities necessary for the secondary recovery operations;
>
> (2) to establish pooled units and cooperative facilities necessary for the conservation and use of gas, including those for extracting and separating the hydrocarbons from the natural gas or casinghead gas and returning the dry gas to a formation underlying any land or leases committed to the agreement.[31]

No Exploratory Units

While at first glance this section seems all inclusive, in effect it is actually quite restrictive. First, it clearly does not authorize voluntary agreements for exploratory drilling or related exploratory activities. Only unit plans necessary to effect secondary recovery operations or to conserve gas are authorized. The act obeyed the strictures of the attorney general's 1947 opinion, though it was by no means clear that this official's reasoning was correct. As early as 1935, the federal government had established a policy of encouraging unitization for exploratory purposes by lessees of public land. Under the authority of the Mineral Leasing Act, almost every lease issued by the Department of the Interior on federal lands since 1935 contained a provision whereby the lessee agreed, within thirty days from demand by the secretary of the interior, to subject his leasehold interest to a cooperative plan.[32] Most of the unit plans approved by the department involved large, unexplored areas of public land interspersed with state or private lands.[33] This policy of unitizing areas for exploratory drilling has many conservation advantages: it insures proper well spacing and production rates from the day of discovery of a reservoir, and minimizes costs and legal risks to operators.[34] Also, if producers can anticipate the early unitization of new discoveries, the expected profitability of exploration is increased, thus promoting more exploratory drilling, a very real public benefit.[35] By 1949, several state statutes authorized exploratory unitization agreements on state lands.[36] Even on private lands, early unitization to effect exploratory drilling had been successfully practiced.[37] Despite these advantages, Texas restricted the scope of the 1949 Act in Section 101.011 to production activities.

Other sections of the 1949 Act confirm that exploratory units are not to be authorized by the commission. Section 101.103(a)(6) requires that the commission find that "the area covered by the unit agreement contains only that part of the field that has reasonably been defined by development." Oklahoma's compulsory unitization statute, enacted in 1945, is similarly restrictive and was clearly used as a model by the Texas lawmakers.[38] The most obvious disadvantage of unitizing a field at an early time is that little information exists to fashion fair and reasonable participation formulas. This disadvantage is particularly serious when compulsory process is being used to force some unwilling owners into a unit.[39] However, if all owners of potential mineral-bearing lands wish to unitize voluntarily and can agree on a participation formula, there is little reason to refuse to authorize approval of such an agreement. Early unitization so strongly serves the public interest of reducing exploratory drilling costs that this would seem to outweigh the possibility of its having an anticompetitive effect. The bias against exploratory units in the 1949 Act may again reflect the political power of independent producers in Texas. Independents drill most of the exploratory wells in the United States.[40] Early unitization can be perceived as reducing the number of drilling and employment opportunities in exploration, even though in the long term, greater efficiency in exploration increases the profitability of this type of activity and would actually encourage more exploration.

No Marketing Agreements

The second restriction imposed by Section 101.011 is that it does not authorize voluntary agreements for marketing oil or gas or for many types of postproduction processing. Any attempt to interpret the authority granted in Section 101.011(1) "to establish and operate cooperative facilities necessary for the secondary recovery operations" so as to encompass joint marketing and processing operations is futile because Section 101.017 of the act later states:

> (b) No agreement authorized by this chapter may provide directly or indirectly for the cooperative refining of crude petroleum, distillate, condensate, or gas, or any by-product of crude petroleum, distillate, condensate, or gas. The extraction of liquid hydrocarbons from gas, and the separation of the liquid hydrocarbons into propanes, butanes, ethanes, distillate, condensate, and natural gasoline, without any additional processing of any of them, is not considered to be refining.
> (c) No agreement authorized by this chapter may provide for the cooperative marketing of crude petroleum, condensate, distillate, or gas, or any by-products of them.

This restriction again reflects the attorney general's 1947 opinion that postproduction cooperative activities are not necessary for conservation. This opinion may have some merit with regard to oil, but gas processing, distribution, and marketing are much more difficult and expensive tasks, and voluntary agreements to perform these tasks could allow greater ultimate recoveries. For example, joint processing of gas condensate in a large plant owned by many operators in the field is much more efficient and low cost than if each operator constructs a separate plant for his own small volumes of gas. This substantial cost saving allows operators to continue to produce the field for a longer period of time, thereby recovering condensate that otherwise would have been abandoned as uneconomic. Joint processing can thus be considered necessary and essential to oil and gas conservation.[41]

This reasoning obviously explains why Sections 101.011(2) and 101.017(b) allow cooperative agreements to extract liquid hydrocarbons from gas as an exception to the exclusion of post-production activities from the scope of the act. This exception allows joint processing of gas in condensate plants overlying gas fields. The same reasoning, however, can justify other forms of joint marketing and processing. For example, joint marketing can enable producers to negotiate better terms with buyers, especially large pipeline purchasers who desire the assurance of large volumes. The better price and terms of sale can again result in larger recoveries of oil and gas than otherwise would occur. Cooperative marketing of a unit's output would also protect correlative rights because pipeline purchasers could no longer discriminate among producers in the field with respect to price or quantity.[42] The 1935 Act had allowed approval of voluntary agreements for gas marketing, and the IOCC's model form also allowed such agreements if a state commission found that it was impractical to deliver the gas or its by-products to the several owners in kind (a situation that would normally obtain).[43] However, the Texas attorney general and lawmakers considered most forms of cooperative marketing and processing a danger to the antitrust laws, and so restricted the operation of the 1949 Act.

Restriction of Unitization to Secondary Recovery Operations

Third, Section 101.011 does not authorize unitization for normal primary production. The first subsection allows pooled units only if "necessary to effect secondary recovery operations for oil or gas." This specific focus on secondary recovery would seem to preclude pressure maintenance operations which often begin during the primary stage of recovery, except that the subsection then proceeds expressly to include pressure maintenance. It seems obvious that the lawmakers were using the term *secondary recovery* as it had been defined and standardized by the API and the IOCC:

" 'secondary recovery' is the oil, gas, or oil and gas recovered by any method (artificial-flowing or pumping) that may be employed to produce them through the joint use of two well bores."[44]

Thus, the lawmakers intended to differentiate primary recovery from secondary recovery and intended that the Railroad Commission authorize only the latter, that is, only operations using injection wells to increase the production of oil and gas. Unitization for the mere convenience of reducing drilling and operating costs during primary production was not to be authorized, even though cost savings were one of the most valued benefits of unitization and even though these cost savings could extend the margin of profitable recovery.[45]

Consider the effect of Section 101.011's limitations on operators of an oil field with a gas-cap drive in a reservoir so shaped that the gas-cap owners have little or no ownership interest in the connected oil zone. This type of reservoir is depicted in figure 4-1. Primary production can proceed without unitization under existing conservation statutes as follows: the operators overlying the gas cap may drill wells into the cap and produce that amount of gas which is the volumetric equivalent in reservoir displacement of the gas and oil produced from the oil well that withdraws the maximum amount of gas in the production of its daily oil allowable.[46] Rule 49(b) prevents the gas-cap operators from producing gas at any greater rate because this would dissipate the gas cap and decrease the ultimate recovery of oil. The oil operators may produce their oil within the allowed gas–oil ratios and prorationing schedules. Eventually the gas cap will expand downward, and the oil wells will produce the migrated gas. In this situation, the correlative rights of the gas-cap producers and royalty interest owners are not well protected, nor is the greatest production of oil achieved because this would require absolutely no production of gas from gas-cap wells. Under these circumstances, the operators in the field may desire to enter into a unitization agreement whereby the gas-cap owners agree to shut in their wells—or better yet, simply do not drill them—and the oil well owners agree that any gas eventually produced from their oil wells will be allocated to the gas-cap owners. All parties benefit: maximum gas-cap pressure is conserved for oil production, and gas-cap owners do not risk losing their gas to the oil well operators. The effect of Rule 49(b) on gas-cap owners often results in uncompensated drainage to the

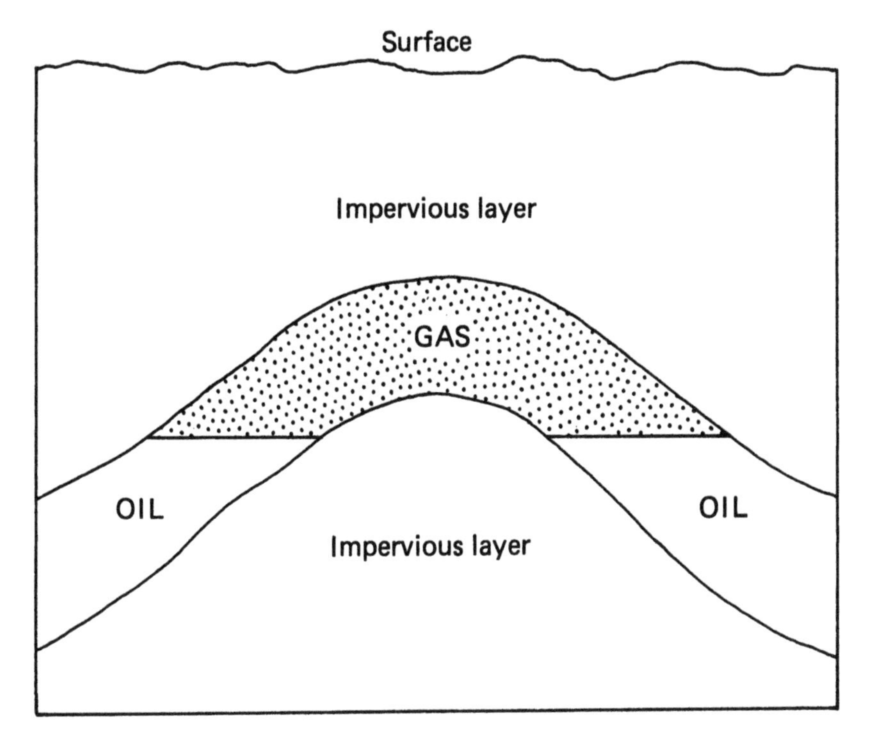

Figure 4-1. Gas-cap reservoir ringed with oil.

oil wells.[47] The unitization agreement is clearly in the public interest of (1) increasing ultimate recovery, (2) reducing drilling costs, and (3) protecting correlative rights.

Yet it is extremely doubtful that the Railroad Commission has the authority to approve such an agreement. No injection wells for any sort of cycling or repressuring are contemplated in the agreement. This is not a "secondary recovery" operation under Section 101.011's nomenclature, *even though* it is necessary to increase the ultimate recovery of oil.

It is difficult to argue that this impact of Section 101.011 is the unintended effect of unfortunate legislative drafting. The Texas lawmakers had the IOCC's model form to assist them which authorized any agreements "reasonably necessary to increase the ultimate recovery or prevent the waste of oil or gas and to protect correlative rights."[48] The 1947 proposed Texas bill used similar language.[49] Texas deliberately eschewed this general phrasing in the 1949 Act.

Types of Secondary Recovery Techniques Authorized

The fourth restriction imposed on voluntary unitization agreements by Section 101.011 is that it does not clearly authorize *in-situ* combustion or tertiary recovery techniques, such as miscible displacement, which do not rely primarily on reservoir pressuring for their success.[50] The enumerated types of secondary recovery operations all relate to reinjecting gas or water into the reservoir to affect reservoir pressure. The word "including" would seem to allow types of secondary operations other than those listed, but a more general phrasing such as used in the IOCC form quoted above would have eliminated any doubt on the subject.[51] The second subsection of Section 101.011 employs a more general phrasing and allows pooled units "necessary for the conservation and use of gas." Because the first subsection does not use this phrasing, it can be argued that the legislators intended to limit voluntary agreements in oil fields to the specific activities listed.

Finally, it should be noted that Section 101.011 does not authorize the formation of pooled units necessary to protect correlative rights. In this regard, the 1949 Act is much narrower than Section 21 of the 1935 Act which it replaced.

Additional Limitations

Other serious limits on the Railroad Commission's authority to approve voluntary unitization agreements appear in Section 101.013, which states that agreements for pooled units are not legal or effective until the commission finds, after application, notice, and hearing:

> (a)(1) that the agreement is necessary to accomplish the purposes specified in Section 101.011 of this code;
>
> (2) that it is in the interest of the public welfare as being reasonably necessary to prevent waste and to promote the conservation of oil or gas or both;
>
> (3) that the rights of the owners of all the interests in the field, whether signers of the unit agreement or not, would be protected under its operation;

(4) that the estimated additional cost, if any, of conducting the operation will not exceed the value of additional oil and gas so recovered by or on behalf of the several persons affected, including royalty owners, owners of overriding royalties, oil and gas payments, carried interests, lien claimants, and others as well as the lessees;

(5) that other available or existing methods or facilities for secondary recovery operations or for the conservation and utilization of gas in the particular area or field concerned or for both are inadequate for the purposes; and

(6) that the area covered by the unit agreement contains only that part of the field that has reasonably been defined by development, and that the owners of interests in the oil and gas under each tract of land in the area reasonably defined by development are given an opportunity to enter into the unit on the same yardstick basis as the owners of interests in the oil and gas under the other tracts in the unit.

(b) A finding by the commission that the area described in the unit agreement is insufficient or covers more acreage than is necessary to accomplish the purposes of this chapter is grounds for the disapproval of the agreement.

Clearly the legislative intent of this section is to ensure that unitization agreements are allowed only as a last resort, that is, only if other methods for conserving oil and gas are inadequate, only if necessary to prevent waste, and only if additional oil or gas will be recovered. The legislative attitude reflected in Section 101.013(a)(5) is remarkably similar to Hardwicke's views concerning compulsory unitization legislation: that such laws were unnecessary and unreasonable because the general conservation laws controlling production rates, well spacing, gas–oil ratios, and the like, could accomplish the same results with less harm to private property rights (see this chapter text at note 17). The state of Oklahoma disagreed with Hardwicke and had passed a comprehensive compulsory unitization statute for both oil and gas fields. Oklahoma's 1945 law and the orders issued thereunder were attacked as unconstitutional in *Palmer Petroleum Corp. v. Phillips Petroleum Co.*[52] The substance of the plaintiff's first complaint in *Palmer* was that the act was unreasonable because it did not require a finding by the commission that unitization would result in greater conservation of oil and gas than could be had under the existing conservation laws.[53] The Oklahoma court upheld the 1945

Act and the compulsory orders issued thereunder despite the absence of such a requirement. In Texas, such a finding is required before a *voluntary* agreement can be approved.

In other areas, the Texas lawmakers were clearly influenced by Oklahoma's statute. The required finding in Section 101.013(a)(4) appears in the Oklahoma statute.[54] This finding is justifiably mandated under a compulsory unitization order to assure that the majority of operators who seek to invoke compulsory process do not force the minority into joining a unit which an impartial, expert tribunal considers an unwise economic proposition.[55] There is much less need for such a finding in the case of a voluntary agreement which affects only those who sign. Indeed, there is a strong public policy interest in encouraging experimentation and the use of new, untested technologies to recover additional oil and gas.[56] This required finding in the Texas statute may reflect the desire for another safeguard against monopoly: if *additional* oil or gas is expected to be produced, the agreement is less likely to restrain trade. But the Railroad Commission must further find that the increased revenues attributable to unitized operations exceed costs. A rational oil or gas producer would not sign a voluntary agreement unless this were expected to be the case, so the legislative purpose of requiring this finding is not clear. It is not needed to protect royalty interest owners who, while much less knowledgeable about oil- and gas-production techniques and economics, do not bear any costs of production.[57] If it is intended to protect small, less knowledgeable producers against signing voluntary agreements that seem uneconomic or unduly risky to the commission, then the function of the finding seems to be to offset the superior persuasive powers of major oil companies who might use voluntary unitization to try to bankrupt independent competitors. This supposition seems very farfetched. More likely, Section 101.013(a)(4) was included simply because it appeared in the Oklahoma statute and because it would restrict the type of voluntary agreements which could be approved by the Railroad Commission. When (a)(4) is read in conjunction with (a)(1) and (a)(2), it is again doubtful that the commission can authorize unitization agreements designed solely to reduce drilling and operating costs, even though such a reduction would indirectly increase ultimate recovery by prolonging the period of profitable production.

Consider the effect of Sections 101.013(a)(1), (a)(2), and (a)(4) on the East Texas field. This reservoir is constantly repressured by the reinjection of almost all the salt water that is produced with the oil. This repressuring is a secondary recovery operation under Section 101.011. The field is expected to recover about 85 percent of its original oil in place at which time the production of further oil will be uneconomic. The East Texas field has thousands of wells which are not needed to recover this oil. The costs of producing oil in East Texas are higher than would exist if the field were unitized and unnecessary wells were shut in.[58] Now suppose that many operators desire to unitize their properties so as to reduce operating costs.[59] Under Sections 101.013(a)(1) and (a)(2) the proposed agreement may not be found "necessary" to accomplish repressuring or conservation since repressuring is already being conducted without the agreement. Further, Section 101.013(a)(4) requires proof of "additional oil and gas so recovered," that is, recovered by the repressuring operation authorized in 101.013(a)(1), not by merely shutting in wells to reduce costs.

Consider also the previous gas-cap example, but assume that some gas recycling will take place to qualify the project as secondary recovery. Assume that the cost of this recycling operation just equals the value of additional oil produced. The oil and gas operators may still wish to proceed with unitization because it better protects correlative rights and can reduce costs by shutting in or not drilling several gas-cap wells.[60] Yet Section 101.013(a)(4) precludes approval of this project because additional revenues do not exceed additional costs. This restrictive interpretation of Section 101.013 is supported by the attorney general's 1947 opinions decrying the authorization of unitization for mere convenience and profit to lessees.[61]

Texas lawmakers in 1949 deliberately shunned the language of Louisiana's compulsory unitization law (enacted in 1940 to apply to condensate gas fields) which authorized that state's commissioner to issue orders "in order to prevent waste and to avoid the drilling of unnecessary wells."[62] Later, in 1960, Louisiana passed another compulsory unitization law applicable to both oil and gas fields. This law requires proof that the proposed unit will prevent waste and unnecessary drilling, and increase ultimate recovery. Under this act, proof of increased recovery is often shown to result from proof that lower operating costs resulting from reduced drilling will allow producers to continue operating for a longer period of time before abandonment of the field.[63] Most scholars agree, however that in states with unitization laws that lack the objective of reducing unnecessary costs, the state's conservation agency does not have the authority to approve unitization plans solely for this purpose even though ultimate recovery is increased in the long run.[64] The Texas law, especially Sections 101.011 and 101.013, seems to preclude such approval.

Correlative Rights and the 1949 Act

Clearly under Sections 101.013(a)(1) and (a)(2) the Railroad Commission has no authority to approve unitization agreements which serve only to protect correlative rights. The commission can approve only agreements which are in the public welfare as necessary to prevent waste and to conserve oil and gas. Yet the public welfare is often served by agreements which protect correlative rights, irrespective of conservation. Waste is often not an issue in natural gas fields, but the protection of correlative rights is of great concern. The 1949 Act does not allow authorization of voluntary unitization agreements in such gas fields, as was once allowed under Section 21 of the 1935 Act.[65]

Still, the 1949 Act does require that the commission find that private rights are being protected before it approves any unitization agreements for secondary recovery or gas conservation. Sections 101.013(a)(3) and (a)(6) require findings that the rights of all owners in the field, whether signers of the agreement or not, are protected, and that all owners have been given an opportunity to join the unit on the same yardstick basis. This latter requirement puts a large administrative burden on the proponents of the unitization plan. Section 101.013(a)(6) requires that the proponents negotiate with every owner of any interest "in the area reasonably defined by development." This area is often the entire field. The unit proponents cannot simply agree to unitize part of the field without offering everyone an opportunity to join. This latter requirement seems to be aimed at preventing operators from gerrymandering unit boundaries in ways which would harm the correlative rights of some of the owners in the field. Section 101.013(b) also requires that the commission take a careful look at the amount of acreage included in the unit to ascertain that it is neither too little nor too much to accomplish the process of secondary recovery or gas conservation.

The required finding in Section 101.013(a)(6) that all owners are given the opportunity to enter the unit on the same yardstick basis does not require a finding that the yardstick itself is fair to all. More than likely, the owners who refused to sign did so because they considered the yardstick unfair as applied to their particular tracts. However, the commission must still find that the rights of these nonsigners are protected under Section 101.013(a)(3). Thus correlative rights do receive much attention under the 1949 Act, even though the term itself is never used in the act.

Another group of provisions in the 1949 Act is designed to assure continuing commission control over the unitized operations. Thus, Section 101.015 subjects any approved agreement to any valid order or rule of the commission relating to location, spacing, prorationing, conservation, or any other matters. Section 101.017(a) prohibits agreements which attempt to contain field rules or to limit the amount of production of oil and gas from the unit area, such provisions being "solely within the province of the Commission." The legislative rationale is obviously to protect the public interest against possible monopolistic practices by retaining public control over the field. However, one of the great advantages of unitizing a field is to free the operators from burdensome state regulations, which are often no longer needed once the field is unitized.[66] The Texas statute does not recognize this advantage to operators or to the state's citizens who must pay taxes to support the agency's work.

Nor is the public interest fully represented in the act. Section 101.013(a)(2) requires a finding that the agreement is "in the interest of the public welfare as being reasonably necessary to prevent waste and to promote the conservation of oil or gas." The act does not authorize the commission to investigate and make findings on all factors which affect the public's interest in secondary recovery projects. For example, it is doubtful that the Railroad Commission has the authority under the 1949 Act to disapprove a voluntary unitization agreement as contrary to the public welfare on the basis that it uses scarce fresh water and will result in large economic losses to farms and agriculture.[67] Growing disputes between farmers and oil operators over scarce fresh water supplies led the Texas legislature in 1983 to grant the commission the authority to condition permit applications for injection wells used in enhanced oil recovery. If the commission finds that a substance other than fresh water is economically and technically feasible to use, the applicant must use this other substance.[68] This new law appears in the Texas Water Code, not in the oil- and gas-conservation statutes, and it applies to all injection wells whether part of a unitized project or not. The new legislation confirms the Railroad Commission's lack of authority to regulate freshwater use when approving unitization agreements under the 1949 Act.

In addition to prohibiting certain provisions in voluntary agreements, the 1949 Act expressly permits others. Section 101.016 reads:

> (a) An agreement authorized by this chapter may provide for location and spacing of input wells and for the extension of leases covering any part of land committed to the unit as long as production of oil or gas in paying quantities is had from any part of the land or leases committed to the unit. However, no agreement may relieve an operator from the obligation to develop reasonably the land and leases as a whole committed to the unit.
>
> (b) An agreement authorized by this chapter may provide that the dry gas after extraction of hydrocarbons may be returned to a formation underlying any land or leases committed to the agreement and may provide that no royalties are required to be paid on the gas so returned.

The issues addressed in this section were frequently bones of contention between lessees and lessors,[69] and it appears that the lawmakers attempted to resolve such disputes by indicating approval of such provisions. In so doing, Section 101.016 is one of the few sections of the 1949 Act which unequivocally encourages voluntary unitization by easing negotiation difficulties.

Another set of provisions in the 1949 Act involves the effect of

Railroad Commission approval of cooperative agreements. In no uncertain terms, Section 101.012 refutes any coercive power or effect:

> Agreements for pooled units and cooperative facilities do not bind a landowner, royalty owner, lessor, lessee, overriding royalty owner, or any other person who does not execute them. The agreements bind only the persons who execute them, their heirs, successors, assigns, and legal representatives. No person shall be compelled or required to enter into such an agreement.

Section 101.018 also protects those who refuse to join a unit agreement by limiting the significance of Railroad Commission approval of a project:

> The approval of an agreement authorized by this chapter shall not of itself be construed as a finding that operations of a different kind or character in the portion of the field outside of the unit are wasteful or not in the interest of conservation.

Thus lessees who refuse to join are not, for this reason alone, to be found in breach of the commission's general conservation laws prohibiting waste or in breach of any implied covenants owed to landowners to develop and operate the land prudently.

Most important, Section 101.004 of the 1949 Act reaffirms the primacy of Texas' antitrust laws (which appear in chapter 15 of the Business and Commerce Code) over voluntary unitization agreements:

> (a) Agreements and operations under agreements that are in accordance with the provisions in this chapter, being necessary to prevent waste and conserve the natural resources of this state, shall not be construed to be in violation of the provisions of Chapter 15, Business & Commerce Code, as amended.
>
> (b) If a court finds a conflict between the provisions in this chapter and Chapter 15, Business & Commerce Code, as amended, the provisions in this chapter are intended as a reasonable exception necessary for the public interest stated in Subsection (a) of this section.

> (c) If a court finds that a conflict exists between the provisions in this chapter and the laws cited in subsections (a) and (b) of this section and finds that the provisions in this chapter are not a reasonable exception, it is the intent of the legislature that the provisions in this chapter, or any conflicting portion of them, shall be declared invalid rather than declaring the cited laws, or any portion of them, invalid.

The IOCC's model form, widely used by other states, included a provision similar to that in Section 101.004(a) above.[70] Texas lawmakers added (b) and (c) to this standard provision.[71] Again, this was a direct reaction to the attorney general's 1947 opinion holding that state approval of cooperative agreements would endanger the vitality of the antitrust laws of Texas. But in adding both (b) and (c) and all the other provisions discussed above, the legislature was surely guilty of overkill. The 1949 Act's restrictions on the Railroad Commission's authority to approve voluntary unitization agreements and the continuing control of the commission over the agreements would alone assure that the agreements were in the public interest of conserving oil and gas. Therefore, antitrust immunity for any agreements which can survive all the required findings of the 1949 Act should be absolute, not limited as in Section 101.004.[72]

The 1949 Act does not speak directly to the effect of not securing Railroad Commission approval of an agreement. Section 101.013(a) states that cooperative agreements are not legal or effective until the commission makes the many required findings discussed above. This language would seem to prohibit operators and royalty interest owners from entering into any voluntary unitization agreements without Railroad Commission approval, except that Section 101.002 states that: "None of the provisions in this chapter restrict any of the rights that a person now may have to make and enter into unitization and pooling agreements." Some persons had entered into voluntary agreements before the 1949 Act,[73] and Section 101.002 seems to retain the right to continue to do so. Juxtaposed to this section is Section 101.014, which states that none of the provisions of the act "shall be construed to require the approval of the commission of voluntary agreements for the joint development and operation of jointly owned property." This section implies that all other types of voluntary agreements require approval of the commission. The more obvious interpretation of Section 101.014, however, is that operators who own land in cotenancy are not to be required to secure Railroad Commission approval to enter into joint operating agreements to develop the property.[74] Thus it would seem that producers who cannot meet the requirements of the act can nonetheless proceed to voluntarily unitize their tracts. They would not receive the "blessing" of Section 101.004, but if they feel that their agreement is secure from the antitrust laws anyway, the act does not prevent them from proceeding. However, given the tenor of the attorney general's 1947 opinion about antitrust immunity for the oil industry, and given the strictness of Texas' antitrust laws until their recent revision in 1983, operators may not want to pursue this course. It is unfortunate that the 1949 Act does not include a provision that failure to submit a cooperative agreement to the commission for approval shall not imply or constitute evidence that the agreement or operation under it violates the antitrust laws. Such a provision appears in other states' laws authorizing voluntary unitization agreements.[75] By expressly limiting antitrust immunity only to the types of agreements which meet the detailed requirements of the act, and by expressly excluding only joint operating agreements from the act's requirements, other voluntary agreements are more susceptible to attack under the antitrust statutes. Ironically, then, the 1949 Act which was to encourage unitization leaves Texas operators with the worst of both worlds: a statute which offers limited antitrust immunity only to certain types of cooperative agreements and which thereby increases the antitrust risks of entering into other types of voluntary agreements.

The Failure of Compulsory Unitization Legislation in Texas

The 1949 Act remains the foundation stone of Texas unitization policy to this day—a carefully chiseled statute honoring voluntary agreements.[76] For the next thirty years following the act's passage, every other major oil- and gas-producing state enacted some form of compulsory unitization legislation. Texas lawmakers added a bit of mortar to the 1949 foundation and nothing more.[77] The next sections of this chapter analyze the reasons for Texas' failure to enact a compulsory unitization law, first treating the period from 1949 to 1972, which was characterized by enormous overcapacity in the oil fields, followed by the period of energy crisis and shortage from 1972 on. The year 1972 is considered the dividing point, because in April of that year the Railroad Commission set a 100 percent market demand factor for crude oil production in Texas, signaling the elimination of excess capacity in Texas fields. The 100 percent factor has been maintained to date.[78]

The 1949–72 Period

The post–World War II economic boom brought a sharp rise in the price of crude oil from 1945 to 1948, but from 1948 to 1952 the price fell, and then remained relatively stable in the 1950s and 1960s. During these two decades, the United States—and Texas in particular—experienced excess producing capacity in the oil fields. The causes of this excess capacity were several: lower-priced imports of crude oil and refined products began to compete with domestic oil; cheap natural gas stole some of the domestic market away from crude oil; new fracturing techniques, secondary recovery technologies, and offshore drilling innovations increased supplies, spurred on by overly optimistic industry forecasts of demand for petroleum products.[79] But above all, the state regulatory systems were to blame.[80] Market demand prorationing insulated the oil and gas industry from market forces by maintaining a relatively stable price of crude oil. Prorationing assured a market to almost every explorer who found a new field. As long as producers expected revenues to exceed costs, they continued to invest in additional producing capacity by drilling more and more wells. Proration formulas which contained per-well factors encouraged this drilling.[81] The end result was that, by 1964, the Gulf Coast states were producing at 56 percent of their capacity.[82] Market demand prorationing in Texas restricted wells to less than 35 percent of their yardstick allowables from 1958 to 1966.[83] As of 1969, the United States had idle, crude-oil producing capacity of 3 million barrels per day, 2 million of this located in Texas.[84]

Eventually, of course, the investment in idle producing capacity reduced the expected profitability of drilling by driving up costs.[85] The number of wells drilled annually in the United States peaked in 1956 at 57,000 and then fell to fewer than 39,000 by 1966.[86] Domestic producers were caught in a cost-price squeeze and their ranks thinned. New reserves were more difficult to find and demanded deeper and more expensive drilling techniques.[87] Producers who were lucky enough to discover good-sized pools found their wells prorated to about one-third of their capacity. This reduced the incentive to explore for new discoveries, and from 1956 to 1966, little change occurred in the level of proved reserves of crude oil in the United States.[88] Crude oil prices could not be increased because of the overhang of excess capacity, the competition from cheaper imported crude, and the explosive political and antitrust problems accompanying such a price rise.[89] The petroleum industry was widely criticized during these two decades for its inefficiency, protectionism, and sheltered tax treatment, especially as these were seen to have anticompetitive effects on independent producers and refiners.[90]

In this economic and political climate, the only way of increasing or maintaining the profitability of oil and gas production was to find ways to decrease costs. Unitization was such a way. It could minimize drilling and operating costs and allow the easier implementation of secondary recovery operations in older fields. Many studies by industry trade groups, government commissions, and academicians reached this same conclusion.[91] By this time, voluntary unitization had been accomplished in many fields.[92] Compulsory process was needed to unitize the more difficult fields and to reduce the costs and delays of the unitization process itself. From 1951 onward, a parade of states enacted compulsory unitization laws for the first time, or amended existing statutes to allow greater use of compulsory process.[93]

Texas lawmakers did not march in this parade. During these same years, many changes were made in the Railroad Commission's regulation of oil fields, often in direct response to the economic conditions facing the industry, but the changes were made within the existing statutory framework.[94] The lack of lawmaking in this area was not the result of a general inability to pass legislation affecting the oil industry. During this time, Texas legislators passed several acts restricting pollution from oil industry activities. Despite the enormous economic pressures pushing the industry toward unitization, compulsory process to achieve this end was a uniquely elusive political goal.

The failure to pass a compulsory unitization statute in Texas during the 1950s and 1960s is more difficult to explain than the 1931 anti-unitization fervor or the cautious approach to voluntary agreements in 1949. The impetus to unitize no longer stemmed from the desire to impose price stability through production controls as had occurred in the early 1930s. Nor could unitization be so easily perceived as an anticompetitive device of the major oil companies seeking to buy out independents. Nor was compulsory unitization subject to the type of reasoning found in the attorney general's 1947 opinions, which expressed grave doubts as to the wisdom of exempting voluntary agreements from the antitrust laws. The typical compulsory unitization statute imposes many conditions and strong state supervision over the unitization process. The quest for unitization in the 1960s was firmly grounded in the need to prevent waste and increase efficiency. Why, then, did Texas fail to enact a compulsory unitization statute?

Ideology is often cited as a reason: that individualistic, entrepreneurial oilmen will oppose government coercion and regulation regardless of its rationale, and that Texas legislators admire and share this attitude. True, the campaign against compulsory unitization has been waged on ideological grounds, just as was the campaign against prorationing in the 1930s. One leader of the well-organized Texas Independent Producers and Royalty Owners Association (TIPRO) called compulsory unitization legislation, "a substitution of force for persuasion . . . another fetter on the step of Freedom, another move down the road towards confiscation, tyranny, and unmorality [whose advocates were] socialists, bureaucrats, self-seeking politicians, fuzzy thinkers of the left, do-gooders, and impractical denizens of ivory towers."[95] While one cannot dispute the sincerity of this belief in freedom versus compulsion, it is most unlikely that ideology alone explains Texas' stance. Just as the independent producers in the 1930s no longer opposed market demand prorationing once they experienced its beneficial effects on prices and revenues, one would expect ideology to give way to economics on the issue of compulsory unitization as well. That it did not implies that a significant number of producers and legislators did not consider that compulsory unitization would benefit either individual balance sheets or the state's economy. Why Texas believed this when twenty-five other states did not remains to be explained.

A review of the factors rendering it difficult to achieve voluntary unitization serves as a starting point for explaining the opposition to compulsory unitization as well. These factors, regrouped here from chapter two, are:

1. Profitable obstructionism
2. Pride of ownership and control
3. Mistrust of the majors
4. Fear of increased legal problems and uncertainty
5. Short-term declines in royalty and other payments
6. Reluctance to take risks, especially when lacking full data on reservoir characteristics
7. Fear of violating antitrust laws
8. The high cost of securing assent from many, even hundreds, of owners.

The seventh factor is irrelevant to the discussion of compulsory unitization, and the eighth factor can be omitted because compulsory unitization is designed to reduce these costs. (Most compulsory unitization laws require the consent of at least 65 percent of the owners of the pool, so negotiating costs are reduced, not eliminated.) The remaining factors are primarily the concern of independent producers and their royalty interest owners. The major oil companies and their trade organizations, such as the API, supported compulsory unitization even before the 1950s and 1960s.

Profitable obstructionism. From the producers' viewpoint, the question is why the economic interest of independents was served by opposing passage of a compulsory unitization bill which offered the potential to decrease costs and increase recovery rates at a time when independents were caught in a severe cost-price squeeze. The factor of profitable obstructionism was discussed earlier in terms of structural advantage in a reservoir: that those lessees and landowners with updip property in a water-drive field or downdip property in a gas-cap field would be unlikely to sign a voluntary agreement unless the unitization formula credited their tracts with more than the amount of recoverable oil and gas in place.[96] There is little evidence to show that independents generally held structurally advantageous tracts of land. In fact, the opposite seems more likely. The major oil companies could afford to purchase the prime tracts and often did so as a matter of corporate policy (see chapter 3 text at notes 90 and 103 to 104). Rather, the incentive for profitable obstructionism arose from the status quo of state regulation, from the legacy of East Texas.

East Texas had bequeathed a rich inheritance to the independent producer. The first bequest was the liberal granting of exceptions to the well-spacing rule so that, as a general principle, every small tract in Texas was entitled to a well permit as a matter of course. As early as 1919, in the first year of its authority over the oil and gas fields, the commission had passed Rule 37, a well-spacing regulation that limited drilling to one well per 10 acres.[97] But Rule 37 also allowed the commission to grant exceptions to the well-spacing pattern if necessary to prevent waste or to prevent confiscation. Confiscation, of course, meant drainage. The East Texas field overlay many townsite areas with tracts of land smaller than the spacing rule. If these tracts could not secure a well permit, they would be drained by adjoining larger tracts meeting the acreage requirements of Rule 37. In the administration of Rule 37 exception requests, the Railroad Commission adopted the position that small tracts in existence at the time of oil and gas discoveries in the area were entitled to Rule 37 exceptions to prevent their confiscation. Indeed, unless the exceptions were granted, the constitutionality of the well-spacing rule was questionable because it would permit confiscation of one person's property for the benefit of another. On the other hand, small tracts that were subdivided after oil and gas had been discovered in the near vicinity or those subdivided by oil and gas leases were not to be protected against drainage and were not entitled to a Rule 37 exception permit for this purpose. These latter tracts are called "voluntary subdivisions."[98] To hold otherwise would render the well-spacing rule ineffective, because every large tract overlying a potential or discovered oil and gas field would be subdivided by its owner into numerous smaller tracts in order to secure more wells and thereby more allowables under the prorationing system, which was East Texas' second bequest to the independent producer.

This prorationing legacy guaranteed the small-tract owner and the marginal well owner that their wells would be profitable to drill and operate. The inheritance was composed of two bequests:

first, the Marginal Well Act and, second, prorationing formulas containing large per-well factors. As noted previously, the Marginal Well Act was passed in 1931 as a direct response to the institution of prorationing in East Texas (see chapter 3 text at notes 136 to 145). By exempting marginal wells from prorationing, the act secured a disproportionate share of production to these wells.[99] Indeed, the exemption of marginal wells from market demand prorationing was so valuable a legal right that producers sometimes falsely classified their wells as marginal wells, and it often became a misfortune to drill a good well rather than a marginal one.[100] Particularly in East Texas, the Marginal Well Act resulted in a prorationing system which granted allowables almost entirely on a per-well basis. The commission realized that the East Texas system created enormous incentives to secure Rule 37 exception permits to drill more wells, and that this incentive was directly contrary to the policies underlying the well-spacing rule, but in their view no other solution presented itself for the extraordinary conditions in the East Texas field.[101] Thereafter, in other oil fields, the commission balanced the two objectives of regulating well spacing and protecting the small tract against confiscation by adopting prorationing formulas that typically allocated one-half of the field's allowable on a per-well basis and the other half on a surface acreage basis.[102] In gas fields, the traditional formula became one-third on a per-well basis and two-thirds on surface acreage. While these formulas better protected correlative rights than those of East Texas by giving some credit for tract size, they nonetheless continued to allow small tracts and wells to produce disproportionate amounts of oil and gas by drainage from neighboring larger tracts of better quality.[103]

The Railroad Commission was constantly criticized by scholars for its encouragement of small-tract drilling and its failure to administer its regulations to promote greater efficiency and fairness in the oil and gas fields. The commission was urged to interpret the Marginal Well Act as exempting from prorationing only those wells which in truth would be abandoned or damaged if production were curtailed, rather than exempting all wells which fell into the specified brackets.[104] Many marginal wells would still be profitable even if they were prorated. The commission was especially urged to consider a tract's surface acreage in applying the

Marginal Well Act, so that a marginal well on a 10-acre tract would be fully exempt from prorationing, while the marginal well on a 5-acre tract would be prorated as long as it was profitable to operate at a lower rate.[105]

The commission was also urged to deny Rule 37 exception requests to all owners who refused to pool their tracts.[106] Pooling was the obvious answer to eliminating unnecessary drilling and yet at the same time protecting the small-tract owner against confiscation.[107] If small tracts were routinely pooled through the Rule 37 administrative process, allowables could be set on the basis of surface acreage or productive acre-feet of sand, and each tract owner, whether large or small, would receive a fair share of the oil and gas underlying his tract. Per-well allowables could be abolished.

This criticism of the commission for its failure to actively pursue a fair and efficient solution to the problem of small-tract drilling seems unduly harsh given the legislative and practical constraints on the commission. It is questionable that the commission could interpret the Marginal Well Act as suggested.[108] Moreover, to administer the Marginal Well Act as proposed would require prorationing on the basis of a cost plus rate-of-return analysis for each of thousands of stripper wells in Texas, a Herculean regulatory task which would necessitate a much larger bureaucracy. To administer Rule 37 exception requests as proposed would amount to compulsory pooling, a policy which the legislature had declined to accept, indeed, which it had condemned in its 1931 declaration that the legislative intent of preventing waste was not to require "that the separately owned properties in any pool be unitized under one management, control, or ownership."[109] This declaration of intent was East Texas' third bequest to the independent producer.

The final bequest by East Texas to the independent producer was judicial approval of this regulatory inheritance. Without such approval, the small-tract owners' legacy was of tenuous value, subject always to the risk that large-tract owners whose oil and gas were being confiscated through drainage would void the inheritance through court review of the Railroad Commission's orders. In 1940 six justices of the U.S. Supreme Court twice gave their blessing to the commission's administration of Rule 37 exception permits, the Marginal Well Act, and per-well prorationing in the East Texas field,[110] viewing it as a rational response to the small-tract problem:

> To deny the holders of these tracts permission to drill might subject them to the risk of losing their oil in place or of being put at the mercy of adjoining holders. . . . If these wells, most of them small, were restricted to production on the basis of an hourly potential formula, it might be unprofitable to operate them at all. Not only are the individual interests of these small operators involved, but their effect on the state's economy is an appropriate factor to be taken into account when plans are devised to keep the wells open.[111]
>
> Small producers have investments in existing wells with low capacities, and these wells need a minimum daily production sufficient to justify their enterprise. In addition to all this, any scheme of proration duly mindful of all those considerations, hardly mathematically commensurable, which constitute the total well-being of a society, must assure the continued operation of a sufficient number of wells for an adequate exploitation of the state's oil resources.[112]

In both of these *Railroad Commission v. Rowan and Nichols Oil Co.* cases, the Supreme Court approved the principle of a "living allowable" for small-tract producers to assure their continued existence.

The Supreme Court then closed the doors of the federal courts to plaintiffs seeking to appeal commission orders.[113] Henceforth, the state courts were to handle oil- and gas-conservation matters so closely tied to the states' public policies and to local conditions requiring familiarity and expertise.

The resulting preeminence of the state courts in prorationing matters did not immediately assure the small-tract owner and independent producer of a firmly vested legacy. In 1935 the Texas Supreme Court had upheld the constitutionality of the well-spacing regulations with this admonition:

> Conditions may arise where it would be proper, right, and just to grant exceptions to the rule so as to permit wells to be drilled

> on smaller tracts than prescribed therein. . . . [B]ut in all such instances it is the duty of the commission to adjust the allowable, based upon the potential production, so as to give to the owner of such smaller tract only his just proportion of the oil and gas. By this method each person will be entitled to recover a quantity of oil and gas substantially equivalent in amount to the recoverable oil and gas under his land.[114]

Under this holding, a large-tract owner might launch a successful attack on the regulatory system that was depriving him of his property. The U.S. Supreme Court's blessing of the East Texas prorationing system in 1940 reduced, but did not remove, the small producers' fear that state courts would invalidate per-well allowable formulas. Indeed, in 1944, the Texas Supreme Court invalidated a prorationing order which was causing uncompensated drainage away from a large-tract operator.[115] However, in 1946 a Texas court eliminated all doubt that this per-well inheritance was not fully vested in the independent producer. The Texas court upheld a prorationing formula in the Hawkins field which allocated production on a 50 percent per-well, 50 percent surface acreage basis, despite a jury finding that this formula would cause drainage of 30 million barrels of oil away from Humble Oil and Refining (the largest producer in the field) toward the Hawkins townsite.[116] In words echoing the U.S. Supreme Court justices, the Texas judge wrote:

> [I]t is held that the owner of an "involuntarily" segregated tract cannot be denied the right to drill at least one well on his tract however small it may be. From which it would seem that his allowable cannot be cut down to the point where his well would no longer produce . . . nor below the point where it could not be drilled and operated at a reasonable profit.[117]

The Texas court also reasoned that the 50 percent per-well factor was fair because large-tract owners could produce oil at a lower cost per barrel than could small-tract owners. It was fair to offset this advantage by allowing some drainage toward the small tracts.[118] Also, the judge reasoned that the restricted allowables imposed by market demand prorationing effected a greater ultimate recovery of oil and a higher price than would have existed without regulation. The large-tract owners received a large part of these benefits. In essence, the judge reminded the majors of the nature of the bargain struck in 1932 when market demand prorationing was authorized: the independent producers with small tracts were not to be injured by the new state conservation legislation which supplanted the rule of capture. Actually, Humble itself acknowledged the 1932 bargain, even while attacking the 50 percent per-well formula. Humble merely sought to reduce the per-well factor in the Hawkins field from 50 percent to 40 percent and to increase the surface acreage factor from 50 percent to 60 percent. The bargain had been struck long ago. Humble was only haggling over the price.

The Texas Supreme Court refused to grant the application for writ of error in the Hawkins field case, thereby approving the result of the lower court's decision, but not necessarily approving all of the reasoning in the opinion. Thereafter the Railroad Commission treated the holding as necessitating continued adherence to per-well allowables.[119] In the same year as the Hawkins field prorationing case, another Texas court of civil appeals decision held that large-tract owners in this same field could not secure exceptions to the spacing rule in order to drill more wells and counteract the drainage by the small tracts.[120] This court told the large-tract owners to seek relief through revision of the allowable formula. In effect, then, no relief was available. The East Texas legacy emerged fully vested from the courts.

The judiciary's failure to invalidate Railroad Commission orders that resulted in confiscation of the property of large-tract owners also reflected the emergence of a new relationship between the courts and administrative agencies. In 1946 the Texas Supreme Court adopted the substantial evidence rule to review administrative decision making.[121] Under this rule, the judge's function is to determine whether there is substantial evidence to reasonably support the commission's decision. The judge is not to weigh conflicting evidence to ascertain which side has the better argument or the preponderance of the evidence. Under the new standard of review, Railroad Commission orders attain a high degree of finality because some evidence almost always exists to support one side of a controverted fact issue. With little chance of successfully appealing any prorationing order, large-tract owners no longer attempted to wrest the riches of the East Texas legacy from the small producer through litigation.[122] Legislative reform was necessary.[123]

Therein arose a circularity difficult to break open. Texas entered the 1930s without compulsory pooling or unitization because such laws were anathema to the independent producers who had discovered and secured an advantageous position in the East Texas field, and to legislators who feared—with some justification under the conditions then existing in East Texas—that such acts would allow the major oil companies to monopolize the oil fields. Henceforth, without compulsory pooling or unitization, small-tract owners and producers in other fields could be protected against confiscation only by being granted Rule 37 exception permits and per-well allowables. Once this regulatory inheritance became vested, no compulsory pooling or unitization bill would be supported by the independent producers and, without this support, no such bill could pass.

Viewed in this light, the U.S. Supreme Court's decisions in the *Rowan & Nichols* cases sealed the fate of compulsory pooling and unitization in Texas for decades to come. In one stroke, the Supreme Court justices both affirmed the Railroad Commission order on substantive grounds and disaffirmed federal court intermeddling in a state agency's administration of oil and gas regulations. These seemingly contradictory holdings derived from the same reasoning: that neither the Court nor the commission could fashion any better formula for East Texas. This reasoning was sound *under existing Texas law*. But in affirming the Railroad Commission order, the Supreme Court endorsed the failure of the Texas legislature to provide conservation laws which protected the correlative rights of all owners of a common reservoir. Had the Court instead affirmed the lower federal courts' rulings that the East Texas prorationing order was confiscatory and wasteful, existing laws would have had to change to allow the Railroad Commission to draft a formula which could pass judicial muster.[124] Compulsory pooling or unitization legislation would probably have been enacted. Instead the Supreme Court treated the existing statutory framework as if it were immutable, and it became so. The circle closed tightly.

Of course, this regulatory system which institutionalized small-tract drilling was the very cause of high costs and inefficient production, and independents were motivated by the cost-price squeeze of the 1950s and 1960s to seek relief from this through unitization. But without compulsory process, independents could bargain their way into a unit from a position of strength. The majors could either offer a favorable unitization formula to independents or forgo the benefits of unitization, in which case the independents kept the relative advantage they enjoyed under the regulatory status quo. The independents seldom risked the complete loss of unitization which could benefit them. If their initial demands were unacceptable to the majors, they could be negotiated downward until mutual accommodation was reached. Moreover, if the increased profitability of unitization was great enough, the large operators in the field who had the most to gain might proceed with the unitized operation on their own. Small producers could then benefit from these repressuring operations without having to bear any of the costs which compulsory process would have imposed on them. Voluntary agreements are most likely to result within small groups whose members differ substantially in size. The largest member will be more willing to bear a disproportionate share of the costs of securing the collective goal because the benefits to him individually outweigh the costs.[125] Thus profit-minded independent producers were not against unitization *per se,* but only against compulsory unitization that would reduce their bargaining power. In sum, the opportunities for profitable obstructionism were enhanced by the regulatory system that gave independent producers and small-tract owners a disproportionate share of the oil and gas in Texas reservoirs. A compulsory unitization statute that would deliver merely a "fair share" of oil and gas to the independents was not to their liking. The legacy of East Texas was long-lasting.

An additional economic rationale for opposing a unitization statute relates to the phenomenon of excess capacity in an ironic way. The cost-price squeeze made unitization attractive as an efficiency measure to lower production costs, but the overhang of excess capacity simultaneously dimmed its allure. After all, if the increased efficiency and output from unitized operations merely lowered oil prices to the benefit of consumers, producers would not be much better off.[126] Market demand prorationing and import quotas on cheaper foreign crude were designed to keep the price of crude oil stable, but both of these programs were under severe attack by powerful economists and politicians. Thus, the changeover to unitized production was risky, especially for independents who had no foreign crude reserves or diversified manufacturing operations to absorb the risk. Indeed, even the most ardent proponents of unitization acknowledged the difficulty of introducing its widespread use because of the price instability and economic dislocations which would probably result as long as excess capacity existed.[127]

Pride of ownership and control. Another factor—pride of ownership and control—also afforded an economic rationale for independents to oppose compulsory unitization legislation. While few producers could take pride in or afford to run a losing operation, as long as the status quo produced adequate profits, the independent could head his own business, hire and fire employees, and enjoy the "perks" of office and the respect of the community. Independents prized their ability to move quickly on shoestring budgets unencumbered by high office overhead and bureaucratic levels of review. (Major oil companies also prize this ability and often "farm out" or assign their leased acreage for independents to drill.)

Mistrust of the majors. Few independents believed that majors could match their own efficiency. Because a major company was most likely to be the unit operator, independents feared they would be forced to pay for the costs of a bloated bureaucracy.

This mistrust of the majors' ability to operate efficiently combined with a long-standing tradition of distrust of the majors' monopolistic motives. Decades of coexistence between the independents and the majors following the East Texas battles in the 1930s had not eliminated the independents' fear of domination. As the number of independents decreased during the 1960s, because of the cost-price squeeze, the majors survived, even increasing their share of domestic crude oil production.[128]

Increased legal problems. Compulsory unitization would also necessitate increased legal costs for independents who wanted to protect their interests through the administrative process. Independents who considered the unit agreement unfair would no longer have the option of refusing to join the unit, but would have to hire considerable technical and legal expertise to testify at commission hearings in an attempt to secure better terms in the unitization order.[129]

For all these reasons, independents could be expected to oppose compulsory unitization on economic versus ideological grounds, and they did. Royalty interest owners with small tracts under lease shared their lessees' preference for the regulatory status quo, and probably shared a basic mistrust of the majors as well. In addition, the fifth and sixth factors listed are important concerns of royalty interest owners, and also explain some of the independents' opposition to compulsory unitization.

Short-term declines in income and reluctance to take risks. Unitization can result in short-term declines in royalties and revenues, even though in the long run total income is increased for all in the unit. Royalty interest owners and small producers often do not share the majors' long-term outlook. This difference in outlook is exacerbated when the unitized operation is technically risky and when the geologic and engineering data are costly to obtain and difficult to interpret. Royalty owners had been roughly treated in the early years of Oklahoma's compulsory statute,[130] so much so that the statute was revised in 1951 to afford them greater protection. In the absence of compulsory unitization, well-informed landowners and royalty interest owners could also bargain their way into a unit from a position of strength. They could demand additional bonus, royalty, or other benefits in return for agreeing to unitize. Compulsory unitization would strip them of this bargaining power. Royalty interest owners often did not view the benefits of reduced costs from unitization as accruing to anyone but the producers, who thus got something for nothing whenever compulsory process was available to force landowners into units. Economists have argued that landowners do share in the benefits derived from compulsory unitization. If lessees are more certain of their ability to unitize a field, they will pay more money for leases in the field and landowners will capture economic rent in terms of larger bonuses and royalties.[131] Landowners seem skeptical of this argument.

Thus it is understandable why independents and royalty interest owners opposed compulsory unitization for self-interested reasons, but it is still not clear why the Texas legislature was swayed by such concerns. What justified support of these special interests against the strong public policy of longer-lived reserves, more secondary recovery, and lower-cost production which could benefit both oil company profits and Texas consumers? Texas is the nation's largest oil and gas consumer, as well as being the largest producer.[132] Lower energy prices would benefit many other Texas industries and the state's agricultural base.

One answer is that no such public interest exists; that the political power of the well-organized and well-financed independents is simply so great that they can wield enormous influence over political campaigns and legislators' votes. This may well be true.[133] But there are also public policy reasons influencing a legislator to vote against compulsory unitization. First is the strong political tradition of populism in Texas that values small-scale operations.[134] Most oil and gas leases are taken from farmers, so oil and gas development by independents simultaneously benefits both family farms and family firms. Oil and gas leases were the salvation of many East Texas farmers during the Depression. These populist values are congruent with the independents' unique risk-taking and entrepreneurial role within the industry. Independents drill most of the nation's exploratory wells which lead to new discoveries of onshore oil and gas fields (see this chapter text at note 40). The fear that independents would be swallowed up by the majors through compulsory unitization was not easy to dismiss. More important, many parts of Texas' economy had flourished under the established state regulatory system. Well-drilling and service companies prospered by drilling and reworking those thousands of unnecessary wells which Eastern politicians and economists so deplored. Employment in the oil fields and in related service industries grew apace.[135] The state's tax revenues were tied to the oil and gas business, and these revenues would fall if prices dropped because fields were unitized at a time of excess capacity. Small towns and communities dependent on oil and gas activities would be seriously disrupted if unit operations run by a single major oil company headquartered in Houston decreased local job opportunities and the local industrial base. Independent producers in Kansas stalled passage of a compulsory unitization bill in that state for many years using these employment-related arguments.[136]

As elected officials, the railroad commissioners could be expected to share this legislative viewpoint. The commissioners clearly viewed their role as "managers of the Texas economy" and favored policies which encouraged drilling and employment.[137] As creators of the regulatory status quo, they could be expected to champion their own handiwork. Commissioner Thompson upheld his 1932 pledge to protect the independents against any domineering by the majors and against any harm from the new prorationing system for the next thirty-three years of his tenure on the commission.[138] His fellow commissioners also honored the pledge. Even so ardent a conservationist as Commissioner Murray, a petroleum engineer by training, refused to push publicly for compulsory unitization of oil fields.[139] While he was quite willing to take unpopular stands against gas flaring and slant well drilling,[140] he only cautiously supported the wisdom of widespread unitization. His attitude is illuminated in the following:

> Unfortunately, there exists a great deal of public misunderstanding as to the purpose of unitization and what may be accomplished by it. Many people seem to feel that the objective of unitization is to obtain the economies of larger scale operation. They attribute to unitization about the same advantages as are said to exist, for example, in combining a number of small independent grocery stores into a chain of grocery stores. It is true that these benefits are generally achieved by the unit operation of a petroleum reservoir which has formerly been operated separately by each of the several diverse owners. There is usually a considerable saving in labor and equipment. However, these advantages also carry with them certain objectionable features. The very fact that unit operation is generally a labor saving device frequently causes unitization to be protested by individual workers, fearful of losing their jobs if unit operation goes into effect, and by labor organizations. In securing the economic advantage of large scale operation, unitization also arouses in the minds of the general public that vague

> fear which they frequently hold toward "bigness" and also a fear of anti-trust violation. Furthermore, the average independent thinks he can operate more cheaply than the average major company. . . .
>
> These alleged advantages and disadvantages become relatively insignificant to the petroleum engineer, because the imperative need for obtaining maximum efficiency of recovery completely overshadows these other considerations. I personally do not believe that I could ever advocate unitization merely for the purpose of obtaining operating economies. It has been my impression that much of the objection to unit operation which exists in the minds of independent petroleum producers, royalty owners, members of the Legislature, and the public in general comes from the fear that unit operation would become a goal per se and that unnecessarily intensive and extensive unit operations could cause considerable abuse.[141]

It was, of course, difficult for Murray to honor both unitization and the commission's pledge to the independents. The following reflects Murray's artful balancing:

> One reason for approaching unitization as a last resort is the fact that Americans have found by experience that competition is a very healthy thing in all types of business. Particularly in the oil industry has it been proven that a system of free, competitive, private enterprise is the only method of effectively developing a nation's underground petroleum reserves.
>
> Consequently, it appears imperative in the national interest that we carefully preserve and protect the independent in all phases of the oil industry but most particularly in wildcatting ventures. . . .
>
> Because the impression exists that unit operations are usually conducted by major companies and thereby the activities of independents in the field of production are reduced, the actual need for unitization should be carefully scrutinized whenever it is proposed. However, enlightened self-interest and public interest dictate we recognize that in many oil fields there is no way which equity can be served and the oil reserves efficiently recovered except through unitization. The person who demands that all oil fields be unitized, regardless of need, and the person who obstructs unitization where it is an actual necessity are equally unpatriotic and are jeopardizing national security as well as the economic future of the oil producing states.[142]

These statements seem to be the most outspoken public support ever given to unitization by the Railroad Commission.[143] Never once did Commissioner Murray mention the words "compulsory unitization" or call for new legislation.

Politically, it was far easier for legislators and railroad commissioners to argue for changes in other types of government regulation which would more surely strengthen Texas' economy. The oil industry's cost-price squeeze could be blamed on the federal government's short-sighted policies, such as its regulation of natural gas prices at low levels, its reluctance to enforce strong import quotas, and the wave of environmental laws enacted in the late 1960s.[144] Texas politicians (and railroad commissioners are elected politicians) argued that the state regulatory system's creation of excess producing capacity was a godsend to the national security of the United States. High-cost marginal wells would be prematurely abandoned if foreign crude imports were allowed to invade the country and drive prices down. The remaining reserves underlying these wells could not be profitably recovered thereafter, and the United States would become dependent on insecure sources of supply for its national defense and economic well-being. Texas' proposed solution to the cost-price squeeze was simple: crude oil and natural gas prices should be increased. Compulsory unitization paled in comparison as a remedy.

Thus the issue of compulsory unitization was not a cut-and-dried one. State legislators interested in promoting the economic wealth and health of Texas may have legitimately opposed compulsory unitization as antithetical to these goals. Of course, from the national perspective, the excessive well drilling and inefficient production practices in Texas oil and gas fields were not in the public interest. The billions of dollars that were needlessly invested in Texas fields could have been placed in alternative investments that would have employed workers and capital more efficiently and allowed consumers to enjoy lower energy prices and greater national economic wealth. Some of these alternative investment opportunities might well have been located in Texas and might have employed more Texas workers in new and growing industries and in the service sector. However, Texas legislators adopted the attitude that a bird in the hand was worth two in the bush. There was no guarantee that alternative investment opportunities would be located in Texas. The major oil companies had the world as their oyster and were fickle in their loyalty to Texas. As a large net exporter of oil and gas, Texans enjoyed most of the economic benefits of the unnecessary well drilling and passed much of the burden of this inefficiency to consumers in other states. As a sovereign state in a government of limited federal powers, Texas could press its own selfish interests at the expense of other states unless the federal lawmakers saw fit to limit this economic sectionalism. Congress never acted to override Texas' legislative choices.[145]

Even in those states which ultimately saw the public interest as meriting the passage of compulsory unitization bills, enactment often came only after fierce opposition and debate. Louisiana, a state renowned for its strong conservation policies, attempted to secure a comprehensive compulsory statute for many years without success until 1960.[146] Mississippi passed a compulsory statute in 1964 which was so narrowly drawn that it was almost never used.[147] It was finally amended in 1972. Oklahoma's original compulsory unitization law did not apply to the state's older reservoirs because the statute was politically unacceptable without this exclusion. Blatant examples of profitable obstructionism or physical waste were often needed to push legislators into enacting compulsory process for unitizing fields.[148] The state statutes required increasingly higher percentages of voluntary agreement before compulsory process could be had. Several states now require 80 to 85 percent of the working interest owners to agree to unitization as a condition precedent to a compulsory unitization order.[149] These statutes may parade falsely as compulsory unitization laws.

Even in Texas the issue was not clear cut. A significant number of independent producers supported a compulsory unitization bill in 1968.[150] The new frontiers of exploration for oil and gas had moved to Alaska, the Arctic, and the Outer Continental Shelf, areas requiring vast amounts of capital to drill and develop. The independents could not compete with the majors in these high-cost, technologically risky areas. From 1956 to 1971, the major oil companies increased their exploration and development expenditures, but the independent producers' expenditures declined more than 50 percent—from $2.45 billion to $1.2 billion.[151] Secondary recovery in Texas' large, old fields could sustain the independents' profits and activity if the fields were unitized. The lack of compulsory unitization in Texas was the petroleum landman's lament.[152] Also, in fields discovered after 1961, independents lost some of their regulatory advantage as compulsory pooling came into play (see this chapter at notes 194 to 218). Finally, some independents were coming to view a compulsory unitization statute which adopted protective procedures and measures to ensure the fairness of unitization orders as preferable to Railroad Commission orders issued under the waste-prevention statutes that had the end result of forcing operators to unitize without any statutory guidelines to govern the process (see chapter 5). Nonetheless, no forceful push for unitization occurred during this period of excess capacity.

The 1972–84 Period

No sooner had the ink dried on many of the studies concluding that excess producing capacity in the oil fields would be a way of life for years to come, than the excess disappeared. In 1969 Texas oil wells were prorated at 52 percent market demand factor, but this jumped to 72 percent and 73 percent in 1970 and 1971, respectively, and then to 100 percent for most of 1972.[153] The causes of this sudden eclipse of idle capacity were many and varied. On the supply side, the unattractive economic situation confronting producers throughout the 1960s had reduced the exploration and development of crude oil reserves, as had environmental laws which, for example, delayed construction of the Trans-Alaskan pipeline for five years and slowed sales of federal offshore leases. The demand for oil jumped unexpectedly in 1972 and 1973 as coal and nuclear power use slowed because of environmental problems. Natural gas could not make up the deficit because the Federal Power Commission regulated its price at too low a rate to motivate additional supplies.

With the overhang gone, oil and gas began to assume the importance of precious commodities, not to be wasted on any account. The risk that unitization would cause oil and gas prices to fall no longer seemed as serious. Between February 1971 and August 1973, the price of foreign crude oil increased from $2.18 to $3.07 per barrel. This increase reduced the threat that cheaper imports would undercut the domestic market price of crude: the domestic price started rising after 1971 as well.[154] The efficient recovery of oil and gas was clearly in the public interest, and if unitization occurred along a rising price trend, economic dislocations to oil and gas producers and royalty owners would be minimized.

The time was right to push for a compulsory unitization bill in Texas, to assault the forty-year tradition of doing without. Some versions of compulsory unitization bills had been introduced in 1969 and 1971, but had failed to even reach the floor for a vote because the Texas Independent Producers and Royalty Owners Association (TIPRO) refused to support them.[155] However, the cost-price squeeze had taken its toll on independent producers. In 1971, TIPRO voted to maintain a neutral stance regarding compulsory unitization in principle. At the same time, TIPRO's executive committee voted to oppose the specific unitization bill proposed in 1971, particularly because it failed to expressly exclude the East Texas, Yates, and Hawkins fields from compulsory process.[156] In late 1972, TIPRO polled its members and found that 60 percent of those responding favored passage of a compulsory unitization law, provided that adequate safeguards protected the rights of the minority interests that were being compelled to join the unit.[157] TIPRO appointed a "watchdog group" to assist in developing a Texas statute having these safeguards. This group listed thirty-nine essential provisions of any compulsory unitization bill which would make it acceptable to TIPRO. In February 1973, by a three-to-one margin, TIPRO's executive committee voted to support House Bill 311, and its twin, Senate Bill 120, because both bills contained the thirty-nine essential safeguards.[158] The majors also organized a large-scale campaign to achieve passage of a compulsory unitization bill in the 1972–73 legislative session. With this type of backing, H.B. 311 passed the Texas House by a vote of 91 to 37 on April 13, 1973.[159] House Bill 311 is as singular a document in the annals of compulsory unitization laws as Texas' 1949 Act is to voluntary unitization laws. A summary analysis of the bill is instructive to understand its subsequent fate.

H.B. 311 authorized the Railroad Commission to approve a compulsory unitization plan only for "unitized operations," defined as "operations related to the production of oil or gas from the unit area known as cycling, recycling, repressuring, waterflooding and pressure maintenance, and tertiary recovery operations for the purpose of increasing the ultimate recovery of oil or gas."[160] Thus H.B. 311 clearly adopted the statutory restrictions of the 1949 Act which disallowed Railroad Commission approval of unitization for exploration or for normal primary recovery operations (see this chapter text at notes 31 to 51). No compulsory order could issue to resolve the conundrum of the gas-cap field discussed earlier. H.B. 311 added tertiary recovery to the 1949 Act's list of permitted unitized activities to eliminate any doubts about the inclusion of this type of operation, but secondary recovery in the sense of operations requiring both injection and output wells was clearly to be the only scope of H.B. 311.[161]

Upon receiving a petition for a compulsory unitization order, Section 4(a) of H.B. 311 stated that the Railroad Commission must hold a hearing to determine if all the following conditions exist:

1. The unitized operation is economically feasible and reasonably necessary to prevent waste *and* to increase the ultimate recovery of oil or gas.
2. The value of the estimated additional recovery will exceed the additional cost.
3. The proposed unitization plan is fair, reasonable, and equitable to all owners *and* reasonable efforts have been made to form a voluntary unit.
4. The expense incurred in establishing the unit is reasonable and necessary.
5. The productive limits of the common reservoir have been reasonably defined by development, *and* the unitized area is reasonably necessary and sufficient.
6. At least 75 percent of the working interest owners have executed the proposed unitization plan.

Obviously many of these findings parallel the statutory traditions of the voluntary unitization act. Nowhere in H.B. 311 was the Railroad Commission authorized to order unitized operations designed primarily to reduce operating and drilling costs, as might be desired in the East Texas field.[162]

A seventh finding required by H.B. 311 was that the unitization plan submitted by the working interest owners meet seventeen additional conditions.[163] Many of these were uniquely designed to protect independents. First, in selecting or replacing a unit operator, the commission must consider all offers to be the unit operator and must consider the offer to operate the unit for the lowest overall cost.[164] In addition, the plan must provide for efficient operations, and no unitization plan could contain a provision for operating charges which included any part of district or central office expense other than reasonable overhead charges.[165] Detailed accounting procedures were required, and working interest owners were assured access to cost and expense data and the right to audit.[166] These provisions were clearly intended to allay the independents' fear that major oil companies would be automatically selected as the unit operator, even though their administrative costs were higher, and that these costs would go unmonitored.[167]

Second, the unit operator must give preference in the hiring of additional personnel to the employees of the working interest owners who had been working within the unit area and whose jobs had been terminated as a result of the unitization. These terminated employees were also guaranteed at least two weeks' severance pay for each year of employment, to be paid as a unit expense.[168] These provisions were clearly inserted to placate workers and labor unions.

Third, H.B. 311 required that the unitization plan provide for the right of any working interest owner to withdraw from the unit at any time by transferring his interest to the other working interest owners,[169] and also for the right of any nonsigning working interest owner to be relieved of liability for unitized operation by assigning his interests in the unit area to the other working interest owners or by offering to sell his interest to the unit operator within sixty days following the effective date of unitization. Elaborate procedures were specified to determine a fair selling price. These provisions guaranteed that an unhappy independent could escape the burdens and denounce the benefits of any compulsory unitization order.

The most important of the plan's seventeen requirements rendered it unlikely that an independent would want to disentangle himself from unit operations. The tract participation factors used to allocate unit production protected the independents' vested advantage in the status quo. Under H.B. 311, each tract's share of remaining primary reserves must be accurately estimated and each tract must be allocated "the share of the estimated remaining primary reserves that such tract or wells thereon can reasonably be expected to produce under the proration plan or allocation formula applicable in the field in the absence of unitization."[170] In the clearest language posssible, H.B. 311 retained the small producers' regulatory advantage in the remaining primary reserves underlying the unit.[171] Those reserves which could be produced only through secondary recovery operations were to be allocated according to a "fair, equitable, and reasonable" formula ascertained by using pertinent "engineering, geological, operating, and other necessary factors,"[172] but only after primary reserves were allocated under the existing spoils system. Under this provision, the East Texas independent was assured a continued disproportionately large share of the East Texas field. The required 75 percent working-interest-owner approval of the unit plan was also to be measured on a "tract participation basis," that is, on the percentage of allocated unit production in the plan. Thus, the independents' voting power also reflected their disproportionate share of primary oil.

Royalty interest owners also had many safeguards under H.B. 311. While they could not initiate the unitization application, the commission could not order unitization until at least 75 percent of the royalty interest owners had approved the unit plan in writing.[173] Nothing in the unitization order could relieve the working interest owners from any obligation to develop reasonably the leases included in the unit area as a whole.[174] Both royalty interest owners and working interest owners had the right to file an application with the commission requesting that geological, engineering, and economic information in the possession of parties owning interests in the proposed area be made available for inspection.[175] Finally, any party at interest aggrieved by a unitization order could appeal to a district court of the county in which the land was located.[176] The advantage to royalty owners and independents of having a nearby forum with jurors chosen from the local communities is self-evident.

Dressed in this protective armor, specially forged to shield independents from too great an assault on traditional vested rights, H.B. 311 passed the House and proceeded to the Senate chamber. Proponents of the bill testified that it would increase the recovery of oil in Texas by 2 billion barrels over and above that which would be recovered without compulsory process, and listed fifty fields in which unitization was either thwarted or delayed for lack of compulsory process.[177] An official of the governor's planning office testified that every additional 500 million barrels of oil produced and processed in Texas would increase economic activity by $1.2 billion, annual taxes by $80 million, and annual employment by 22,000 workers.[178] TIPRO also testified on behalf of the bill.[179] A majority of senators seemed pledged to the cause, but Senator Peyton McKnight, an independent producer from East Texas, delayed final action on the bill by requesting that the transcripts of all of the testimony on H.B. 311 be typed.[180] The completion of this lengthy task and the opposition's purposeful filibustering of the bills that preceded H.B. 311 on the legislative docket left only a few days remaining in the legislative session. A majority of senators voted to suspend the regular order of business to permit consideration of the bill, but the vote was three short of the required two-thirds, and the bill was tabled.[181] Thus, despite the fact that the bill had passed the House and seemingly had the support of a majority of senators, compulsory unitization remained a uniquely elusive political goal.

The bill's defeat in 1973 was clearly not attributable to any particular provisions which unduly favored the major oil companies or prejudiced the independents.[182] Rather, the reasons for the bill's failure are largely the same as those discussed in the pre-1972 era. The main opposition derived from the perception that the bill would upset the allocation of oil in large fields that had old-style proration formulas. Despite the bill's express declaration that small-tract owners and producers would continue to receive shares of primary oil measured by the existing allocation formula in the field, opponents characterized the bill as an "end run" to change allowables, particularly in the East Texas, Hawkins, and Yates fields.[183] Even though both the proponents and opponents of the bill testified that the East Texas field would probably never be unitized,[184] the bill made lawmakers from East Texas "nervous."[185] Also, the new economic scenario of energy scarcity and the many safeguards included in the bill reduced, but did not eliminate, the fear of major oil company domination,[186] of decreased drilling, employment, and tax revenues,[187] and of lower short-term royalty payments.[188] Additionally, the irony of regulation provided another barrier to deflect the bill from passage. The brighter economic picture facing producers in 1972 and 1973 increased the attractiveness of profitable obstructionism under the regulatory status quo. With rising oil and gas prices and emerging shortages, wells based on Rule 37 exceptions were profitable to drill again.

The number of such wells approved by the commission soared from 1,562 in 1970 to 3,031 in 1980.[189] Gas producers could find ready markets for their gas, reducing the incentive to recycle it for pressure maintenance or secondary recovery operations. "Going it alone" was good business again.

The defeat of H.B. 311 in 1973 spelled the end of any further organized effort to adopt compulsory unitization in Texas.[190] Even the massive havoc wreaked on the U.S. economy by the Arab oil embargo a few months later in October 1973 did not inspire another effort to pass a compulsory unitization bill. Governor Briscoe called a special session of the legislature to pass a 55-mile-per-hour speed-limit bill, but refused to add the compulsory unitization bill to the agenda.[191] Nor did the subsequent decade of energy travail and woe provide enough inspiration. Texas governors and legislatures established formal councils to study and propose energy policies to cope with the short- and long-term effects of the energy crisis, but these councils, unwilling to renew hostilities on a recent battlefront, gingerly dodged the issue of compulsory unitization. After four years of intense research on energy policy, the Governor's Energy Advisory Council in 1977 developed the following postembargo policy on unitization:

> It is recognized that both crude oil and natural gas production are declining in Texas; that a significant potential exists for increasing recoverable reserves and production from the application of enhanced recovery techniques; that some producers stand to gain production and income at the expense of other producers in enhanced recovery operations unless unitization is instituted; that the difficulty encountered in unitizing a reservoir is a significant deterrent to the initiation of enhanced recovery operations; that the opportunity currently exists for unitization of reservoirs in Texas on a voluntary basis; and that many reservoirs in Texas are now unitized under the voluntary rules.
>
> It is also recognized that . . . the potential exists for one or more of the mineral estates which enter a production unit to receive less from the proceeds of such unit than it would have received had it been produced separate from the production unit.
>
> Therefore, it is recommended that the state maintain a statutory and regulatory framework which will facilitate the unitization of oil and natural gas reservoirs where such unitization

> would result in an increased total ultimate recovery from such reservoirs. While it is necessary to place primary emphasis on maximizing the long-term recovery of oil and natural gas, the state must make every effort to assure the equitable division of production from unitized reservoirs.[192]

The railroad commissioners and Texas politicians continued to blame the federal government for the energy crisis. The major oil companies were too busy defending themselves against massive antitrust attacks and fending the onslaught of federal price controls, crude oil allocation laws, windfall profits tax legislation, and divestiture bills to resume the fight for compulsory unitization. Even the unveiled threat of federal intervention into the Railroad Commission's regulatory preserve failed to stir Texas legislators into implementing compulsory unitization as assurance to federal officials that the ultimate recovery of the nation's precious oil and gas resources underlying Texas soil could be entrusted entirely to the state of Texas.[193] No source of momentum could be found to overcome the inertial effect of the 1973 defeat and the East Texas legacy.

Compulsory Pooling in Texas

Given these reasons for the conclusive defeat of compulsory unitization in Texas, it is puzzling that Texas lawmakers passed a compulsory pooling bill in 1965. The objective of compulsory pooling is to reduce unnecessary drilling costs and to maintain well-spacing rules without violating the correlative rights of any of the owners of tracts within the spacing unit. As we have seen, Texas politicians did not consider excessive drilling to be evil; indeed they took pride in the number of wells drilled and the ensuing benefits to the Texas economy. The correlative rights of independents and small-tract owners were being amply protected under the existing regulatory system, so little impetus for change arose on this account.[194] Indeed, in many ways it would seem that compulsory pooling would suffer even greater legislative disfavor than compulsory unitization. Unitization for secondary recovery purposes generally results in greater ultimate recovery of oil and gas and longer-lived fields, thus providing additional employment and drilling opportunities, especially in water-flooding operations that use closely spaced injection wells. Pooling generally does not have these beneficial effects.[195] The mystery then is why a compulsory pooling act passed and compulsory unitization failed.

The answer is simple. On March 8, 1961, the Texas Supreme Court found invalid the commonly used one-third per well, two-thirds surface acreage, gas prorationing formula on the basis that it confiscated the property of large-tract owners.[196] Under this formula, in the case at hand, the owner of a three-tenths-of-an-acre tract overlying about $7,000 worth of gas in the Normanna field would be allowed to drain about $2.5 million worth of gas from the neighboring large-tract owners during the life of the field. The supreme court found the prorationing order unreasonable. Lest oil and gas attorneys, judges, and the Railroad Commission viewed this 1961 decision as limited only to its facts, the supreme court in 1962 invalidated the same one-third–two-thirds formula in the Port Acres field upon a showing of uncompensated drainage from large-tract to small-tract owners.[197] The court, not the legislature, broke open the circle.

What caused this reversal in judicial opinion which previously had approved allocation formulas with large per-well factors? More fundamentally, what led the plaintiffs to contest a prorationing formula which had become commonly accepted throughout Texas gas fields? It is not difficult to read between the lines of the court's opinions to see that the economic pressure of the cost-price squeeze on producers due to excess capacity in the oil and gas fields was having a serious effect. The wells in these fields were deep and costly to drill.[198] A well drilled under a Rule 37 exception would need to drain more and more from neighboring tracts in order to receive a living allowable. If the principle of a living allowable to small-tract owners remained inviolate, it was easy to foresee the need for even greater per-well factors in future allowable formulas to maintain the profitability of small-tract drilling. For almost thirty years, the large-tract owners had accepted the political compromise that market demand prorationing required some per-well factor in the prorationing formula, but the price of this deal had escalated beyond all expectations.[199] In addition, the conflict was not strictly between large tracts and small tracts. It was between pooled or unitized tracts and small tracts. In both fields, the economic pressures of the 1960s had encouraged many small producers to voluntarily pool their small tracts into larger tracts in order to save drilling costs.[200] The economic benefits of pooling were considerably reduced when those who refused to pool could drain from those who did. Under the economic conditions of the 1960s, with costs mounting, crude prices falling, and drilling on the decline, the allocation of shares of production in a discovered field was more important than it had been during an expanding market when all could share in a growing pie.[201] Third, deeper drilling in the Port Acres field had led to the problem of retrograde condensation, and the field required pressure maintenance and cycling in order to achieve efficient recovery of the liquids with the gas. The plaintiffs sought not only to invalidate the prorationing formula which drained gas from them, but also sought to have the Railroad Commission enter a compulsory cycling and pressure maintenance order for the entire field.[202]

The justices thus had a clear choice before them: to encourage efficiency and greater ultimate recovery in the gas fields, or to allow the old order to reign even in new fields. In deciding to invalidate large, per-well allowable factors, the court clearly chose to encourage pooling: first, by protecting the voluntarily pooled tracts from drainage; and second, by eliminating the small tract's incentive to "go it alone."

The court's judicial activism in inducing pooling by no means divested the independents entirely of their East Texas legacy however. The supreme court did not overrule the 1946 decisions in the Hawkins and Yates fields which had upheld the 50 percent per-well, 50 percent acreage formula for prorationing crude oil. The court distinguished these prior cases with the following language:

> We are in agreement with the reasoning of the courts in the Humble and Standard Oil Company cases in holding that where producers have acquiesced in and have failed to complain of the Commission's proration orders for a long period, during which time other operators have expended vast sums in exploration and drilling operations, such producers should not be heard to complain.[203]

The old order would remain in the old fields. This point was forcefully made in the subsequent *Railroad Commission v. Aluminum Co. of America* case,[204] in which the court held that laches prevented the plaintiffs from invalidating the one-third–two-thirds gas prorationing formula in the Appling field. No one would deprive the East Texas independent of his disproportionate share of the East Texas reservoir.

Judicial activism was also very limited in the Rule 37 exception process. The plaintiffs in both the *Normanna* and the *Port Acres* cases had first attempted to prevent the small-tract owners from securing permits for Rule 37 exceptions.[205] They argued that the small-tract owners had been offered a fair opportunity to pool and did not require a Rule 37 permit to prevent confiscation. The courts upheld the commission's granting of the exception permits and refused to condition the permits on proof of an inability to pool, as had been suggested by Hardwicke. The large-tract owners' *only* recourse was to attack the prorationing formulas.

Nor would judicial activism stretch to force the Railroad Commission to enter a compulsory cycling and pressure maintenance order, as requested by the plaintiff in the *Port Acres* case.[206]

Despite these limits to the courts' decisions involving the Normanna and Port Acres fields, two supreme court justices dissented in the prorationing cases, arguing that the court was creating compulsory pooling, directly contrary to the legislative declaration of intent that the commission was not authorized to require that separately owned tracts be unitized under one management and control.[207] Semantically, the majority opinions did not compel pooling. They simply declared the existing prorationing formulas unreasonable and confiscatory. But the essence and effect of the majority's decisions were obvious. Small-tract owners in the future could either voluntarily pool into larger tracts or suffer uncompensated drainage under new prorationing formulas which rendered small-tract drilling uneconomic.

Following the *Normanna* decision, the Railroad Commission entered a new prorationing order for the field based entirely on surface acreage. However, a special allowable would be accorded the small-tract owner who could prove that drilling a well was not economically feasible under this new formula and that neighboring tract owners refused to pool with him on fair terms. This special allowable was designed "to encourage a reasonable attitude in such neighbors so that they would endeavor to work out this common problem."[208] Compulsory pooling had come to Texas, not in the statute books, but in the Railroad Commission's administrative regulations. With *de facto* pooling in effect, the Texas legislature enacted the Mineral Interest Pooling Act in 1965, applicable to reservoirs discovered or produced after March 8, 1961, the date of the *Normanna* decision.[209]

Perhaps this history of the passage of compulsory pooling legislation in Texas supports criticism of the Railroad Commission for failing to administer its regulations at an earlier date to encourage pooling.[210] If the Railroad Commission had displayed initiative and activism, it could have adopted this new prorationing system which brought greater fairness and efficiency in the oil and gas fields long before the supreme court forced the issue.[211] The Railroad Commission in fact was extremely concerned about the depressed economic conditions confronting producers in the 1960s. But the commissioners perceived their role more passively. In Commissioner Langdon's words: "It has long been a policy of the Commission merely to administer the matters delegated to it by the law, not to foster legislation or write the law."[212]

Of course, this attitude can be viewed as a defensive screen used to shield the elected commissioners from any active endorsement or administrative promotion of pooling which ran the substantial risk of alienating the politically strong independents. But it is also a solidly defensible position. In our tripartite system of coequal branches of government, the legislature, not the executive, is to write the law. Texas lawmakers had not passed a compulsory pooling statute. They had not delegated pooling authority to the Railroad Commission. Moreover, the makeshift use of the commission's prorationing authority to administer pooling was not a satisfactory solution to the small-tract problem. Without a compulsory pooling statute, the Railroad Commission faced the tremendous task of administering the details of the pooling–prorationing system without any legislative standards as guidance. Many legal uncertainties plagued this approach.[213] Ultimate responsibility for compulsory pooling policy lay with the legislature.

With the tables turned, the independents now supported a compulsory pooling bill and drafted a version favorable to their interests. This version was enacted in 1965 with few changes.[214] The independents' imprint on the bill is readily apparent in the following provisions:

1. The act applies only to post-*Normanna* reservoirs and only to developed fields. Exploratory drilling is excluded. Thus independents who discover a new field cannot be forced to share their new discovery immediately.
2. The Railroad Commission cannot initiate pooling on its own motion, but must await an application from a private party. If no application is ever made, operators are free to play the rule of capture and Rule 37 exception game.
3. Pooled units cannot exceed 160 acres for an oil well or 640 acres for a gas well, plus 10 percent tolerance. Thus pooling cannot be used as a fieldwide unitization device.
4. The act applies only to separately owned tracts. Cotenants cannot be forcibly pooled.
5. The commission has no authority to enter a pooling order unless the applicant has made fair and reasonable offers to voluntarily pool with other owners in the proration unit. This provision for "compulsory voluntary" pooling reduces the need to pursue administrative channels and the attendant bureaucratic costs and time. It also requires that small-tract owners be bargained with, giving them a sense of participation and control over their property.
6. The act specifically allows small-tract owners to "muscle in" to tracts which are of standard size.
7. The act provides that production will generally be allocated on a surface acreage basis. If any other formula is used, a nonconsenting owner shall not receive less than he would receive under a surface acreage allocation. In effect, the nonconsenting small-tract owner who has edge tracts of poor quality is guaranteed a relatively favorable surface acreage formula. The general use of surface acreage also reduces the need to hire costly experts to analyze and evaluate more complicated formulas.[215]

8. The act prohibits certain provisions in a pooling agreement which increase the unit operator's power and control and which independent producers particularly dislike, such as a preferential right of the operator to purchase mineral interests in the unit; a call on or option to purchase unit production; operating charges that include unreasonable district or central office expense; and prohibitions against nonoperators questioning the operation of the unit.[216]

9. The act provides that owners can elect not to pay drilling costs in advance, in which event the parties advancing the costs can be reimbursed only out of production. Thus if the drilling parties hit a dry hole, the independent who has elected not to contribute any front-end capital to the drilling operation cannot be charged for any costs of the dry hole. If the well is productive, the parties advancing the costs can charge the carried producer a risk factor on his proportionate share of the costs, but not exceeding 100 percent. Private joint operating agreements often provide for risk factors double or triple this percentage.

10. Venue for appeals of pooling orders is the district court of the county in which the land is located rather than the Travis County courts located in Austin, the state capital, where most appeals of commission orders must be filed. The small producer can thus appeal on friendly turf and avoid travel time and expense.

In all these respects, the Mineral Interest Pooling Act favored the independent producer. By contrast, royalty interest owners were almost completely ignored. Indeed, the language of the original act prohibited royalty interest owners from even applying for a pooling order.[217] The act was amended in 1971 to remedy the inequity that arose from this situation.

The net effect of the pooling act has been to encourage wider spacing and more efficient production in oil and gas fields discovered after March 8, 1961. Presumably, these fields are more easily unitized than fields in which the East Texas legacy lives on, because fewer operators in the newer fields have a vested interest in excess wells or living allowables. However, compulsory pooling into drilling units is not an effective substitute for optimal well placement and production rates on a fieldwide basis (see chapter 2).

Moreover, the many limitations on the power of the Railroad Commission to force pooling under the Texas act still result in less pooling than would be desirable from an efficiency standpoint. The *Normanna* and *Port Acres* cases did nothing to disturb prorationing formulas in old fields or to disturb the small-tract owner's right to receive a Rule 37 exception permit. When crude oil prices rocketed upwards after 1973, drilling for oil and gas under small tracts became profitable again, and the number of Rule 37 exception permits soared. No doubt some of the permits were granted to recover oil and gas that could not be produced from existing wells, but many represent unnecessary drilling. The Railroad Commission has no authority to initiate pooling or to refuse a Rule 37 exception permit on the basis that the applicant's correlative rights could be protected by pooling. Nor does the Railroad Commission monitor the Rule 37 applications in conjunction with pooling act applications. Thus the large-tract owner who is seeking a pooling order to force a small tract into a unit may find that the small-tract owner has in the interim received a Rule 37 exception permit and started to drill on his own.[218]

Summary of Texas Legislation Affecting Unitization

This historical analysis of Texas' legislative framework of oil- and gas-conservation policy shows a strong and sustained antipathy to the very concept of unitization, whether voluntary or compulsory. This statutory framework consists of:

(a) State antitrust laws that from 1889 until 1983 essentially condemned voluntary unitization agreements as *per se* violations of the antitrust laws.

(b) Oil and gas conservation statutes that grant the commission very broad, strong powers to prevent waste, with the specific proviso that these waste-prevention powers are not intended to authorize the commission to require unitization or repressuring.

(c) Oil prorationing statutes that allow market demand prorationing and grant the commission vast discretion to allocate oil allowables "on a reasonable basis." For many years, this discretion resulted in allocation formulas with large per-well factors which reduced the incentive for small-tract owners to pool or unitize.

(d) The Marginal Well Act, enacted in 1931 and revised in 1933, that exempts stripper wells from market demand prorationing, thus promoting excessive investment in these wells and creating serious inequities in the distribution of a field's allowable. These effects reduce the attractiveness of either voluntary or compulsory unitization to the many owners of marginal wells in Texas.

(e) The 1949 Act that begrudgingly grants limited antitrust immunity only to certain types of voluntary unitization agreements, and only after the commission has ascertained that unitization is necessary as a last resort to increase oil and gas recovery. The 1949 Act supplanted a provision in the 1935 gas act which had allowed voluntary unitization agreements in gas fields. This 1935 provision was passed in an effort to encourage unitization in order to protect the correlative rights of independent producers who lacked pipeline connections.[219]

(f) The belated enactment of the Mineral Interest Pooling Act in 1965 which allows compulsory pooling, but only in fields discovered after March 8, 1961.

The Texas legislature has consistently refused to define correlative rights or economic waste in terms of protecting producers from having to drill costly and unnecessary wells in order to obtain a fair share of oil and gas. Excessive well drilling is not a sin to a Texas politician.

Because Texas' conservation legislation came of age in the 1930s with the discovery of the East Texas field, the fate of unitization became intertwined with the battle over market demand prorationing. At that time, both unitization and prorationing were understandably viewed as monopolistic, price-fixing devices, masquerading as conservation measures in the major oil companies' campaigns for control of domestic and world oil markets. However, even had there been no proof of the majors' use or abuse of their pipeline power and umpire systems to regulate the oil fields, the political power of the numerous Texas independents and the antimonopoly ideology of the Texas populace may yet have been enough to secure this anti-unitization policy framework. The independents had secured a disproportionate share of the oil in the East Texas field and would fight to institutionalize this advantage in the statute books. They had enormous political appeal in this fight against the oil industry's unpopular Goliaths. The Marginal Well Act best demonstrates this political stroke. It was passed in 1931 on an emergency basis to precede the scheduled institution of prorationing in the East Texas field. It was amended in 1933 to accommodate a well-potential prorationing formula in East Texas that would pass judicial muster and preclude a formula based on any acreage or reserves factor. Humble Oil and Refining's well-argued defense of unitization as a low-cost, efficient, and equitable method of increasing ultimate recovery and preventing the waste of large amounts of oil and gas was no match for the independents' political clout and prowess.

Texas entered the 1930s without a compulsory pooling or unitization law, and the events in East Texas assured that no such laws would thereafter be passed, absent extraordinary pressure from the judiciary. A costly, inefficient, and inequitable system of prorationing developed in the East Texas field, but it was declared rational by the highest court of the land, and was passed on, in somewhat diffused form, as a legacy to all Texas oil and gas fields. This legacy vested a rich regulatory advantage in small-tract owners, small oil and gas producers, marginal well owners, well drilling and service companies, and many towns and communities. The Texas legislature would never willingly divest this strong constituency of its inheritance. The continued legislative antipathy to unitization after the 1930s, as reflected in the restrictive 1949 Act and the failure to pass a compulsory unitization statute in 1973, was no longer credibly based on a view that unitization was an antitrust menace. Rather it reflected the political power of those who had inherited the East Texas legacy.

Few scholars have accurately explained the events which doomed the concept of unitization in Texas. Most scholars have viewed the birth of Texas' per-well prorationing system and the consequent death of unitization and pooling as the result of mistake, ignorance, or unconscious design.[220] Some quirks of fate did indeed affect the development of Texas' conservation framework: that the East Texas oil field, the largest in the United States, was discovered at the start of the greatest economic depression ever to befall the nation; that it underlay populated areas subdivided into hundreds of small tracts; that it was a water-drive field for which no ready engineering knowledge existed to guide proper production practices; that it was located near transportation facilities; and that it was discovered at a time when the Railroad Commission was composed of sleepy politicians with no expertise or control over the oil industry. Nonetheless, the framework of conservation law that developed in Texas was anything but an unconscious accident. In the 1930s the Texas lawmakers faced a clear choice between two methods of preventing the large-scale physical waste of oil and gas and the economic ruin of the industry—unitization or the statewide control of production rates. The latter was the only politically acceptable course of action.[221] Unitization would place control of a field in the hands of a major oil company as unit operator. The enforced prorationing of oil under martial law showed producers the beneficial price effects which could result from state control of production rates. The majors won the hard-fought battle for market demand prorationing, but victory was premised on the pledge that prorationing would not injure the independents. This translated into allowable formulas that closely approximated the rule of capture by being based on per-well factors. Humble acknowledged the Devil's bargain it had struck when it sought only to reduce, but not to eliminate, the 50 percent per-well factor used in the Hawkins field in 1946.

Market demand prorationing and the commission's emerging competence and expertise in regulating gas-oil ratios and similar production practices substituted for unitization as methods of preventing the physical waste of oil and gas in Texas. These substitutes were effective in eliminating much of the flagrant waste that had characterized the oil and gas industry under the rule of capture. By one estimate, this conservation framework increased the recovery of oil and gas in Texas by 50 percent and thus decreased the cost of petroleum products to consumers in the long term, as well as supplying the nation with vast dependable underground stocks of oil and gas.[222]

Texas' framework of conservation laws grants the commission the power to prevent waste and protect correlative rights. The commission has no statutory authority to control the price of oil and gas, to provide employment for oilfield workers, or to preserve and maintain competition in the oil and gas industry.[223] But the authority to regulate the oil fields naturally carries with it the power to manage the Texas economy. Commissioner Thompson and subsequent commissioners have recognized and used this power. Market demand prorationing was used to raise the price of oil to acceptable levels. Rule 37 exception permits and per-well allowables were used to spread the underground wealth of the reservoirs over a larger number of citizens, particularly farmers, and to protect the independent producer. There is a natural congruence between protecting correlative rights and protecting competition in an oil field because of the migratory nature of oil and gas. A large oil company with ample capital for drilling and an integrated pipeline system can drain less fortunate co-owners of a common oil and gas pool without liability under the rule of capture. Because of this, the individual oil and gas pool is easily viewed as the relevant market area for defining and preventing monopoly power.[224] Competitive struggles over correlative rights in individual fields assume an antitrust character. The Railroad Commission could not and would not divorce its power over correlative rights from this larger antitrust context.

It is not strange or unexpected that the commission acted beyond its statutory authority in considering employment, income distribution, and effects on competition in its administration of the conservation laws. The commissioners were elected officials, operating in the same environment as the lawmakers who created the policy framework. It is the function of the courts to check unauthorized actions of administrative agencies and unconstitutional legislative enactments. However the Great Depression and the New Deal had revolutionized the relationship between the courts, the legislature, and executive branch agencies. The U.S. Supreme Court was no longer interested in striking down state legislation on due process grounds, especially legislation which spurred drilling, raised prices, and created employment and wealth. The Texas Supreme Court's subsequent adoption of the substantial evidence rule rendered commission orders almost impregnable to attack. Thus the judicial system also played an important role in the selection of a highly imperfect prorationing system as a substitute for unitization.

It is within this imperfect framework of conservation law and policy that the Railroad Commission's subsequent administrative decision making concerning unitization must be analyzed. This is the next subject at hand.

5

The Railroad Commission and Unitization: Field Rules and Statewide Orders

Chapters 5 and 6 address the Railroad Commission's administration of the oil- and gas-conservation laws, specifically focusing on those regulations and orders which have affected the implementation of unitization in Texas oil and gas fields. The objective of these chapters is to assess the degree to which the Railroad Commission has exercised its administrative discretion to either encourage or retard unitization in Texas within the legislative framework described in chapters 3 and 4.

This framework is classifiable into two types of statutes: those which prohibit waste; and the 1949 Act authorizing voluntary unitization agreements upon approval by the Railroad Commission. As we have seen, the Texas legislature adopted strong statutes against waste as early as 1919 and strengthened them throughout the legislative battle over prorationing in the 1930s.[1] However, these waste statutes include a legislative declaration of intent that the Railroad Commission not require repressuring or unitization in its efforts to prevent waste. For many years, this declaration of intent was codified as Article 6014(g) of Texas' civil statutes, and all references to the declaration either by the commission or by scholars were in terms of Article 6014(g). This nomenclature will be retained throughout the book to ease identification of this legislative enactment.[2] As we have seen in chapter 4, the 1949 Act restricts the Railroad Commission's power to approve many types of voluntary unitization agreements. Despite these limitations on its powers, the commission has actively used its administrative authority to effect unitization in oil and gas fields, particularly after World War II. This chapter analyzes the administrative decision making inherent in the commission's field rules and statewide orders. A field rule is applicable only to a specific oil or gas field. It is adopted after notice and hearing procedures and is based on the particular field's conditions. Every field has at least four rules to govern the four fundamentals of well spacing, proration unit size, allocation formula, and well casing. A statewide order is applicable to all fields in the state unless superseded by a special field rule.[3] Chapter 6 analyzes the commission's interpretation and implementation of the 1949 voluntary unitization act.

Field Rules Affecting Unitization

Five types of field orders issued by the commission have had a significant impact on unitization and pressure maintenance operations. These five are (1) orders preventing waste in condensate gas fields; (2) "no flare" orders directed at preventing the waste of casinghead gas in oil fields; (3) "no waste" orders directed at preventing the underground waste of oil; (4) the special allowable field rules granted to secondary recovery and pressure maintenance operations; and (5) the unique field rules and special allowables adopted in the East Texas field. After documenting the extent and impact of these field rules, this chapter assesses the degree of administrative discretion demonstrated by the Railroad Commission in its promulgation of these field orders, including an analysis of whether the commission's zeal for preventing waste and encouraging unitization may have carried it beyond the limits of its statutory authority.

Pre-1949 Orders in Condensate Gas Fields

The Railroad Commission was largely a reactive agency in the early 1930s, buffeted by legislative winds, judicial thunder, and forceful pressures within the oil and gas industry. Because its members had little professional expertise in oil and gas operations or their regulation, the commission lacked firm ground for its recommendations. Only after the storm over market demand prorationing subsided—both in the East Texas oil field and the Panhandle gas field—could the commission tend to other matters. After the passage, in 1935, of comprehensive bills prohibiting the waste of gas from gas wells (H.B. 266) and from oil wells (H.B. 782), the Railroad Commission's major tasks were the enforcement of these acts and deciding Rule 37 exception cases. In the latter half of the 1930s, the revolution in administrative law wrought by the Roosevelt era had increased the power of the agency, and the forceful presence of Commissioner Thompson had increased its prestige, but it was still too poorly staffed to undertake vigorous administrative action on its own motion. Moreover, for most of the decade, the regulatory problem was an excess of oil and gas. Conservation above and beyond that induced by market demand prorationing was not of pressing importance.

The one important exception to this state of affairs involved the Railroad Commission's actions to prevent waste in the newly discovered, condensate gas fields. The statutory framework prohibited the flaring of gas from gas wells,[4] but allowed the flaring of casinghead gas which was produced along with oil from oil wells. During the 1930s, operators in condensate fields produced the gas and passed it through a wellhead separator which dripped out a light liquid called natural gasoline or condensate. The remaining gas was piped to a gasoline plant to recover additional liquids, after which millions of cubic feet of tail gas were flared into the air. The operators claimed that the light liquid was crude oil and that their wells were oil wells so they could legally flare the residue gas.

In 1934 the Railroad Commission issued an order in the Agua Dulce field prohibiting this gas flaring on the basis that the wells were gas wells. This classification was made after the commission undertook scientific tests and studies which determined that the light liquid existed originally in a gaseous state in the reservoir.[5] Unless the producers marketed or recycled the gas from these wells, the wells would be shut in. The commission's order was upheld by the courts.[6] Producers with wells located far from natural gas pipelines and markets were thus forced to reinject the gas. The Agua Dulce producers began to return the residue gas from their stripper plant to the producing horizon.[7] The technology for cycling was experimental in 1934, but by 1938 a commercial cycling plant was operating in northeast Texas in the Cayuga field, and from then on the cycling industry grew apace. By 1942, over twenty-nine cycling plants existed in Texas.[8] Cycling can only be accomplished in unitized fields, because injection wells must be placed at one end of the field and gas-producing wells at the other, with no wells in the center. It is almost impossible to perform cycling economically without a fieldwide unit. Thus the 1934 Agua Dulce field order set a precedent which could force operators in other fields to unitize or else shut down. In 1939 the Railroad Commission passed a statewide order classifying condensate wells as gas wells throughout Texas.[9] By 1948, forty voluntary unitization agreements had been approved by the attorney general under Section 21 of the 1935 Act, most of them involving condensate fields.[10]

In its investigation and regulation of condensate gas fields, the commission displayed initiative and vigor. Still, this example was the exception and not the rule before World War II. During the early 1940s, the crucial importance of crude oil and gas supplies to the war effort focused national attention on conservation and on the state agency which controlled so much of the nation's oil and gas resources. A 1944 study of the Railroad Commission's administrative control of oil and gas production in Texas showed that the agency still suffered from abysmal personnel policies and deficient procedural practices,[11] yet the commission was seldom criticized by the industry it regulated. This dearth of criticism was attributed to the fact that the substantive programs of the Railroad Commission were by and large accepted by the industry. Indeed, much of the agency's agenda was sponsored by the oil companies. The commission still played a largely reactive role. The initiative for statewide or fieldwide hearings came primarily from oil and gas producers experiencing particular problems under the existing regulations. The commission seldom conducted its own investigations or advocated a particular position at hearings. The administration of oil and gas production in Texas was "government with the consent and cooperation of the governed."[12]

This is not to say that the war years were placid ones for the commission. The agency labored mightily with the federal government and industry to increase production from the oil fields to meet military needs. To raise output without damaging the fields required the agency to shift from prorationing based on market demand factors to prorationing based on scientific evidence of each field's maximum efficient rate (MER) of production. This meant estimating individual MERs for hundreds of oil fields.

The war years did bring significant efficiencies in oil and gas conservation, especially wider well spacing and cycling in condensate fields. However, much of this was accomplished by federal directive under the Petroleum Administration for War (PAW), not by the exercise of the commission's administrative discretion. For example, the large Katy gas field was unitized on the insistence of the PAW that a cycling operation be initiated to recover the maximum amount of liquid from the field.[13] The PAW never issued a direct order requiring unitization, but it refused to authorize materials for a field's development until the operators agreed to an efficient plan. Liquid hydrocarbons were essential to the war effort, and the PAW especially insisted on cycling operations.

The push to produce additional liquid hydrocarbons from condensate fields for the war effort did result in some dissension within the commission regarding its authority to require cycling. In July 1944, the commission entered an order requiring pressure maintenance in the Lake Creek condensate field. The order recited the feasibility of cycling to increase ultimate recovery, and directed that "the reservoir pressures . . . be maintained at such a point as will insure the efficient recovery of the natural resources constituted of the hydrocarbons contained therein to the end that preventable waste of said natural resources shall be prevented and the ultimate recovery thereof increased."[14] Commissioner Culberson dissented, citing Article 6014(g) as prohibiting the agency from entering such an order, regardless of its merits in preventing waste.[15] The Agua Dulce condensate field was subject to a similar order in 1944 which, in no uncertain terms, required pressure maintenance.[16] The order clearly was not a popular one. Shortly after its effective date, some of the operators in the field applied to exempt certain producing horizons from the required repressuring. At the hearing on the application, the evidence showed that the operators had a firm, long-term sales contract to market the gas. The commission denied the application, finding that the waste of liquid hydrocarbons by retrograde condensation would occur without repressuring.[17] The legislative declaration of intent in Article 6014(g) seemed not to exist.

The dissension on the commission over the legality of these repressuring orders in condensate gas fields ultimately caused the commissioners to break with tradition and to support a 1946 bill which provided for compulsory cycling of gas.[18] This seems to be the only time that the commission publicly supported any form of compulsory unitization statute. The tenuous legality of its repressuring orders juxtaposed against the essential need for efficient recovery of condensate during the war years undoubtedly caused this break. The Railroad Commission also feared, with some justification, that the Federal Power Commission would step into its regulatory preserve in the postwar years if the state agency could not or would not act to prevent large-scale waste of those oil and gas resources which had proved, once more, to be strategically critical to our national defense. Despite the commission's support, the compulsory cycling bill did not pass the Texas legislature.

No-Flare Orders

Any characterization of the Railroad Commission as a passive agency ended abruptly on March 17, 1947, when the commission issued an order prohibiting the production of oil and gas from the entire Seeligson field until the casinghead gas from the wells was put to a beneficial use rather than being flared. This order marked the commission's adoption of an active stance to prevent waste in oil fields, a stance which would inevitably encourage unitization. The commission's new attitude was the result of several factors. First, the lessons of World War II underscored the importance of conserving oil and gas resources for national defense. The huge increase in oil production for the war effort was accompanied by large-scale flaring of casinghead gas. Because of material shortages during the war, few casinghead gas plants or pipelines were constructed. Liquid hydrocarbons were more important to the war effort than gas, and the PAW directed scarce materials to plants capable of producing large amounts of condensate rather than to casinghead gas plants. By 1944, 70 percent of the casinghead gas produced in the United States was flared.[19]

The massive postwar gas flaring was difficult to hide from the general public and reflected poorly on the commission's reputation as a conservation agency. At the same time, the pent-up consumer demand for petroleum products and the lapse of federal price control authority led to sharp increases in the price of crude oil and natural gas, providing private incentives to conserve these resources. In 1946 the Federal Power Commission held hearings on the subject of gas waste, and the Railroad Commission—anxious to forestall federal intervention into its regulatory preserve—took the initiative to reduce the waste.[20]

Also, the Texas courts and legislators were still embroiled in the controversy over the proper scope of judicial review of agency decision making. The courts' adoption of the substantial evidence rule increased the agency's power and its corresponding duty to act responsibly. Federal lawmakers had recently enacted the Administrative Procedure Act of 1946 in an effort to settle the raging controversy about the proper control of administrative discretion in federal agencies. The ferment on this issue at the state level was just as real, and the commissioners could not but be affected by criticism of their agency as an industry-dominated institution, operating less for the public interest than for the benefit of private concerns.[21] Finally, William Murray was appointed to the Railroad Commission in 1947. A petroleum engineer, he had served with the Petroleum Administration for War and was an ardent conservationist. He had written a highly publicized report in 1945 estimating that 1.5 billion cubic feet of casinghead gas was flared daily in Texas.[22] The Seeligson order was issued two months after his arrival at the agency.

The Seeligson order was the first of its kind in the history of oil and gas regulation in Texas.[23] While it was similar to the 1934 Agua Dulce order in that it prohibited production until the gas was put to a beneficial use, the Seeligson order flexed far more administrative muscle than had the Agua Dulce order. Once the wells in the Agua Dulce field had been classified as gas wells, the statutes required that the commission prevent gas flaring. In the Seeligson field, the wells were clearly oil wells; they were being operated within the gas–oil ratio rules set by the commission, and no statute specifically prohibited the flaring of casinghead gas. Indeed, such flaring had been commonly practiced and accepted in all oil fields for years, and in the Seeligson field since 1941.

Before the Seeligson order was to become effective, the producers in the field obtained a temporary injunction against its enforcement until final determination of its legality. On appeal, the Texas Supreme Court affirmed the temporary injunction but reversed the trial court's conclusion of law that the Railroad Commission had no power to promulgate such an order. The supreme court found ample statutory authority to support the commission's waste-prevention orders. The statutory enumeration of specific types of waste to be prohibited was obviously meant to be inclusive, not exclusive. The court wrote: "Whatever the dictates of reason, fairness, and good judgment under all the facts would lead one to conclude is a wasteful practice in the production, storage, or transportation of oil and gas, must be held to have been denounced by the legislature as unlawful."[24]

Whether reason and good judgment dictated the no-flare order in the Seeligson field would await development of the facts in a trial on the merits under the substantial evidence rule. The operators were prepared to argue that the commission's finding that a market currently existed for the casinghead gas was arbitrary and that compliance with the order was a physical impossibility.[25] However, no trial on the merits was ever held; the casinghead gas found a market.[26] While the processing and marketing of this gas does not always require that a field be unitized, the construction of a large-scale casinghead gas plant which can efficiently service a field does necessitate cooperation among operators. The commission's administrative muscle in the Seeligson order dealt a blow to prideful individualism. Further, if a casinghead gas plant cannot market the residue gas, it must be reinjected, and this will generally require a coordinated poolwide program. In the aftermath of the Seeligson order, the operators sought approval of a private agreement for the cooperative development of the field. Because of the limitations of Section 21 of the 1935 Act, the Texas attorney general declared that he could approve only those parts of the Seeligson unitization agreement involving production from gas areas. The attorney general could not grant antitrust immunity to the provisions in the agreement concerning the sale of casinghead gas or the construction of a plant for processing casinghead gas.[27] The Seeligson order sparked serious consideration of the need for a statute authorizing voluntary unitization agreements in oil fields.

With its broad authority to prevent waste validated by the supreme court, the commission embarked on a statewide campaign against gas flaring. On November 22, 1948, the agency issued no-flare orders for sixteen oil fields.[28] The operators again brought suit to enjoin these orders. In the Heyser field, the producers were not flaring casinghead gas at the wellhead as in the Seeligson field. They had constructed a casinghead gas plant in 1938, only two years after the field's discovery. However, the residue plant gas, comprising about 95 percent of the casinghead gas, was being flared. The producers argued that compliance with the no-flare order was impossible, that negotiations to market the gas to a nearby pipeline had foundered because some of the operators in the field refused to install compressors, that agreements and facilities to pipe the gas would be completed in a few more months, and that a shutdown would cause irreparable damage to certain wells.[29] The court looked at the commission's long record of hearings held to eliminate flaring in the Heyser field and upheld the order. The operators' long delay in agreeing on marketing facilities, while on notice for three years that the flaring was to end, was a problem of their own making. The court found that the commission was quite willing to grant applications for exceptions to the shutdown order to operators who could show their wells would be irreparably damaged.[30] The commission was reasonable; the order was reasonable. Only one justice dissented, unable to agree that the order was reasonable since profit-minded operators did not consider it so.[31]

Similarly situated operators in the Flour Bluff field were also unsuccessful in attacking their no-flare order in court.[32] In the Flour Bluff case, the Railroad Commission seemed interested not just in the marketing of the residue gas from a processing plant, but also in its reinjection to increase the ultimate recovery of oil in the field. A petroleum engineer for Humble Oil testified that recycling was not financially practical since the increased oil recovery from reinjection would not pay for the large investment required. The court was not sympathetic. Conservation of the state's natural resources was more important than immediate profits to producers.[33]

Humble Oil then argued that the shutdown order would cause a greater waste of oil than would be saved by the cessation of flaring. Humble presented data showing that 275,000 barrels of oil would be permanently lost if the field were shut in for five months, because the water drive would push the oil into the reservoir space that had been filled with gas. This low-saturated sand would not release the oil easily. The oil equivalent of the gas saved by eliminating flaring for five months was only 217,000 barrels. The court was equally unsympathetic. Even though the data were not disputed, the court credited the commission with expert knowledge of the field and of the public interest in conservation, and then stated that the anticipated waste could be minimized by gas injection to retard the water drive.[34] In essence, the commission's discretionary powers were immune to attack on economic *and* conservation grounds. No operators in the fourteen other shutdown fields proceeded to court. Operators in twenty-six other fields were subject to show cause hearings in late 1948,[35] but they also seem to have accepted their fate without litigating the commission's authority.

While the Flour Bluff case was proceeding through the courts in 1949, the huge Spraberry oil field was discovered. By March 1953, hundreds of oil wells were flaring casinghead gas in this field. On March 17, 1953, the Railroad Commission held a hearing on the flaring, and eight days later entered an order shutting down all the wells until the casinghead gas was put to a valid use. At that time, 468 wells were marketing their gas to a pipeline for light and fuel. The operators of 1,800 other wells were either flaring casinghead gas at the wellhead or as residue gas from gasoline plants.[36] The Railroad Commission shut down the 1,800 flaring wells under its waste-prevention authority and shut down the 468 nonflaring wells to protect correlative rights and prevent the confiscation of oil and gas from the flarers to the nonflarers. The facts showed that the facilities to use and market all of the field's casinghead gas were under construction but would not be complete until January 1, 1954, nine months after the order issued. The delay was not caused by a lack of diligence, but by the slowness of the Federal Power Commission's procedures to approve construction of the facilities and the actual construction time required to service this large field.

The supreme court held that the Railroad Commission had no statutory authority to shut down the 468 nonflaring wells which were not wasting oil or gas. But, the court continued, the statutes did give the commission authority to regulate the flow of these wells in order to protect correlative rights. The court cited those sections of H.B. 266 which had been enacted in 1935 to protect the correlative rights of operators in gas fields.[37] These statutes had never before been held to apply to oil fields.[38] The supreme court also declared that the commission had the power to shut down the flaring wells, even though the operators had exercised great diligence in attempting to market their gas. In previous court opinions involving shutdowns, the commission's orders had been upheld on the basis that the operators lacked diligence. Now the defense of diligence was gone. The court then voided the entire Spraberry order, at the same time praising the commission's "courageous effort" to prevent waste and hinting that administrative control of allowables could accomplish the twin objectives of preventing waste and protecting correlative rights.[39] In a subsequent order, the commission shut down the flaring wells and restricted the field allowable to a small percentage of the nonflaring wells' capacity.[40] This created space in the pipeline for gas from unconnected wells. As casinghead gas facilities were completed and more wells were connected, the field allowable was increased. Superficially, the Railroad Commission had lost the court battle. In fact, the commission had won court approval of a more powerful and flexible regulatory tool—the use of fieldwide prorationing of oil to protect correlative rights independent of waste prevention. The court's approval was important because even the commission's authority to prorate gas solely to protect correlative rights had been the subject of considerable legal controversy.[41] The court's holding allows the commission to prorate oil fields on the basis of local market demand for casinghead gas, in addition to the statewide basis authorized in the oil conservation statutes.[42] Furthermore, the Spraberry case now established that great diligence on the part of producers was as irrelevant in defining waste as economic infeasibility. The commission's anti-waste orders were virtually impregnable, especially under the substantial evidence rule.[43] With each appeal of the Railroad Commission's no-flare orders, the courts had broadened the power of the commission to prevent waste. No other appeals were taken. The commission's campaign against flaring succeeded in reducing the percentage of casinghead gas flared from 58 percent in 1946 to 30 percent in 1950.[44]

No-Waste Orders

In the cases discussed above, the Railroad Commission had selected fields which were located close to existing pipelines. In most instances the gas was ultimately marketed without the need for unitized production practices.[45] In other fields, however, efficient recovery required gas reinjection to maintain reservoir pressure. This type of operation generally requires unitization. The Railroad Commission boldly attacked such fields also. The Old Ocean field which produced oil, gas, and condensate was the subject of a Railroad Commission hearing regarding gas flaring and waste on May 13, 1947. The evidence showed that cycling would increase the ultimate recovery of oil by 50 percent and that of condensate by 20 million barrels.[46] In August 1948, the Old Ocean field was unitized.[47] The Anton–Irish field was the object of similar attention from the agency,[48] and was unitized in July 1951 so that pressure maintenance operations could begin.[49] Later in 1951, the commission issued orders for the Pegasus and Fort Chadbourne fields prohibiting production from these reservoirs until nonwasteful techniques were used. The orders did not actually require pressure maintenance or unitization.[50] In fact, this was their specific intent and effect. Commissioner Murray described the situation as follows:

> The Texas Railroad Commission has recently embarked upon a policy of requiring oil fields which were wasting great quantities of oil underground to be shut in until more efficient production practices are put into operation. When issuing such orders, the Railroad Commission does not require or even suggest that unit operations be followed. All the Commission desires is that improved recovery practices be adopted. It is up to the operators, with the advice of their lawyers, to determine whether unitization represents the best approach for maintaining equity while accomplishing conservation.[51]

The Fort Chadbourne field was unitized in May 1953 and the Pegasus field in June 1954.[52] No operators appealed these orders in court. yet these orders were materially different from the no-flare orders in their less than subtle coercion of repressuring and unitization. The gas from the Fort Chadbourne field could have been marketed, but the commission wanted the gas recycled, not sold, so that the ultimate recovery of oil would be increased.[53] The field was shut in for many months until the gas-reinjection program was started, and this must have caused considerable economic hardship to some operators. Still, the commission was careful to select fields in which repressuring would be profitable to the operators.[54] For example, Humble Oil and Refining, the largest operator in the Fort Chadbourne field, had made studies showing the desirability of reinjecting gas to recover millions of barrels more oil. The commission was anxious that Humble include the smaller producers in the repressuring operation, and given the no-waste order, these producers became equally anxious to join.[55]

The orders in these fields were but warm-up exercises compared to the administrative muscle displayed in the Kelly–Snyder (SACROC) field. This gigantic oil field was discovered in late 1948, and by 1951, more than 1,000 wells had been completed. The field was a solution gas drive, and primary production was dropping the reservoir pressure precipitously. By July 1954, it was clear that less than 19 percent of the original oil in place would be recovered unless pressure maintenance operations were begun.[56] The operators themselves were alarmed at the rapid decline in reservoir pressure, and formed an engineering committee in February 1950 which determined that oil recovery would double if repressuring were instituted. The Railroad Commission called a hearing in November 1951 to discuss the pressure decline, and promptly thereafter reduced the top well allowable from 250 barrels per day to 100 barrels daily in order to prevent waste.[57] Throughout 1952, negotiations proceeded to unitize the giant field. The field had many small tracts vested with the legacy of large per-well allowables which would bequeath them a disproportionate share of reserves. This share would be quite different from that accorded under any rational unitization agreement. In addition, the engineering committee had determined that a unique, center-to-edge water-injection program was the best way to repressure the field; that is, water would be injected in the thickest part of the reservoir's center, and oil would be captured by wells along its edges. Some small operators were alarmed at this untraditional approach to water flooding and feared that it would cause rather than prevent waste.[58] Both of these factors made it difficult to secure unanimous approval of unitized operations. By late 1953, an agreement was signed by 1,900 different persons comprising 96.58 percent of the working interests and 83.84 percent of the royalty interests in the field. The gargantuan SACROC unit embraced 47,400 acres; the estimated cost of the secondary recovery project was $60 million; and an additional 720 million barrels of oil were to be recovered.[59] After twenty-three days of hearings, the commission approved the SACROC unit's secondary recovery plan on January 18, 1954. However, three months later, a group of operators and owners located at the edge of the field overlying thin sands brought suit to enjoin the order approving the unit. The suit never came to trial because the largest oil companies in the unit succeeded in buying out the disgruntled owners. Throughout the long months of engineering studies and unitization negotiations, the Railroad Commission kept the field's allowable controlled at a low level. In September 1954, the SACROC unit commenced its repressuring program, one of the most advanced and most successful in the nation.[60]

The Railroad Commission's active display of muscle in these fields owed much to William Murray's leadership. A petroleum engineer, he was able to assess the feasibility of gas marketing and reinjection programs and to select good targets for the Railroad Commission's exercise of authority. Yet he never advocated compulsory unitization, knowing that neither of his fellow commissioners would ever go along with it because of their pledge to the independents.[61] Instead Murray pursued "permissive unitization," that is, permitting the majority of operators who had concluded that unitized operations served their private interests to proceed with their unitization plans as long as they also served the commission's conservation goals.

However, Murray's policy was something more than permissive passivity. To ensure that enough operators joined the unit to make it feasible, Murray applied what he termed the "doctrine of equal coercion."[62] The commission would call a "show cause" hearing for the targeted field.[63] Producers would have to appear and show cause why the commission should not shut the field down to prevent waste. The producers were quite aware that the commission would make life difficult for them if they did not reach agreement on a plan to maximize the field's recovery. Voluntary unitization agreements were much easier to achieve when all parties were equally coerced into bargaining fairly with each other in order to forestall a shutdown order. Landmen also found it easier to persuade royalty interest owners to sign up under these conditions.[64]

The precedent set by these no-flare and no-waste field orders carried through even after Commissioner Murray resigned. In 1965 efforts to unitize the Fairway field were stalled. Several years of unsuccessful negotiations had already passed since the field's discovery in 1960. The commission reduced the field's allowable from 28 percent (or 12,000 barrels of oil daily) to 5 percent (or 2,000 barrels daily).[65] The field was unitized eight months later.[66]

The Conroe field, discovered back in 1931, required pressure maintenance in the 1970s. The gas from the field had long been marketed, and gas prices were rising, increasing the attractiveness of continuing these sales. But without recycling, oil was migrating to parts of the reservoir where it would be forever unrecoverable. The commission initiated an investigation of the field and spurred the major operator into action.[67] The field was unitized in late 1977.[68] The Yates field—that paragon of efficient production in 1926—required unitization and pressure maintenance by 1970. Its old-style prorationing formula made voluntary unitization difficult to achieve.[69] The operators in the field wanted an increase in the field's MER allowable. The commission made it known that a larger allowable depended on the operators' ability to agree to an efficient plan of operation.[70] In July 1976, Yates was unitized, and its allowable was doubled.[71]

Special Allowable Field Rules

As the Yates field example shows, the commission used carrots in addition to the stick to encourage unitized operations. The field rules for the original SACROC unit granted additional allowables for water injected into the field and permitted the transfer of allowables from wells with high gas–oil ratios to wells with lower ratios.[72] The right to transfer allowables is valuable. Wells which bring up much gas (or water) along with the oil can be shut in, yet the field's total allowable will not decrease because the shut-in well's allowable can be transferred to another well with a lower gas–oil (or water–oil) ratio. This second well is then allowed to produce the oil allowable for both wells. This reduces operating costs. Also, because the second well produces less gas (or water) with the oil, the costs of reinjecting gas (or water) and repressuring the reservoir are less. These carrots helped to ensure that the unitized operations would be profitable, and this also explains the absence of court challenges to these fieldwide orders.

Water flooding in depleted reservoirs was commonly served the fattest carrot of all: capacity allowables completely exempt from market demand prorationing.[73] The Railroad Commission reasoned that the technology and expense of water flooding required this special treatment. Efficient recovery of oil in a water flood requires that the output wells produce at capacity just as the oil reaches them in front of the advancing water. If the oil migrates past the well, it is more difficult and expensive to recover. Obviously, such wells cannot be subject to shutdown days under market demand prorationing. Because water flooding often requires large front-end costs, the special allowables also let operators recover income sooner. This favorable treatment directed considerable capital into water flooding during the postwar years. Indeed, the system was heavily criticized in later years because the number of wells exempt from prorationing under this policy and under the Marginal Well Act resulted in increasingly severe restrictions on nonexempt wells. Many of the capacity water-flooding projects involve small fields with low-volume, high-cost wells. Low-cost, high-volume wells were penalized by the statewide prorationing system. After 1957, when excess capacity resulted in drilling declines and reduced profits, the majors complained about the liberal granting of exempt capacity allowables in water-flood fields. At the instigation of several large companies, the Railroad Commission called a hearing in 1960 to explore this controversy. The commission did not revoke its rules, but it began to examine water-flooding projects more carefully to see if they were *bona fide* efforts to increase recovery rather than devices to circumvent prorationing restrictions.[74] The exemption of capacity water floods from market demand prorationing clearly induced more secondary recovery, but this was not always accompanied by more unitization. Many of these water floods covered a single operator's property, and many probably represented a step backward in terms of efficiency. Economists have argued that by favoring high-cost oil at the expense of low-cost oil, the exemption of capacity water floods distorts investment decisions and raises the long-run cost of oil.[75]

Cycling and pressure maintenance activities were not accorded exempt status and were not generally granted special allowables except upon a showing that the allowable was required to make the project profitable.[76] The commission reasoned that these projects usually generated additional oil and gas at significantly reduced costs, and this was its own reward. Still, bonus allowables for water injection were sometimes granted as incentives to perform pressure maintenance as in the SACROC example. The equivalent incentive for pressure maintenance using gas injection was the grant of a net gas–oil ratio; that is, any gas produced from oil wells and then injected into the reservoir to increase recovery efficiency was not counted in measuring the well's permissible gas–oil ratio, thereby allowing additional oil to be brought up.[77]

In almost all requested cases, the commission granted secondary recovery and pressure maintenance projects the right to transfer allowables on a lease basis. This allowed producers to shut in wells with high gas–oil or water–oil ratios and transfer their allowables to better wells on the same lease tract, thus reducing operating costs and conserving reservoir pressure without suffering a loss of total allowable.[78]

Bonus allowables for pressure maintenance operations and capacity allowables for water flooding are still granted by the Railroad Commission, but they are of much less importance in guiding private investment decisions of producers now than they were in the 1960s. Since 1973, the statewide market demand factor has been set at 100 percent and all oil fields in Texas have been operating at their MERs, or at 100 percent of their capacity. Hence, exemptions from market demand prorationing no longer bring a favored status to projects. In 1968 the commission passed a statewide rule allowing the transfer of allowables to wells anywhere on a lease, so this special status has also disappeared by becoming the norm rather than the exception (see this chapter text at notes 115 to 121).

Special Allowable Rules for the East Texas Field

Amazingly, the precedent for using special allowable field rules to induce cooperative pressure maintenance arose from East Texas, that bastion of independent power. This example of the Railroad Commission's use of administrative discretion commands further attention, both because the field rules here preceded statewide action by many years, and because the prohibition in Article 6014(g) against compulsory repressuring or unitization originated out of conditions in the East Texas field. This field seemed the least likely in all of Texas to achieve any sort of cooperative action for repressuring.

As output expanded in the East Texas field, large quantities of salt water were brought up with the oil. As early as 1935, operators realized that the disposal of increasingly large amounts of salt water would be a serious problem.[79] By 1938, some of the larger operators had started to experiment with the use of disposal wells to inject the water back into the reservoir. By this time, the pollution hazards of storing brine in surface pits or flushing it into streams were more than evident. In 1939 a suit was brought by the Texas attorney general against 155 oil operators in the southern part of the East Texas field to enjoin saltwater pollution of the watershed. Judgment was entered perpetually enjoining the pollution as a public nuisance.[80] It was obvious that something had to be done.

On March 29, 1940, the Railroad Commission issued an order permitting the oil allowable of a well which was converted into a saltwater injection well to be transferred to other producing wells on the same lease. This order provided some incentive for operators to convert producing oil wells into disposal wells, because without the order, the loss of a producing well would decrease the leasehold's allowable under East Texas' infamous per-well proration formula. By the end of 1940, thirty-one injection wells returned 50,000 barrels of salt water a day to the reservoir, but this was far less than the 200,000 barrels produced. The existing saltwater injection systems were owned by large oil companies with large leases. Small-tract operators could not afford to drill their own injection wells and install costly treatment facilities. Because of the lawsuits enjoining pollution, the small operators were threatened with well shutdowns. Several of the smaller operators in the field sought the Railroad Commission's cooperation to solve the disposal problem.

At a series of commission hearings in 1941, a group of small operators proposed that the commission grant an additional oil allowable to operators who reinjected salt water. It was also suggested that operators of any well producing more than 100 barrels of salt water daily be permitted to shut in the well and take its allowable from other wells on the lease without the necessity of converting the shut-in well to an injection well. On November 20, 1941, a commission order for the East Texas field granted a bonus oil allowable of 1 barrel of oil for each 50 barrels of salt water injected. On February 20, 1942, an amendment to the bonus-allowable rule permitted wells producing more than 100 barrels of salt water daily to be shut in and their allowables transferred to other wells on the same lease. To avoid fraud and protect correlative rights, wells shut down under this order had to be tested every six months to determine if they were still capable of producing an allowable. Obviously, a flooded-out well should not be allowed to transfer a nonexistent allowable. The testing of so many wells imposed such a large administrative burden on the commission that an order was promulgated in March 1943 establishing a decline curve for production from shut-in oil wells. The decline curve was used to estimate the amount of allowable transfer. In 1942 and 1943, additional orders penalized wells producing large amounts of salt water by restricting the amount of bonus that could be earned. The denial of carrots (that is, bonus allowables) became a stick.

Simultaneously with the promulgation of these orders, the operators and the commission actively discussed the possibility of forming a fieldwide, saltwater disposal organization. Testimony at the 1941 hearings showed that saltwater injection would effectively prevent further declines in the reservoir pressure of the field, and that this would substantially increase the ultimate recovery of oil and also reduce production costs by prolonging the period of natural lift. Under the Railroad Commission's administration of the bonus-allowable rules, the recovery and sharing of expenses for the fieldwide system was easy to negotiate: the commission would vary the amount of bonus (originally 1 barrel of oil for 50 injected barrels of water) so that the value of the bonus oil just equaled the cost of saltwater injection charged by a fieldwide disposal company. This gave assurance to the disposal company that its costs would be recovered and also ensured that operators had no profit incentive to produce more salt water. The East Texas Saltwater Disposal Company was quickly formed in January 1942, and in October the company started injection operations. By 1947, the company was reinjecting over 400,000 barrels of salt water a day. The company served all operators in the field without discrimination, whether or not they were stockholders. Efforts were made to distribute the stock widely among all operators, roughly in proportion to the number of wells owned.

A 1947 Railroad Commission order made the transfer allowable rule more flexible by allowing wells producing more than 100 barrels of salt water per day to shut down and transfer their allowables to wells on other leases rather than on the original lease. This provided the incentive to shut in many large water producers and decreased both costs of production and reinjection. Those lessees and lessors who agreed to such a transfer in effect created a mini-unitized area of the field. The arrangement was completely voluntary. Nonetheless, this 1947 order was fairly controversial. Small producers who did not own several leases in the field obviously could not benefit from the more flexible transfer rule to the same degree that large operators could. As a conservation measure, however, the order was an immediate success. Within a year, 298 oil wells with high water–oil ratios were shut down, and their total allowable of 2,700 barrels per day was transferred to other wells. The expense of producing and then reinjecting 103,000 barrels of salt water was saved.[81]

In sum, the Railroad Commission's field rules for East Texas encouraged a large-scale pressure maintenance program, which has stabilized the field's reservoir pressure for decades and increased the recovery of oil by at least 600 million barrels.[82] The East Texas injection program often is cited as the model of voluntary cooperation by producers to prevent waste. The paradox is that the same field is also cited as the grossest example of wasteful and excessive drilling as a result of the same producers' absolute intransigence to pooling or unitization. That pressure maintenance could be successfully accomplished in light of the latter defies common sense. Pressure maintenance operations normally change the amount of oil and gas that each tract can capture. The independents in East Texas fought unitization because they wanted to keep the disproportionate share of production which the field's allowable formula accorded them.

Yet again, the unique setting of the East Texas field explains the paradox. First, the injection program was forced on producers by lawsuits seeking injunctions against saltwater pollution. The Railroad Commission did not dishonor its pledge to the independents by initiating the move to repressure. Second, at heart, the bonus-allowable rules were designed to help the little operator. Small producers needed a large-scale, low-cost disposal system more than the large producers, who could finance and construct such facilities themselves. The independents approached the commission with their problem, and the agency was anxious to help them secure nondiscriminatory facilities. Third, the reinjection program involved fieldwide collection and disposal of salt water, but left each operator free to produce his own oil wells and run his own business. The bonus and transfer allowables encouraged reinjection and shutting in of large water-producing wells but did not require either. Plus, these two allowable programs were still rooted on a per-well allocation basis. Fourth, the water-repressuring operation was used to supplement the natural water drive in the field. No artificial gas caps or water drives were created that would have drastically affected correlative rights. The repressuring generally kept everyone in the same relative position, but kept everyone's wells producing longer.[83] The increase in additional oil recovered was shared by all producers. Lastly, the pollution problem appeared shortly before the United States entered World War II. The oil from East Texas was desperately needed to fill the Big Inch pipeline carrying crude to the East Coast to fuel the war effort. If the field were wastefully produced, the specter of compulsory unitization or of federal intervention hovered as a patriotic solution.

For these reasons, the East Texas field—that bastion of aggressive individualism—accepted cooperative pressure maintenance.

Still, this acceptance had definite limits, as will be seen when a statewide, transfer-allowable rule was proposed in 1968 (see this chapter text at notes 115 to 121).

After the East Texas experience, bonus- and transfer-allowable rules were used in other fields as well, largely to encourage secondary recovery operations rather than to prevent pollution. It was not until 1967 that the Railroad Commission promulgated a statewide order prohibiting the surface disposal of salt water.

Summary and Assessment of the Legality of the Field Orders Used to Induce Unitization

It is evident from this review of the Railroad Commission's field orders affecting unitization that the agency has actively used an array of special field rules, both carrots and sticks, to encourage unitization. The commission's bold use of such field rules first appeared when the agency sought to prohibit gas flaring in condensate fields in the 1930s, and to maximize the recovery of condensate in these fields during World War II. After the war, the commission intensified, rather than disarmed, the campaign against waste. The agency's no-flare orders succeeded in eliminating the surface waste of casinghead gas in oil fields and survived numerous court challenges. These orders also pushed the legislature into passing the 1949 Act authorizing the commission to approve voluntary unitization agreements in oil and gas fields. Strengthened by these victories, the commission launched a similar campaign against underground waste. Targeted fields such as Anton–Irish, Pegasus, Fort Chadbourne, and Kelly–Snyder (SACROC) were vanquished and unitized in the early 1950s, with such weapons as show-cause hearings, shutdowns, and reduced allowables. Other campaigns resulted in the unitization of large oil fields in later decades such as Fairway (James), Conroe, and Yates. The granting of bonus allowables and transfer allowables for injection operations often led producers to surrender peacefully to unitization. In difficult fields, olive branches in the form of increased MERs led warring factions to accept unitization gracefully.

Carrots and olive branches aside, the administrative muscle flexed and sometimes used by the commission to shut down or severely curtail production from entire fields until nonwasteful operations were adopted was the Railroad Commission's most powerful weapon for effecting unitization. It is impossible to measure the exact extent of this muscle power in terms of the number of fields unitized after commission arm-twisting. It is also unnecessary to do so. It is clear from this history that the Railroad Commission's actions to induce unitization were more than sporadic forays launched for a year or two to deflect a prying federal bureaucracy or to win popular support of particular commissioners up for reelection. The no-flare and no-waste campaigns were long-term Railroad Commission policy in furtherance of the conservation of oil and gas. Operators in fields other than those mentioned in this book did not need show-cause hearings directed at them to understand this fact. Undoubtedly, the widespread public knowledge of the commission's policy and practices resulted in the voluntary unitization of other oil fields by operators and landowners who foresaw and accepted the inevitable.

It is also evident that the agency's biggest stick—the shutdown or severe prorationing of all wells in a field—has been used to force operators to repressure and unitize condensate gas fields and oil fields, despite the legislative declaration in the very statute used as authority for the orders that "it is not the intent of this Act to require repressuring of an oil pool or to require that the separately owned properties in any pool be unitized under one management, control, or ownership."[84] Commissioner Murray admits that the agency paid mere lip service to this declaration by phrasing the orders in terms of requiring the prevention of waste rather than requiring repressuring or unitization.[85] Even this small but polite gesture of respect to the legislative branch seems to have been ignored or forgotten at times. The commission orders in the Lake Creek and Agua Dulce condensate fields unequivocably required pressure maintenance (see this chapter text at notes 14 to 17).

During these years of activity, the commission never asserted that it had the implied authority to issue orders requiring operators to unitize.[86] On the contrary, the commission often publicly stated that it did not have the power to force either pooling or unitization.[87] There is no record of the Railroad Commission pressing an official interpretation of Article 6014(g) that would narrow its scope. The commission has not urged an interpretation of this provision as a hortatory legislative message rather than a law with substantive content. It has not interpreted the statute as permitting, but not requiring the Railroad Commission to issue compulsory unitization and repressuring orders, as A. W. Walker suggested in 1938.[88] The commission has not attempted to hold that Article 6014(g) applies only to fieldwide unitization but not to pooling;[89] or that it prohibits mandatory repressuring orders in oil fields, but not in gas fields. Nor is there any evidence that the commission used sections of the conservation statutes other than the waste definition in Article 6014(g) as authority for its orders.[90] The commission seems to have acknowledged the plain meaning of Article 6014(g) and then ignored it.

It well may be that internal memoranda of the commission argued some of these points, and that the commissioners who signed the no-waste orders and those requiring pressure maintenance believed that Article 6014(g) did not limit their actions. Because no operators ever challenged these orders in court, the commission's position on these points has never surfaced. Court challenges of the no-flare orders, which mainly sought to have casinghead gas marketed, did not directly involve Article 6014(g)'s strictures against required repressuring and unitization.

None of the interpretations proposed above to limit the effect of Article 6014(g) deserves much respect. The legislative history of this law, documented in chapter 3, shows that it was an integral and substantive part of the 1931 Anti-Market Demand Act. The statute's own words defy Walker's permissive interpretation. The prohibition against requiring that separately owned properties be unitized under one management, control, or ownership describes pooling as neatly as it does unitization. The only possible limitation on the scope of Article 6014(g) which seems at least arguable is that it does not prohibit the Railroad Commission from requiring repressuring in gas fields. The omission of gas pools from the first phrase of the article seems deliberate when contrasted with the second phrase, which forbids the required unitization of "any pool." This interpretation could support orders such as those issued in 1944 for the Lake Creek and Agua Dulce condensate-gas pools which unmistakably required repressuring. Commissioner Culberson's dissent in the Lake Creek order shows that the officials were aware of Article 6014(g) (see this chapter text at notes 15 and 18). It seems extraordinary that the other two commissioners may have been willing to press the argument that Article 6014(g) did not prohibit mandatory pressure maintenance orders in condensate-gas pools.[91]

However, these were extraordinary times. Liquid hydrocarbons were essential to the war effort, and the federal government, through the PAW, was a strong force in the regulation of Texas oil and gas fields. The exigencies of war may also explain the failure of operators to challenge these orders in court. Some of the affected producers were clearly unhappy with the orders, which prevented them from selling their unrecycled gas to waiting pipelines (see this chapter text at notes 16 to 17). One major deterrent to challenging these orders publicly was the prospect that quick legislative reform, either state or federal, would follow such litigation, especially if a court found that Article 6014(g) did, indeed, prohibit regulatory agencies from serving the public interest by maximizing the recovery of liquid hydrocarbons.[92] Also, the cycling operations no doubt were profitable in the longer term, especially after the war when oil and gas prices rose. Patriotism may have prevented litigation during the war years, but profitability explains the absence of litigation after the war ended.

To those schooled in the legislative history of Article 6014(g), the omission of gas pools from the provision against repressuring does not evidence a deliberate signal that mandatory cycling orders to prevent waste in gas fields would meet with the lawmakers' approval. In 1931 deep condensate-gas fields were only just being discovered. Their scientific nature and engineering characteristics were still unknown. The Railroad Commission had to conduct lengthy studies and investigations in 1933 and 1934 to determine the reservoir characteristics of the fluids from these new fields (see this chapter text at notes 4 to 6). To lawmakers in 1931, repressuring only existed in the context of oil fields. Had the legislators known that pressure maintenance and unitization could be applied to gas fields in the future, it seems likely that the first part of Article 6014(g) would have included gas pools as well. The lawmakers clearly understood that repressuring generally requires unitization, and the second part of the article specifically prohibits compulsory unitization of any pool.

Whatever legal support buttressed the commission's compulsory pressure maintenance orders in gas fields, no such justification would exist for orders requiring pressure maintenance in oil fields. The commission often worded these orders carefully so that they did not require repressuring, but stated that the field would be shut in if wasteful operations were not abated. However, in at least one oil field, the commission's edict read: "NOW THEREFORE IT IS ORDERED . . . that all operators in the Shafter Lake Devonian Field, Andrews County, Texas proceed immediately to inaugurate a pressure maintenance program by returning to the Devonian reservoir all water produced and all gas produced in excess of lease requirements."[93] This order clearly violates Article 6014(g) and exceeds the commission's statutory authority.

Even those orders which were more carefully phrased to avoid the express prohibitions of Article 6014(g) seem to have exceeded the limits of the commission's powers. To argue their validity is to exalt form over substance. Admittedly, it is difficult to read the other sections of Article 6014 and the many other statutes enacted which place a strong and continuing duty on the commission to prevent waste of all types without sensing a paramount legislative concern for the conservation of oil and gas. Article 6014(g) is irreconcilable with this concern. Certainly the commission served the public interest in conservation by removing this conflict with aptly phrased orders. Just as certainly, Article 6014(g) was inserted into Texas' conservation laws to assure that any such conflict would be resolved in favor of individual freedom of action, not the prevention of waste through coerced repressuring or unitization. The legislative history of Article 6014(g) shows an antipathy to waste-prevention measures which would necessitate that producers in a field unitize under the management and control of the dominant oil company in the field, usually a major oil company. Of course it can be argued that the commission's no-waste orders retain the individual producer's freedom of action: he can choose to be shut in or to produce at only 5 percent of the field's capacity if he does not want to join a unit. But these alternatives would bankrupt a fair number of independents, and this was surely not the intent of the waste-prevention statutes, especially in 1931. Faced with this choice, there is no real difference between a no-waste order and a compulsory unitization order.

The only public record of the Railroad Commission's position on the relationship between Article 6014(g) and the no-waste orders appears in a legal challenge to the agency's prorationing rules. In the Port Acres field case,[94] the plaintiffs sought court review of two issues: (1) the commission's promulgation of an inequitable and confiscatory one-third per-well, two-thirds surface-acreage prorationing formula for the gas field, and (2) the commission's failure to promulgate an order enjoining the wasteful production of gas in this condensate field. Halbouty and the other plaintiffs won the first issue and ushered in the new era of pooling. As to the second issue, the Texas Supreme Court ruled that it lacked jurisdiction on direct appeal to hear lawsuits whose purpose was to compel official agency action.[95] Thus the court did not address the substantive merits of Halbouty's position. Halbouty had argued in commission hearings and in the lower court that 7 million barrels of condensate worth $20 million would be forever lost unless the commission issued an order preventing waste in the field.[96] The commission refused to do so, even though it had issued such orders in the past, such as the Agua Dulce order. In fact, Halbouty and the other plaintiffs had proffered a proposed order to the commission for its consideration and signature. The language of the proposed order was nearly identical to that actually used in the Agua Dulce order.[97] Halbouty argued that Article 6014(g) did not apply to gas pools and that the oil and gas statutes placed a strong duty on the agency to issue a mandatory cycling order, citing the no-flare cases as precedent. Halbouty also argued that Article 6014(g) was a simple expression of legislative intent, not a prohibitive law and that, because it was repugnant to the general intent of the conservation laws, it should be narrowly construed to apply only to oil pools. The Halbouty brief for the appellants continued:

> The Appellees have, in their pleadings, stated that a pressure maintenance program would require unitization which they say the Commission cannot compel. The subject of unitization is a "straw issue". Appellants have never asked the Commission to unitize the field.[98]

In response, the commission stated:

> [T]he statutes of the State of Texas, particularly Article 6014(g), Vernon's Texas Civil Statutes, prohibit the Railroad Commission from issuing orders to require the repressuring of the reservoir or to require that the separately owned properties in any pool be unitized under one management, control or

> ownership, and that as a matter of common knowledge, it would be impossible to carry out any program of cycling, recycling, or pressure maintenance without a unitization of the separately owned properties in said reservoir under one management, control or ownership.[99]

The commission's own interpretation of Article 6014(g) would seem to estop it from asserting the legality of its earlier no-waste orders.[100]

The conclusion seems inescapable that the administrative discretion displayed by the commission in some of its field orders went beyond its statutory authority. Yet no operators challenged these orders in court. This requires some explanation. The legal underpinning of the no-waste orders exalted form over substance, but good form is sometimes quite enough to win an argument.[101] Most no-waste orders did not expressly require repressuring or unitization, and a court could rationally uphold the orders on this semantic basis, especially with such a strong public interest in conservation at stake. Operators also may have been deterred from challenging these orders because every court challenge of the no-flare orders had only resulted in strengthening the commission's authority and duty to prevent waste. Under the substantial evidence rule, form could carry the commission a long way.

More important, the Railroad Commission took care in selecting its targets and aimed at fields that offered ripe profits for repressuring and unitization. If profitability was at all doubtful, the commission often granted bonus allowables, transfer allowables, and increased MERs to guarantee that all producers would benefit from the new order. Undoubtedly, this is the most important reason for the lack of court review of these no-waste orders.

The question remains, Why did the Railroad Commission refuse to issue a no-waste order in the Port Acres field? A full assessment of the commission's use of administrative muscle to induce unitization requires an account of those instances in which it refused to act. It is even more difficult to find evidence of orders not issued than to find those which were promulgated. It does appear that operators sometimes requested that the Railroad Commission reduce a field's MER, particularly in condensate fields. Applicants seeking reduced MERs had to prove that the lower allowable was necessary to prevent waste. Brooks Peden, a commission hearing examiner for many years, reports:

> The amount of proof required varies substantially with the unanimity of the request among the operators in the field. If all operators are in agreement, the amount of proof required may be fairly small. If the application is protested, however, the proof offered should be substantial, because the Commission realizes quite well that a reduction in allowable can make a lease inoperable and can provide the leverage for a most effective squeeze.[102]

It can be inferred from this report that the Railroad Commission was not always willing to issue orders for reduced MERs to prevent waste, especially if this could result in financial losses or coercive "squeezes" on small producers. This description of Railroad Commission decision making has characterized many other episodes in the agency's history[103] and would explain the commission's reluctance to issue a zero MER (a shutdown) in the Port Acres field, which contained many small tracts and was already subject to enormous controversy and litigation on the prorationing-pooling issue.

Commissioner Murray recalls that the Port Acres field was not particularly large or rich in condensate.[104] As a general rule, the Railroad Commission did not apply as much pressure on operators to unitize gas fields as oil fields because the expected additional recovery from the former was usually not as large or as certain. Those cycling projects which were technically feasible were usually profitable to operators because condensate was exempt from the restrictions of market demand prorationing. The gas could not be flared, and if no market existed, cycling was the only action open to operators to recoup income from their gas wells. Thus, the commission did not normally have to arm-twist operators into cycling operations. Unless a condensate field was very large or extraordinarily rich in liquids, the commission left such fields alone and concentrated on large oil fields, such as Kelly–Snyder (SACROC), where expected additional recovery from pressure maintenance totaled in the hundreds of millions of barrels, not a mere seven million. Also, if the commission forced cycling on all small fields like Port Acres, the supply of gas flowing into existing pipelines would have been curtailed, causing market disruptions and inevitable lawsuits. Because of the tenuous legal status of the commission's orders requiring repressuring, the agency did not risk court litigation about its statutory authority for small returns from small fields.[105]

Thus, while the commission's use of field orders to induce unitization attained the status of a well-recognized policy, it remains less than a full substitute for an organized routine of unitizing oil and gas fields as a matter of course. The Port Acres field was never unitized, and Halbouty estimates that 30 percent of the recoverable hydrocarbons have been left in the now-depleted field because of the lack of an efficient, fieldwide pressure maintenance program.[106] Commission action in the East Texas field, and in SACROC, resulted only after a number of operators sought the agency's help. The commission's near shutdown of the Fairway (James) field came only after two years of delay in the voluntary unitization process. The commission does not conduct annual reviews of Texas fields to determine which of them would benefit from repressuring or unitization. If the use of field orders were a complete substitute for a compulsory unitization statute, one would not have expected such a strong push to pass a compulsory unitization statute in 1973. A statistical assessment of the degree to which the commissioner's administrative decision making has substituted for such a statute appears in chapter 9. Before such a final assessment can be made, the commission's statewide orders and those promulgated under the 1949 Act must be analyzed.

Statewide Orders and Unitization

Under its statutory authority to prevent waste, the Railroad Commission has often promulgated statewide orders which have indirectly facilitated unitization. Many statewide orders enacted to conserve reservoir energy and increase the efficiency of production assist in unitizing a field by reducing the differential between a producer's vested interest in the regulatory status quo and his expected interest under a unitization agreement. The greatest agitation for changes in statewide rules to achieve greater efficiency occurred in the early 1960s because of the cost-price squeeze on producers (see chapter 4 text at notes 79 to 89). The resulting statewide rule changes with the most direct impact on unitization are the 1962 spacing rules, the 1965 allowable yardsticks, and the 1968 lease-allowable transfer system. The fourth statewide rule having an impact on unitization was Rule 8, the "no pit" order, promulgated in 1967 because of pressure from federal authorities to reduce water pollution caused by oil operations.

The 1962 Spacing Rules

By 1960, deeper and costlier drilling led producers to seek wider spacing rules. A lessee with a 40-acre tract often could not afford to drill two wells under the existing 20-acre statewide spacing rule. Even when allowable formulas used large per-well factors, the severe restrictions on production rates caused by market demand prorationing made dense drilling unprofitable. One oil well can typically drain 40 acres, and the commission had increasingly sized prorationing units at 40 acres in setting field rules. Even 80-acre and 160-acre spacings were being used.[107] In 1962, Rules 37 and 38 were changed to establish 40-acre units statewide.[108] In the same year, the commission amended another statewide rule to allow operators to request temporary field rules after one well had been drilled into a newly discovered field.[109] The previous rule required the operator to wait until five producing wells had been drilled. Predictably, this resulted in a clustering of wells around the initial discovery well and made subsequent orderly development more difficult. These two reforms now allow the commission to establish large drilling units through temporary spacing rules applied early in a field's life. This pattern of development accords much better with the pattern that operators would seek in a unitization agreement.

New Allowable Yardsticks in 1965 and 1966

Of course, the allowable system is a key factor in determining the profitability of drilling deeper and spacing more widely. The economic pressures of the 1960s led the Railroad Commission, aided by a committee of oil industry representatives, to devise a new statewide allowable yardstick which would reduce incentives to drill unnecessary wells. The new yardstick became effective on January 1, 1965.[110] It assumed a standard spacing unit of 40 acres, strongly penalized tracts of lesser size, and granted larger allowables to larger units. The yardstick eliminated most incentives to overdrill fields.[111] During these same years, the courts outlawed the use of large per-well factors in allowable formulas, and the legislature passed the Mineral Interest Pooling Act. These events, combined with the new yardstick, reduced the profitability of small-tract drilling and the financial motive to "go it alone" in new fields. Unitization is easier to accomplish in fields with wide spacing and prorationing formulas based on productive acre-feet of sand or surface acreage. Indeed, Commissioner Murray viewed the large per-well allowable formulas as the single greatest obstacle to unitization in Texas.[112] Thus, these changes in the spacing and prorationing rules facilitated voluntary unitization agreements in fields discovered after 1965.

In 1966 the commission also amended the allowable yardstick for discovery wells. Under the old rule, discovery wells were exempt from market demand prorationing for eighteen months, or until six wells were completed, whichever came first. The new order exempted discovery wells for twenty-four months or until eleven wells were completed.[113] The change was obviously made to motivate more exploratory drilling. This drilling is particularly risky and the exemption allows successful wildcatters to recoup costs more quickly. However, from a conservation viewpoint, discovery wells should be produced slowly to conserve reservoir pressure while the field is explored and the most efficient drilling pattern and production rate established. On this score, the new discovery yardstick is not conducive to unitization.[114] On the other hand, the new rule placed exploratory drilling more on a par with marginal and capacity water-flood wells. The exemption of these latter wells from market demand prorationing was directing capital away from exploration, and the new discovery allowables sought to neutralize this effect. This illustrates the compounding inefficiency which can result from using special and exempt allowables as "carrots" for certain categories of production. Even though some of these exemptions are designed to encourage cooperative pressure maintenance and secondary recovery projects, they may skew investment decisions into inefficient channels.

The 1968 Lease-Allowable System

The third major change in the statewide rules occurred in May 1968, when the Railroad Commission adopted an order permitting use of the lease-allowable system in all fields in Texas except the East Texas field.[115] Before this time, allowable transfers were customary in field rules for secondary recovery projects, but had not been adopted on a permanent statewide basis largely because of their controversial effect on the correlative rights of oil producers in older, established fields. The right to transfer allowables from poor wells to good wells is of greater value to operators with large tracts of land and many wells under lease than to small-tract producers who only have a well or two on a few acres.[116] Therefore, the Railroad Commission had shied away from adopting allowable transfers as a routine matter in all fields. Unlike the new spacing and yardstick orders and the Mineral Interest Pooling Act, which operated prospectively only, the lease-allowable system only made sense if it applied to old and new fields alike, and this fact made it even more controversial. The external shock of the Arab–Israeli conflict in 1967, during which the Arab oil exporters suspended shipments of crude oil to the United States, impelled the Railroad Commission to adopt a statewide lease-allowable system on a temporary emergency basis for five months. To make up the deficit in crude supplies, the Railroad Commission raised the Texas market demand factor from 33.8 percent in May 1967 to 54 percent in August.[117] In order to avoid producing oil from wells with high water–oil and gas–oil ratios, the commission allowed operators to combine the allowables of all the wells on the lease and produce this total only from the most efficient wells. Very real savings in operating costs were achieved with this system, in addition to greater conservation of reservoir pressure. Humble Oil and Refining Co. saved 70 million cubic feet of gas per day, cut saltwater production by 75,000 barrels per day, and reduced operating costs by 5 percent. Annual savings for the entire oil industry in Texas were estimated at $3.7 million.[118] Also, the 1967 crisis came at the same time that the federal pollution control laws put enormous pressure on the oil industry to eliminate saltwater storage pits. The best way to reduce the cost of reinjecting salt water is never to bring the water up, and allowable transfers assisted this solution.

The new lease-allowable system indirectly facilitated unitization agreements by demonstrating vividly the cost savings that could be accomplished by changes in the prorationing rules which permitted each lease to be treated as a mini-unitized area without penalties in the form of reduced allowables to operators who shut in unnecessary wells on each lease.

The 1968 statewide order making lease allowables permanent also merits attention because of its express exclusion of the East Texas field. This exclusion once again demonstrates the remarkable political power of the East Texas independents and their ability to successfully oppose any regulation which could possibly affect their correlative rights. The independents sought exclusion from the statewide order because they expected that it would disproportionately benefit the larger oil companies who had more acreage and more wells and who therefore could transfer allowables more freely.[119] The independents also argued that East Texas already had its own lease-allowable system dating from 1942 and that no additional flexibility in production was required to produce efficiently. The larger oil companies countered this argument with evidence showing that operating costs could be significantly reduced and an additional 6 million barrels of oil might be recovered if the statewide rule was applied in the East Texas field.[120]

The independents won the debate: East Texas' 1942 lease-allowable order became a vested legacy, just as the per-well allowable formula had. The 1942 order became a limit to—not a precedent for—more efficient operating practices in the field.[121] The exclusion of the East Texas field from the 1968 statewide lease-allowable order foreshadowed the defeat of the compulsory unitization bill a few years later in 1973.

The 1967 No-Pit Rule

Lawsuits against saltwater pollution in the East Texas field led to a fieldwide cooperative injection program which eliminated most surface brine storage pits in this field in the early 1940s. It was not until 1967 that Rule 8 was adopted banning surface pits statewide and, even then, the ban was not effective until January 1, 1969.[122] Rule 8 was passed when it became evident that the federal government would enter the arena of water quality protection if the states failed to act. The Texas legislature had given the Railroad Commission sole responsibility for the prevention of pollution and control of waste from oil and gas activities.[123] Yet this task conflicted with the Railroad Commission's ingrained philosophy of promoting and nurturing the oil industry of the state, especially the independents. Not surprisingly, then, Rule 8 allowed exceptions for producers who could prove that their operations would be uneconomic if they could not use surface pits or flush their wastes into streams. The liberal granting of exceptions and lax enforcement of the no-pit rule brought heavy criticism of the commission and of the oil industry.[124]

The commission was caught in desperate straits, forced to choose between shutting down small producers or abdicating their pollution authority to the federal government or another state agency. The best solution to the predicament was for producers to cooperate in constructing efficient fieldwide injection systems. Commissioner Langdon, hewing to the credo that the agency's function was to enforce the law—not to propose new laws—ended a 1966 speech on the subject of preventing saltwater pollution with the following "anticipatory questions":

> (3) Under the present laws, does the Railroad Commission have the power or authority to require operators in a common field to engage in disposal of produced oil field brines and wastes by means of a cooperative organization?...
>
> (4) To what extent could the Statutory Pooling Act be used to implement a common salt water disposal project?
>
> (a) should the Railroad Commission consider the disposal aspects of an operation involved in an application to pool under the statute?...
>
> (5) Since our Legislature found it reasonable and necessary to enact the Statutory Pooling Act, could it not be proposed in the public interest and to the end of conserving our fresh water resources that compulsory cooperative disposal systems be employed?[125]

The tremendous federal pressure on Texas to clean up the oil fields forced the commission to push the industry into cooperative reinjection ventures.[126] The pollution laws thus induced significant pressure maintenance, sometimes accompanied by unitization.

Summary of the Effectiveness of Statewide Rules in Encouraging Unitization

In contrast to the commission's bold use of field rules, this account of the agency's statewide rules shows that the commission more often lagged than led. External pressures such as the Suez Canal crisis and the federal enforcement of water pollution laws were needed to spark the commission into passing statewide rules for transfer allowables and underground injection disposal. The cost-price squeeze on the domestic oil industry pushed the commission into adopting wider spacing. The 1962 spacing order based on 40-acre units followed years of using this standard in field rules. The 1965 allowable yardstick followed the court decisions in the Normanna and Port Acres fields, which invalidated the commission's traditional methods of setting prorationing formulas. The statewide rules that were finally promulgated contained liberal procedures for granting exceptions, and the 1968 lease-allowable transfer rule excluded the entire East Texas field.

It is not surprising that statewide rules are a less effective mechanism for inducing unitization through administrative discretion than field rules. Field rules can be tailored to fit the particular needs of an individual reservoir's physical characteristics and of its operators' financial positions, location to markets, and willingness to cooperate. Across-the-board statewide rules cannot attain such flexibility even with exception procedures and exclusions.

Nonetheless, the net effect of these statewide rules was to increase the efficiency of producing and drilling operations in Texas. In this way, the new rules reduced the differential regulatory advantage of "going it alone," and made voluntary agreements easier to negotiate. The elimination of the prorationing system's incentives to overdrill was especially important, although the courts deserve much of the credit for pushing through this reform.

6

The 1949 Voluntary Unitization Statute and the Railroad Commission

As we have seen, Texas passed a statute in 1949 authorizing the Railroad Commission to approve voluntary unitization agreements. The great controversy surrounding the bill's passage resulted in a very restrictive law (see chapter 4 text at notes 31 to 75). Only certain types of secondary recovery operations can be approved by the agency. Many findings must precede agency approval, each of which is a potential source of protest, delay, and litigation. Despite the legislature's clear objective of limiting the Railroad Commission's power to approve unitization agreements freely, the act still leaves room for the exercise of administrative discretion by the Railroad Commission, which can choose either to promote voluntary unitization agreements by interpreting ambiguities in favor of applicants and by purposefully lowering the statutory hurdles, or to interpret the act narrowly, by demanding strict proof of the many detailed findings, and by generally treating unitization as a disfavored last resort.

As will be seen in this chapter, the Railroad Commission clearly has taken the first course of action.[1] Its administration of the 1949 Act has sought, in almost all instances, to encourage those voluntary unitization plans that can show an expected increase in the ultimate recovery of oil and gas. The statutory hurdles are sometimes placed so low as to be virtually nonexistent.

Requiring Railroad Commission Approval

Section 101.013 of the 1949 Act states that "agreements for pooled units and cooperative facilities are not legal or effective" until the commission makes certain findings after application, notice, and hearing. Taken literally, this would preclude landowners and lessees from entering into any voluntary unitization agreement without first obtaining Railroad Commission approval. Because of the restrictions of the 1949 Act, many types of unitization agreements cannot meet the statutory requirements for approval, and Section 101.013 implies that all such contracts are illegal. The need to secure agency approval gives protestors a chance to delay, and possibly thwart, a proposed agreement. Also, the time and expense involved in participating in a hearing in Austin may inhibit operators in some fields from conducting small-scale recovery projects which, despite their limited size, would be more successful if unitized.

Juxtaposed against Section 101.013 is Section 101.002 which states that "none of the provisions in this chapter restrict any of the rights that a person now may have to make and enter into unitization and pooling agreements." This section can be interpreted as allowing landowners and producers to continue their pre-1949 right to unitize voluntarily without securing Railroad Commission approval. This interpretation would ease the administrative burden imposed on small projects and also would allow lessees and royalty interest owners to join units which do not meet the statute's restricted purposes and required findings.

The commission's first act of discretion was, therefore, to decide whether it would require approval of all cooperative agreements in Texas, or only those which operators desired to bring forward. The commission chose the latter course. Operators who want to secure the antitrust immunity benefits of the 1949 Act may apply for commission approval of their unitization plan, but approval is not required.[2] The commission has even suggested to operators whose projects do not meet the strictures of the act that they proceed without the agency's approval (see this chapter text at notes 7 to 8). Almost all operators involved in large, fieldwide projects do, in fact, seek this approval. If the plan is the least bit controversial, the major oil companies, who have the most to fear from antitrust charges, are especially insistent on the need to obtain the Railroad Commission's stamp before a unitization plan becomes effective. Yet the Railroad Commission's interpretation still leaves the operators of small water-flooding projects free to conduct their activities without bureaucratic entanglement. Many of the larger oil companies in Texas have unitized reservoirs without securing commission approval, especially when only one or two leases were involved or when 100 percent of all the parties consented to the unitization.[3]

The Railroad Commission's Definition of Secondary Recovery Operations

Section 101.011 of the 1949 Act authorizes the Railroad Commission to approve voluntary agreements "necessary to effect secondary recovery operations for oil and gas." The commission has readily approved voluntary agreements for enhanced or tertiary recovery, such as miscible displacement, which are not commonly classified as secondary recovery operations in current terminology and which were unknown technologies in 1949.[4] The commission has shown a willingness to approve experimental repressuring and water-flooding techniques in order to further the growth of technology.[5] As long as the unitized operation's technology uses injection wells, the Railroad Commission considers it a secondary recovery project.

Unitization projects not involving the use of injection wells cause the commission greater difficulty because the wording of the 1949 Act derives from the IOCC's and API's standard definition of secondary recovery as operations which produce additional oil and gas through the use of injection and output wells (see chapter 4 text at notes 44 to 45).

Chapter 4 discussed one example of an operation that would require unitization in order to increase the ultimate recovery of oil, better protect correlative rights, and reduce unnecessary drilling costs, but which would not qualify as a "secondary recovery" operation (see chapter 4 text at notes 46 to 49). The example involved a gas-cap field ringed with oil. Most of the operators desired to unitize the field to prevent the drilling and production from gas-cap wells. However, no injection wells were required to operate the unit efficiently. This example confronted the commission in early 1980, when the Quintana Petroleum Corporation sought the commission's advice as to whether the agency could approve a unitization agreement for a project that did not use injection wells. The General Counsel's Office determined that the statute prohibited the Railroad Commission from approving such an agreement, but also noted that at least two hearing examiners at the agency were aware of past instances in which the commission nonetheless had approved such "pseudo-secondary recovery" projects.[6] These same examiners "were not bothered by what Quintana planned to do and would be receptive to such a plan if they sat as examiners."[7] Despite this past precedent, the General Counsel's Office advised Quintana that an injection program was required by the statute in order to secure the Railroad Commission's blessing of the proposed unitization plan. The general counsel recommended that Quintana proceed with the plan despite the Railroad Commission's inability to approve it. Indeed, the legal memorandum questioned why Quintana even desired to receive Railroad Commission approval: the agency's order could not compel the unsigned owners to join the unitization agreement.[8]

Quintana's own reading of the statute had obviously led it to the conclusion that injection wells would be necessary to secure commission approval. Therefore, the company simultaneously sought advice on an alternative unitization plan which would involve drilling one injection well to recycle gas produced from the oil wells back into the gas cap. This injection program was clearly not necessary to recover additional oil. Indeed, it would simply increase the cost of the project by many thousands of dollars without any offsetting gain in the amount of oil recovered. Section 101.011 allows approval of voluntary agreements for cooperative facilities "necessary to effect secondary recovery" or "necessary for the conservation and use of gas." The only purpose of the injection well was to qualify the unit for commission approval. Under these circumstances, the General Counsel's Office advised that Quintana might also find it difficult to establish all the findings required by Section 101.013 of the 1949 Act, especially the fourth finding that the value of additional oil and gas produced from the project exceed the project's cost. The General Counsel's Office was concerned that a project, so frankly acknowledged to be for pseudo-secondary recovery, could not make this showing.

Quintana and its major partner in the field, Exxon Co. U.S.A., treasured Railroad Commission approval of the project despite the commission's assessment of its dubious value to them.[9] Almost a year later, Quintana applied to the commission for approval of a voluntary unitization agreement for this field with its one injection well. During the interim, Quintana had obtained the consent of over 99 percent of both working interest and royalty interest owners in the field.[10] After notice and an uneventful hearing, the Railroad Commission approved the unit, effective March 9, 1981.[11]

The approval of Quintana's unitization project illustrates the commission's dedication to the goal of conserving oil and gas. The Quintana operation was expected to yield an additional 750,000 barrels of oil through conservation of the gas-cap pressure.[12] With virtually 100 percent voluntary agreement achieved, no reason except legal technicality existed to disapprove the unit. However, because of the restrictions of Section 101.011 of the 1949 Act, one more unnecessary well was drilled in the Texas oil fields.

The commission has also stated that the 1949 Act does not authorize the agency to approve pooling agreements for drilling and primary production.[13]

The Required Findings

The legislative heart of the 1949 Act is Section 101.013, which requires many findings before a unitization agreement can be approved. Most of the findings are designed to assure that voluntary agreements are approved only after unitization is proved to be the necessary last resort for the conservation of oil and gas and that the unitization plan is fair to signers and nonsigners alike (see chapter 4 text at notes 52 to 66). These findings are a breeding ground for controversy, contested hearings, lengthy delays in securing approval, and litigation. Yet the one outstanding characteristic of Railroad commission hearings on unitization is that they are usually uncontested.[14] The hearings are brief; the application is unopposed; and the final Railroad Commission order approving the unit is generally issued a few weeks after the hearing. Despite all the required findings, few protests are ever made on the basis of a failure to meet statutory provisions.[15] Most applicants seeking commission approval of a unit plan have secured the consent of at least 90 percent, and often 95 percent, of the working interest and royalty interest owners in the area to be unitized (see table 6-1). The commission has an informal rule requiring that applicants have the consent of 85 percent of the working interests and 65 percent of the royalty interests before applying. These percentages are often thought to be mandatory,[16] but they are not a statutory requirement for securing Railroad Commission approval. The commission could consider an application which did not meet these percentages.[17] The 65 percent–85 percent "rule" originated at the first commission hearing to approve a unitization proposal in 1949. One commissioner proposed it as a guideline, reasoning that without these minimum percentages, the unit was unlikely to be economically feasible. The rule has existed ever since. No standard measurement has ever been specified but the oil companies generally report it as a percentage of the total acreage in the proposed unit, not as a percentage of the number of owners.

Table 6-1. Review of Twelve Unitization Agreements Approved by the Railroad Commission in 1976

Field	No. of acres in unit	Percentage of working interest consent	Percentage of royalty interest consent
Tom O'Connor	15,000	99.6	98.28
Hollow Tree	337	100.0	100.0
Forest Hill	273	99.78	88.15
Kildare	613	85.6	68.2
West Tyler	584	99.77	78.83
Stephens Co.	3,932	93.86	96.61
Tippett, W.	2,214	97.02	96.47
Cowden, N.	480	100.0	100.0
Goldsmith, W.	840	100.0	100.0
Yates	26,400	99.996	90.68
Ackerly	800	100.0	99.98
Engle	671	100.0	74.72
Average	4,345	97.97	90.99

Source: *Secondary Recovery Railroad Commission Application Summaries, 1976* (Austin, Texas State House Reporter, Inc., 1976) pp. 15–16, 47, 70, 72–73, 80, 106, 128–129, 137, 150–151, 176–178, 187, and 243–244. Data for the other two units applied for in 1976 were not reported.

The actual percentages on the application forms are usually much higher than the rule requires, and the hearing process itself often induces some of the unsigned interests to sign up. In medium-to-large fields, a small fraction of the royalty owners can never be located, and this explains the failure to achieve 100 percent consent in some fields. In those fields whose owners simply refuse to sign, the application usually passes through the hearing process unopposed. The nonsigners are silent. The hearings seldom last more than two hours. The examiner has a standard list of twenty questions, and most are quickly disposed of with "yes" and "no" answers by the applicant's landman and petroleum engineer.[18] There are no specific provisions in proposed unitization agreements which the Railroad Commission will either prohibit or require other than those mandated by statute. The unitization agreements presented to the commission almost never are modified.

In practice, the only finding of concern to the commission is whether *additional* oil or gas will be produced by the proposed unitization plan. The 1949 Act requires that "the estimated additional cost, if any, of conducting the operations will not exceed the value of additional oil and gas so recovered."[19] The unitized project must promote the public interest of conservation by this showing. The commission does interpret the word "additional" as requiring that the applicant show a net increase in projected production over and above that which would be achieved without unitization.[20] It is not sufficient to simply project future production from the unitized area. The relevant finding requires a projection of future production beyond that which would be expected by normal primary production. Evidence to satisfy this finding is established through engineering studies and expert testimony. Most of the time spent in a unitization hearing involves discussion of the technical and scientific issues underlying the estimates of additional recovery.

The commission does not, however, require that the additional recovery be proved with absolute certitude. Such an attitude would thwart technological progress and experimentation with new injection additives and with reservoirs having unique geological conditions. The evidence simply must show that additional recovery can be reasonably expected.[21] The commission has actively encouraged the testing and use of new technologies. SACROC is the best example of this, both in its unique center-to-edge secondary recovery program in the 1950s and its current carbon dioxide tertiary recovery program (see chapter 5 text at notes 56 to 60). The Railroad Commission approved SACROC's voluntary unitization agreement on January 30, 1953, when only the general nature of the proposed water-flooding operation had been determined. A number of landowners who opposed the unit's formation protested that the unitization application was premature and too indefinite. The commission responded that waste was already occurring in the field, that more serious waste was imminent, and that the application was timely.[22] Three months later the Railroad Commission held several weeks of hearings to discuss the specifics of the proposed water flood. The commission's order approving the actual secondary recovery operation was issued in January 1954, almost a year after its approval of the unitization agreement.[23]

Section 101.013 also requires that the estimated additional cost of the unitized operation will not exceed the value of the additional oil and gas recovered. Little time is spent reviewing the cost estimates. Quintana's application for its pseudo-secondary recovery project went through the hearing without a problem, despite the existence of an earlier agency memorandum questioning the project's ability to meet this required finding (see this chapter text at notes 6 to 11). The agency assumes that the operators would not voluntarily agree to a project that did not promise profits. Moreover, if the secondary recovery operation is not expected to make a profit—but the commission wants it to proceed because of its conservation benefits—bonus allowables are awarded to ensure its profitability. SACROC's carbon dioxide tertiary recovery project was clearly uneconomic without these bonus allowables.[24] Thus it appears that the Railroad Commission will simply lower the hurdle of this fourth finding so that any desired project can make the jump.

The current commission's liberal attitude toward approving unitization projects follows a long tradition. The commission seems never to have disapproved a unitization proposal which could make the finding that additional recovery would be obtained and which had the support of most of the operators and a large majority of royalty interest owners. Commissioner Murray's "permissive unitization" is an accurate description of Railroad

Commission policy (see chapter 5 text at note 61). Most of the findings required by Section 101.013 are cursorily treated except for proof that additional oil and gas will be recovered. Once this is shown, Railroad Commission approval is certain.

The only possible exception to this liberal tradition appears in Commissioner Murray's remark that he would never personally advocate unitization "merely for the purpose of obtaining operating economies" (see chapter 4 text at note 141). This remark was made in 1952 in the middle of the commission's no-flare and no-waste campaign. His statement seems clearly intended to placate producers throughout the state that the commission's orders were aimed at preventing the physical waste of hydrocarbons, not at the elimination of economic waste, that is, the unnecessary drilling costs incurred because of the sacrosanct prorationing system and the lack of compulsory pooling. But when the producers' balance sheets felt the cost-price squeeze in the 1960s, reforms such as wider well spacing, new yardstick formulas, and the lease-allowable system (which also required a Suez Canal crisis) were instituted to decrease the number of wells in operation. Because unitized operations generally achieve both additional oil recovery and operating efficiencies, the commission seldom has been presented with a case in which unitization approval was sought primarily for the latter purpose.[25] It is possible, however, that such situations will confront the commission more often in the future as older fields with densely spaced wells and injection programs approach their economic break-even point. The applicant seeking approval of such a unitized project, which mainly involves shutting in existing wells and transferring allowables fieldwide, can be expected to argue that a reduction in operating costs will prolong the life of the field and thus increase recovery by widening the spread between producers' revenues and costs. If this showing is made, the commission seems disposed to approve such a unit.[26] While the 1949 Act lacks clear authority for the agency to bless such an agreement (see chapter 4 text at notes 54 to 61), the Quintana example illustrates that the commission is apt to interpret the statute broadly in order to sustain any agreement which will recover additional oil and gas.

The sixth finding required by Section 101.013 is that "the area covered by the unitization agreement contains only that part of the field that has reasonably been defined by development." A narrow interpretation of this phrase would prevent approval of unitization plans that are proposed early in the life of a reservoir. The 1949 Act clearly prohibits unitization for ordinary primary production, but pressure maintenance is allowed and its early implementation is often essential for conservation, especially in condensate fields. The commission has had few occasions to interpret this phrase because most unitization proposals involve well-developed fields that had been discovered many years earlier (see chapter 9 text at notes 1 to 23). The attitude reflected in the SACROC case would seem to apply to this issue also. If an area is sufficiently developed to sustain a finding that waste now is occurring, or is imminent, then the area "has reasonably been defined by development." Certainly the commission will not require drilling on every lease or spacing unit in order to find reasonable development.[27]

While the sixth finding of Section 101.013 limits units to parts of the field reasonably defined by development, it does not require that the unit encompass the entire developed area of the field. The Railroad Commission may, and often does, approve units which cover only part of the developed acreage in the field (see chapter 9 text at notes 24 to 32). The ability to unitize on a less-than-fieldwide basis undoubtedly contributes to the explanation for the very high percentage of consenting owners brought to the commission by unit operators. If most of the owners in one area of the field refuse to join the unitization agreement, this area may be excluded from the application for commission approval. Thus the fact that 95 percent of the owners of interests in the unitized area have signed does not mean that 95 percent of the owners in the field have agreed to unitization. A large number, particularly those favored by a structural advantage in one part of the field, may have refused to join. If, however, so much acreage in the field is excluded from the proposed unit that the unit is unlikely to accomplish the purposes of the 1949 Act, that is, an increase in the ultimate recovery of oil and gas, then Section 101.103(b) states that insufficient acreage is "grounds for the disapproval of the agreement." Of course, few unit proponents would ever seek commission approval of an agreement that was unlikely to succeed because a large amount of acreage was excluded from the unit. However, when a group of operators does seek approval of a less-than-fieldwide unit, the matter of protecting the correlative rights of the owners who have not signed comes to the fore. The unit's operations often have the potential to drown out prematurely the wells on adjacent nonunitized tracts, to replace valuable oil and gas on adjacent tracts with salt water or other injected substances, or to increase drainage from the adjacent tract to the unit area. The commission's protection of correlative rights under the 1949 Act's required findings merits special attention because so many unitization agreements in Texas are for less-than-fieldwide units.

Correlative Rights and the Required Findings

Two findings required by Section 101.013 are intended to protect correlative rights. The first, appearing in Section 101.013(3), requires the commission to find that "the rights of the owners of all the interests in the field, whether signers or not, would be protected." The second, appearing in Section 101.013(6), requires a finding that all owners are given an opportunity to join the unit on the same yardstick basis. These statutory provisions are potentially the most fertile ground for protest. Most protests of unitization applications do involve the issue of correlative rights, such as whether the unit's participation formula is fair and whether the amount or value of the protestor's productive acreage has been accurately determined.[28] However, few unitization applications are contested and, in practice, the findings appear on the examiner's list of twenty questions and are simply answered "yes" by the applicant. Only if a protestor appears at the hearing to dispute the applicant's testimony regarding these findings does the examiner look into the issue. The agency will not do so on its own initiative.[29] The commission requires that a copy of the unitization agreement be filed with the application, but the hearing examiners often do not even look at the participation formula in the unit agreement. They generally do not ask why the nonsigners have refused to join. Nor does the commission monitor the unitized operation after approval to assess its actual effect on correlative rights. The commission maintains jurisdiction over the unit, but unless someone protests that his or her correlative rights are being injured, the commission will not concern itself with the unit's operation for any reason other than waste prevention.

The provision that all owners be given an opportunity to enter the unit on the same yardstick basis is obviously designed to prevent the gerrymandering of unit boundaries in such a way that some owners in the field are purposefully left out of the unit. The 1949 Act envisioned the authorization of less-than-fieldwide units, and to counteract the possibility that inequitably drawn boundary lines would adversely affect the correlative rights of those outside the unit, Section 101.013(a)(6) requires that "the owners of interests in the oil and gas under each tract of land in the area reasonably defined by development are given an opportunity to enter into the unit on the same yardstick basis as the owner of interests in the oil and gas under the other tracts in the unit." Thus, if the "area reasonably defined by development" is an entire field—as it normally would be in a secondary recovery project involving a field which has been under production for many years—all owners in the field would have to be offered an opportunity to join the unit. The necessity of negotiating with what may amount to hundreds of different owners is expensive and time consuming, but it does protect against "squeeze outs" of some owners in the field.

Despite the language of the act, the commission does not require that all the owners in a field be offered an opportunity to join the unit. Rather, the agency requires that the unit proponents extend an offer to all owners of adjacent tracts bordering the proposed unit. Once the proposed unit area is ringed by nonconsenting tracts or by nonproductive land, the commission does not require that unitization offers be made to owners of tracts beyond this ring.[30] The commission interprets the "area reasonably defined by development" in Section 101.013(a)(6) to mean the part of the field that is developed *and* covered by the proposed unit agreement. In line with this interpretation, the commission requires that the unit applicant give notice of the hearing which seeks approval of the unit plan, first, to the lessees and unleased mineral interest owners of the tracts adjacent to the proposed unit, and, second, to *all* the owners (including royalty interest owners) of any tract within the area to be unitized.[31] Notice is not extended to all owners in the field unless the unitization agreement is fieldwide. The commission's reading of the 1949 Act seems at odds with the plain language of the statute, but it does ease procedural delays and negotiating expense in establishing less-than-fieldwide units. Very often, unit proponents do try to achieve fieldwide unitization because this maximizes the efficiency and recovery of the unit's operations. In this event, the offer to join will be extended fieldwide. However, a group of operators desiring to institute a waterflooding operation on one part of a field may be able to escape the burden of fieldwide offers because of the commission's interpretation of the statute.

The commission's willingness to ease the formation of less-than-fieldwide units probably reflects the voluntary nature of the agreements that it is authorizing. The agreements cannot bind anyone who does not sign them. The common law continues to govern the relationship between the unit and those outside the unit. As we shall see later, the common law is probably still effective to protect those whose property rights are injured by the unit's activities, even though the unit has received commission approval (see chapter 7 text at notes 125 to 161). Thus, those operators who form partial units and secure commission approval under the agency's rather relaxed procedures do so at the risk that the unit will be liable for any injury to the tracts of nonsigned owners.

Of course, the commission should never approve a project which fails to protect the correlative rights of all owners in the field. Section 101.013(a)(3) requires that the commission find that "the rights of the owners of all interests *in the field,* whether signers of the unit agreement or not, would be protected under its operation" (italics added). Despite this language, for many years the commission's orders approving unitization agreements contained a phrase in the findings section which stated that "the Commission in no way passes upon the equity of said basis of participation, that being a matter of contract between parties executing said agreement."[32] This curious phrase appeared despite the statute's clear requirement that the commission affirmatively find that the rights of the signers are protected. It is true that the commission generally has no authority to adjudicate contract or title disputes between private parties,[33] but this does not prevent the legislature from conferring on the commission a mandatory duty to review particular types of contracts. The legislative antipathy to the very concept of unitization arose from the fear that the majors would coerce small producers into joining units on unfair terms. In this light, it is not strange that the lawmakers would require the commission to affirmatively find that the rights of the signers and nonsigners of voluntary unitization agreement are protected. The most important right in such an agreement is the participation factor. During the lengthy hearings on the contested application for the SACROC unit, the protestors argued that the commission had a statutory duty to examine the fairness of every term of the unit agreement to protect both signers and nonsigners, and alleged that many people signed because they were "uninformed, ill-advised and misinformed."[34] The commission replied that the parties to the unit were capable of handling their own business, and that the 1949 Act did not give the commission authority to assess the fairness of the agreement as to the signers.[35] The latter argument flies in the face of the clear wording of the statute, but as support, the agency cited dictum from a Rule 37 case holding that the commission had no authority to compel pooling, and therefore the right to pool was controlled by the private freedom to contract. The absurd use of this dictum in the context of a statute requiring Railroad Commission approval of voluntary unitization agreements simply illustrates that the agency was (and is) utterly unwilling to let this mandatory finding delay the hearing and approval process.[36]

The rationale for the commission's laissez-faire attitude regarding correlative rights is that the agreement before the commission is a voluntary one. Those who signed it did so of their own free will. Those who did not sign will not be forced into the unit by the Railroad Commission's order. The self-interest of the nonsigners who fear that the unit's operation will adversely affect their property rights is relied on to bring questions concerning correlative rights to the commission's attention.[37]

The commission's posture seems perfectly justified for agreements which are, in truth, negotiated voluntarily. However, we have noted numerous occasions when the commission has used its waste-prevention powers to induce—if not coerce—unitization. In these cases, the small operator often faced either a complete shutdown of his wells or submission to a unitization plan prepared by the larger operators holding the majority interest in the field. Commissioner Murray's "doctrine of equal coercion" will induce small operators favored by a regulatory advantage to bargain with others, but the result is sometimes difficult to describe as purely "voluntary." The small operator may not have the legal and financial resources to secure amendments to the proposed unitization formula at the bargaining table, at commission hearings, or in court. He may simply accept what seems inevitable and sign the agreement proffered by the majority of larger operators. Indeed, it seems that a significant number of producers and landowners believe to this day that the Railroad Commission has compulsory unitization powers.[38]

Even in fields not targeted by the Railroad Commission for special orders designed to effect unitization, most small operators are probably quite aware of the commission's attitude favoring unitization plans that conserve oil and gas. The 1949 Act does not specify the measure of the nonsigners' correlative rights which the commission is to protect. In the 1954 order approving the SACROC unit, the commission stated that its order must "as far as practical afford each of the several property owners in a field a fair opportunity to produce the recoverable oil underlying his properties."[39] The order further stated that if protecting correlative rights conflicted with the duty to prevent waste, the commission "must use its discretion in balancing the two problems." These statements do not seem to offer protection to the nonsigners' right to produce more than the oil and gas in place under his tract by draining from other tracts under the existing prorationing and spacing rules. If this was to be the commission's measure of protection, the nonsigner could be stripped of the regulatory advantage which formed the very basis of his refusal to sign. Without a statutory definition of correlative rights to constrain the commission's discretion, the certain value of participating in the unit may outweigh the uncertain value of the nonsigners' correlative rights. For all these reasons, many individuals—particularly small producers and royalty interest owners—may sign a unitization agreement which they consider unfair. They simply feel they have no other choice.

This seems to be what happened to Francis Oil & Gas, Inc., a small producer in the Yates field in 1975. As previously noted, several larger operators in this field had sought a change in its prorationing formula and an increase in its MER since 1967 (see chapter 5 text at notes 69 to 71). The Railroad Commission refused to act on all such applications until the field was unitized, thus depriving all operators of the benefits of both increased production and increased prices under the postembargo federal pricing rules. This finally induced the major operators in the field to unitize. The unitization formula—according to the federal judges of the Tenth Circuit Court of Appeals—was "facially inequitable" to Francis Oil and other, similarly situated independents who mainly produced stripper well oil, that is, oil produced from property with wells pumping an average of ten barrels or less per day.[40] Under the federal rules, the price of stripper well oil was uncontrolled and was substantially higher than the price of other crude. The unitization formula did not credit Francis Oil for the higher value of this oil, which it contributed to the unit in a disproportionately large amount relative to its tract participation factor. Francis Oil sued Exxon, the purchaser of the unit's crude, for the difference between the value received under the unit formula and the value of the stripper oil contributed to the unit.[41] The federal court described the unit's formation as follows:

> In negotiating the Unit Agreements, Marathon limited itself to the larger working interests of the field and after obtaining the consent of these working interests offered the final draft to the small interests, including Francis, on a take it or leave it basis. Needless to say, Francis, although wishing to leave it, opted to take it.[42]

The facts showed that Marathon, the unit operator, held a 49 percent share in the unit, and six other major oil companies—Chevron, Gulf, Amoco, Shell, Getty, and Exxon—held 43.5 percent, leaving only a small minority of 7.5 percent to be bargained with.[43] The rank injustice of the unit's participation formula to the small producer in Francis Oil's position obviously shocked the appellate justices. Yet the Railroad Commission approved this unitization agreement, finding that the rights of all the owners, including the signers, were protected. The hurdle of Section 101.013(a)(3) seems nonexistent.

This case raises a number of disturbing issues. If voluntary units are often formed this way in Texas, then the independent producers' allegations and fears of overreaching and dominance by unit operators are quite real. The Railroad Commission's failure to honor its statutory mandate to protect the rights of signers allows the forced unitization of minority-interest owners on "facially inequitable" terms. The Railroad Commission's laissez-faire attitude regarding correlative rights derives from its assumption that the agreements before it for approval are voluntarily made. If Francis Oil's experience in the Yates field is typical, one must wonder why independent producers in Texas do not press for a compulsory unitization statute with adequate safeguards to protect correlative rights.

However, the behind-the-scenes story of Francis Oil's plight casts a different light on the Tenth Circuit Court's description. Francis Oil's experience is in many ways atypical of the unitization process in Texas. In the opinion of the company's lawyer, Michael Medina, and its president, the behavior of the major operators in the Yates field was uncommon.[44] Most often, the large operators make every effort to negotiate fairly with all interested owners and the result is usually a compromise formula that all owners can accept. The large operators generally bargain fairly in order to avoid such administrative hassles and legal entanglements as occurred in this instance. The extraordinary events in the Yates field occurred because of the unique timing of the field's unitization. In 1975, when the unit agreement was negotiated and signed, the issue of allocating stripper-well pricing benefits in the participation formula was nonexistent. Under the 1975 federal pricing rules, all tracts in a unitized area were treated as a single property; and any well originally producing less than 10 barrels per day of stripper oil that produced more than 10 barrels after unitization (a likely occurrence in the Yates field, where producers were to receive a 100 percent increase in the MER following unitization) lost its pricing benefit.[45] Under this 1975 rule, Francis Oil had no complaint with the proposed unitization formula. Francis would still benefit by joining the unit because every barrel of new oil produced from the doubled MER resulted in 1 barrel of "released oil," which could be sold at uncontrolled prices under the federal rules.

The intricate interplay between the federal pricing rules, the unit participation formula, and the existing field prorationing rules made negotiations particularly complex in the Yates field. The Railroad Commission's promise of a doubled MER was clearly the key to the unit's success in securing 99.99 percent consent of the working interests and 89.5 percent of the royalty owners. Of the 1,109 individual interest owners in the field, 770 would receive a decrease in their fractional shares of the field's production under the unit agreement, compared with the existing field rules.[46] Nonetheless, almost all owners signed because, under the proposal, the MER increase ensured that no owner would receive fewer barrels of allowable than he received under the September 1, 1974, proration schedule. Also, before unitization, the average composite price of crude oil from the Yates field under the pricing rule was $5.79 per barrel, based on 87 percent "old" oil and 13 percent "new" oil. With the doubled MER, the proportion of new oil would expand to 57 percent, and the average composite price would rise to $8.79 per barrel.[47]

The 1975 federal pricing rule for stripper-well oil in unitized projects was widely acknowledged as providing a disincentive for producers to unitize. Therefore, in February 1976, the federal government devised a new regulation called the "imputed stripper-well exemption" to remove this disincentive. At this point, the issue of allocating the pricing benefits of stripper oil became important to Francis, but the unitization agreement was already signed. After years of haggling over the participation formula, none of the major operators wanted to start all over again with the 199 working interest owners and 962 royalty owners. The Railroad Commission held hearings on the Yates unitization proposal in April 1976. Francis Oil did not raise the issue of the fairness of the agreement at that hearing. It was so obvious to the company that the commission wanted the field unitized and that any objections to the agreement would be disfavored that Francis Oil did not protest publicly. Nonetheless, according to Michael Medina, it did protest the unit's formula privately with Railroad Commission staff members. However, it was rumored that the Railroad Commission staff had told operators in the field in 1975 that any nonsigners to the unit would not receive an increase in the MER. Only the unit would get additional allowables. Under these circumstances, Francis Oil was much better off joining the unit and receiving the increased allowable than being a holdout. Francis Oil also feared its tract would be drained if it refused to join and all the wells on adjacent tracts in the unit produced at double the rate of Francis's wells. Of course, Francis Oil would have been even better off joining a unit with a more favorable formula.[48] Had the federal government promulgated its imputed stripper-well exemption rule earlier, the unit formula would undoubtedly have incorporated the rule that Francis Oil desired.[49] In sum, Francis Oil is really quite satisfied with the way fields are unitized in Texas.

In fact, the most common criticism of Texas unitization practices is that holdouts and nonconsenters are treated too well. Commissioner Murray characterized the final Yates participation formula as "absurd," "outlandish," and "unfair."[50] The Yates unitization formula is based 50 percent on the production allowables established for the field as of September 1, 1974, and 50 percent on gross acre-feet of pay adjusted for quality. Thus more than half the benefits of the unitized operations are allocated under the old legacy. Many of the unitization agreements for the old fields use split formulas that allocate the remaining primary production under the existing prorationing rules and the secondary recovery production under a formula more rationally related to each tract's remaining reserves of oil. The use of a time-split formula to allocate unit proceeds is a sure sign that the proration formula in the field is not an equitable basis for permanently allocating the increased recovery from the unit.[51] Murray's characterization of the agreement that SACROC finally negotiated with the nonconsenters who brought suit against the Railroad Commission's approval of the unit is even more blunt: "extortion" and "blackmail."[52] In his opinion, the SACROC unit purchased the nonconsenters' tracts for $25 million more than they were worth. This is an enormous sum of money to pay for avoiding further administrative and legal contests. It also reflects, however, the value of the freedom to place injection wells and output wells optimally.[53] State Representative Milton Fox, the petroleum engineer who for so long has advocated compulsory unitization, believes that holdouts in Texas receive much more than their fair share of the unitized production.[54] After the defeat of the proposed compulsory bill in 1973, the major oil companies redoubled their efforts to unitize fields voluntarily. They often succeeded, but only by accepting formulas which, in Fox's opinion, perpetuated the inequitable distribution of the reservoir's bounty.

Thus despite the harsh light in which the judges of the Tenth Circuit Court of Appeals viewed the process of "voluntary" unitization in Texas, it seems that those who sign the agreements generally receive fair treatment. The lack of compulsory process allows owners of tracts with structural or regulatory advantages to capture the economic rents of their favorable position. The applicant must secure the agreement of most of the owners in order to operate the unit efficiently and to avoid legal battles. Negotiated formulas which perpetuate vested rights to some degree are inevitable. The Railroad Commission's laissez-faire attitude toward protecting the rights of all signers seems justified. Even in those instances where the commission itself has put enormous pressure on the operators to unitize and to do so quickly under threat of a general shutdown, the small producers' easy access to the Railroad Commission staff in informal, prehearing discussions is likely to ensure fair bargaining by the unit proponents and a fair result.[55] Also, the major operators who generally propose and implement the large fieldwide unitizations are clearly constrained from overreaching by political concerns.[56] Thus, by the time that a unit agreement reaches the formal hearing stage, it reflects the end product of months, and even years, of arms-length fair bargaining. There is little reason for the commission to initiate an independent assessment of the unit's fairness to those who signed, except for the fact that the statute seemingly requires this.

The 1949 Act also requires that the Railroad Commission find that the rights of all nonsigners are protected by the unit agreement's operation. This finding is treated just as cursorily, unless a nonsigner appears at the hearing to protest the unit's approval. Most nonsigners do not do this. Even the Yates field hearing, which lasted for three days, had no protestors.[57] The nonsigners are happy with their lot. Some even attend the hearings and testify to this effect.[58] The existing field rules remain in effect as to all the nonsigned interests, and their regulatory or structural advantage is preserved.[59] In addition, the unit applicant has often appeased all nonsigners by agreeing to place and operate injection and output wells in such a way as to eliminate dissent. Two methods of protecting nonsigners are standard: (1) injection wells are placed a minimum distance from the lease lines of unsigned tracts to avoid drowning out the nonsigners' producing wells; and (2) the unit's lease-allowable transfers are controlled so that output wells near the unsigned tract do not produce at more than twice their top allowable rate;[60] this avoids drainage from the unsigned tract to the unit. Unit applicants routinely incorporate these standard controls in their proposed plan. If they did not, the Railroad Commission would probably require these two items.[61] Other reasonable requests of nonconsenters are also privately agreed to.

Many of the fields remaining to be unitized in the 1970s were those which had eluded cooperative operations for decades because of their large size, the large number of owners involved, and their old-style prorationing formulas. With the defeat of compulsory unitization legislation in 1973, the unit proponents in these fields turned to increasingly sophisticated and tempting participation formulas to lure parties into unitization. In the Webster field, Exxon offered all parties a choice of participation formulas: either each tract's proportionate share of productive surface acres or the share of acre-feet of oil pay.[62] In the Hawkins field—a notoriously difficult field to unitize because of its 50 percent per-well allowable factor—Exxon agreed to protect the unsigned owners of fractional interests in tracts which the co-owners had agreed to unitize by guaranteeing that the unsigned owners' total payment over the remaining life of the field would not be less than the amount of oil and gas they would have received had the unit not been formed.[63] Thus even though a unit applicant approaches the hearing with only 85 percent of the owners signed up, the hearing is usually uncontested.[64] The Railroad Commission lets sleeping dogs lie.

On the rare occasion when nonsigners do protest the unit's approval, the hearing process itself will often produce a mutually agreeable solution.[65] If not, the Railroad Commission's attitude toward the nonsigners is not sympathetic or solicitous. The burden of proof is on the protestor to show that his correlative rights are not being protected. He must present positive, geological evidence to sustain his position and cannot simply rely on weaknesses in the applicant's case or on broad allegations and vague worries.[66] To protect unsigned owners of interests within the unitized area, the agency will not allow unit operators to convert the *last* producing well on the tract to an injection well, and the production from any wells on the tract must be measured and reported separately from unit production.[67] For the nonsigned owner with property adjacent to, but not in the unit, limitations on distances and allowable transfers are considered sufficient protection. The commission may also require that the unit operator allow nonsigners to join the unit on reasonable terms at a later time.[68] These rules define the limits of protection that a nonsigned protestor can expect from the commission.[69]

Nonetheless these rules may seriously interfere with the operation of the unit. The requirement that production be separately metered from any tract with even a minute, unsigned interest costs thousands of dollars' worth of extra equipment. Restrictions on allowable transfers to wells near unsigned tracts mean that one, small unsigned "window" tract in the middle of the unitized area can impose production restrictions on many surrounding unit wells.[70] These are the inevitable consequences of the lack of compulsory process.

In sum, the Railroad Commission plays a very inactive role in protecting correlative rights in the unitization approval process. The agency's posture is unsympathetic to correlative rights issues that might delay or conflict with the conservation of oil and gas attainable through unitization. Yet correlative rights seem to be well protected by the inherent nature of voluntary unitization.

The Effect of Railroad Commission Approval

Sections 101.015 and 101.017 of the 1949 Act require that voluntary agreements remain subject to any Railroad Commission rules and orders relating to spacing and prorationing and also prohibit agreements from containing the field rules or limiting oil and gas production. The legislative intent of these sections is to ensure public control over the unitized project so that the antitrust laws are not abrogated.[71] The 1949 Act does not define the benefits to be achieved by securing agency approval, other than the limited antitrust immunity conferred by Section 101.004. The SACROC protestors used this silence and the legislative phrasing of the above sections to argue that the Railroad Commission had no authority to approve a unitization agreement containing a participation formula that differed from the field rules. The Railroad Commission vehemently rebutted this interpretation of the 1949 Act by citing the section which states that nothing in the act shall restrict any existing rights to enter into unitization agreements.[72]

In practice, the commission's approval of a unitization proposal results in two sets of final orders: one approves the unit agreement as meeting the requirements of the 1949 Act (although the order itself says nothing about its effect on the unit's antitrust liability); and the second approves specific field rules for the secondary recovery operations in the unitized area. Unless antitrust liability is viewed as a substantial risk, the second order is substantively more important to the unit's success. This order permits the unit to transfer allowables throughout the unitized area. Wells can be shut in without decreasing the unit's total allowable, and producing wells can be clustered in the most advantageous locations (subject to the restrictions imposed to protect correlative rights). Generally, however, the total unit allowable is still calculated under the existing field prorationing formula.[73] The unit operation benefits primarily from being able to distribute this total to wells anywhere in the unitized area. In some cases, after a production increase is realized, the unit applicant may also request capacity allowables or bonus allowables for water or gas injection.[74] Thus, if a unitization plan cannot meet the statutory requirements of the 1949 Act, the unit proponents can still request this second type of Railroad Commission order allowing transfers of allowables and special allowables for the unitized area, and the agency would probably grant the request if an increase in recovery were expected.

Summary of the Commission's Administration of the 1949 Voluntary Unitization Act

In all respects, the Railroad Commission's administration of its authority under the 1949 Act shows an active, positive regard for unitization agreements that promise to recover additional oil and gas. The 1949 Act is consistently construed in favor of unit applicants. Despite the opportunities for delay, obstruction, and contention provided by the 1949 Act's many required findings, unitization hearings are handled with dispatch by the commission. The primary focus is proof of an expected increase in ultimate recovery. The commission will lower the hurdles of particular findings, most notably by granting special allowables to assure the feasibility of the project and by assuming that correlative rights are protected unless protestors prove otherwise. Operators who cannot or do not want to meet the requirements of the 1949 Act need not do so. Their private unitization agreements still will be considered valid contracts without Railroad Commission approval and the limited antitrust immunity it secures. The commission will grant the benefits of transfer allowables to cooperative projects even if the producers do not also file a unitization application under the 1949 Act. The commission even will allow "pseudo-secondary recovery" projects to clear the hurdles.

Summary of the Effect of the Railroad Commission's Decision Making on Unitization

Chapters 5 and 6 have looked at the Railroad Commission's administrative decision making with the objective of assessing whether the agency has encouraged or discouraged unitization in its rule-making and adjudicative functions. Clearly, the Railroad Commission has used its discretionary powers to effect unitization, often pressing and sometimes even exceeding the limits of its statutory authority.

The agency has had three sets of tools at its disposal to implement unitization. The most powerful one is the individual field order. The commission first used this weapon during World War II to require pressure maintenance and cycling in condensate gas fields. After the war, the agency's no-flare and no-waste campaigns directly resulted in the unitization of a significant number of oil fields. In steps, the Railroad Commission first forbade the flaring of casinghead gas from the wellhead, then gas flaring from casinghead gas plants, after which they prohibited the marketing of casinghead gas in fields where its reinjection was necessary to increase the ultimate recovery of oil.

Field orders can be tailored to fit the precise needs of a particular field, whether for carrots (increased MERs, transfer allowables, and bonus allowables), or for sticks (reduced MERs and shut-downs), or both. In issuing these field orders, the Railroad Commission simply ignored Article 6014(g)'s proscription against required repressuring and unitization, or paid it lip service by careful phrasing of the orders. Commissioner Murray frankly acknowledges replacing Article 6014(g) with the "doctrine of equal coercion."

In its administration of the 1949 Act, the commission has shown a similar disregard for the legislature's intended restrictions on unitization. The commission does not make a searching inquiry into many of the act's required findings. Indeed, for many years, the commission's unitization orders disavowed any authority over the correlative rights of signers despite the statute's express mandate that the agency find that these rights were protected in the voluntary unitization process. In the one area in which the Railroad Commission does make a serious investigation of the proposed unitization plan—that is, whether it will increase the ultimate recovery of oil and gas—the commission can and will issue special allowables to assure that desired projects meet the act's requirement of economic feasibility.

The Railroad Commission's use of statewide orders to induce unitization displays much less initiative. While the 1960s saw a number of statewide rules passed that moved in the direction of encouraging unitization, these rules were often imposed by forces external to the agency: the cost-price squeeze of cheaper oil imports; the court decisions in the *Normanna* and *Port Acres* cases; the Suez Canal crisis; and the federal water pollution acts. Some of this foot-dragging illustrates the Railroad Commission's traditional concern with protecting the small producers. Liberal exceptions to Rule 8, the no-pit order, were granted for many years to producers who could show that without surface disposal of brine, they would be unable to operate their marginal wells profitably. The least successful field rules, in terms of inducing unitization, were the capacity allowables granted to water-flooding operations in nearly depleted fields. These allowables often helped small producers with marginal wells stay in business. The granting of capacity allowables to such fields became so standardized as to achieve the status of a statewide rule. This liberality induced secondary recovery, but not much unitization, and probably contributed to more inefficient and high-cost production.

The political influence of East Texas has figured prominently in both the field rules and the statewide rules. When the independents in East Texas needed the Railroad Commission's help and the cooperation of the majors to dispose of salt water in a nonpolluting way, they secured bonus allowables for saltwater injection and a complex system of transfer allowables tailored to their needs. This explains East Texas' acceptance of a large-scale pressure maintenance program. The East Texas field set the precedent for the agency's use of special allowables to induce pressure maintenance and the use of transfer allowables to increase the efficiency of production and lower operating costs. However, when a more liberal statewide lease-allowable system was proposed, the East Texas independents were powerful enough to exclude their field from the order. A few years later, this same power defeated a compulsory unitization bill which even TIPRO supported.

The one tool that the Railroad Commission seems never to have actually used is a change in the field's prorationing formula. The commission either shut the entire field down or lowered the MER for everyone. This approach penalizes both those who support and those who oppose unitization. Nonconsenters and consenters suffer equally. The commission has not issued field orders revising the existing prorationing formula so that the nonconsenters would lose the regulatory advantage which makes it profitable to be a holdout. The unit's total allowable is still based on the field's existing prorationing formula. The two mechanisms used to adjust the correlative rights of the unit and the nonconsenters are rules that limit allowable transfers to boundary wells and rules that prohibit the conversion of the last well on a nonsigner's tract into an injection well. New prorationing formulas designed to prevent nonconsenters from benefiting from the unit's repressuring operations through net migration of valuable fluids are not promulgated.[75] The courts have affirmed, and perhaps mandated this commission policy of refusing to revise per-well formulas that have been accepted for many years.[76] The legislature has also affirmed this policy by making the Mineral Interest Pooling Act prospective only. Thus the greatest obstacle to unitization—the per-well factor in allocation formulas—remains in older fields. The Yates field had to be granted a doubled MER before unitization could proceed. This was the only way to ensure that no producer would suffer a decline in production by shifting from the old prorationing formula to the unit's participation formula. The Railroad Commission is often blamed for creating this obstacle of large per-well allowables and for refusing to revise allowables to encourage unitization more directly. However, as we have seen, the legislature must bear the ultimate responsibility for its failure to pass a compulsory pooling bill early in the development of the Texas oil and gas industry.

The Railroad Commission's furtherance of unitization through its administrative practice is of sufficient strength, determination, and permanence to attain the status of policy. However, the commission is still constrained by the lack of a compulsory unitization statute. In the controversial setting of the Port Acres field, the commission refused to issue a requested order for pressure maintenance and cycling, unwilling to test its precarious authority in what was surely to be a court-reviewed case. It is impossible to measure the number of instances in which the Railroad Commission refused to enter requested orders which would have induced unitization. Even if such a list were available, it would not account for the instances in which operators were dissuaded from ever bringing controversial projects to the Railroad Commission because it lacked compulsory unitization and repressuring authority. Much useful information on the extent to which the legal system has actually induced unitization can be derived from a statistical analysis of the degree and nature of unitization in Texas oil and gas fields today. Such a statistical assessment appears in chapter 9. It is premature to undertake this review at this point because the third group of players in the legal system has not yet been heard from. The courts can have a powerful influence on the course of unitization. Through the common law, through judicial review of Railroad Commission decision making, and through decisions in antitrust cases, the courts can either encourage or retard the progress of unitization. As we have seen, the gap between the Texas legislature's acts and the Railroad Commission's orders is often a large one. Whether the courts will allow such gaps to exist or create more on their own initiative is the next subject at hand. An analysis of the court's role in this drama will also explain some features of the voluntary unitization process in Texas, such as the remarkably high percentage of consent achieved without compulsory process.

7

Unitization and the Courts: The Common Law

The courts are the third in the triumvirate of legal institutions that influence the course of unitization in Texas. Indeed, it is difficult to imagine a law more hostile to unitization than the judge-made rule of capture. This rule allows one landowner to drain oil and gas from underneath an adjoining landowner's tract without liability. The adjoining landowner's only recourse is to drill an offset well to drain back the oil and gas. The rule creates an enormous incentive to drill wells, many of which are unnecessary from a conservation standpoint, and encourages a landowner to deplete the reservoir as rapidly as possible. Nor is the rule much of a guarantor that each landowner will receive the value of the oil and gas in place beneath his tract. But when the courts announced the rule in the early 1900s, the science of petroleum engineering and geology was only nascent, and the concept of unitization was unknown. In this context, the rule of capture was a rational response of judges to cases alleging drainage.

Other judicial doctrines have also influenced the implementation of unitization in Texas, and these will be discussed in this chapter. These common law doctrines include the rule of nonapportionment, the cross-conveyancing theory of unitization, judicial interpretation of contractual provisions and implied covenants in oil and gas leases, and tort liability of a unit to adjoining nonconsenters. Because many of the common law concepts that govern pooling are equally applicable to unitization, this chapter analyzes both pooling and unitization cases.

Chapter 8 then continues to assess the judicial influence on pooling and unitization in Texas by analyzing the courts' review of Railroad Commission orders promulgated under the conservation legislation enacted in Texas from 1919 onward. Much of this legislation was passed to limit the wasteful effects of the rule of capture and the frenzied, competitive drilling that it spurred. The Railroad Commission was granted broad discretion and strong powers under this legislation to prevent waste and protect correlative rights, but it was specifically forbidden the one power that would promote both goals simultaneously—compulsory unitization. Despite this prohibition, the commission, after World War II, actively pursued unitization of the largest Texas oil fields. Chapter 8 assesses the extent to which the courts constrained the commission's actions through judicial definition of the agency's statutory authority and through review of the reasonableness of its orders. Some of these cases have already been discussed as an integral part of the formation of Texas' legislative policies (such as the compulsory pooling act) and of the Railroad Commission's use of administrative orders to effect unitization. However, many other important cases of judicial review of commission decision making remain to be analyzed.

Chapter 8 also reviews court decisions involving unitized operations and the antitrust laws. Its concluding section then summarizes whether the courts have been a positive or negative factor in the unitization of oil and gas fields in Texas.

No Equitable Pooling in Texas

At an early date, Texas courts adopted the concept that the owner of land overlying oil and gas owned these minerals in place beneath his land, even if oil and gas were not actually being produced and brought into physical possession by the landowner.[1] This "ownership-in-place" or "absolute ownership" concept allows corporeal estates in realty to be created in the minerals, much like any other estate in land. At the same time, Texas adopted the rule of capture as the basic common law defining correlative rights to oil and gas. The inconsistency of creating absolute property rights to oil and gas in place and then allowing this same oil and gas to be drained without liability under the rule of capture was reconciled by the Texas Supreme Court as follows: "If the owners of adjacent lands have the right to appropriate, without liability, the gas and oil underlying their neighbors' land, then their neighbor has the correlative right to appropriate, through like methods of drainage, the gas and oil underlying the tracts adjacent to his own."[2]

From 1930 onward, the rule of capture became a less important determinant of property rights than Railroad Commission orders specifying where wells could be placed and at what rate they could be produced. The courts largely upheld the validity of this conservation legislation, thus relegating the common law rule in many instances.[3] Still, the rule of capture retained much of its vitality in Texas, even after conservation laws rendered the rule irreconcilable with the concept of absolute ownership of oil and gas. The most striking example of this occurred in *Ryan Consolidated Petroleum Corp. v. Pickens.*[4]

Ryan and Pickens were lessees of adjacent, small town lots that had been subdivided after oil and gas had been discovered in the vicinity. Thus, under the spacing rule, neither lessee could get a Rule 37 exception to drill on his individual lease. However, one lessee could secure a well permit to drill on the combined acreage leased by both. Pickens obtained the permit and drilled a productive well on his leased tract. Ryan sued for an accounting for his fair share of the oil and gas coming from Pickens's well.[5] In its initial opinion, the Texas Supreme Court held for Ryan on the basis that Pickens's application stated that the requested well would produce from both tracts and, therefore, Pickens was judicially estopped to deny the fact that drainage from Ryan was occurring.[6] However, after a rehearing, the court withdrew this opinion, and a majority of six justices denied Ryan any accounting. The final opinion emphasized that the rule of capture was a well-settled property rule in the state and that equitable principles could not defeat it.[7] In the absence of compulsory pooling legislation, neither the commission nor the court could authorize pooling of the two tracts. Thus, the supreme court closed the door to the concept of equitable pooling, a concept that had been adopted in Mississippi.[8] Three justices dissented, arguing that the constitutionality of the rule of capture was at issue when Rule 37 no longer allowed a landowner the equal right to drill and drain back. The dissent viewed the rule of capture as anachronistic because of the growth in knowledge of reservoir dynamics and geology. Stressing the paramount public interest in maintaining the well-spacing rule, the dissent expressed fear that Rule 37, rather than the rule of capture, would be adjudged unconstitutional. Without equitable pooling, the enforcement of Rule 37 would be difficult because the denial of well permits to small tracts like Ryan's violated the legal concept of absolute ownership and a fundamental sense of justice.[9] As we have seen, before the 1960s, the Railroad Commission seldom denied Rule 37 exception requests (see chapter 4 text at notes 97 to 98). Thus, the supremacy of the rule of capture, as epitomized in *Ryan v. Pickens,* probably induced unnecessary drilling, rendering subsequent unitization of oil fields more difficult to implement.

In another line of court decisions, the rule of capture was also held superior to equitable principles. In 1925 the Texas Supreme Court in *Japhet v. McRae* adopted the rule of nonapportionment: that absent an express provision, royalties would not be shared among the owners of separate tracts created by conveyances made after the original, larger tract was leased as a whole.[10] In this case, a 15-acre tract was leased by Keeble, its owner. Subsequently, 10 acres of the 15 acres were sold to Japhet. A producing well was drilled on this 10 acres. McRae and Keeble owned the remaining 5 acres and sued for five-fifteenths of the royalty from the well on the 10 acres. The court denied their demand. Again, the court premised its decision on the rudimentary state of knowledge about petroleum reservoirs, which prevented any presumption that the well on the 10 acres was necessarily draining the 5 acres. But even if drainage were to be proved, the court found no injustice to Keeble and McRae. Keeble "traded with his eyes open" and should have known that if a well was drilled on land not owned by Keeble, the rule of capture would apply.[11] The court considered its denial of relief to Keeble fairer than allowing Keeble to "wait for others to develop a certain portion of the tract and then ask for a division of the developed tract."[12] The nonapportionment rule has been strongly criticized,[13] but it has survived much subsequent litigation, although developments in reservoir science subsequently obviated the court's first premise. Because the injustice to the landowner of the drained tract in *Japhet* did not derive from state regulations, the constitutionality of Rule 37 was not questioned. In subsequent cases involving a *Japhet*-type situation, the spacing rules prevented the lessee from drilling on both tracts. Still, the court refused to apportion royalties.[14] The *Japhet* case and its progeny demonstrate an early and sustained judicial hostility to implied rights to pool in Texas.

The one exception to this judicial hostility is the common law of the community lease. In *Parker v. Parker,*[15] the Texas court adopted the view that owners of separate, contiguous tracts of land, who execute a single lease covering all the land, have created a community lease as a matter of law. The royalty will be apportioned among all the lessors on a surface-acreage basis, so that the owners of a tract without a well may, nonetheless, secure their proportionate share of royalties, and the lessee may develop the land as a single tract without regard to interior tract lines and offsets. The presumption that the landowners intended to pool their royalties can be rebutted only by express provisions in the lease, not by parole evidence of the lessors' intent not to pool. In *Parker,* the court was clearly concerned with the "manifest injustice and hardship" that would result to some landowners if royalties were not shared.[16] The case involved a 66-acre tract and a 178-acre tract leased together as a 244-acre tract. Wells were drilled on the 178 acres, but not on the 66. That the courts would find an intent to apportion royalties in *Parker* as a matter of law and no such intent in *Japhet v. McRae* seems anomalous. The court in *Parker* simply stated that "we find that *Japhet* v. *McRae* . . . is not on point."[17]

Despite their inconsistency, both lines of decisions have been adhered to in subsequent cases, although the Texas Supreme Court on one occasion did express some misgivings about the *Parker* holding.[18] The community lease is the only example in the Texas common law of the creation of an implied intent to pool based on equitable principles. As we will see in the section on the common law of cotenancy, the Texas courts have generally refused to apply equitable principles in the context of fieldwide unitization as well as in the pooling context.

The Pooling of Nonexecutive Interests

The judicial hostility to implied rights to pool is further manifest in the rule that the owner of the executive right—that is, the right to lease—does not have the right to pool nonexecutive interests such as royalty interest owners. This rule was announced in 1943 in the case of *Brown v. Smith.*[19] The Smiths owned two adjoining tracts of land totaling 62.75 acres. Brown wanted to purchase a community lease on the entire 62.75 acres. Twenty acres of this 62.75 acres were subject to a one-thirty-second royalty interest owned by Mrs. Lee. Brown had agreed to purchase the lease on the entire 62.75 acres, pending completion of a title search. When the search revealed Lee's nonexecutive royalty interest, Brown refused to follow through with the purchase. Lee had neither signed nor ratified the lease. The Smiths sued to force Brown to purchase the lease. The court refused to grant specific performance to the Smiths on the lease purchase contract. While the Smiths as mineral interest owners had the right to lease the 20 acres burdened by Mrs. Lee's royalty, the court held that the Smiths did not have the right to pool her interest into the larger 62.75-acre tract. If the Smiths did not have this right, their lessee could not acquire it either. Therefore, Brown—the would-be lessee—could not be forced to purchase a lease that did not fully merge the two tracts for development purposes as Brown had desired.[20]

The effect of *Brown v. Smith* is that lessees who desire to pool or unitize tracts subject to nonexecutive interests must secure the consent of all such interests in addition to the consent of the person with the right to lease. Separate negotiations must ensue, resulting in higher transaction costs, greater delays, and—if the consent is not secured—greater difficulty and higher costs in actually operating the unitized area. Indeed, the law invites nonexecutives to be holdouts and to delay signing or ratifying pooling or unitization agreements involving their tracts until they are sure that the economic benefits of joining the unit are greater than those of remaining unpooled.[21]

The Texas rule that executives cannot pool nonexecutives without express authority to do so is not followed in all other states.[22] The rule has been vigorously condemned, and prominent scholars have urged the court to overrule *Brown v. Smith.*[23] Instead, the court has retained the general rule and has attempted to confine some of the harmful consequences of the holding by using tortuous reasoning to distinguish other cases.[24] Such a judicial posture renders this area of the law more uncertain and invites further litigation. In states which follow the *Brown v. Smith* rule, compulsory pooling or unitization laws that would vitiate the need to attain the assent of every nonexecutive interest owner are especially important to the efficient development of oil and gas resources.

The Cross-Conveyancing Theory

The court's holding in *Brown v. Smith* was based on a concept of the nature of unitized title, which had been developed only a year earlier in *Veal v. Thomason.*[25] In this case, the court held that all persons having an interest in a unit's production were necessary and indispensable parties to a lawsuit involving the unit. Thus when the plaintiff, Thomason, sought to recover the title and possession of a 197-acre tract from Veal, the latter successfully argued that Thomason had to join all the other owners of interests in the 6,000-acre unit into which Veal had placed the 197-acre tract. The court held that the unitization had made all the lessors of land in the unit joint owners of all royalties produced from the unit. Because royalties are real property in Texas, the lessors were held to have cross-conveyed interests in realty. If Thomason won the title dispute and withdrew the 197-acre tract from the unit, the royalty owners of other tracts in the unit would no longer receive royalties from wells on the 197 acres. The unit would comprise only 5,803 acres, and all the owners' fractional interests in unit production would change. Because their interests would be vitally affected by the outcome of the title dispute, the court held that all the owners should have "their day in court," and required joinder.

This rule of indispensable parties can significantly influence a landowner's or lessee's motivation to join a unit. Once unitized, the landowner and lessee may not be able to litigate an issue concerning the validity of the unitization or the proper interpretation of the unitization agreement without joining all other owners in the unit.[26] In large units this may sometimes require the joinder of hundreds of parties. The expense and difficulty of joinder in effect render the unitization agreement almost impregnable.[27] While this rule may thereby advance the progress of unitized operations after the unitization agreement has been negotiated, it also may retard an owner's willingness to join. Maurice Merrill, a leading scholar and advocate of landowners' rights, commented as follows on the *Veal v. Thomason* rule:

> If I owned land in Texas, I should not enter into a joint lease, which in that state has a unitizing effect. Neither would I execute a lease empowering the lessee to unitize, a power by which he might involve my interests with those created by strangers in favor of a multitude of other strangers.[28]

In some jurisdictions, unitization is not viewed as having the effect of cross-conveying interests in real property. Rather, the unitization agreement is simply treated as a contractual method of determining the amount of royalties due to each owner. No owner acquires an interest in the oil and gas produced from another's property. Some writers have urged the courts to adopt this "contract" or "allocation" concept to avoid joinder and other problems created by the cross-conveyancing theory, and some lessees have inserted clauses in unitization agreements that expressly negate any intent to cross-convey interests.[29] However, the mere conceptual classification of the nature of unitized title would not appear to alter the necessity of joining those owners whose interests will, in truth, be materially affected by the outcome of a lawsuit in which one party is seeking to invalidate or to modify the unitization agreement. Williams and Meyers conclude that the best solution to the joinder problem is to rule that the owners of nonoperating interests need not be joined as indispensable parties.[30] Typically, the nonoperating parties are royalty owners whose interests will coincide with those of the unit operator or of the party who originally committed the land to the unit. These royalty owners will be adequately represented by their own lessees, by the owner of the executive right to their tract, or by the unit operator. Unfortunately, this solution may be difficult for courts to adopt in Texas because of the ruling in *Brown v. Smith,* which refutes the principle that the executive can represent the nonexecutive in pooling decisions and which requires the joinder of nonexecutives if a unitization agreement is to affect their interests. It might be difficult for the Texas courts to reconcile requiring joinder in the one instance and not in the other. Thus the two sets of cases—*Brown v. Smith* and *Veal v. Thomason*—may have created a situation in which Merrill's advice to royalty owners is quite sound. Compulsory unitization legislation would seem the best way to solve some of the problems of joinder in lawsuits involving title or other disputes on unitized land. The resolution of disputes concerning a tract of land which has been compulsorily unitized will not affect the interests of other persons in the unit because the tract will continue to be bound by the compulsory order.[31]

In one other context, a jurisdiction's view of the nature of unitized title may influence the implementation of unitization or pooling. One case implies that a pooling or unitization clause in a lease would violate the rule against perpetuities in a jurisdiction which follows the cross-conveyancing theory of unitization, but not in a contract allocation jurisdiction.[32] While some scholars doubt that such a lease clause or the unitization agreement itself could be construed as violating the rule against perpetuities,[33] the uncertainty in this area has caused some lessees to impose a twenty-one-year time limit on the exercise of their pooling and unitization powers.[34] Such a time limit on unitization may not be long enough in Texas. Most unitization in Texas occurs in older fields that have been under lease for many years before unitization is initiated (see chapter 9 text at notes 17 to 23). Here again, the common law of cross-conveyancing is inimical to unitization in Texas.

Cotenancy

Mineral estates are often owned in cotenancy. This fact makes development of the oil and gas resources more difficult, because if any of the cotenants refuses to lease, the lessee who proceeds to drill under lease from the consenting cotenants must account to the unsigned owners for their proportionate share of the profits from the operation on the leased tract.[35] If the lessee encounters a dry hole, the unsigned cotenants cannot be forced to bear a share of these costs. Thus, the economics of drilling on land burdened by unsigned cotenants is skewed against the lessee who wishes to develop: he must share any discovered fortune with owners who do not bear any drilling risks or contribute any front-end capital, yet he cannot force them to share any bad luck. Also, the developing cotenant's lessee must be prepared for disputes and possible litigation over the reasonableness of the costs that may be charged against the nondeveloping cotenant's share of production.[36] Similarly, when a cotenant undertakes the expense of a secondary recovery project, the law in Texas will allow him to recover only those costs which are necessary and beneficial to the mineral estate.[37] The costs of unsuccessful reworking of wells or ineffective water flooding cannot be recovered, even if such costs resulted from a sound business judgment. The common law of cotenancy is thus another reason why compulsory unitization laws are often needed to reduce the risks and costs of operating a unit without the consent of all owners.

The Texas rule which allows a cotenant to lease and develop the tract without the consent of all other co-owners is at least more progressive than the rule in a minority of states which holds that a lessee cannot develop a tract unless all cotenants are joined in the lease.[38] In Texas, a cotenant also can pool or unitize his undivided fractional share of the mineral estate even though the remaining co-owners refuse to do so.[39] This rule certainly promotes unitization. However, it may also subject the owners of small fractional shares of a mineral tract to a "squeeze play" if they refuse to lease.

The potential for this is illustrated in the case of *Superior Oil Co. v. Roberts.*[40] In this case, Roberts, the plaintiff, owned an undivided one-half interest in a tract of 1.5 acres. The owners of the other one-half interest executed an oil and gas lease to Superior Oil that purported to cover the entire 1.5 acres. This lease authorized Superior Oil to place the small tract into a large unit covering several hundred acres. Production was obtained from the unit, but no wells were drilled on the 1.5 acres. The plaintiff then sought an accounting from Superior Oil for his proportionate share of the profits from the unit. The court of civil appeals ruled that, because Superior Oil had leased the entire 1.5 acres, entered all 1.5 acres in the unit, and then paid itself profits on the basis of the full 1.5 acres, Superior held the plaintiff's *pro rata* share of the profits in trust for plaintiff.[41] The supreme court reversed the decision of the appellate court and refused to impose a constructive trust on the unit's profits from Roberts's fractional interest. In denying the accounting, the supreme court noted that the plaintiff had refused Superior's offer to lease his undivided one-half interest and had made no attempt to ratify or adopt either his cotenants' lease to Superior or the unitization agreement. Thus, there was no contractual relationship, express or implied, between the plaintiffs and Superior or the unit. The rule of capture and the common law of cotenancy applied: Roberts could receive nothing from a well not on his tract. The court refused to recognize any principle of equitable unitization just as they had refused to recognize equitable pooling. The court wrote:

> In claiming an interest in [the unit's] mineral production, plaintiffs in effect seek to claim the beneficial provision of such contracts and repudiate the unfavorable portions thereof which would limit their claim to the same rights as those held by their cotenants who joined in the Superior lease, namely, royalty rights only. In our opinion, this position is untenable.[42]

Obviously, the cotenant of a small tract who refuses to lease or unitize, is in a precarious situation. It is unlikely that his small tract will be drilled on and very likely that it will be drained. The unit operator will prefer to drill on adjacent tracts in which all the cotenants have agreed to lease and unitize, so as to avoid the problems of accounting to the unsigned cotenant. The court's decision in *Superior Oil* certainly favors unitization, and in this sense is good public policy, but it may do so at the expense of both vested property rights and justice. The court made no finding that the lease and unitization offers to Roberts from Superior Oil were fair. Such a finding is crucial to the reasonableness of a rule which allows fractional interests in small tracts to be drained by unit operations on surrounding land. Moreover, the Texas court seems to deny Roberts the option of ratifying the unit agreement and participating in the unit proceeds as a profit-sharing mineral interest owner. Should Roberts seek to ratify the unit agreement, the court limits him to securing "royalty rights only" from the unit, even though he has never leased his acreage.

The holding that Roberts lacks an option to share in profits is difficult to reconcile with Texas decisions in two analogous types of situations. First, the courts have held that the cotenant of a tract who refuses to join in a lease signed by the other owners, which purports to cover the entire interest in the tract, may choose to either ratify the lease and receive a royalty interest or refuse to ratify and receive instead a proportionate share of the profits from operations on the tract.[43] Furthermore, it appears that the cotenant can delay making this choice until it is obvious which decision is most remunerative.[44] This cotenancy rule does not facilitate the development of oil and gas, but this fact has not prompted the courts to modify the rule. The second line of cases involves the unauthorized pooling of a nonexecutive's interest by the owner of the executive right. The Texas courts have held that the nonexecutive may ratify this unauthorized pooling and secure a share of the pooled production, sometimes many years after the pooling agreement has been made and the adjoining tracts have been fully developed.[45] Thus it is not clear why under Texas law, Roberts, as an unleased mineral interest owner, does not have the option of ratifying the unitization agreement which has purported to cover the entire 1.5 acres, thereby securing a profits share from the date of ratification.[46]

Certainly the analogous cotenancy cases can be distinguished as involving only production from the jointly owned tract. In contrast, Roberts was seeking a share of the proceeds from a well on an adjacent tract. Likewise, the analogous cases authorizing the nonexecutive to ratify a pooling agreement can be distinguished on the basis that the nonexecutive has no power to lease and develop the tract in which he owns an interest; that is, the nonexecutive has no self-help remedy.[47] In contrast, Roberts has the right, in theory, to lease his undivided half-interest in the 1.5 acres and secure a well on this tract.[48] In fact, however, his chances of obtaining a lease or drilling his own well are virtually nil, and his position is that of a nonexecutive cotenant whose fractional interest has been leased and unitized by the other cotenants. The Texas courts will not imply an intent on the part of a nonexecutive cotenant to limit his production to a royalty share when the executive cotenant develops the tract by his own drilling rather than by leasing to another party.[49] This zealous safeguarding of the nonexecutive cotenant's right to a profits share in the self-development situation is difficult to reconcile with the holding in the *Superior Oil* case that limits *executive* owners of mineral interests to a royalty share only.

In denying Roberts the option to ratify the unitization agreement as an owner of a mineral interest, the Texas justices did not attempt to reconcile or distinguish their holding with analogous cases in Texas law. Nor did the court even mention public policy considerations that would justify the possible sacrifice of individual rights for a greater societal good.[50] Undoubtedly, the decision makes life easier for unit operators who need not carry the working interests of unleased cotenants. The profitability of a unit operation may be considerably enhanced by the *Superior Oil* holding. The opinion marks a strong judicial posture in favor of unit participants, but its lack of any accompanying oratory or analysis to this effect is remarkable. Indeed, on close examination, the decision may not be as supportive of unitization as it first appears. At heart, the decision is akin to that of *Ryan v. Pickens*: the Texas Supreme Court is as unwilling to adopt equitable unitization as it is to adopt equitable pooling. Were the shoe on the other foot, and were a small-tract owner draining a unit and unjustly enriching himself at the unit's repressuring expense, the rationale of *Superior Oil v. Roberts* would deny the unit relief. The *Superior Oil* opinion forces Roberts to enjoy his mineral estate either by ratifying the lease to Superior Oil (the same lessee chosen by his cotenants) or by seeking partition of the jointly owned tract. If partition in kind is granted and the resulting separate tract is large enough to secure a drilling permit under the spacing rules, the unitized operation may be thwarted or rendered less efficient and profitable by a well on this unsigned tract.[51] Hence, a holding that allowed an owner of an unleased, fractional mineral interest to ratify the unit and receive a profits share might promote unitized operations more than the *Superior Oil* holding.

In many instances, however, partition in kind will create a voluntary subdivision that is unable to obtain a Rule 37 exception permit to drill. Thus, even with the cotenancy aspects of the case removed by a partition decree, the owner of a small tract may be squeezed out by the unit operator.[52] This is why it is important that either judge-made law or statutory law requires that such owners be offered a fair opportunity to join the unit before they are denied any relief against drainage by the unit.

The unitized operations in the *Superior Oil* case covered several hundred acres in the Altair field. Most owners in the field had signed or ratified a unitization agreement dated April 10, 1947.

The agreement preceded passage of the 1949 voluntary unitization act and therefore the commission was not called upon to approve it.[53] Section 101.013 of the 1949 Act requires that the Railroad Commission make a finding that the rights of the owners of all interests in the field, whether signers or not, are protected under the unitized operation, and that the owners of all interests are given an opportunity to enter into the unit on the same yardstick basis. Were the owner of a fractional mineral interest or a separate small tract to protest commission approval of a proposed unit on the basis that the unit operator's offer to join was conditioned on leasing to the unit first, Section 101.013 of the 1949 Act would seem to require that the commission deny approval of the unit. The unit proponents might argue that the rights of all owners of unleased minerals interests in the field are adequately protected by a fair offer to lease to the unit on the same yardstick basis as other owners in the unit have leased, but this interpretation of the 1949 Act is of doubtful validity. The purpose of the 1949 Act is to provide antitrust immunity to voluntary units, not to expand the power of unit operators to alter the property rights of nonsigners. Most compulsory unitization laws in other states do not force owners of unleased mineral interests to convert to a royalty owner's share of unit production.[54] Nor does Texas' Mineral Interest Pooling Act force unleased owners to exchange mineral interests for royalty shares.[55] Further, private voluntary unitization agreements typically provide that owners of unleased mineral interests receive a proportionate share of both the royalty and the working interests.[56] In the event that the commission and the courts were to construe the 1949 Act in such a questionable manner, Section 101.013 would certainly require that the Railroad Commission find that the offer to lease was a fair one before approving the unit. This would provide some protection to owners of small tracts, which is lacking in the supreme court's decision in the *Superior Oil* case. But the small-tract owner must be alert and active in his own defense. As we have seen, the Railroad Commission relies solely on the self-interest of protestors to bring allegations of unfairness to its attention at the unitization hearings (see chapter 6 text at notes 28 to 32). While the commission's unitization orders no longer expressly disavow any authority to pass upon the equity of the participation factors in the unitization agreement, the commission views its duty to protect correlative rights as being fulfilled by providing adequate notice of the hearing, so that unhappy owners can appear and protest the unit's formation. As previously noted, few such protests occur. Despite the lack of protection to unsigned interests under the common law of *Superior Oil v. Roberts* and despite the commission's laissez-faire attitude toward correlative rights under the 1949 Act, the very nature of voluntary unitization seems to protect the interests of both signers and nonsigners.

Judicial Interpretation of Pooling and Unitization Clauses in Leases

Pooling clauses in oil and gas leases now are common. Generally, such clauses authorize the lessee to pool all or any part of the acreage covered by the lease with other land in the immediate vicinity when, in the lessee's judgment, it is necessary or reasonable to do so in order to develop and operate the premises. The pooling clause often limits pooled units for oil to 40 acres in size and gas units to 640 acres. Few leases contain clauses authorizing unitization of the tract.[57] Judicial interpretation of these pooling and unitization clauses has an important influence on the ability of lessees to use these lease provisions to develop land efficiently.[58]

The typical pooling clause vests great discretion in the lessee to pool at his option. Pooling has large consequences for the lessor whose royalties from a well on the leased tract may be greatly diminished by the lessee's act of pooling. When lessors have sought to invalidate pooling clauses on the basis that they are too broad and vague to be enforced, the Texas courts have upheld the clauses. For example, in *Tiller v. Fields*,[59] the court upheld a pooling clause which had no limitations on either the time for exercising the pooling authority or the size of the unit to be formed with the statement that:

> Anticipatory provisions in leases for the commitment by the lessee of such leases to unitization, of necessity must be in general terms. Neither the lessor nor the lessee has any way of knowing at the time the lease is taken the facts with respect to which it will be necessary for the lessee to apply his power. It is not practicable for the lessee to await the ascertainment of such facts. He knows from experience that because of the possibility of many changes in ownership of the lessor's interest as

> time goes on, it may be difficult to effect an agreement if the right to unitize is not included in the lease itself.[60]

The courts have often rationalized their decisions to uphold broad pooling clauses with public policy considerations. Thus, in *Texaco Inc. v. Letterman,*[61] the court ruled that the pooling clause authorized the lessee to exercise the pooling power a second time after the termination of the original unit, and stated the following:

> That pooling or unitizing of oil and gas leases is a standard practice in the industry cannot be questioned. It is equally recognized that unitization is often a more feasible method of operation from an engineering and scientific point of view. Unitization can be said to be advantageous to both lessors and lessees. We think these facts lead to the conclusion that in the absence of clear language to the contrary, pooling clauses should not be construed in a narrow or limited sense.[62]

To protect landowners against the abusive use of this broad pooling power, the Texas courts simultaneously declared that the lessee must exercise the pooling authority in good faith. The determination of whether a lessee has acted in good faith is a question of fact to be decided on a case-by-case basis. Thus, in *Amoco Production Co. v. Underwood* the jury found that the lessee's formation of a gas unit was in bad faith,[63] and the court canceled the unit designation and declared certain leases terminated for lack of production. In this case, the lessee was found to have gerrymandered eight leases covering 2,252 acres into a gas unit of 688 acres in an attempt to hold all 2,252 acres by production from the one unit well. The unit, which was formed only two days before the leases were to expire, included some unproductive acreage and excluded other productive land.

The good-faith standard does not create a fiduciary relationship between the lessee and lessor.[64] The lessee may consider his own self-interest in deciding whether to pool. Thus, pooling often has been found to be done in good faith, even though the landowner's royalty is thereby diminished.[65]

In addition to imposing a good-faith restraint on pooling, the Texas courts have interpreted the express provisions of pooling clauses very narrowly. Thus, a pooling clause allowing the lessee to "enlarge or change the shape of an existing unit" was construed as not authorizing the lessee to reduce the size of an existing unit.[66]

A lessor whose pooling clause provided for payment of royalties on an acreage basis was held not to be bound by a pooling agreement entered into by her lessee on a productive-sand basis.[67] When a lessee executed, but failed to record, a pooling agreement before the primary term of the lease ended, the court declared the lease terminated for lack of production, because the pooling clause expressly required that the lessee "shall execute in writing and record . . . an instrument identifying and describing each such unit or units so created."[68] This judicial inclination to construe pooling clauses narrowly can hinder sound conservation practices, as illustrated by the case of *Jones* v. *Killingsworth*.[69] The pooling clause in landowner Jones's lease read as follows: "Units pooled for oil hereunder shall not substantially exceed 40 acres each in area, . . . provided that should governmental authority having jurisdiction prescribe or permit the creation of units larger than those specified, units thereafter created may conform substantially in size with those prescribed by governmental regulations." The Railroad Commission's proration field rules for the Fairway (James Lime) field permitted only one well to each 80-acre proration unit, but also provided that operators "may elect to assign tolerance of not more than eighty acres of additional unassigned lease acreage to a well on an eighty (80) acre unit and shall in such event receive allowable credit for not more than one hundred sixty (160) acres." The lessee formed a unit of 160 acres, including all of Jones's land. The Texas Supreme Court held that the pooling clause in the lease only allowed oil units this large if *prescribed* by regulatory authority. The field rules permitted, but did not prescribe, 160-acre units. Therefore, the pooling was invalid, and the lease terminated for lack of production at the end of the primary term, notwithstanding that the lessee had acted in good faith.

The pooling provision interpreted in the *Killingsworth* case was widely used throughout Texas. The majority's opinion could invalidate many other pooled units and lead to great confusion and uncertainty in titles to land so pooled. Citing these considerations, three justices dissented,[70] arguing that a reasonable construction of the pooling clause would uphold the 160-acre unit. The dissent wrote:

> (T)he authority granted by the lessor for pooling necessarily had to be stated in broad and general terms because it could not

> be foreseen what the circumstances in the future might be, what the regulations of the Railroad Commission might be nor in what terms they might be stated. For that reason a liberal interpretation should be given to the pooling provision to accomplish the purpose for which it was intended, that is, to promote conservation beneficial both to the lessor and the lessee.[71]

The difficulties of unitizing the Fairway (James Lime) field have been mentioned in chapter 5 (see chapter 5 text at notes 65 to 66). Only after the commission drastically reduced the field's allowable rate of production would enough owners in the field agree to unitize. The regulation permitting 160-acre prorationing units was intended to minimize the number of wells drilled into the reservoir and to allow early definition of the reservoir's boundaries at reduced costs. The unit proponents actively sought the wide, 160-acre spacing rule and considered its adoption by the commission to be a major victory in their efforts to secure early unitization of the Fairway field.[72] The supreme court's strict construction policy toward pooling clauses in leases made fieldwide unitization even more difficult to negotiate and implement.

The dissenting justices' prophesy that the ruling would cause further litigation became reality. Another group of landowners who held leases with the same pooling provision brought suit for title and possession of the mineral estate in a 66.5-acre tract of land that had been pooled into a 160-acre unit and then placed in the fieldwide Fairway unit.[73] In this case, the court held that the lease on the 66.5-acre tract was void because of the unauthorized pooling. Furthermore, the court held that the landowners' subsequent ratification of the fieldwide unit was as mineral interest owners rather than as royalty interest owners.[74] Litigation like this is expensive.[75] In addition, it is difficult for unit proponents to assess the expected profitability of unit operations when the ownership status of a significant number of landowners and lessees in the field is uncertain.

In some instances, the harshness of the court's strict construction policy toward pooling clauses is mitigated by judicial rulings that the landowners have ratified the unauthorized pooling and thereby revived the underlying lease. Thus, when a lessee pooled a tract into a large gas unit greatly exceeding the 40-acre pooling limit authorized in the lease, the court nonetheless held that the lease on the tract was still in effect, because the landowners had accepted royalty checks covering their share of the gas unit's production and had executed a royalty deed reciting that the lands deeded were at that time under an oil and gas lease and subject to it.[76]

The courts' narrow construction of pooling clauses does not seem to reflect any inherent hostility to pooling, but seems to be a corollary of the basic rule of contract interpretation that contracts are to be construed against their drafters.[77] Oil and gas leases are usually drafted by the lessee and so the lessor often wins cases which involve the construction of express limitations on the lessee's pooling authority.

Implied Covenants

The doctrine of implied covenants is an established feature of Texas oil and gas law. From an early date, the Texas courts have ruled that implied promises exist between the lessee and lessor of an oil and gas lease, absent express provisions to the contrary, and have used these implied promises to resolve disputes between the two parties. Implied covenants entered the common law because the courts recognized that it is impossible to specify in advance in a lease contract all of the details of exploration, production, and development that a lessee should attend to. Despite scientific advances in exploration technology, it is still not possible to know whether a drilling operation will result in a "gusher" or a "duster." For this reason, lessees do not want to commit themselves to a specific development program in the lease that subsequent information may indicate to be unprofitable. Lessors are also wary of express lease provisions that might unduly limit the lessee's obligations. Thus, the typical oil and gas lease passes to the lessee the broad and exclusive right to explore, produce, and develop the property, with little else specified.

This leaves the lessor in a disadvantageous position. The lessor's primary economic return for granting the lease is usually the right to receive royalty. If the lessee does not develop the property, the lessor receives little benefit from the contract. Both parties share a strong mutual interest in finding and producing oil and gas, but the lessee must bear all the costs of these operations while the royalty interest owner's share is cost free. Thus, the two parties are apt to have different views of the proper timing and implementation of development decisions, and conflicts between the two parties are inevitable.

These inherent characteristics of the lessee–lessor relationship gave birth to the doctrine of implied covenants. Judges resolve disputes between the two parties by filling the silent spaces of the lease with an implied promise on the part of the lessee to do "whatever, in the circumstances would be reasonably expected of operators of ordinary prudence, having regard to the interests of both lessor and lessee."[78] Using this broad principle, the courts have obligated lessees to protect their landowners' tracts against drainage; to drill additional wells once production has been obtained; to market the oil and gas on the best possible terms; and to manage and administer the lease with reasonable care.[79]

Implied covenant law has been vilified almost as much as the rule of capture for its promotion of additional drilling, which redounds to the benefit of self-interested lessors, but which is inefficient and wasteful from society's viewpoint.[80] No doubt some of this criticism is warranted, and the doctrine of implied covenants has retarded unitization because lessors can use this common law to force lessees to drill offset wells or additional development wells on their tracts which make subsequent pooling or unitization more difficult—if not impossible—to achieve.[81] Indeed, even if all the lessees in a field are united in their desire to perform cooperative repressuring, lessors who refuse to join may defeat the plan. The lessee who joins a unit, but who cannot persuade his lessor to do so, may find himself liable to his lessor for drainage if the unit does not drill a well on the tract and the lessor can prove that it would be profitable for the lessee to drill a protection well. Furthermore, at the end of the primary term, the lease without a well on it will expire and the lessee will be stripped of any ownership interest whatsoever in the oil and gas. If the lease does have producing wells on it, the lessee must account to the nonjoined landowner for a full royalty from these wells, but the lessee will receive only a unit's share of production from them.[82] The unsigned lessor whose tract experiences an increase in production resulting from the pressure maintenance activities of the unit cannot be forced to share this increase with the unit.[83] If the nonconsenting landowner can prove that a reasonably prudent operator would drill additional wells on the tract, the lessee also may be found liable for breach of the implied covenant to develop.[84]

Of course in many instances, the interests of the lessee and the lessor are mutual, and both will either join or not join a unitization plan based on a comparative economic analysis of the benefits of cooperating versus going it alone. The owners of tracts favored by nature or by the existing regulatory system generally will have an incentive to hold out, and the owners of disfavored tracts will want to cooperate. However, because lessors do not bear the costs of developing the oil and gas resource and because they may be more risk averse and therefore value current income more than future income,[85] sometimes fewer lessors than lessees agree to unitize. The statistics show that the percentage of royalty interest owners consenting to unitization is often lower than that of owners of working interests.[86] Some lessees in the unit therefore must contend with nonjoined landowners and the law of implied covenants.

Although implied covenant law is generally inimical to unitization because it works to maximize the number of wells drilled while unitization seeks to minimize the number, this body of common law does not always operate at cross-purposes to pooling and unitization. In certain circumstances, a lessee may have an implied obligation to perform secondary recovery operations to increase the ultimate recovery of oil and gas from a leased tract.[87] This implied covenant may impel lessees to consider the merits of unitization in order to achieve better recovery. When lessees do instigate secondary recovery operations, the courts have shown a willingness to recognize the unique nature of these operations and assess an operator's conduct in light of the practical necessities of secondary recovery. Thus, in one case, the court refused to find that a unitization agreement had terminated because the operator had failed to pump water into the field for a period in excess of ninety days.[88] The court found that secondary recovery operations did not always require the continuous injection of water. Prudent operators would, in certain circumstances, cease water injection and wait several months for the injected water to push the oil into the producing well bores.

More recently, the Texas Supreme Court has propelled producers toward unitization by holding that lessees have a duty to protect their landowners from fieldwide drainage. In *Amoco Production Co. v. Alexander,* the court wrote:

> The duties of a reasonably prudent operator to protect from field-wide drainage may include (1) drilling replacement wells, (2) re-working existing wells, (3) drilling additional wells, (4) seeking field-wide regulatory action, (5) seeking Rule 37 exceptions from the Railroad Commission, (6) seeking voluntary unitization, and (7) seeking other available administrative relief.[89]

The Alexanders, plaintiffs in this case, owned downdip tracts of land in the Hastings field. They had leased to Amoco for a one-sixth royalty. Amoco owned about 80 percent of the field's production and had leased many updip tracts in return for a one-eighth royalty. The Hastings field is a water-drive reservoir. As oil is produced from the field, the water moves upward and the wells on downdip leases are drowned out first. The downdip oil naturally migrates to the updip leases, resulting in fieldwide drainage.

The Alexanders contended that Amoco had deliberately slowed its production rate on downdip leases and increased production rates on updip leases to hasten this natural migration. They contended that Amoco had a duty to drill additional wells and rework existing wells on their downdip tract to capture additional oil before it was drained away.

Amoco claimed that the implied covenant to protect against drainage extended only to local drainage caused by wells on adjoining tracts. If a lessee had to protect against regional drainage affecting all tracts in the field, each lessee would be required to drill offset wells on each lease. Every additional well drilled would increase the fieldwide migration and set off a self-defeating chain reaction requiring more offset drilling that would, in turn, accelerate drainage. Amoco contended that the only solution to fieldwide drainage was a change in the field prorationing rules. Otherwise, Amoco faced conflicting obligations to its several lessors. If downdip production were increased to protect the Alexanders, the updip lessors would bring suit, and vice versa. Liability for fieldwide drainage would place lessees like Amoco in an untenable position.

The Texas Supreme Court upheld the lower court's decision in favor of the Alexanders. The court found that if Amoco had sought permits for Rule 37 exceptions from the Railroad Commission to drill replacement wells at the uppermost edge of the Alexanders' tract, the commission would have granted the request.[90] The court also found that the wells would have been profitable to drill. The Alexanders were awarded almost $2 million in damages, representing the additional royalty that they would have received had the replacement wells been drilled.[91] The court was unimpressed by Amoco's argument that the lower court's decision imposed impossible and conflicting obligations on Amoco. Amoco's conflict of interest as a common lessee to both updip and downdip landowners was of its own making. Amoco clearly had no economic incentive to increase production on the Alexanders' tract, with its one-sixth royalty, when the Alexander oil would be swept updip and produced with only a one-eighth royalty. Even without the royalty difference, money invested updip would have a greater economic return because the wells there would last longer. The court wrote:

> These conflicts would not occur if Amoco was not a common lessee. . . . If the Alexanders were the only Amoco lessor, their interests would more nearly coincide. Amoco's interest would be to capture the most oil possible from the Alexander leases before they watered out.
>
> Amoco's responsibilities to other lessors in the same field do not control in this suit. This lawsuit is between the Alexanders and Amoco on the lease agreement between them and the implied covenants attaching to that lease agreement. The reasonably prudent operator standard is not to be reduced to the Alexanders because Amoco has other lessors in the same field.[92]

Because prorationing rules, once established, are almost immutable, the only way that a lessee in Amoco's position can avoid conflicting implied obligations to both updip and downdip lessors is to unitize the field. Amoco owned 80 percent of the field production, and its only major competitor seemed to be Exxon. Exxon owned downdip leases and would probably have been willing to agree to a reasonable unitization offer. The unitization agreement would eliminate the need to drill replacement wells on each tract as the water line rose, and both Exxon and Amoco would gain from these large cost savings. If Amoco could not secure the consent of the updip lessors to a unitization agreement, failure to unitize probably would be excused under implied covenant law.[93]

The *Alexander* opinion provides an incentive for the updip lessors to consent. If they do not, a lessee in Amoco's position will be forced to drill additional wells downdip, perhaps reducing the amount of oil obtained by the updip lessors to less than what they would have received under unitization. Or, the next set of Alexander-type plaintiffs might argue that Amoco had a duty to reduce the entire field's production rate to allow the downdip wells to produce for a longer period of time and recover their fair share of oil.[94] The updip landowners would then have to accept either a reduced MER or a unitization agreement. Armed with the *Alexander* opinion, an Amoco-type lessee would have some persuasive arguments to attract updip lessors to unitization. The supreme court's express mention of voluntary unitization as a means of fulfilling the implied covenant to protect against fieldwide drainage is well reasoned and deliberate.

The *Alexander* opinion breaks new ground, for it was by no means obvious that a Texas court would ever hold that an implied covenant to pool or unitize existed. The courts had so often eschewed equitable pooling that the "fair dealing" basis for implying a covenant to pool seemed doubtful (see this chapter text at notes 1 to 13). Also, some lower court decisions had hinted that no such duty existed.[95] And many cases had been decided in East Texas holding that Rule 37 exception wells would not be granted to protect tracts against fieldwide drainage.[96] These cases involved appeals of Railroad Commission orders that had either granted or denied Rule 37 exception permits, rather than implied covenant litigation, but the drainage effect was the same, and the court had made no effort to encourage pooling or unitization in these cases. The courts' stock answer to lessees and landowners hurt by fieldwide drainage was to assign them the virtually impossible task of seeking to amend the field's prorationing rules. The language in the *Alexander* case bespeaks judicial recognition and appreciation of fieldwide unitization as a solution to fieldwide drainage.

Still, in many cases, a lessee in Texas will be able to successfully defend against any breach of the implied covenant to unitize by showing that the high costs of negotiating and the impossibility of securing enough signatures under voluntary unitization prohibit a reasonably prudent operator from unitizing. The fact remains that the Alexanders won damages, not for Amoco's failure to unitize, but for its failure to drill Rule 37 exception wells.[97] In fields where a reasonably prudent operator cannot secure voluntary unitization because of high costs or too many holdouts, the approved method of protecting individual landowners from fieldwide drainage is to drill these costly, inefficient, and unnecessary wells.[98] The *Alexander* opinion can result in a wasteful race to drill Rule 37 exception wells. Without compulsory unitization legislation, the East Texas legacy lives on. The *Alexander* case encourages producers with leases from multiple landowners in a field to seek passage of a compulsory unitization bill which would reduce the risk of expensive, implied covenant litigation and also reduce negotiating, drilling, and producing costs. But a major oil company, not an independent producer, is more likely to be the common lessee serving several landowners. The major companies have supported compulsory unitization for years. The small producer who holds only one lease on a small part of a field's acreage is unlikely to be converted to the cause of compulsory unitization merely because of the *Alexander* case. Judicial appreciation of unitization is no substitute for compulsory legislation.

Those large producers who do face the dilemma of serving both updip and downdip masters in the same field will undoubtedly seek to unitize the field. Those updip lessors who are better off "going it alone," even if more wells are drilled downdip in the absence of unitization, may not sign the agreement. If such landowners can then also succeed in an implied covenant action against their lessee for failing to maximize the potential production of the updip tracts because of downdip drilling, the *Alexander* case has had the perverse result of fueling the incentive to be a holdout.[99] If, on the other hand, the lessee can defend against such a suit by proving that he made a fair and reasonable unitization offer to the nonconsenting landowner, and this offer is deemed to satisfy the lessee's implied covenant duties to the updip landowner, the motivation to hold out subsides. In a federal case involving Texas land, the Fifth Circuit Court of Appeals held that the lessee's implied covenant obligations to a group of landowners who had refused to join a unitization agreement and who were damaged by the same lessee's cycling operations on adjacent tracts, were satisfied by making a fair offer to unitize. In *Tide Water Associated Oil Co. v. Stott,*[100] the lessee, Tide Water, had unitized several of its leases in the Long Lake condensate gas field. The plaintiff-lessors refused to join the agreement. Tide Water continued to operate three wells on the plaintiffs' tracts, extracting condensate in separators at the well and selling the remaining dry gas to a pipeline. Tide Water's cycling operations on the surrounding tracts drove the condensate-rich "wet" gas away from the plaintiffs' land, replacing it with less valuable dry gas. The plaintiffs sued for breach of the implied covenant to protect against drainage, and the lower court awarded damages. On appeal, the Fifth Circuit judges reversed the decision, holding that Texas law required that the landowner prove that the lessee could have drilled additional wells on the landowner's tract with an expectation of receiving a reasonable profit for the gas so produced. The court continued:

> It is conceded that a reasonable and prudent operator would not have drilled an additional well upon any of the appellees' three tracts; that the lessees were producing from these tracts all of the mineral products which could be produced in the absence of recycling; and that recycling was not practicable in the absence of unitization, which the lessors had refused. The appellants, therefore, have fulfilled their implied covenant to protect the premises from drainage.[101]

The plaintiffs alleged, however, that Texas law imposed an additional implied covenant on lessees who were draining their own lessors by operations on adjacent tracts: an implied covenant not to injure their own lessor's estate. The court found no support for this principle in Texas law. In fact the Texas case of *Hutchins v. Humble Oil & Refining Co.*[102] had repudiated any such principle a few years earlier. In the *Hutchins* case, Humble Oil held leases on 28 acres of the lessor's land and was draining the acreage with two wells located less than 200 feet away. Humble Oil had little economic incentive to drill on the lessor's acreage because the company owned the adjacent tracts in fee and paid zero royalty thereon. The jury in *Hutchins* found that a reasonably prudent operator would not drill an additional well on the lessor's 28 acres and also found that the express clause of the lease limited Humble Oil's offset obligation to wells that were located within 150 feet of the lease.[103] If Texas law did not allow Hutchins to recover damages, it also would not allow the Stotts to do so.

The *Tide Water* holding is obviously of great assistance to a lessee who proceeds to conduct unitized operations without the consent of all of his landowners. And while it is a federal case, it seemed well-based on Texas law. However, nine years after the *Tide Water* case, the Texas Supreme Court, in another case involving drainage by a common lessee, upheld a judgment in favor of the landowner and in so doing clouded the precedential value of the *Tide Water* rule. In this case, *Shell Oil Co. v. Stansbury,*[104] the landowner had proved that a reasonably prudent operator could have drilled a profitable offset well. However, an express offset clause in this lease would have negated any implied duty to protect against drainage if the court followed the *Hutchins* precedent. The lower court refused to find that the express offset clause limited the lessee's obligations to drill when the lessee was the very one doing the draining. On appeal, the Texas Supreme Court refused the application for writ of error, finding no reversible error, and added a brief *per curiam* opinion, stating that:

> We disapprove of any language in the opinion of *Hutchins v. Humble Oil & Refining Co.,* . . . which conflicts with the principle that a lessee is under a duty to protect his lessor against depletion of the lessor's minerals by the affirmative act of the lessee upon adjacent land.[105]

This opinion seems to support the implied covenant not to injure the lessor's estate, which the lessors in *Tide Water* had unsuccessfully sought to impose. If so, the *Tide Water* case would no longer reflect Texas common law. The unfortunate lessee in a *Tide Water-* or *Alexander*-type situation would be liable under implied covenant law for any injury to the lessors who refused to unitize and who were injured by the lessee's drilling or unitized operations conducted on other tracts. If, because of this probable liability, the lessee refused to unitize and thus failed to maximize the recovery of oil and gas from the tracts of those lessors who were willing to join the unit, the hapless lessee would face the equally serious risk of implied covenant litigation brought by this set of lessors.[106]

For many years, this brief *per curiam* opinion caused consternation as to whether it made the common lessee holding adjacent leases from different landowners strictly liable as an insurer against drainage, regardless of what a reasonably prudent operator would do.[107] An alternative reading of the *per curiam* opinion was that it simply overruled that part of *Hutchins* involving the effect of express offset clauses in common-lessee drainage cases, but that implied covenant liability still required a showing that drilling offset wells or taking other protective measures would be profitable to the lessee as a reasonably prudent operator. Oil and gas producers often seek to acquire large blocks of leases in a field to achieve economies in exploration, drilling, and producing operations and to avoid drainage by offset operators. Especially in high-risk areas, producers desire to drill on large blocks of land so that if oil or gas is found, the risk-taking operator can reward himself with additional successful development wells and recoup his investment (which probably included several prior dry holes) before rival operators move into the area. If a common lessee is a guarantor against drainage between his separate landowners, the law chills this industry practice which enhances efficiency and the willingness to take risks.

The Texas Supreme Court in the *Alexander* opinion clarified its previous *per curiam* opinion, as follows:

> This Court in *Shell Oil Co. v. Stansbury* . . . expressly overruled the *Hutchins* case, *supra,* and held that an express offset provision does not limit the lessee's obligation to protect from drainage when the lessee is the one causing the drainage. In drainage cases, Texas courts place upon the lessor the burden to prove that substantial drainage has occurred and that an offset well would produce oil or gas in paying quantities.[108]

The court expresssly removed Texas from those jurisdictions that held a lessee strictly liable for drainage caused by his own acts on nearby tracts. In so doing, the Texas court allowed the *Tide Water* holding to continue to reflect Texas law: a lessor who refuses to accept a fair offer to unitize will not have veto power over operations of his lessee on other tracts, even if such operations cause some loss to the lessor. This judicial clarification of implied covenant law in Texas provides even greater encouragement to unitization than the court's statement that an operator's duty to protect against fieldwide drainage may include seeking voluntary unitization. The landowner who refuses to accept a fair offer to unitize will probably not succeed in an implied covenant action against the lessee who continues operations on other tracts which injure the nonconsenter's tract. The incentive to hold out is considerably reduced.

Powerful as is the *Tide Water* case in protecting unitized operations, three factors constrain its usefulness to unit proponents. First, The Fifth Circuit judges seem to have been influenced by the fact that other operators in the Long Lake field had started cycling operations before Tide Water.[109] Tide Water and its other lessors would suffer noncompensatory drainage if Tide Water could not engage in cycling because of the veto power of a nonconsenting landowner. It is not clear that Tide Water would be free from liability to a nonconsenting landowner if Tide Water were the first or only operator in the field to do cycling.

Second, the *Tide Water* case still requires that the lessee act as a reasonably prudent operator on the nonconsenting landowner's tract. This duty will often interfere with the efficiency of the unitized operation. Thus, Tide Water had to continue to operate the three wells on the Stotts' tract and to sell as much gas as possible from these wells. Tide Water had to incur the expense of installing and operating condensers at the wellheads of these three wells. Tide Water's lease would have ended had no wells been on the tract; and had the lessor been able to prove that additional wells would have been profitable to drill, Tide Water would be subject to damages for breach of the implied covenant to protect against drainage and, perhaps, to develop as well.[110]

Third, the court's reasoning in *Tide Water* depended to a large extent on the conclusion that the plaintiffs had been offered a fair opportunity to join a fair unitization plan.[111] The plaintiffs argued vigorously that the offer and plan were not fair. Any lessee who unitizes without the consent of all interested landowners knows that the fairness of the offer may be litigated in court. If the lessee's judgment of fairness is not accepted by the court, the lessee will be liable in damages. This threat of litigation is a serious obstacle to the introduction of sound conservation measures by lessees under voluntary unitization agreements that are not signed by all the landowners. However, if the common law did not require a finding that the unitization offer was fair and that the lessee act as a reasonably prudent operator on the nonconsenter's tract, the rights of nonconsenting owners could be seriously abused by unit operators. Thus, the *Tide Water* holding seems to be as good a balance as possible between the rights of the lessee who unitizes and the rights of landowners who refuse to join. The judges in the *Tide Water* case also required that the unit operator allow the losing landowner to participate in cycling operations in the future, if desired.[112] The decision still is not a complete substitute for a compulsory unitization law that would allow lessees to force landowners into a unit on fair and reasonable terms.

Texas' 1949 voluntary unitization act can provide some but not all of the missing protection to lessees who cannot use compulsory process to unitize holdouts and who face conflicting implied covenant duties to different lessors, some of whom have signed and some of whom have not signed a unitization agreement. The 1949 Act requires that the Railroad Commission find that the rights of all signers and nonsigners in the field are protected and that the owners outside the unit are given an opportunity to enter the unit on the same yardstick basis as the owners in the unit.[113] The fact that a lessee has secured commission approval of the unit agreement with these two findings can be introduced as evidence of the fairness of the unitization offer in any subsequent implied covenant litigation. This evidence may well be persuasive to judge or jury. This persuasive effect is valuable to lessees caught in a *Tide Water* situation, because a finding of fairness will allow the lessee to escape liability for injury to the nonconsenting owners as long as he or she fulfills his other implied covenant duties. This may be one reason that lessees inevitably seek commission approval of their unitized operations. However, it is unlikely that the mere fact of commission approval of the unit plan would be held to bar a lessor by collateral estoppel from bringing a cause of action under implied covenant law, or that commission approval would be considered conclusive evidence of the fairness of the offer. (These matters are discussed fully in the next section on "Tort Liability of the Unit.")

As we have seen, in Texas many unitization agreements have the consent of a larger percentage of lessees than of lessors. Therefore, some lessees in the unit have nonconsenting royalty-interest owners and may be at risk under implied covenant law. Despite the potential for controversy in this situation, very little litigation has ensued. It appears that the lessee and the unit operator are careful to observe the dictates of implied covenant law. The Railroad Commission does not actively pursue the rights of the nonconsenters. The agency's protection of the unsigned lessor's interest extends only to forbidding the conversion of the last producing well on the lessor's tract to an injection well; to restrictions on the unit's transfer of lease allowables to wells close to the unsigned tract in order to protect it against drainage; and to restrictions on the unit's placement of injection wells.[114] Unless the unsigned lessor protests that these standard mechanisms do not adequately protect his royalty interest, the commission will routinely approve the unitization agreement. Few unsigned lessors ever contest the unitization application; the lessee and unit operator have privately accommodated their demands.

On the rare occasion that a dissident lessor protests the unit's formation, the commission will not deny authorization of the unit merely because the lessor alleges that the lessee is breaching an implied covenant obligation. The Railroad Commission will issue the order as long as it finds that the correlative rights of all the owners in the unit area are reasonably protected by the mechanisms listed above. The commission will deny any jurisdiction over the private implied covenant controversy between the lessor and his lessee. The unhappy lessor can then seek to invalidate the commission's order approving the unit by appeal to the courts, but if substantial evidence exists to support the agency order, the lessor will lose the suit. The odds are against the lessor in this forum.[115]

The unhappy lessor can also file suit in court against his lessee for breach of implied covenant. The burden of proof is on the lessor to show that a breach occurred and to prove the amount of damages therefrom. In Texas, damages will ordinarily be measured by the amount of royalty that the landowner would have received had additional wells been drilled on his tract to prevent drainage. This is often difficult to establish and will require that experts in geology and petroleum engineering be hired at significant expense to the lessor. The lessor who is being drained may not be able to prove damages because of these factors. The dissident lessor may argue in this instance that equitable relief, such as an injunction against the unit's operations, should be granted.[116] The Texas courts will probably deny such injunctive relief to the lessor; the nonconsenting lessor has then lost any veto power over the unit.

The case of *McLachlan v. Stroube* illustrates many of the points discussed here.[117] The case involved a cause of action for breach of contract between an overriding royalty-interest owner and the operator of an assigned oil and gas lease, but it is analogous to a suit between a lessor and a lessee for breach of an implied covenant. The McLachlans assigned a leasehold interest to the Stroubes in return for an overriding sliding-scale royalty. The Stroubes placed the tract in a unit, but the McLachlans refused to join the unit until the Stroubes agreed that the unitization would not adversely affect the McLachlans' contractual rights to a seven-sixteenth royalty under the sliding scale. The Stroubes would not agree. At the Railroad Commission hearing to approve the unitization agreement, the McLachlans testified that they were not willing to have any of the sixteen oil wells on their tract converted to water-injection wells. The unit operator at that time did not propose to convert any of these wells. The Railroad Commission approved the unit, but granted the McLachlans' request that they be afforded an opportunity to protest any future well conversions on their tract, if and when requested by the operator.

A year later, the unit operator applied to the Railroad Commission for authority to convert three of the McLachlan tract's wells to injection wells. The McLachlans vigorously protested the application. The Railroad Commission granted the conversion request, reciting that the agency had no jurisdiction over the private controversy between the McLachlans and the Stroubes.[118] The commission order also provided that the allowable production from the converted wells could not be transferred off the McLachlan tract until the McLachlans joined the unit. The McLachlans then brought suit in the district court of Travis County to set aside the commission's order. The district court granted a temporary injunction, restraining the conversion of the oil wells pending trial on the merits. However, at this trial a few weeks later, the court upheld the commission order, and conversion of the wells was allowed to proceed even without the McLachlans' consent.

At the same time that the suit attacking the Railroad Commission order was being litigated, the McLachlans sued the Stroubes in a district court in Scurry County for breach of contract, alleging that the royalty clause in the lease assignment required the payment of a seven-sixteenth, not a seven-thirty-second royalty. While this lawsuit was proceeding and after the McLachlans had lost the court battle against the commission order authorizing the well conversions, the McLachlans ratified the unitization agreement. The ratification expressly stated that it was without prejudice to the McLachlans' lawsuit contending that the royalty clause required payment of a seven-sixteenth royalty, but that the McLachlans could no longer controvert Stroube's right to have executed the unit agreement without the McLachlans' consent or the unit operator's right to convert the three wells on the tract to injection wells.[119] At the trial on the royalty clause, the jury found that the McLachlans had been forced by the Stroubes and the unit operator to commit their interest to the unit because their tract was going to be flooded out under the Railroad Commission's order. In the Stroubes' own words, "The McLachlans decided to accept the inevitable."[120]

Ultimately, the McLachlans won the lawsuit on the royalty issue and recovered a large monetary judgment. But the fact that the Travis County court upheld the commission's conversion order and refused to permanently enjoin the conversion of the wells suggests that a nonconsenting lessor who brings an action for breach of contract, express or implied, against a lessee who has placed the tract in a unit over the lessor's objections, cannot enjoin the unit operations on his tract but is limited to seeking damages for the breach. The secondary recovery operations will advance unabated. The lessor who cannot prove damages is left without a remedy. This pro-unitization stance by the courts seems a fair accommodation of the public interest in increasing recovery through unitization and the private rights of litigants under implied covenant law, although the courts should probably be careful to find that the unitization agreement which the lessors refused to sign was a fair one before the unit operator is allowed to benefit from the denial of injunctive relief to the nonconsenting lessor in his implied covenant action against his lessee, especially if the lessee is also the unit operator.[121]

Because the McLachlans ratified the unitization agreement, the question of the lessee's and the unit operator's liability for using the tract for secondary recovery without the consent of the royalty interest owner was not litigated. Had it been, it is likely that the conversion of the wells would have been allowed as long as it resulted in increased oil and gas production on the tract.[122] If, however, the tract suffered net drainage, the lessee would be at risk under implied covenant law,[123] and the unit operator would be at risk under tort principles of trespass. Thus, it is not surprising that the unit operator in the *McLachlan* case was anxious to secure a ratification that would give him the freedom to place injection wells optimally and also to transfer allowables freely throughout the unitized area.[124] Had enough lessors refused to join the unit, the unit might never have been formed. Judicial efforts to interpret implied covenant law so that unitized operations are not thwarted still do not substitute fully for compulsory unitization legislation.

Tort Liability of the Unit

Unitized operations cause large-scale migrations of salt water, oil, and gas in a reservoir, which often would not have occurred naturally. These deliberately induced migrations may cause injury to the tracts of nearby nonconsenting landowners and thereby give rise to the issue of tort liability. The issue of tort liability arises whether the injections are pursuant to a unitization plan or are conducted by a single operator. Because repressuring activities often require cooperative efforts, the ensuing discussion assumes that the offending intrusions are caused by a unit operator. The most common form of injury to the nonconsenter's tract is the invasion of salt water from water-flooding operations on adjacent property. This salt water may prematurely flood out producing wells on the nonconsenter's tract and displace valuable oil and gas from underneath the tract. Similarly, in cycling operations, injected dry gas from unitized operations elsewhere in the field may displace higher-valued wet gas. The judicial response to causes of action brought by nonjoiners seeking to impose liability for tortious conduct on the unit is an important determinant of the progress of voluntary unitization. If the unit operations can be enjoined or if damages—including punitive damages—can be recovered against the unit by nonconsenters on the ground that the unit's activity constitutes trespass or nuisance, the unit proponents may be dissuaded from water flooding or cycling without the consent of all or a very large majority of owners in the field.

The early common law in Texas suggested that trespass liability would exist for saltwater invasions. A 1927 case between two lessees on adjacent tracts held that damages could be recovered when the plaintiff's oil well was flooded with salt water following the defendant's negligent use of nitroglycerin to "shoot" his nearby well.[125] The court stated that the defendant had a duty to avoid physical invasion of the plaintiff's land and disturbance of the plaintiff's right of possession and enjoyment of his property. In a 1948 case involving damages to a landowner resulting from a well blowout on neighboring land, the Texas Supreme Court wrote that "[e]ach owner of land owns separately, distinctly and exclusively all the oil and gas under his land and is accorded the usual remedies against trespassers who appropriate the minerals or destroy their market value."[126] One usual remedy accorded against trespassers is an injunction.[127]

Again, in 1961, the Texas Supreme Court affirmed the granting of a temporary injunction against a defendant's proposed sandfracing operation on a small tract just 30 feet wide that adjoined the plaintiff's tracts. Delhi-Taylor, the plaintiff in *Delhi-Taylor Oil Corp. v. Holmes,*[128] alleged that the sandfracing would be a trespass because the great hydraulic pressure used to fracture the producing sands would cause subsurface cracks to extend beyond the boundaries of the defendant's tract into Delhi-Taylor's mineral estate. These cracks were analogous to a pipeline leading to the well bore and would allow the defendant to drain huge quantities of gas from the plaintiff. Delhi-Taylor contended that trespass liability for these cracks should exist notwithstanding the rule of capture, which imposes no liability for drainage in the normal situation where the adjacent lessee has not caused a physical invasion of another's tract. The trial court had granted the temporary injunction to preserve the status quo pending a trial on the merits, but this opinion was reversed by the court of civil appeals on the basis that primary jurisdiction to decide this issue lay with the Railroad Commission, not with the court.

In a companion case, *Gregg v. Delhi-Taylor,*[129] the Texas Supreme Court held that the courts had the power to determine whether sandfracing would result in a subsurface trespass. None of the conservation statutes expressly delegated this issue to the Railroad Commission, and the commission itself asserted no power to resolve the dispute. Indeed, the commissioners refused to be drawn into the controversy, even though the defendant had requested that they issue rules on the subject.[130] The supreme court ruled that the direct, intentional invasion of the plaintiff's subsurface estate by sandfracing raised the issue of trespass and that courts had the power to enjoin such conduct.[131]

The probability that courts would find sandfracing to constitute a trespass spelled doom for many small-tract producers. Without fracturing, some of their wells would be unprofitable to produce and could not be transformed into prolific wells. The technology of sandfracing was not advanced enough that the small-tract owner could control the extent and direction of the fractures. Therefore, the small-tract operator would have fewer economically attractive projects to buy or sell. The operators on large tracts suffered no such constraints on their sandfracing activities. Thus, the *Delhi-Taylor* cases pushed small-tract producers to support passage of a compulsory pooling bill. Unless the owner of a small tract that required sandfracing to produce gas profitably could pool his acreage with the adjoining land, his tract would be worthless for development purposes because fracturing might be enjoined as a trespass. The *Delhi-Taylor* cases were decided at the same time that the supreme court invalidated allowable formulas with large per-well factors in the *Normanna* decision.[132] Small producers had two reasons to press for passage of compulsory pooling legislation from 1961 onward.

The defendant in *Gregg v. Delhi-Taylor Oil Corp.* also contended that if the Railroad Commission authorized the sandfracing, it would not be a trespass and would not be subject to injunction.[133] Because the Railroad Commission had not authorized fracturing, the supreme court held that it was not necessary to decide this issue at the time. An *amicus curiae* brief advised the court that the commission had promulgated rules authorizing secondary recovery operations such as water flooding and gas recycling, which caused subsurface invasions of the land of others. The court decided that "[t]he validity and reasonableness of the rules and orders involved in those [secondary] operations may be passed upon when and if they reach this court."[134]

Only a year later, the reasonableness of a Railroad Commission order authorizing the placement of a saltwater injection well used for water flooding at a location only 206 feet from another tract reached the court. In *Railroad Commission v. Manziel,*[135] the Manziels sued to cancel a Railroad Commission order that permitted the Whelan brothers to drill and inject water at this irregular location. A regular location would have placed the injection well 660 feet away from the Manziels' tract. The Whelans had unitized all of the properties under lease to them and were water flooding the entire acreage. The Manziels were the only other operator in the field and had partially unitized and water flooded their leased acreage. The effect of the water-flooding projects was to push oil from the Whelans' tract toward the Manziels' acreage. The irregularly located injection well had been granted to protect the Whelans' correlative rights by minimizing the amount of oil which would cross boundary lines. However, salt water from this well would cause one of the Manziels' wells to water out in three and one-half to eight months instead of in thirty-two months. The court held that the Manziels' pleadings gave rise to the issue of trespass, and then stated that:

> Secondary recovery operations are carried on to increase the ultimate recovery of oil and gas, and it is established that pressure maintenance projects will result in more recovery than was obtained by primary methods. It cannot be disputed that such operations should be encouraged. . . . It is obvious that secondary recovery programs could not and would not be conducted if any adjoining operator could stop the project on the ground of subsurface trespass. As is pointed out by amicus curiae, if the Manziels' theory of subsurface trespass be accepted, the injection of salt water in the East Texas field has caused subsurface trespasses of the greatest magnitude.
>
> . . . If the intrusions of salt water are to be regarded as trespassory in character, then under common notions of surface invasions, the justifying public policy considerations behind secondary recovery operations could not be reached in considering the validity and reasonableness of such operations. . . . Certainly, it is relevant to consider and weigh the interests of

> society and the oil and gas industry as a whole against the interests of the individual operator who is damaged. . . .
>
> We conclude that if, in the valid exercise of its authority to prevent waste, protect correlative rights, or in the exercise of other powers within its jurisdiction, the Commission authorizes secondary recovery projects, a trespass does not occur when the injected, secondary recovery forces move across lease lines, and the operations are not subject to an injunction on that basis. The technical rules of trespass have no place in the consideration of the validity of the orders of the Commission.[136]

The language in the *Manziel* case prompted one commentator to declare that Railroad Commission approval of a secondary recovery or pressure maintenance project immunized the operators from tort liability as long as they complied with the commission orders and operated nonnegligently.[137] However, such a conclusion does not necessarily follow from the court's holding in *Manziel*. The "technical rules of trespass" might still be considered in a private cause of action brought by one landowner against another for damages rather than in a lawsuit for an injunction against a Railroad Commission order. Thus, had the Manziels sued the Whelans directly for trespass damages to their oil and gas interests, tort liability might still be found. Indeed the supreme court specifically left open this possibility by stating that:

> [I]n the case at bar we are not confronted with the tort aspects of such practices. Neither is the question raised as to whether the Commission's authorization of such operations throws a protective cloak around the injecting operator who might otherwise be subjected to the risks of liability for actual damages to the adjoining property.[138]

Both the commission and the Texas courts have long held that the agency has no authority to determine issues of tort liability such as trespass and conversion,[139] or to adjudicate private disputes over the interpretation of a contract or title disputes over land,[140] or antitrust suits brought by private plaintiffs.[141] Simply put, the commission has no jurisdiction to license a trespass, and the *Manziel* case does nothing to disturb this principle. Thus, even though a unit operator's injection activities are authorized by a Railroad

Commission order approving the unit under the 1949 Act, the risk of tort liability to nonsigners still exists. Despite the fact that the commission order approving the unit under the 1949 Act recites a commission finding that the rights of all owners of interests in the field—whether signers or not—will be protected under the unitized operation, a party injured by the unit's acts may nonetheless proceed to sue the unit and recover trespass damages if proved. This administrative finding will not bar a private cause of action in trespass under the principles of *res judicata* or collateral estoppel, nor will this finding be conclusive evidence of lack of injury.

That tort liability to nonconsenters can still exist despite agency approval of the unit is vividly demonstrated by a sequence of cases arising in Kansas. Because these Kansas cases reflect many established principles of Texas law, the cases are apt to be used as solid precedents in the event that a lawsuit alleging a private trespass cause of action is brought against a unit operator in Texas. Thus these Kansas cases bear close attention.

In *Jackson v. State Corporation Commission,*[142] the Kansas Supreme Court upheld a commission order which allowed Tidewater to drill injection wells only 12 feet from the Jackson brothers' tract. The injected salt water flooded eight of the Jacksons' wells within a matter of hours and days. The Jacksons complained to the commission, which entered a cease and desist order against the water flooding pending a full hearing before the commission. After the hearing, the commission concluded that no waste was being committed and that the placement of injection wells close to lease boundaries was the only way to prevent the movement of oil from lease to lease and thereby protect correlative rights. But the commission was careful to point out that its hearing was limited to considering evidence involving its statutory authority to prevent waste and protect correlative rights and that "an action for damages for the flooding of producing wells may be available to the owners of such wells, but the forum for such a damage action is in the courts, not this commission."[143] The Kansas Supreme Court affirmed the commission's decision to allow the water flooding to proceed on the basis that substantial evidence existed to uphold the commission order.

The Jackson brothers had also brought a second lawsuit in federal court directly against Tidewater for damages to their oil and gas operations.[144] The trial court awarded actual damages to the Jacksons of $620,700 in lost profits and punitive damages of $25,000 for Tidewater's reckless and wanton disregard of the plaintiffs' property rights. The Tenth Circuit Court of Appeals affirmed the judgment against Tidewater, holding that "though a water flood project in Kansas be carried on under color of public law, as a legalized nuisance or trespass, the water flooder may not conduct operations in a manner to cause substantial injury to the property of a non-assenting lessee-producer in the common reservoir, without incurring the risk of liability therefore."[145] Tidewater's primary defense to this second lawsuit was the principle of *res judicata* or collateral estoppel: that the federal court was bound by the Kansas commission's determination, made after a full hearing and affirmed by the Kansas Supreme Court, that Tidewater's secondary recovery project prevented waste and protected correlative rights. The Tenth Circuit judge conceded that the doctrine of collateral estoppel could apply to determinations made by administrative agencies just as fully as it applied to determinations made by the courts. However, whether the agency determination of a critical issue was to have a conclusive effect in subsequent judicial proceedings depended on whether the agency had the authority to determine the issue. Because the Kansas commission had no jurisdiction to determine Tidewater's tort liability, collateral estoppel did not bar the Jacksons' lawsuit nor preclude Jacksons' ultimate success in proving that their rights were not protected by the water flooding. The appellate court noted that the trial court's findings that Tidewater had deliberately injected water at unreasonable and excessive rates were "irreconcilably inconsistent" with the conclusive findings of the Kansas commission that the water flooding was lawful,[146] but held that it was unnecessary to reconcile the two conclusions in this separate and different cause of action. A majority of judges on the Tenth Circuit did, however, reverse the lower court's award of punitive damages because Tidewater's acts were committed under color of law and in accordance with the state's interest in conservation. A dissenting judge would not have allowed the commission's administrative order even to shield against punitive damages.

The Jacksons' ultimate victory in the Kansas courts is an important lesson for operators conducting water-flooding operations in Texas in partially unitized reservoirs. As we have seen, the Railroad Commission, like the Kansas Corporation Commission, claims no authority and has no authority to resolve disputes over private causes of action. Furthermore, the Texas courts have absolutely denied that *res judicata* applies to make an agency finding conclusive in a subsequent judicial proceeding.[147] Thus, Texas law supports the holding of liability in *Tidewater Oil Co. v. Jackson* even more than does Kansas law. The Railroad Commission's finding that the rights of all owners of interests in the field are protected by the unitization plan approved by the commission will not, in and of itself, shield an operator from liability for actual damages to an adjacent tract.

The only way in which the Texas courts might part from the Kansas law applied in the *Jackson* case is in the choice of a standard of liability to be imposed on the unit operator. In the *Jackson* case, the federal court found that Kansas law would impose a duty of strict liability on Tidewater.[148] Therefore, the reasonableness of the defendant's acts or his lack of negligence were irrelevant to the judgment.[149] It is doubtful that the Texas courts would impose a standard of strict liability on secondary recovery activities. The *Manziel* opinion clearly portends that the court will consider the reasonableness of the conduct of the unit operator in assessing tort liability because of the great public interest in secondary recovery.[150] The established rules of trespass do not require more than a finding that the defendant has intentionally invaded the property owned by another.[151] Invasions of salt water from adjacent, secondary recovery operations are usually intentional and would thus constitute a trespass, against which injunctions would freely issue. The Texas court in *Manziel* disapproved of these "technical" results.

In the context of commission-approved secondary recovery operations, the court seems anxious to replace traditional trespass law with more flexible rules such as those used to resolve disputes over the tort of nuisance. Liability for a private nuisance requires a finding of substantial and unreasonable interference with another's use and enjoyment of his property. The nuisance approach gives the court considerable leeway in assessing whether the defendant's conduct is unreasonable and therefore actionable.[152] Moreover, under nuisance principles, even if the invasions of salt water are found to be unreasonable because of the substantial harm caused to another's property, it is most unlikely that an injunction would issue against the unit operator who is proceeding under a commission order that recites the additional recovery of oil and gas expected under the unitization plan.[153] The agency approval would weigh as evidence of the reasonableness and social utility of the defendant's conduct.[154] Such evidence would probably be highly persuasive to many judges and juries.[155] Also, the Texas court would undoubtedly follow the *Tidewater Oil Co. v. Jackson* precedent and find that the commission order approving the unit's operations bars recovery of punitive damages by carrying a presumption that the authorized activity is reasonable.

Nonetheless, despite all of these pro-unitization effects which flow from the *Manziel* opinion, it still appears that an injured holdout in Texas can overcome the presumption of reasonableness that attaches to the commission order and recover damages upon proof of the unit's intentional and unreasonable disregard of his property rights. The commission order is not an impenetrable shield to tort liability.[156]

On the other hand, it is at least arguable that some of the language in the *Manziel* case stands for the broader principle that secondary recovery operations, because of their public benefits in increasing oil and gas recovery, are immune from all trespass liability, whether for damages or an injunction and whether in an action attacking the validity of a commission order or in a lawsuit between private plaintiffs. The court in *Manziel* approvingly quoted the following from the foremost treatise in oil and gas law:

> "What may be called a 'negative rule of capture' appears to be developing. Just as under the rule of capture a land owner may capture such oil or gas as will migrate from adjoining premises to a well bottomed on his own land, so also may he inject into a formation substances which may migrate through the structure to the land of others, even if it thus results in the displacement under such land of more valuable with less valuable substances."[157]

A negative rule of capture would encourage secondary recovery and pressure maintenance operations even more than the narrower reading of *Manziel,* which would shield the unit only from injunctions and possibly from strict liability for damages. Surely such a rule of nonliability for injury to holdouts would also require that all owners in the field are offered a fair and reasonable opportunity to join the unit, and that the unit operations are conducted nonnegligently. With these conditions appended, the adoption of a "negative rule of capture" as the governing common law would encourage unitization and also arguably protect correlative rights.[158] Fewer owners in a field would refuse a fair offer to unitize if, as a consequence of their refusal, oil and gas could be displaced from their tract without liability. A state adopting such a common law rule would have less need for a compulsory unitization statute.[159] This very fact is probably why the prevailing weight of judicial opinion is that tort liability for damages does exist.[160] Freedom from such liability so transforms the common law that it must be legislatively procured.[161] Even then, such legislation is sure to face court challenge under the takings clauses of both the state and federal constitutions.[162]

Still, even the narrower holding in *Manziel* that pressure maintenance operations pursuant to a Railroad Commission order cannot be enjoined as a trespass greatly facilitates unitized operations. Because of the public value of secondary recovery operations, the law allows the unit to condemn the nonconsenter's property, pay a fair price for it, and continue to operate. The unit has a private right of eminent domain as long as it has secured official authorization from the commission that it is operating in the public interest of increasing ultimate recovery. This official authorization also bars any punitive damages against the unit and probably precludes the imposition of a standard of strict liability for damages in tort actions. It is small wonder, then, that unit operators inevitably seek Railroad Commission approval. The legislated antitrust immunity thereby secured may be of less importance than the benefits bequeathed by the courts in suits brought by nonconsenters alleging tort or breach of implied covenant liability.

The court's measure of damages in trespass cases also encourages owners to join the unit and discourages holdouts from suing a unit operator. Damages will require proof of net drainage.[163] If outmigration of oil or gas from the nonconsenter's tract is matched by in-migration, no trepass and no damages will result. Likewise, had the Manziels sued the Whelans directly for trespass damages stemming from their prematurely flooded well, the Whelans would have undoubledly countersued for the same type of damages to one of their producing wells, which was being watered out by a nearby injection well on the Manziels' tract.[164] Few holdouts may be able to prove net drainage or net damages when the certainty of countersuits against them is factored into the decision to sue for trespass.

Even if net damages can be shown, the courts have been notably reluctant to award damages which would encourage tract owners to hold out. The usual measure of damages for trespass or nuisance is the difference in the fair market value of the property before and after the tort occurred. This value can include the economic rent obtained from structural or regulatory advantage, but it will not include more than this. Punitive damages will probably not be granted against a commission-approved unit, nor will damages for a willful trespass which, if allowed, would permit the nonconsenter to recover the value of the oil and gas drained away without deduction for the reasonable costs of producing it. Most voluntary unitization agreements recognize and credit regulatory or structural advantages (see chapter 6 text at notes 50 to 56), so nonconsenters will not gain by seeking trespass damages versus joining a unit.

Several cases from other jurisdictions illustrate these points. In *Baumgartner v. Gulf Oil Corp.*,[165] the Nebraska Supreme Court refused to allow a nonconsenting lessee, who sued a unit for sweeping oil off his tract, to recover any damages greater than the profits which he could have secured had he operated alone, without the assistance of the unit's surrounding repressuring activities. Few lessees on small tracts will ever be able to prove that it is economic to repressure their tract if the nearby tracts do not repressure. The small-tract owner's solo repressuring performance will most often push oil off his own tract onto his neighbors' land. The Mississippi Supreme Court in *California Co. v. Britt*[166] refused to allow a nonconsenting mineral-interest owner to recover any damages whatsoever after he refused to accept a fair opportunity to join for many years while he was under lease to a producer who had joined the unit. The Texas courts have not had an occasion to rule on the proper measure of damages to be awarded for a trespass or tort caused by secondary recovery operations, but the Texas Supreme Court's recitation in the *Manziel* case of the strong public interest in secondary recovery foreshadows the same judicial antipathy to holdouts as appears in the law of other jurisdictions. Indeed, the Texas Supreme Court used the *Britt* case as precedent to refuse any relief to the plaintiff in *Superior Oil Co. v. Roberts*[167] whose land was drained by the surrounding unit's operations. The *Roberts* case was not brought as a tort case against the unit operator, but as a cause of action by one cotenant against his fellow cotenant seeking an accounting for a fair share of the jointly owned tract's unit proceeds. Therefore, it is not a firm precedent for the latter situation. But the Texas court's ready willingness to use the Mississippi case should cause some anxiety among holdouts.

The Nebraska case of *Baumgartner v. Gulf Oil Corp.* seems firmly rooted in the same principles that govern Texas' existing common law, and it is quite likely that the Texas courts would adopt this case as precedent for measuring damages. In this case, the lessee of a small tract refused to join a unit unless he was exempted from paying any costs whatsoever. The unit declined to so favor him, and thus excluded his tract from the unit, there being no compulsory process available in Nebraska at that time to force the plaintiff to join on reasonable terms. The plaintiff sought a well permit to drill a well on his tract, but the commission delayed granting this permit for such a long time that his tract was swept dry by the unit at the time he finally secured his permit.[168] The lower court awarded the plaintiff damages of $90,000 for a willful trespass. This recovery was measured by the value of the oil and gas drained away by the unit without deduction for the costs that the plaintiff would have incurred to produce it. On appeal, the Nebraska Supreme Court refused to find a trespass because the unit was authorized by the Nebraska commission under a statute passed to encourage voluntary unitization, and because secondary recovery was so crucial to the state's oil industry and economy. The court quoted extensively from the *Manziel* case in its decision to disregard trespass law.[169] However, because Nebraska's conservation statutes required the protection of correlative rights, the court held that the plaintiff could recover damages measured by what he would have obtained through his own efforts on his property as if no unitized operations existed next door.[170] The evidence in this case showed that the plaintiff could have made a profit of $27,000 if he had joined the unit. If he had drilled his own well, he would have secured a maximum profit of $12,000. Even this $12,000 was not guaranteed to the plaintiff because it depended on the unit's conducting secondary recovery simultaneously with the plaintiff. The evidence seemed to show that the plaintiff would have sustained nothing but losses if he had attempted to do his own secondary recovery without the nearby unit doing the same. The case was remanded for a determination of damages. Because few small-tract owners can conduct profitable operations alone—especially by sweep methods which require fieldwide migrations—few owners will find it beneficial to hold out.[171]

Additional uncertainty in the law defining correlative rights in Texas may further dissuade persons from refusing to join a unit and also dissuade nonconsenters from litigating. The Texas courts have often defined a landowner's correlative right as the opportunity to recover the amount of oil and gas in place underneath his tract.[172] To further the public interest in secondary recovery operations as expounded in the *Manziel* case, the Texas courts might restrict a nonconsenter to precisely this amount which does not recognize a regulatory or structural advantage. Indeed, in two cases involving judicial review of the reasonableness of commission orders, the Texas Supreme Court refused to recognize that the plaintiffs (both of whom were holdouts to a unitization agreement) had a correlative right to a structural advantage.[173] The commission order for the SACROC unit spoke of the correlative rights of each of the protesting, nonjoined owners in terms of "a fair opportunity to produce the recoverable oil and gas underlying his properties."[174] While it seems more likely that the measure of the nonconsenters' correlative rights under the common law will include the value of a regulatory or structural advantage (especially for primary production) as in the *Baumgartner* case, this additional uncertainty in the measure of damages to holdouts in Texas may encourage owners to join a unit.

These judicial limits on a nonjoiner's damages are still not a complete substitute for a compulsory unitization statute, however. It is quite possible for a landowner's tract to be so favored by a regulatory or structural advantage that joining a proposed unit is economically unattractive, particularly in the early life of the field when both primary production and the rule of capture are in full force. Thus, even though Texas jurists can be expected to follow their Nebraska brethren in a case involving a unit's liability to an injured holdout, the common law still protects the right of these favored owners to secure the profit or royalties that they could have made by "going it alone."[175] In fields with prorationing formulas based on large per-well factors, these profits will lure owners away from any cooperative venture that does not credit fully this regulatory advantage in the participation formula for primary production. Moreover, lessees who are able to obtain their drilling permits without the delay that was imposed on plaintiff Baumgartner will seldom have cause to bring suits in tort or suits to protect their correlative rights. They will drill their wells and sweep up a disproportionate share of the reservoir's oil and gas, often benefitting from the unit's nearby repressuring operation without paying for it.[176]

The Nebraska legislature passed a compulsory unitization statute during the litigation involved in the *Baumgartner* case. The political power of those favored by Texas' old-style prorationing formulas has consistently defeated this legislation in Texas. As might be expected, then, very little voluntary unitization occurs in Texas early in the life of a field when primary production exists (see chapter 9 text at notes 17 to 23).

The fact that approval of the voluntary unitization agreement probably does not shield the unit from tort liability also explains why unit proponents are careful to protect the rights of nonconsenters. As we have seen, few nonconsenters ever appear at the Railroad Commission hearings to protest the unit's formation (see chapter 6 text at notes 57 to 64). If they appear at all, it is to state that they do not oppose the unit's formation. Obviously, the nonconsenters believe that their property rights are protected. The commission is not particularly vigilant about considering correlative rights issues. The lack of protest from nonconsenters stems from the unit proponents' own vigilance in these matters. The nonjoiner in Texas has three forums in which to argue unfair treatment by unit operators: (1) the commission hearing, (2) judicial review of the reasonableness of the commission's subsequent orders in protecting correlative rights, and (3) court litigation directly against the unit or the joined lessee for tort or implied covenant liability. The unit's advocates prefer to settle with nonconsenters privately rather than risk lengthy and costly litigation in any or all of these forums.

The obvious judicial antipathy to holdouts evidenced by the *Roberts* and *Manziel* cases in Texas and by cases from other jurisdictions provides some motivation to recalcitrant owners to settle or join the unit rather than risk reliance on the common law to protect their correlative rights. Still, the nonconsenters can generally secure much of the economic rent due to their privileged position of regulatory advantage.[177] This fact strongly suggests that most operators in Texas expect that tort liability to nonconsenters exists and that the nonconsenter's correlative rights are measured by what the nonconsenter would receive by drilling and producing on his own under the existing field rules. As we have seen, this expectation is well founded.

If a large number of owners refuse to join a proposed unit, the unitization allocation formula will often be renegotiated to attract the holdouts. If only a few owners refuse to join, restrictions on the unit's activities will be privately negotiated with the nonconsenters who then become silent. Because unitization can increase oil and gas production so much, the unit can pay these economic rents to the nonconsenters and still make a profit. Of course, if enough owners in the field refuse to unitize, their considerable power to restrict well location, to prohibit or limit transfers of allowables, and to subject the unit to possible tort liability may defeat the unit's formation. This is another reason that case law regarding the tort liability of a unit is scarce: the unit does not exist.

Liability to the Surface Estate

The final body of common law that significantly affects the progress of unitization is the unit's liability for injury to the surface estate. A unit operator is no less liable than an individual producer for injury to the surface estate of an adjoining landowner caused by saltwater pollution, oil spills, explosions, and the like. Liability exists on theories of nuisance, negligence, breach of a duty imposed by statute or regulation, or strict liability.[178] As we have seen, common law liability for polluting surface lands and streams led producers in the East Texas field to seek a cooperative saltwater disposal program for the large amounts of brine produced in this field (see chapter 5 text at notes 79 to 83).

Of equal significance to the course of unitization in Texas is the common law regarding the dominance of the mineral estate owner's use of the overlying surface estate. When ownership of the minerals and the surface is combined in the same landowner, the lessee's use of the surface will be guided by both the express terms of the lease and by implied easements to use the surface. In Texas, as in most jurisdictions, the mineral estate is dominant and can use as much of the surface as is reasonably necessary to conduct oil and gas operations.[179] Liability for excessive or negligent use of the surface still exists. If the lease contains a pooling or unitization provision, the lessee may use the surface to conduct operations on the leased premises and on lands pooled or unitized with it.[180] However, if no pooling or unitization authority exists in the lease, the surface can be used only to conduct operations which benefit the leased premises as a separate unit.

The dominance of the mineral estate greatly assists oil and gas operations generally and secondary recovery operations in particular. Secondary recovery often requires more use of the surface estate than does primary recovery of oil. Pipelines for both the repressuring substance and the oil and gas produced, pumping stations, closely drilled injection and output wells, and connecting roads will take up significant amounts of surface space. Secondary recovery operations often will require the use of vast quantities of the surface owner's water resources.[181]

Quite often, the mineral estate is severed from the surface estate, and different owners have title to these separate property interests. The oil and gas operator will take a lease from the mineral estate owner, without ever contacting the surface owner. The triangular relationship between the surface owner, the lessor of the mineral estate, and the oil and gas lessee is a fertile ground for conflict respecting the dominance of the mineral estate, especially when secondary recovery operations are instigated.

This conflict is well illustrated by the case of *Sun Oil Co. v. Whitaker.*[182] Gann originally owned both the surface and the mineral estates. In 1946 he leased the minerals to Sun Oil, and in 1948 he conveyed the surface to Whitaker, subject to Sun's lease. Whitaker engaged in irrigated farming, which required the use of fresh groundwater pumped up from the Ogallala aquifer. After several years of diminishing primary recovery, Sun Oil started to water flood the tract and used 100,000 gallons a day of this scarce groundwater to repressure the reservoir. These large withdrawals of groundwater would decrease the expected life of Whitaker's farm by at least eight years. Sun Oil sought a permanent injunction enjoining Whitaker from interfering with its production of fresh water, and Whitaker filed a cross action seeking to enjoin Sun Oil from using any fresh water for water flooding. The lower courts granted judgment for Whitaker, and the Texas Supreme Court, in its first opinion, upheld this judgment on the basis that Sun Oil's withdrawals constituted an excessive and unreasonable use of the surface estate. Sun Oil had a reasonable alternative of purchasing water off the tract at a cost of $42,000 compared with the $3.2 million value of additional oil expected to be produced from the tract by the water flood. However, on motion for rehearing, the supreme court reversed its position, and in a five-to-four opinion, granted judgment for Sun Oil. Whitaker had stipulated that the water flood was a reasonable and proper method of producing oil from the tract and was being properly conducted. The majority of justices wrote:

> We have concluded that there is no evidence to support the jury's finding that it is not "reasonably necessary" for Sun to use the water underlying the Whitaker farm for its waterflood project. As pointed out above, efforts to use available saltwater for the waterflood project have failed, and there is no other source of usable water on the leased Whitaker tract which is available to Sun. To hold that Sun can be required to purchase water from other sources or owners of other tracts in the area, would be in derogation of the dominant estate.[183]

The force of the *Sun Oil v. Whitaker* decision is that secondary recovery operations in Texas can proceed without having to pay for significant external costs imposed on the surface estate owner. The private value of secondary recovery oil does not reflect its social costs. The common law encourages secondary recovery, but this may not redound to the benefit of society at large.[184] The landowner who owns both the surface and the minerals can protect against this common law result by inserting express provisions in the lease which prohibit the lessee from using fresh water for secondary recovery activity. However, the owner of a severed surface estate has no such option and is at the mercy of the mineral estate owner's lease.[185] For many years, the Railroad Commission disavowed any power to consider whether the use of fresh water was reasonable when it approved unitization agreements under the 1949 Act. In 1983 the Texas legislature finally authorized the commission to require operators to use substances other than fresh water for injection purposes, if other substances are economically and technically feasible (see chapter 4 text at notes 67 to 68).

The *Whitaker* case did not involve the question of the lessee's right to use the surface estate of one tract for operations on other premises. The water withdrawn by Sun Oil was used exclusively to benefit the mineral estate underlying Whitaker's tract. Sun Oil's right to pump water from this tract and inject it into wells on other tracts in order to increase their recovery of oil and gas would depend on whether the lease to which Whitaker was subject authorized pooling or unitization. If it did, Sun Oil would be able to use the surface of Whitaker's tract to benefit the unitized area.

Thus, in *Miller v. Crown Central Petroleum Corp.*,[186] the Millers acquired title to the surface of a tract subject to a prior oil and gas lease that authorized the lessee to pool the leased acreage with other lands in the immediate vicinity. Subsequently, all the owners of the mineral estate—both working and royalty interests—joined in a water-flooding program. The unit operator piped salt water across the Millers' land to inject into wells on adjoining tracts. The Millers sought to enjoin the transportation of this water on their surface and also sought damages. The court rendered judgment against the Millers because the mineral estate was dominant, the water flooding was a reasonable and necessary method of increasing production from the mineral estate underlying the Millers' tract, and the lease authorized pooling.

On the other hand, if the oil and gas lease does not authorize unitization, the owners of the mineral estate cannot use the severed surface estate to enhance oil and gas recovery on other tracts. Thus, in *Robinson v. Robbins Petroleum Corp.*,[187] the owner of the surface estate, Robinson, was allowed to recover damages from the unit operator who was using a well on Robinson's tract to produce salt water, which was then injected into wells on other tracts. Robinson's deed to the surface was subject to the prior oil and gas lease, but nothing in this lease or in Robinson's deed authorized the mineral estate owner to increase the burden on the surface estate for the benefit of other lands. Nor could the Railroad Commission order approving the secondary recovery alter Robinson's title or rights in the surface estate.[188] The unit operator had the free use of salt water from Robinson's tract to produce additional oil and gas from this tract, but the operator would have to pay Robinson for the salt water used to benefit other tracts in the unitized area.

The *Robinson* case provides symmetry to the *Whitaker* holding: the unit operator must pay the surface owner for water obtained from the tract and used elsewhere, but the surface owner cannot force the operator to purchase alternative water from sources off the leased premises for use on the mineral estate underlying the surface owner's tract. The common law thus draws clear lines around the surface owner's separate tract. This result is hostile to unitization, which progresses most effectively only if such lines are erased.[189] Few oil and gas leases contain unitization provisions, although some pooling clauses may be broad enough to allow the use of the surface of one tract for unitized operations on other tracts. The result of the common law rule is that unit operators will sometimes have to pay the severed surface-estate owner for the right to use the surface for unitized operations. If no reasonable amount of money will satisfy the surface owner, the unit proponents may be forced to place injection wells and other facilities at suboptimal locations especially if the surface owner can enjoin their use on his land. The courts may well deny injunctive relief to the surface owner in this situation for the same public policy reasons discussed in the preceding section on tort liability. The dearth of case law in this area suggests that most surface owners' demands for compensation are met through private negotiation.

In one other situation involving surface rights, the common law has encouraged small-tract drilling and thus discouraged pooling and unitization. In *Atlantic Refining Co. v. Bright & Schiff,*[190] Bright & Schiff had leased such a small tract that they did not have sufficient surface acreage for their pits, pumps, and drilling equipment. Bright & Schiff obtained a surface lease from the adjoining landowner and sought to use this surface in its drilling operation. Atlantic had leased the minerals underlying this adjoining surface estate and had pooled this lease with several others into a larger unit. Atlantic moved onto the surface estate leased by Bright & Schiff and started erecting a warehouse to prevent them from using this same surface to drill their well. Needless to say, the prorationing formula in this gas field had a large per-well allocation factor and Bright & Schiff's well would drain large quantities of gas from Atlantic's tract.[191] Both lessees sought injunctions against the use of this surface acreage by the other. Atlantic argued that its mineral lease granted it certain surface rights, including an implied covenant that prohibited its own lessor from using the surface in a manner which would diminish the value of Atlantic's mineral estate. The court enjoined Atlantic from using the surface for the sole purpose of preventing Bright & Schiff's drilling operations. Atlantic had not made a satisfactory showing to the court that its use of the surface estate for a warehouse was reasonably necessary to the use and enjoyment of its own mineral estate. This holding has been criticized because it allows lessors to decrease the value of their own lessee's mineral estate by selling surface easements to rival operators.[192] Lessees have a duty to maximize oil and gas recovery to the benefit of the lessor, and lessors should have a reciprocal duty not to interfere with their lessees' efforts to do so. When the surface estate is severed, there is an even greater incentive on the part of the surface owner to grant easements to competitors, because the surface owner has no interest in maximizing the recovery of the oil and gas under his land. The common law of the *Bright & Schiff* case is hostile to pooling and unitization because it facilitates small-tract drilling by holdouts on acreage adjacent to a pooled unit. Once the holdout's well is drilled, it can drain the unit without liability under the rule of capture.

Summary of the Effect of the Common Law on the Progress of Pooling and Unitization in Texas

The common law principles of the rule of capture, of implied covenants, and of tort are all inherently hostile to unitization. The rule of capture and the doctrine of implied covenants cause unnecessary well drilling and promote the rapid production of oil and gas—exactly opposite to the aims of unitization. Because tort liability exists for injury to adjacent, unsigned mineral interests, the unit that cannot secure unanimous voluntary agreement must place its injection and producing wells in suboptimal positions and produce them at suboptimal rates or else risk having to pay damages to the injured estate. The unit must also suffer drainage to the holdout's tract under the rule of capture. Even if the 1949 Act did not require the commission to protect the rights of all the nonsigners in the field before approving a unit, the common law provides protection to these rights. This common law can significantly decrease the efficiency of a unit's operations or thwart the unit altogether.

The three common law doctrines of the rule of capture, implied covenants, and tort are universal in oil and gas jurisprudence. The Texas courts have added certain other doctrines that make Texas' jurisprudence even more inimical to unitization and pooling. The worst offender in this regard is the rule that the owner of the right to lease (the executive right) does not have the right to pool the owners of nonexecutive interests in the land. Because of this rule, the proponents of unitization must negotiate with a much larger number of owners in the field. If some nonexecutive owners refuse to join, the unit's risk of tort liability increases, and the optimal placement of wells and transfer allowables is further hindered. The economics of drilling on a tract with an unsigned interest makes unitization difficult to effect. Under the rule of capture, unsigned owners must be paid their full share of production from any well on their tract, undiluted by the tract's participation factor in the unit agreement. The common law invites royalty interest owners to wait and see how the field is developed before deciding whether to ratify a unitization agreement or to remain outside the unit.

The rule of nonapportionment does not impair unitization efforts as much as the rule against pooling nonexecutives, but it does have the effect of creating injustice to the owner of a small, subdivided tract that does not have a well and often cannot receive one under the spacing rules. The supremacy of the rule of capture over any principles of equitable or implied pooling creates an environment in which the Railroad Commission will want to grant Rule 37 exception wells freely rather than allow the injustice to rankle their collective conscience. While litigation may invalidate some of the exception well permits illegally granted by the commission, and thus prevent the drilling of wells which are unnecessary for conservation, the injustice and inequity will continue, absent compulsory pooling or unitization laws.

A third inimical feature of Texas common law is the holding that allows surface owners of unitized tracts to grant surface easements for drilling operations to adjacent small-tract owners who have refused to unitize because it is more profitable to drill on their own and drain from the unit.

Texas' cross-conveyancing theory is often cited as another common law rule which is offensive to unitization because it creates such grave joinder problems that owners of interests in a field are reluctant to join a unit which would preclude them from litigating issues that might vitally affect their interests. However the root of the cross-conveyancing theory lies, not in Texas' common law, but in rules of procedure which are necessary to assure fairness and due process to all owners with an interest in a commonly owned source of supply. Moreover, once a unit is formed, the cross-conveyancing theory can work in the unit's favor by diminishing the likelihood that anyone will sue the unit over an issue that could require the joinder of hundreds of other persons.

Except for the surface-easement case, these anti-unitization graftings onto Texas common law appear to have resulted unintentionally rather than because of any judicially perceived public interest opposed to unitization or pooling. The nonapportionment rule was born in 1925 before pooling and unitization were established concepts in scientific, industry, or legal circles. The cross-conveyancing theory derives from rules of procedural fairness, not from judicial antipathy to unitization. The seminal case establishing the rule that executives do not have the power to pool nonexecutive interests did not arise in the context of an executive attempting to create and enforce a unitization agreement against the desire of a nonexecutive holdout. The issue in the case was whether two landowners could force a lessee in a suit for specific performance to take a community lease that was not signed by a nonparticipating royalty-interest owner. It is not surprising that the case occasioned no great comment when it first was decided. Few observers foresaw the holding's grave impact on lessees' pooling and unitization efforts. Similarly, the many Texas cases that strictly construe pooling clauses in leases—especially those which wreaked havoc in the efforts to unitize the Fairway field during primary recovery—do not evidence a judicial hostility to pooling, but rather embody the application of a general rule of contract construction.

When faced squarely with situations that forced consideration of the virtues of pooling in order to protect correlative rights as exemplified by the *Ryan v. Pickens* case, the court clung to the rule of capture rather than creating a new common law of equitable pooling. In the court's view, it is the function of the democratically elected legislature, not the judiciary, to create the law.

Having denied equitable pooling, the court could not then adopt equitable unitization. Thus, the unleased cotenant in *Superior Oil Co. v. Roberts,* whose fractional interest was leased and placed into a large unit by his fellow cotenant, was not allowed to receive an equitable accounting from a well not on his tract. The *Superior Oil* case appears as strikingly pro-unitization because it allows a unit operator to drain all the oil and gas from an unsigned cotenant without liability. However, the court's refusal to recognize any principle of equitable unitization can just as often work against a unit operator. Were the unit in the position of being drained by a nearby nonjoined owner who was benefitting from the unit's re-pressuring activity without paying any of the costs, the lack of equitable unitization in Texas jurisprudence would deny relief to the unit. The *Superior Oil* case was decided without any mention of the public interest in secondary recovery, and thus it is difficult to view this case as a deliberate effort by the court to encourage unitization.

Two cases brought the conflict between unitization and the common law of implied covenants and torts to the fore—*Railroad Commission v. Manziel* and *Amoco Production Co. v. Alexander.* In both cases, the Texas Supreme Court struggled to shape the common law in ways which would assist unit operators and promote secondary recovery. In the *Manziel* case, the court refused to enjoin the water flooding by a unit operator who had been granted an irregular injection-well location pursuant to a secondary recovery operation approved by the commission. The technical rules of trespass were held to have no place in determining the validity of the commission's grant of this well permit. The broad language used in *Manziel* and its clear statement of the public interest in secondary recovery strongly suggest that the technicalities of trespass law will also be disregarded in a cause of action brought by a holdout alleging tort liability directly against the operator of a commission-approved unit. In this event, the unit operator will probably not be subject to an injunction, strict liability for damages will probably not be imposed, and punitive damages will not be assessed, even though the operators' acts were intentional. Notwithstanding these beneficial results, tort liability still survives the *Manziel* case. The commission's finding under the 1949 Act that the rights of all nonsigners have been protected will not operate to collaterally estop the holdout from bringing suit, nor will it be conclusive evidence that no harm has occurred or that the operator's conduct is reasonable.

In *Amoco Production Co. v. Alexander,* the court held that lessees with multiple tracts under lease have an implied duty to protect each of their landowners against fieldwide drainage. One method of doing this is to unitize the field, and the court expressly mentioned this alternative. Thus, an implied covenant to unitize now exists in Texas common law. More important, the *Alexander* court resolved a longstanding ambiguity in Texas law, which until that time suggested that the lessee who held leases from several owners—some of whom had joined a unit and others of whom had not—would be strictly liable in damages for drainage to the nonconsenters, even though a reasonably prudent operator would not find it profitable to drill protection wells or adopt any other measures to prevent this drainage. The *Alexander* opinion makes it clear that the federal precedent in the case of *Tidewater Associated Oil Co. v. Stott,* which refused to find a strict duty to protect lessors who refused to consent to a cycling operation, is good Texas law.

A third case, *McLachlan v. Stroube,* also demonstrates a strong pro-unitization stance by the courts. Although the case itself involves only an issue of contract interpretation, the history of the litigation in the lower courts shows that the courts will not enjoin authorized secondary recovery operations on a tract, even though the tract is embroiled in a fierce controversy between a lessee who has joined the unit and a lessor who has not. The project will be allowed to advance, like a juggernaut, and the nonconsenting lessor is left to a remedy in damages, if damages can be proved.

Despite the courts' obvious concern with shaping the common law to encourage unitization, the actual results of these pro-unitization cases are as follows: the Alexanders won damages equal to the royalties that they would have received had their lessee drilled Rule 37 replacement wells higher up on their tract. In *Tidewater v. Stott,* no liability to the nonconsenting lessors was found, but only because the lessee had already drilled three wells on their tract and was producing and selling gas from this tract rather than recycling it. These three wells undoubtedly decreased the efficiency of the lessee's cycling operation on adjacent tracts. In the *Manziel* case, the unit operator was allowed to proceed with his Rule 37 exception well, a well which was completely unnecessary for conservation purposes.

Almost all of the problems imposed on unit operations by the common law would be eliminated or greatly alleviated by compulsory unitization legislation. Such legislation would allow the owner of the executive right to force nonexecutive interest owners into a unit on fair terms. Such legislation would eliminate the injustice which flows from the lack of equitable pooling or unitization. Such legislation would make irrelevant the cross-conveyancing theory because the compulsory order would bind all the owners of the tracts in the unit, regardless of the outcome of title or other disputes. A compulsory unitization law would make obsolete litigation over the construction of pooling clauses because one fieldwide unit would replace the many smaller drilling units as the relevant geographic area from which production must be secured to hold a lease in effect. Such a law would ease the burden of cotenancy by providing a predetermined basis of participation and accounting to cotenants who do not willingly join the unit, and possibly by providing compensation to the unit participants for carrying any working interest owners who refuse to contribute front-end capital to the project. Such legislation would greatly ease the formation of fieldwide units and thereby erase all interior lines around separate tracts, voiding all implied covenant or tort liability of the unit to nonconsenters.

Without compulsory unitization legislation, one would expect to find that little unitization occurs in the early life of fields when the rule of capture and primary production are in full force and lessees with a regulatory or structural advantage can produce profitably on their own. One would expect to find partial rather than fieldwide units, and to find a large number of unnecessary wells being drilled and produced in Texas oil and gas fields. Without compulsory process, one would expect to find unitization agreements allowing the owners of tracts with a regulatory advantage to secure the economic rents from their favored positions. Before looking at the statistics which bear out some of these expectations, the court's role in reviewing the Railroad Commission's orders must be analyzed. These orders were promulgated under legislation designed to mitigate the pernicious effects of the common law. Whether these orders survive judicial review can significantly affect the progress of unitization in Texas.

8

Judicial Review of Railroad Commission Orders

Because much of the common law is inherently hostile to sound conservation practices in the production of oil and gas, states enacted legislation to override this law and appointed regulatory agencies to administer and enforce the new statutory regimes. Court decisions reviewing a conservation agency's orders are another extremely important determinant of the course of unitization. Judicial review of an agency's acts generally centers on two issues: (1) whether the agency is acting within the statutory authority delegated to it by the legislature, and (2) whether the agency's order is reasonable. In both cases, the courts have a mighty influence on the amount of power and discretion accorded to the regulatory commission in its efforts to prevent waste and to protect correlative rights.

This chapter analyzes the role of the Texas courts in performing their duty to constrain Railroad Commission decision making within lawful limits on matters affecting the progress of unitization. The chapter looks first at judicial review of the commission's statutory authority to prevent waste and to protect correlative rights. Second, the chapter looks at judicial review of the reasonableness of the commission's orders, especially prorationing orders, Rule 37 orders, no-flare and no-waste orders, and orders promulgated under the 1949 voluntary unitization act. A third section of this chapter reviews the court decisions involving unitization agreements and the antitrust laws. The concluding section summarizes the courts' influence on unitization in both the common law sphere (discussed in chapter 7) and the sphere of judicial review of agency actions.

Judicial Review of the Commission's Statutory Authority to Prevent Waste

From the earliest time forward, the Texas courts have been willing to interpret the commission's statutory authority to prevent waste in a broad and expansive manner. In the midst of the rancorous debates over prorationing in the East Texas field during the early 1930s, the Texas court of civil appeals, in the case of *Danciger Oil & Refining Co. v. Railroad Commission,*[1] upheld the commission's power to prorate production in the following words:

> . . . [A]ny order of the commission bearing a reasonable relationship to the general duty imposed upon the commission, which is not unreasonable nor unjust, and which is reasonably calculated to prevent waste, comes, if not within the express powers granted to the commission, clearly within those necessarily implied.[2]

At the time of this decision, Texas conservation laws did not expressly authorize the agency to prorate production. Nonetheless, the court upheld the commission's statewide prorationing order on the basis that it was reasonably calculated to prevent physical waste. The extant conservation act expressly prohibited the commission from preventing "economic waste," but this also did not deter the court from supporting the prorationing order, even though it was unequivocally based on market demand. The court wrote:

> Just what the Legislature meant by "economic waste" is not clear. It is obvious we think that physical waste of such resources must of necessity result in economic waste. But it is equally true that economic waste by producers, such as expenditure upon a given well or lease, in bringing in production

> thereon, of a sum in excess of what the well or lease would return to such producers financially, does not necessarily mean physical waste of the natural resource. In the latter case there would be economic waste of the resources of the producer; but, if his wells were properly operated under regulation, there would be no physical waste of the natural resource itself, which is the only matter in which the state and the public are interested. That sort of waste, or economic loss in the production, sale, use, or disposition by the owners or operators of oil properly produced by them without physical waste of the resource itself... is a character of waste the commission was without authority to prevent.... But such limitation is not a denial to the commission of power to take into consideration an economic standard, or economic conditions, if such conditions bear a direct or reasonable relationship to physical waste.[3]

The judges found evidence that tremendous physical waste was occurring in the Panhandle field where the plaintiff Danciger operated, and upheld the commission's order.[4] By the time this decision reached the Texas Supreme Court on appeal, the specific order at issue had expired by its own terms and the legislature had materially changed the conservation statutes by expressly authorizing the commission to prorate oil and gas on the basis of market demand. The question on appeal was therefore moot. The Texas Supreme Court seemed unwilling at this time, however, to embrace the lower court's sweeping view of the commission's powers. Rather than dismissing the case for mootness, which would have left the lower court's opinion intact, the supreme court reversed the judgments of the lower courts because of mootness and then dismissed the case.[5] However, subsequent supreme court decisions show that the *Danciger* court's opinion accurately forecast a strong judicial attitude of according the commission very broad powers to prevent waste.

Nowhere is this more evident than in the judicial decisions upholding the Railroad Commission's authority to issue "no flare" and "no waste" orders. The Texas Supreme Court was a full partner in the commission's campaign to eliminate waste in Texas oil and gas fields at the end of World War II. Much of the story of this partnership has already been told in chapter 5. The court's expansive interpretation of the commission's power to prevent waste was an integral factor in the commission's decision to extend its arm-twisting from no-flare orders to no-waste orders and from condensate-gas fields to oil fields. The supreme court's interpretation of the commission's statutory authority to prevent waste bears repeating here:

> Whatever the dictates of reason, fairness, and good judgment under all the facts would lead one to conclude is a wasteful practice in the production, storage or transportation of oil and gas, must be held to have been denounced by the legislature as unlawful.[6]

Thus, even though the conservation statutes did not specifically prohibit the flaring of casinghead gas, the courts upheld the commission's power—indeed its duty—to prohibit this flaring.

Just as important as the court's broad interpretation of the commission's statutory power to prevent waste was the court's studious avoidance of any mention of the one, express legislative limitation on the commission's power to prevent waste: Article 6014(g)'s crystalline pronouncement that "it is not the intent of this Act to require repressuring of an oil pool or that the separately owned properties in any pool be unitized under one management, control, or ownership." By ignoring this provision, the supreme court surpassed even the *Danciger* court's previous wide embrace of the commission's waste-prevention powers. The court of civil appeals in the *Danciger* case was interpreting the commission's powers under the 1929 conservation laws, which did not yet contain Article 6014(g). Thus the *Danciger* court could truthfully state that, other than the exclusion of economic waste, "nowhere in these statutes . . . do we find any express limitation upon the powers of the commission in ascertaining what constituted waste, nor upon methods it might use to prevent it."[7]

Article 6014(g) was enacted in 1931 precisely to limit the commission in its methods of preventing waste. Never once in all of the later challenges to the commission's power to prevent waste did the supreme court mention this limitation on the agency's actions. This silence was not due to ignorance of the intended effects of the commission orders to force repressuring and unitization. In the litigation involving the Flour Bluff field, the court was presented with evidence showing that the shutdown order would result in a waste of oil greater than the equivalent savings in casinghead gas. The court responded that "[t]he evidence suggests that the loss anticipated . . . could be minimized by the injection of gas to retard the water drive."[8] The court's strong support of the commission's waste-prevention orders and its silent disregard of Article 6014(g) were persuasive deterrents to further lawsuits questioning the agency's power to prevent waste, even when the commission's orders were blatantly designed to force repressuring and unitization.

The full force of the Texas Supreme Court's encouragement of unitization through judicial sanction of the commission's sweeping statutory authority to prevent waste is illuminated by contrasting its decisions with that of the Colorado Supreme Court. In 1952 the Colorado Oil and Gas Conservation Commission issued a no-flare order for the Rangely field, reciting that "no gas shall be produced from the Weber Sand Reservoir unless all gas so produced shall be returned to said reservoir."[9] At the time of the order, the Rangely field accounted for almost two-thirds of Colorado's oil production. The wells in this field produced 39 million cubic feet of casinghead gas daily, of which 20 million cubic feet were being flared. The commission's order sought to prevent this flaring and force reinjection of the gas, which would increase the amount of recoverable oil from the field by 30 to 87 million barrels. Colorado's conservation statutes prohibited the waste of oil and gas in as broad terms as Texas'; indeed, in even broader terms, because the Colorado laws did not contain the equivalent of Article 6014(g).[10] The Colorado order was challenged by three operators who argued that the conservation act prohibited only the unreasonable or excessive production or flaring of gas; that the act conferred no power on the commission to compel operators, under threat of shutdown, to take affirmative action and expend costly sums for gas-injection wells, compressor equipment, and pipelines; that the authority for such a drastic and far-reaching order could not be supported by implication but must be expressly granted in the act; and that successful gas reinjection required unitization and the commission had no authority to compel unitization which the operators were unable to achieve voluntarily. The Colorado Supreme Court agreed with the operators on all accounts and held that the Colorado commission had exceeded its statutory authority by issuing the order. Despite the Colorado act's broad definition of waste, the court found that the legislature had delegated authority to the commission "stintingly, sparingly, and almost grudgingly."[11] The court wrote:

> We recognize that there may be great advantages to the operators and substantial additional recovery of oil in the Rangely Field if repressuring secondary recovery mechanisms are adopted. We also appreciate the fact that substantial benefits might accrue to the operators in the field and to the State of Colorado if unitization of Rangely Field were to be accomplished. Under section 12 of the 1951 Act, either or both of these objectives might be had by voluntary agreements between the operators. A valid compulsory program to accomplish those ends must have definite authorization from the legislature.[12]

Had the Texas Supreme Court desired to restrain the Railroad Commission's issuance of no-waste orders, it would have been a simple thing to look beyond the wording of the orders to their coercive unitizing effect on operators and to void the orders, even without resort to Article 6014(g). The Colorado commission's order by its terms did not require unitization; it required reinjection of gas. The Railroad Commission's shutdown orders also did not require unitization, but simply required that the operators prevent waste. Even if the Colorado commission had veiled its Rangely order in the broader phrasing used by the Texas commission, it seems likely that the Colorado Supreme Court would have voided the order.[13] It also seems likely that, had one of the Railroad Commission's no-waste orders been appealed to the Texas Supreme Court as being outside the commission's authority, the order would have been upheld as long as it did not, by its express terms, require repressuring or unitization. The Texas judges would not unveil their partner's nakedness; instead they cloaked the commission with power.

The case of *Pickens v. Railroad Commission* further illustrates Texas' judicial posture.[14] Pickens, the plaintiff in this case, sought to invalidate a commission prorationing order on the precise grounds that would arise in challenging a no-waste order: that the order was designed to force Pickens to join a unitization agreement and therefore it was beyond the agency's statutory authority because the commission "sought to do by indirection what it could not do directly."[15] Pickens asserted that the commission had abdicated its duty to make an independent determination of what constituted a fair prorationing formula by blindly adopting the 50 percent acreage, 50 percent acre-feet of sand formula used in the voluntary unitization agreement signed by 88 percent of the producers in the Fairway (James Lime) field. The unitization committee had reached this formula as a compromise after many months of bargaining among the owners in the field. Pickens refused to join the unitization agreement because his land overlay the thickest part of the reservoir and the half of the participation formula based on surface acreage did not credit him for the thickness of his underlying sands.

The court acknowledged that "this formula will encourage unitization,"[16] but the justices did not therefore consider it invalid as beyond the commission's authority. Indeed, the court recounted that "since the earliest days of the oil industry, it has been customary for interested people to recommend orders to the Commission,"[17] and cited the early experience of the Yates field as an example. Without ruling that the courts would never seek an inquiry into the motives of an administrative agency, the justices found no evidence that the commission had abdicated its duty or exceeded its power by promulgating that prorationing formula which would best encourage voluntary unitization.

Similarly, in the *Normanna* and *Port Acres* cases that invalidated the one-third–two-thirds prorationing formula in gas fields, the majority of justices ignored the obvious fact that their ruling would force small-tract owners to pool rather than to drill on their own.[18] These cases differ from the *Pickens* case in that the commission was seeking to uphold its one-third–two-thirds formula despite its deleterious effects on the efforts of some operators to cycle gas and to produce efficiently with fewer wells. Yet even in this situation, the majority of justices made no reference to Article 6014(g), even though the commission cited the article in its briefs as evidence of the commission's lack of power to promulgate prorationing formulas that indirectly coerced operators to pool or unitize. Instead, the justices cited the 1949 Act for its express authorization of voluntary unitization as being in the public interest.[19] Following the *Normanna* decision, the Railroad Commission allocated production in this field on the basis of surface acreage but established a special allowable for small-tract owners who could show that drilling a well on their acreage was not economically feasible and that all lessees on adjacent tracts had refused to pool on fair reasonable terms. No court challenge was ever brought against these field rules on the basis that the commission had no authority to force pooling, but had judicial review been sought, the same rationale used later in the *Pickens* case would probably have been used to uphold the Normanna field prorationing order.[20] Thus, commission orders that strongly but indirectly encourage unitization are judicially sanctioned as long as they do not expressly require unitization or repressuring.

Rule 37 Orders and Waste

Judicial review of the scope of the commission's statutory authority to prevent waste also occurs in the context of Rule 37 orders. Rule 37 fixes minimum distances between wells and lease lines, provided that the commission—in order to prevent waste or to prevent the confiscation of property—may grant exceptions to permit drilling within shorter distances.[21] Because the Texas statutes do not define waste in general terms, but simply list noninclusive examples of waste, the Texas courts have had to define waste for purposes of reviewing the commission's Rule 37 decisions. In a 1939 case, the supreme court defined waste as "the ultimate loss of oil."[22] If a well at an irregular location would recover a substantial amount of oil that would otherwise be left in the ground, a commission order granting a Rule 37 permit would be upheld by the courts. Thus the mere fact that a well, if drilled, would produce oil did not suffice to prove waste. If the same oil could be recovered from already existing wells, a Rule 37 waste exception was not justified.[23] In subsequent cases, the court elucidated the conditions under which a waste exception was justified. An applicant had to show that unusual underground conditions existed in a localized area of a reservoir so that an exception to the field's spacing rule was necessary to produce oil that would not be otherwise recovered.[24] For many years, proof of unusual reservoir conditions was the only method of acquiring a Rule 37 waste exception. Then in 1978, in the case of *Exxon Corp. v. Railroad*

Commission,[25] the supreme court upheld a commission order granting a Rule 37 exception permit based on waste, even though the applicant admitted that there was no difference in reservoir conditions between the requested Rule 37 exception location and a regular location. The applicant, BTA Oil Producers, requested the Rule 37 exception location because it had an existing well bore at this spot, which could be plugged back and recompleted in the new producing zone at substantially less cost than if BTA had to drill an entirely new well at a regular location. BTA presented evidence that a well at either location would produce oil that would otherwise not be recovered, but the amount of oil was not sufficient to cover the cost of a new well. The commission granted the Rule 37 permit. Exxon, the lessee of an adjoining tract, was producing in the same zone and protested the grant. On appeal, the supreme court upheld the commission's authority to consider economic factors in its Rule 37 decision. The court wrote:

> Consideration of reasonable economic factors upon which operators must act is one of the underlying bases for Rule 37 itself. . . . The adoption of the spacing rule represents an economic decision that the density of development should be regulated and restricted, at least in part to prevent physical or economic waste from the drilling of wells which are not reasonably necessary to drain a reservoir adequately.
>
> In the present case, common sense dictates that the economic waste that would result from BTA's drilling a completely new well, simply so as not to crowd its existing well, is a most relevant consideration.[26]

The court's opinion in this Rule 37 case has led to speculation that the commission now has the authority to promulgate orders to prevent economic waste, not just the physical waste of oil.[27] The debate on this issue is reminiscent of the great controversy over the commission's power to curtail production in excess of market demand. The U.S. Supreme Court ultimately upheld market demand prorationing laws on the ground that they prevented the *physical* waste of oil.[28] That such laws incidentally raised prices and prevented "economic waste," namely financial losses to producers, did not defeat their validity.[29] For decades, the Railroad Commission's broad grant of authority to prevent waste has been considered in terms of preventing the ultimate loss of oil. The definitions of waste appearing in Texas conservation laws do not include the economic waste of drilling or operating unnecessary wells.[30] It was not until the enactment of the Mineral Interest Pooling Act in 1965 that Texas conservation statutes even mentioned the objective of preventing the drilling of unnecessary wells. The Texas attorney general's vehement denunciation of the proposed voluntary unitization bill in 1947 was based on the fear that it empowered the commission to authorize unitization agreements which merely saved money for oil and gas operators, but which did not necessarily benefit the public by increasing the ultimate recovery of oil and gas (see chapter 4 text at notes 23 to 27). Prior to 1979, no court decision involving waste had ever suggested that the commission had the authority to issue orders on the sole basis that the orders would allow producers to operate more profitably.[31]

In *Exxon Corp. v. Railroad Commission,* the Rule 37 exception well prevented both physical and economic waste. BTA's evidence that no other producing well could recover the oil around the existing well bore was accepted by the commission and the court. However, in a subsequent Rule 37 docket, the commission granted an exception permit even though the applicant, Hughes and Hughes, admitted that the same gas could be recovered by drilling a new well rather than by using an existing well bore.[32] The applicant could save $900,000 by recompleting the existing well instead of drilling a new one, but the physical recovery of gas at this location would be unaffected, as would the total recovery in the field.[33] The commission granted the Rule 37 permit to prevent confiscation, even though the evidence also showed that the correlative rights of Hughes and Hughes would be unaffected by the location of the well. Hughes and Hughes could profitably drill a new well at a regular location, which would protect them from drainage by the adjacent operator, Texas Oil and Gas, just as nicely as would the Rule 37 well.[34] Thus, the only basis for granting the Rule 37 permit to use the existing well bore was to let Hughes and Hughes save $900,000. The commission held that the rationale of *Exxon Corp. v. Railroad Commission* allowed the agency to consider economic costs in a Rule 37 permit to prevent confiscation and therefore granted the permit. Texas Oil and Gas vigorously protested the Rule 37 application throughout several days of hearings, arguing that the exception was not necessary either to prevent waste or to protect correlative rights. However, Texas Oil and Gas did not seek judicial review of the final order granting the permit, so it is not yet known whether the court will uphold the commission's grant of a Rule 37 permit on grounds of preventing economic waste alone.[35]

The commission's obvious concern for economic waste, and the supreme court's acceptance in the *Exxon Corp.* case of the relevance and importance of economic factors in the commission's decision making auger well for future unitization efforts which may be instigated largely to reduce operating costs by shutting in unnecessary wells. As discussed in chapter 4, the commission's statutory authority to approve such unitization agreements under the 1949 Act is uncertain. The *Exxon Corp. v. Railroad Commission* case suggests that the court today, like the earlier *Danciger* court, will uphold the commission's statutory authority to issue unitization orders as long as the commission can trace a reasonable nexus between economic waste and physical waste. Reduced drilling and operating costs can prevent physical waste by extending the margin of profitable operations; wells can be pumped longer, and additional oil recovered.[36] If there is substantial evidence that greater economic efficiency will result in greater ultimate recovery of oil and gas, the Texas courts seem willing to support commission orders designed to save operators' costs and increase profitability.

In a different context involving Rule 37, the Texas Supreme Court has interpreted the Railroad Commission's authority to prevent waste in another way that also encourages unitization. In *Railroad Commission v. Manziel,*[37] the Whelans had received a commission order that permitted them to drill and inject water into a well only 206 feet from the Manziel property. The field rules provided that wells be located at least 660 feet from lease lines. If the irregular well location were allowed, one of the Manziels' wells which had an estimated life of thirty-two months would be flooded with salt water within three to eight months. The Whelans had been granted the irregular location in order to protect their correlative rights by minimizing the amount of oil which would be pushed by their saltwater injections across to the Manziels' land. The Whelans had unitized the lands leased to them and had secured commission approval of their unitization project under the 1949 Act.

The Manziels asserted that the commission had no authority to grant this Rule 37 exception: first, because the commission could not license a trespass, and, second, because Article 6029(4) of the conservation statutes prohibited such a result. Article 6029(4) states that the commission shall require wells "to be drilled and operated in such manner as to prevent injury to adjoining property."[38] As we have seen, the court refused to find that a trespass existed in this situation. The supreme court also dismissed Article 6029(4) as having no application to authorized, secondary recovery projects. By enacting the 1949 voluntary unitization act, the court reasoned that the legislature "has made it the policy of this state to encourage the secondary recovery of oil."[39] The legislators must have known that successful, secondary recovery operations would eventually flood out all the wells in a field, including those on tracts outside the unitized area. Citing the great public interest in secondary recovery, which could often double recovery rates in oil fields, the court stripped Article 6029(4) of any power to interfere with the commission's statutory authority to approve well locations for water flooding. The economic waste caused by prematurely flooding out the Manziels' well was condoned by the court because of the necessity to protect the unit operator's correlative rights. If the commission's power to protect these unit operator's rights were constrained, fewer lessees would perform secondary recovery, and the physical waste of oil and gas would result. The prevention of physical waste is a more important public concern than the prevention of economic waste. Therefore, the court upheld the commission order and allowed economic waste to an individual producer to proceed by interpreting the commission's power under the 1949 Act as superseding the agency's duties under other sections of Texas conservation laws.[40]

Judicial Review of the Commission's Statutory Authority to Protect Correlative Rights

Judicial interpretation of the scope of the commission's statutory authority to protect correlative rights is as important to the course of unitization as the scope of the agency's power to prevent waste.

At first blush, the protection of the private rights of owners to a fair division of a reservoir's bounty would seem to be a subsidiary, and perhaps even inappropriate, issue to command the attention of a regulatory commission. Decisions about correlative rights determine who is to get what share of the pie. The public interest in increasing the size of the pie does not seem to be served by this process. However, the prevention of waste and the protection of correlative rights are inseparably related. A conservation order issued by an agency would be struck down as a taking if it failed to protect adequately the private property rights of some of the owners in a field. Furthermore, few regulatory commissions would initiate efforts to prevent waste if the unavoidable effect of their orders would be to inflict serious harm on a significant segment of the oil and gas industry. As we have seen, the Railroad Commission's inability to protect the correlative rights of small-tract owners through compulsory pooling led the agency to adopt a per-well prorationing system, which produced tremendous economic waste and contributed to physical waste by discouraging unitization and raising production costs.

The inseparable link between waste prevention and the protection of correlative rights is aptly illustrated in the case of *Corzelius v. Harrell.*[41] Harrell had leased almost the entire Bammel gas field in 1939 when he applied to the Railroad Commission for a permit to install a recycling plant to produce wet gas, process the condensate, and pump the dry gas back into the reservoir. In 1941 leases belonging to Harrell on 4 percent of the field lapsed, and Corzelius acquired them. Corzelius drilled a well on his newly acquired acreage and began producing and selling 5 million cubic feet of gas daily to two distributing companies owned by him that supplied light and fuel to a city. Harrell's recycling plant had a capacity of 35 million cubic feet per day, of which 33 million were reinjected into the ground. The other 2 million cubic feet were consumed in the cycling and repressuring operation. Harrell applied to the commission for an order prohibiting permanent withdrawals from the field until the recycling was concluded; in the alternative, he argued that withdrawals should be prorated. After a hearing, the commission found that production of 20 million cubic feet daily would not result in waste and issued a prorationing order that allowed Corzelius to continue to withdraw and sell gas from the field. Harrell appealed this order, and the lower courts permanently enjoined the commission from continuing this order in effect on the basis that the order did not adequately protect the correlative rights of Harrell.[42] Corzelius, with only 4 percent of the field's acreage, was withdrawing 5 million cubic feet per day, while Harrell, with 96 percent of the field, was only withdrawing a net amount of 2 million cubic feet per day. On appeal, Corzelius argued that Sections 10 and 11 of Article 6008 authorized the commission to prorate gas to adjust correlative rights only as an incident to preventing waste.[43] The statute declares that it is the duty of the commission to prorate gas "for the protection of public and private interests: a) In the prevention of waste. . . ; b) In the adjustment of correlative rights. . . . " Because the commission had found that 20 million cubic feet of gas could be withdrawn nonwastefully, Corzelius contended that the commission could not curtail his production of 5 million cubic feet. The Texas Supreme Court interpreted the statutes to give the commission power to prorate gas either to protect correlative rights or to prevent waste, each power standing independently.[44] Before this opinion, the commission had made no attempt to enforce compulsory proration of gas fields because it was almost impossible to show that such prorationing was necessary to prevent waste, and the agency was unsure that it had the authority to prorate for any other purpose. The *Corzelius* decision was hailed by many because it seemed to give the commission a powerful tool which could be used to discourage owners from refusing to sign a unitization agreement for recycling operations: the production from the wells of holdouts could be prorated to ensure that they were not unjustly enriched by the cycling operations of others that kept gas pressure in the field high.[45]

The scope of the commission's authority to protect correlative rights under Sections 10 and 11 of Article 6008 was broadened further in *Railroad Commission v. Rowan Oil Co.*,[46] the Spraberry field case. In this case, the commission had shut down all of the 2,400 wells in the entire field. Flaring wells were shut in to prevent waste, and nonflaring wells were shut in to protect the correlative rights of the flarers, who would otherwise be drained by those fortunate producers having pipeline connections for their casinghead gas. The court invalidated the commission order because the justices could find no statutory authority in the oil-conservation laws which allowed the commission to shut in nonwasteful wells. However, the court transposed Sections 10 and 11 of Article 6008 relating to natural gas into the statutes regulating oil wells, and suggested that the agency could accomplish the twin objectives of preventing waste and protecting correlative rights under the "broad powers" to regulate the flow from nonwasteful wells in order to protect correlative rights.[47] This was a bold move by the court because Article 6008 clearly applied only to gas wells and the Spraberry field was an oil field. But a bold move was required if the waste of casinghead gas was to be averted in this field. The attorney general, speaking on behalf of the commissioners, had asked the court to either uphold the shutdown order in its entirety or strike it down in its entirety.[48] The commission was clearly unwilling to shut in the flaring wells if it could not protect the owners of these wells from drainage. By interpreting Sections 10 and 11 to apply to oil fields, the court cleared the way for the commission to prorate the nonflaring wells on the sole basis of the agency's broad authority to protect correlative rights. The non-flaring wells would be allowed to produce, but at such low rates that they would not drain much oil from the wells that were shut in awaiting the completion of casinghead-gas facilities. As these facilities came on-stream, the field's allowable was increased.

Thus, the *Spraberry* case gave the commission a more powerful tool to use than the blunt bludgeon of a complete field shutdown. It also allowed the agency to prorate oil fields on the basis of the localized market demand for casinghead gas rather than on the statewide demand for oil. The commission could now fine tune the oil-prorationing system on a field-by-field basis, both to prevent waste and protect correlative rights. Thus, an opinion that ostensibly constrained the commission's power to prevent waste actually enlarged it by broadening the scope of the commission's power to protect correlative rights. A subsequent supreme court opinion, *Railroad Commission v. Sample*,[49] also accorded the commission a broad statutory authority to adjust oil well allowables to protect correlative rights under the oil-conservation laws that require the agency to prorate oil "on a reasonable basis."

However, despite the court's expansive view of the commission's authority to prorate oil and gas wells to protect correlative rights in the *Corzelius* and the *Spraberry* cases, a subsequent opinion took a much narrower view of the amount of discretion delegated to the commission in the operation of the gas-prorationing system, and this later decision effectively negates much of the commission's power to protect correlative rights in gas fields. In *Railroad Commission v. Woods Exploration and Producing Co.,*[50] the Texas Supreme Court held that the commission does not have the authority to limit gas production from a reservoir to less than reasonable market demand, except to prevent waste. This case involved the vitriolic battle between large- and small-tract gas producers in the Appling field, a battle which was waged on many fronts: (1) through attacks on the commission's grant of Rule 37 permits to the small tracts;[51] (2) through attacks on the commission's one-third per well, two-thirds surface acreage allocation formula for the field, which resulted in massive drainage toward the small tracts;[52] (3) through an attack on the commission's authority to protect the correlative rights of the large-tract owners by issuing the prorationing order involved in the case under discussion;[53] and (4) through attacks on the large-tract owners for conspiring to destroy the small-tract producers in violation of the antitrust laws, both state and federal.[54]

The large-tract owners had lost the first two battles and were being drained by the small-tract producers. The latter were tremendously favored by the one-third–two-thirds formula, but not content with this advantage, they played with the gas-prorationing system to receive even greater allowables. The game worked as follows: to determine the monthly market demand for gas from a field, the commission required that producers file forecasts stating the volume of gas that each producer expected to be able to market from his well the following month. An operator could forecast any amount as long as it did not exceed the delivery capacity of his wells. The small-tract producers followed the practice of filing forecasts at the full delivery capacity of their wells. These forecasts, when added to those of other producers in the field, resulted in the assignment of very large monthly allowables for the total reservoir. The allowables were so great that the large-tract wells were incapable of producing the allowables representing their fair share of the gas under the one-third–two-thirds allocation formula. The large-tract wells were allowed to produce at capacity and the remainder of their allowable was reallocated to the small-tract wells, which therefore received even more allowable than the one-third–two-thirds formula accorded them.[55] A large-tract producer requested that the commission amend its statewide rule for determining market demand as applied to the Appling field. The commission agreed and established a formula for computing the maximum quantity of gas that might constitute market demand. In effect, the commission's new prorationing order for this field established a ceiling on reasonable market demand, which was fixed to permit the large-tract wells to receive allowables that they would be capable of producing. No allowables would remain for transfer to the small-tract wells.

The commission based its authority to promulgate this order on Sections 1 and 22 of Article 6008, the 1935 natural gas conservation act, which was enacted "in recognition of past, present, and imminent evils occurring in the production and use of natural gas, as a result of waste in the production and use thereof in the absence of correlative opportunities of owners of gas in a common reservoir to produce and use the same," and which expressly vested in the commission "a broad discretion in administering this law."[56] Yet a majority of the Texas Supreme Court voided the Appling field order based on a narrow interpretation of Section 12 of the act, which states that the monthly reservoir allowable of gas shall be fixed "at the lawful market demand for the gas or at the volume that can be produced from the reservoir without waste, whichever is the smaller quantity."[57] Because of this express provision, the agency had no statutory authority to prorate gas below market demand in order to protect correlative rights. The majority suggested that the commission could end the small producers' practice of nominating fictitious forecasts of market demand for natural gas by relying on other data to project reasonable market demand, such as pipeline purchasers' forecasts.[58]

Three justices dissented and, in lengthy opinions, set forth the legislative history of the gas statutes and the case precedent, including *Corzelius,* which would affirm the commission's power to consider correlative rights in establishing a reasonable reservoir allowable.[59] This history showed that the very purpose of Article

6008 was to protect correlative rights. It was enacted primarily to protect the small producers in the Panhandle who had no pipeline connections and who were being drained by large, integrated producers. One dissenting justice warned:

> The Court holds that the Commission may not consider correlative rights as a factor in the determination of market demand. . . . If Woods Exploration Company and a small pipe line can destroy correlative rights in the reservoirs in question it is unthinkable what larger pipe lines can do to gas fields all over this state wherever there are small tract owners located.[60]

The majority, in contrast, looked at the legislative history of the *oil*-conservation statutes, which showed that the lawmakers had been extremely reluctant to authorize market demand prorationing at all. Citing Hardwicke's legal history of oil prorationing in Texas, the majority wrote:

> Against this background, it is rather difficult to believe that when the Legislature finally authorized the Commission to limit production to reasonable market demand, it meant that an even lower figure might be set whenever the Commission concluded that the same would constitute a reasonable reservoir allowable under all the circumstances.[61]

The majority's reliance on the legislative history of the oil-prorationing statutes to defeat the commission's gas-prorationing order in the Appling field seems particularly inappropriate because the court's previous opinion in the Spraberry field case authorized the commission to prorate this oil field far below its market demand in order to protect the correlative rights of the shut-in producers. Also, in an even earlier case, the court had held that the commission could prorate an oil field below the field's market demand even though this prorationing was not required to prevent waste. In *Railroad Commission v. Continental Oil Co.*,[62] the commission had distributed the statewide allowable among the various oil fields so that production from the Conoco-Driscoll field was limited to 1,300 barrels daily. The undisputed evidence showed that a firm market demand for 4,000 barrels a day existed for this field's oil and that 2,300 barrels a day could be produced without waste.

Continental sought to overturn the prorationing order limiting output to 1,300 barrels, but the court upheld the agency's order by construing the oil-prorationing statutes to allow the commission to set field allowables based on the field's fair share of the state's total, even if this share was less than the field's market demand.[63]

Of course there is a very real public interest to be served by meeting market demand. If the Railroad Commission often restricted fields below market demand solely to protect correlative rights, the price of oil and gas products to consumers would rise as shortages developed. In contrast, prorating fields below market demand to prevent waste may increase prices to consumers in the short term, but by maximizing the recovery of resources, the public is benefited by lower prices in the long run. The court's suggestion to the commission to prorate the nonwasteful wells in the Spraberry field in order to protect the correlative rights of the shut-in producers was done at a time of excess capacity in the oil fields. The market demand for the oil which no longer flowed from either the shut-in wells or from the severely prorated wells could be met by oil supplied from other fields. Alternative transportation to the affected market area via pipeline or rail shipments is usually possible in the case of oil. This is not always the case with natural gas. If gas fields are prorated below their existing market demand, alternative sources of supply may not be easy to arrange. This is one reason why the commission prorates gas fields differently from oil fields.[64] This may also explain why the Texas Supreme Court was unwilling in the *Woods* case to interpret the commission's authority to prorate gas broadly, even though the court had shown no such compunctions in the *Spraberry* case. However, the court did not enunciate any such public policy rationale in the *Woods* case to distinguish it from the Spraberry field situation.[65]

The result in the *Woods* case must give one pause to consider the commission's authority to prorate wells in oil fields at a time when all oil fields are operating at 100 percent capacity. In such a situation, would the court continue to support the commission's broad authority to prorate nonwasteful oil wells in order to protect the correlative rights of the owners of wasteful wells that have been shut in, when the result is to reduce the field's output even further below market demand and cause even greater shortages? Or would the court transpose Section 12 of Article 6008, as interpreted in the *Woods* case, into the oil statutes, just as it transposed Sections 10 and 11 in the *Spraberry* case? More fundamentally, would the commission even issue a no-waste order in a field at a time of energy shortage if the unavoidable effect of such an order is either to decrease supplies greatly (because both flaring and nonflaring wells are curtailed), or to allow drainage from shut-in wells to nonwasteful wells? In a unitized field, the commission would not face such a difficult choice. The shut-in wells would receive a fair share of the production from the producing wells. Moreover, some of the wells producing a large amount of casinghead gas would probably never have been drilled.

On the surface, the majority's decision in the *Woods* case simply denies to the commission an easy method of establishing reasonable market demand to prevent abuses in the gas-nomination system,[66] but the holding does not overrule the broad authority granted to the commission under Sections 10 and 11 of Article 6008 to prorate gas in order to protect correlative rights independent of its waste-prevention powers, as upheld in the *Corzelius* case.

However, consider the effect of the rule that the commission cannot prorate gas below market demand, except to prevent waste, as applied to the facts in the Bammel field which resulted in the litigation between Corzelius and Harrell. The commission found that 20 million cubic feet per day of gas could be withdrawn from the reservoir without causing waste. Corzelius, with 4 percent of the acreage, was withdrawing and selling 5 million cubic feet per day, which represented the market demand for the field. Harrell, with 96 percent of the acreage, was withdrawing only 2 million cubic feet daily for use in his cycling operation. The majority opinion in the *Woods* case means that the commission cannot curtail production below the 5 million cubic feet representing the market demand. Indeed, were market demand to equal 18 million cubic feet per day, Corzelius could withdraw this amount as long as Harrell continued to withdraw only 2 million cubic feet daily. Corzelius could not withdraw more than this amount even if market demand far exceeded 20 million cubic feet. The commission's authority to prorate gas to prevent waste still acts as a brake on the ability of noncyclers to withdraw and sell gas from a condensate reservoir, but the commission's lack of authority to curtail production below market demand prevents the agency from being able to protect correlative rights fully, especially when the market demand factor is truly 100 percent, without any fictitious nominations.

Correlative rights are also difficult to protect in depletion gas fields that do not contain much condensate. Prorationing in these fields is not required to prevent waste; the maximum recovery of gas is independent of its rate of production. Therefore, the reservoir must be produced at the rate which satisfies market demand, even if this rate causes uncompensated drainage between tracts. Ever since the energy crisis of the 1970s, many gas fields in Texas have had 100 percent market-demand factors. The owners of large tracts of land in these gas fields are often drained by small-tract wells in the field, even when the allocation formula for the field is based on net productive acre-feet of sand, rather than the notorious one-third–two-thirds formula. Whenever the delivery capacity of the large-tract wells is insufficient to produce the allowable based on net acre-feet of sand, the small-tract wells are assigned the unused allowable of the large-tract wells in order to meet market demand. The only way that the large-tract owners can increase the delivery capacity of their tracts and prevent this drainage is to drill additional wells. This is the context in which Hughes and Hughes was seeking the Rule 37 exception permit discussed earlier in this chapter. Hughes and Hughes needed a third well to increase the delivery capacity of its tract to match its ownership of 90 percent of the reserves in the field and its right to produce 90 percent of the field's allowables under the allocation formula based on reserves. Without the third well, Hughes and Hughes would be drained by Texas Oil and Gas, which held less than 10 percent of the gas reserves, but whose one well could deliver 35 percent of the gas from the field.

In effect, the gas-prorationing system in Texas results in a per-well allocation of allowables—regardless of the actual field-allowable formula adopted—in any gas field with wells of roughly equal delivery capacity and a 100 percent market-demand factor. The *Woods* decision forbids the commission from prorating a field's production below market demand in order to protect correlative rights. Thus, without compulsory unitization of these gas fields, correlative rights can only be protected by drilling wells that are unnecessary from the viewpoint of increasing the ultimate recovery of gas.[67] However, even if H.B. 311, the proposed compulsory unitization bill, had passed in 1973, it would not have authorized the commission to approve any unitization whose sole purpose was the protection of correlative rights rather than an increase in the ultimate recovery of oil or gas (see chapter 4 text at note 160). Nor does Texas' current, voluntary unitization act authorize such approval (see chapter 4 at notes 31 to 65). Only the now discarded Section 21 of the 1935 natural gas act recognized the role that unitization could play in protecting correlative rights in gas fields.

Conservation Versus Correlative Rights

At times, the two objectives of preventing waste and protecting correlative rights will conflict. Judicial review of the commission's orders that have by necessity selected one objective over the other, can also affect the progress of unitization in Texas. If conservation orders are frequently overturned because the courts find that they do not protect correlative rights and therefore are a confiscatory taking, the commission's hands would be tied in effecting waste-prevention measures, absent a compulsory unitization statute that could accomplish both goals simultaneously.

The case of *Texaco Inc. v. Railroad Commission* involved a serious conflict between the commission's twin goals.[68] In 1954 the commission had authorized bonus allowables to encourage reinjection of salt water in the Fig Ridge field, a water-drive field. The reinjection kept pressure high in the field and increased the ultimate recovery of oil. The program continued from 1954 to 1983 without incident. But after twenty years the reinjection had created a high-pressure area along the western edges of the field, and oil was being pushed eastward toward the low-pressure area. Oil migrated away from Texaco's leases toward Sun Oil's property. Texaco applied to the commission to suspend the bonus-allowable rule. The commission denied the application on the basis that the bonus rule, if continued, was expected to result in the production of 800,000 additional barrels of oil. The supreme court upheld the commission's decision even though it would result in large, net uncompensated drainage from Texaco. Texaco argued that the *Normanna* and *Port Acres* decisions prohibited the commission from issuing a prorationing order that allowed confiscation of one person's oil by another. The supreme court found these cases not controlling because they had not involved the issue of waste prevention. The court wrote:

> The business of producing, storing and transporting oil and gas is a business affected with a public interest and subject to regulation by the state. . . . Between protecting correlative rights and protecting the public interest of preserving our state's natural resources, the prevention of waste has been held to be the dominant purpose.[69]

Many other cases exist in which the courts have upheld conservation orders against the contention that they failed to protect private property rights.[70] The *Texaco* case is especially important because it is fairly recent and because it involves a prorationing order. The bonus-allowable rule was not changed because it increased ultimate recovery. In the converse situation, if the commission changed an existing prorationing formula in a field because a new one was necessary to prevent waste, the court's rationale in *Texaco* could be used to uphold this changed formula against the complaint of owners whose share of allowables decreased under the new formula relative to the old.

The Commission's Unused Authority

We have seen that the courts have bequeathed the commission almost unlimited authority to prevent waste. The agency's authority to protect correlative rights independent of waste has also been construed broadly except in the area of gas prorationing. All of the cases discussed so far in this chapter suggest that the commission has the power to encourage unitization by changing existing allowable formulas which render voluntary unitization especially difficult to achieve.

The commission uses many carrots and sticks to effectuate unitization, but it does not change the allocation formula in a field so that producers favored by the regulatory status quo would lose the incentive to go it alone and decide to join the unit instead.[71] The commission's loyalty to the old prorationing formulas may be partly a result of judicial decisions that sanctify the stability of property rights to produce. The *Normanna* and *Port Acres* decisions, which invalidated prorationing formulas with large per-well factors, did so only prospectively. The court in the *Normanna* case refused to authorize a wholesale reform of the allowable system that had been accepted and acquiesced in for many years.[72] The court's policy in this regard was strengthened in litigation over the one-third per well, two-thirds surface acreage formula in the Appling field. A large-tract owner sought judicial review of this formula several years after it had been adopted in the field. The court held that the company was barred by unreasonable delay from seeking to overturn the formula, notwithstanding the fact that the average per-well allowable for small tracts was forty-three times the average per-acre allowable for standard-sized units.[73] The court wrote:

> There are many reasons why stability in respect to proration formulas is vital to the well being of the industry as a whole, to the property owners in the field and to the public at large. . . . Individuals and institutions have invested in royalties and in other oil and gas interests. Loans have been made with these properties as security, and taxes have been levied by various municipal and school authorities. It is well known that the economy of the whole state rests to a large extent on the oil and gas business.[74]

This language makes it almost impossible to modify established prorationing formulas for the sole purpose of better protecting correlative rights. However, in the same case, the court affirmed the commission's power to change its regulations and orders in light of changing circumstances:

> Unquestionably the Commission's power to regulate oil and gas production in the interest of conservation and protection of correlative rights is a continuing one and its orders are subject to change or modification where conditions have changed materially, new and unforeseen problems arise or mistakes are

> discovered [citations omitted]. It is equally true that no landowner or operator acquires any vested right to continue to produce the same amount of oil or gas during the life of such well as was fixed by the proration schedule in force at the time he drilled his well.[75]

The court found no material changes, new information, or unforeseen problems that would justify amending the one-third–two-thirds formula in the Appling field.[76] But in many cases involving unitization, the commission can argue that changed conditions merit a change in the prorationing formula. For example, if almost all the primary oil in a field is depleted, the need for secondary recovery operations to increase production would certainly be a changed circumstance justifying an alteration in the field formula. An amended prorationing formula could be based on the percentage of secondary oil reserves in place beneath each tract rather than on the percentage of primary oil recovered by each tract under the field's original proration order. Such an amended formula would induce owners of small tracts to join the unit because their wells might not be able to produce enough secondary oil to operate profitably under this formula. Voluntary unitization is fairly easy to achieve in fields depleted of their primary oil because most owners realize that, without repressuring, no one will be able to produce profitably. This is not the case in fields requiring pressure maintenance when primary recovery is still in effect. If the field's original prorationing formula hinders efforts to achieve pressure maintenance, then it would seem that the new occurrence of waste would be a changed condition that would justify a new prorationing formula. For example, after many years of production in the Conroe field, oil migrated into spaces previously occupied by gas, and this migration forever precluded the recovery of some of this oil. The emergence of this waste is certainly the type of "new and unforeseen problem" that is subject to the commission's continuing duty to regulate and prevent by modifying existing prorationing formulas.

Thus, it appears that if waste is occurring in a field and many owners refuse to unitize mainly because they are better off under the existing per-well allowable formula, then the commission has the power to amend the prorationing formula to one which better prevents waste and simultaneously provides a greater incentive to join the unit. If the holdouts argue that the new formula violates their correlative rights because it accords them a smaller percentage of the reservoir's bounty than the old formula, the *Texaco* case from the Fig Ridge field holds that their rights may be sacrificed to the greater public interest of preventing waste. In addition, if the unit proponents decided to perform the repressuring without the joinder of all owners, the commission would seem to have the authority to change the field's prorationing formula due to the changed condition of newly instituted secondary recovery or pressure maintenance in order to assure that the holdouts recover only their fair share of the oil.[77] Had waste not been at issue in the *Texaco* case, changed conditions in the field's pressure pattern may have justified elimination of the bonus-allowable rule so that correlative rights could be better protected.

The commission has not manipulated allowable formulas in this manner despite its dedication to increasing the ultimate recovery of oil and gas in Texas. The agency prefers to encourage holdouts to unitize by lowering the maximum producing rate or the gas–oil ratio for the entire field, by offering sufficient economic incentives to assure that all owners in the field gain from unitization, and by granting Rule 37 exception wells to the unit to protect its correlative rights to a fair share of the oil and gas in the pool. The commission's refusal to manipulate prorationing formulas to induce unitization is as much a testament to its success in securing unitization by other means as it is a reflection of its reluctance to engage so openly in "squeeze plays" against the owners and operators of small tracts. The current commission seems wholly dedicated to the prevention of waste. There seems to be little doubt that if the only method of preventing waste in a field were to change the old-style prorationing formula, the commission would do so.[78] In this event, the courts have ample precedent to uphold the commission's new order.

Summary of Judicial Review of the Commission's Statutory Authority

In general, only a handful of judicial opinions have invalidated commission orders on the basis that the agency exceeded its statutory authority.[79] The courts have granted the commission strong and all-inclusive powers to prevent waste, even when the exercise of this power conflicts with the protection of the correlative rights of some oil and gas owners and operators. The courts have also interpreted the statutes to allow the commission to prorate non-wasteful oil wells in order to protect correlative rights. Without this authority, it is doubtful that the commission would have shut down the flaring wells in the Spraberry field to prevent the waste of casinghead gas, because the shutdown order then would have allowed the nonflaring wells with pipeline connections to drain oil from the flarers—an unjust and politically unacceptable result. Thus judicial support of a broad power to protect correlative rights facilitated the commission's efforts to prevent waste. The court's silence regarding Article 6014(g)—the legislative prohibition against issuing waste-prevention orders that would require repressuring or unitization—was so deafening that producers did not even attack the shutdown orders as illegal on this ground.

Judicial construction of the natural gas conservation statutes has similarly accorded the commission strong powers to prevent waste in natural gas fields. However, the justices' narrow reading of these same statutes has tied the commission's hands in the performance of its duty to protect correlative rights in gas fields. By denying to the commission the statutory authority to consider correlative rights in the ascertainment of the market demand for gas, the court has precluded the agency from prorating gas on an equitable basis. Ironically, the commission's only tool for correcting inequities in the gas-prorationing system is to grant Rule 37 exception wells to those operators whose gas is being confiscated. The East Texas legacy lives on, even in gas fields!

Clearly, the Texas Supreme Court has been a full partner in the Railroad Commission's efforts to prevent the waste of oil and gas in Texas fields. While this partnership has often succeeded, judicial cloaking of the commission with broad powers is not a full substitute for compulsory unitization legislation. A compulsory unitization law would allow the commission to prevent waste without sacrificing the protection of correlative rights. Were the Fig Ridge oil field unitized, both Sun Oil and Texaco would share fairly in the increased oil recovery of 800,000 barrels resulting from the bonus-allowable rule. Were the Bammel gas field unitized, condensate recovery would have been maximized, and the dry gas and condensate shared fairly between Corzelius and Harrell. Indeed, had the field been unitized, Harrell's leases on 4 percent of the acreage probably would have never lapsed and fallen into Corzelius's hands. The partnership of the court and the commission cannot always correct shortcomings in the statutory law.

Judicial Review of the Reasonableness of the Railroad Commission Orders

The Substantial Evidence Rule and Prorationing Orders

Most parties seeking to overturn a Railroad Commission rule or order do so on the basis that the order is unreasonable, not that the commission has acted outside its statutory authority. For many years, until the passage of the Administrative Procedure and Texas Register Act of 1976 (APTRA), judicial review of Railroad Commission decisions occurred under the substantial evidence rule. Using this rule, the courts inquired whether the agency's orders were supported by substantial evidence, and, if so, the order was sustained. The judges did not review the record made at the commission hearing, but heard the case anew, and either party could introduce evidence at trial that had never been considered at the administrative hearing. The scope of judicial review under the substantial evidence rule was described as follows by the Texas Supreme Court:

> This does not mean that a mere scintilla of evidence will suffice, nor does it mean that the court is bound to select the testimony of one side, with absolute blindness to that introduced by the other. After all, the court is to render justice in the case. The record is to be considered as a whole, and it is for the court to determine what constitutes substantial evidence. The court is not to substitute its discretion for that committed to the agency by the Legislature, but is to sustain the agency if it is reasonably supported by substantial evidence before the court.[80]

The substantial evidence rule was often criticized.[81] Few commission orders involving controverted fact issues could be overturned under this test. Because conflicting evidence usually exists in contested cases, the substantial evidence rule gave commission decisions a high degree of finality. Also, because the court judged the commission's orders on the basis of evidence presented at trial, the record produced at the agency hearing was a virtual nullity and the commission had little incentive to develop fair hearing procedures.

The Administrative Procedure and Texas Register Act of 1976 (APTRA) amends the substantial evidence rule by requiring that courts conduct their review on the basis of the record made before the agency.[82] APTRA does not significantly change the scope of judicial review afforded commission orders however. The new act states that "the court may not substitute its judgment for that of the agency as to the weight of the evidence on questions committed to agency discretion,"[83] and directs that the court shall reverse or remand a case if the administrative decision is "not reasonably supported by substantial evidence in view of the reliable and probative evidence in the record as a whole" or is "arbitrary or capricious."[84] Thus, Railroad Commission orders still carry a high degree of finality.[85]

The effect of the scope of judicial review on the commission's substantive decision making is illustrated by contrasting two prorationing cases: *Marrs v. Railroad Commission*[86] and *Railroad Commission v. Humble Oil & Refining Co.* (the Hawkins field case).[87] The *Marrs* case in 1944 was the first prorationing case to reach the Texas Supreme Court, all others having been litigated in the federal courts.[88] The court invalidated the commission's prorationing order for the Gulf–McElroy field because it allowed drainage from the less densely drilled area of the field to the more densely drilled areas. In deciding the case, the court did not seem to use the substantial evidence rule, which would have resulted in an affirmance of the commission order.[89] The *Marrs* case created considerable confusion over the proper scope of judicial review of agency orders, so much so that in 1946 the court seized the occasion of a Rule 37 case to expound upon and commit itself fully to the substantial evidence rule.[90] In the same year, the Texas court of civil appeals decided the Hawkins field case which institutionalized small-tract drilling and made compulsory pooling legislation impossible to enact in Texas. The jury in this case had found that the 50 percent per well–50 percent surface acreage allocation formula for oil allowables in this field would result in drainage of 30 million barrels of oil from Humble Oil, the largest tract owner with 76.5 percent of the recoverable reserves, to the small-tract owners in the Hawkins townsite. The district court had therefore invalidated the commission order. The appeals court reversed the district court despite this evidence, and sustained the commission's order as reasonable under the substantial evidence rule. The appeals court reasoned that, as a corollary to Rule 37 exception permits which were granted to small tracts to prevent confiscation, the allowable from these exception wells could not "be cut down to the point where (the) well would no longer produce nor below the point where it could be drilled and operated at a reasonable profit."[91] The court then justified the disparity in withdrawals between large and small tracts on the basis that large-tract owners had lower drilling and production costs from their more widely spaced wells and also benefitted greatly from the overall effects of market demand prorationing which kept the price of oil high.[92]

The Texas Supreme Court refused the writ of error in the Hawkins field case, finding no reversible error. This refusal may have been because Humble Oil had partially acquiesced in the 50–50 formula initially so that many wells had been drilled by small-tract owners relying on this formula, rather than for the reasons given by the appellate court to sustain the order.[93] Whatever the reason, the result was to sustain "living allowables" to small-tract producers as a matter of course for years to come.[94]

The court's change of mind in 1961 regarding living allowables in the *Normanna* and *Port Acres* cases has already been chronicled in chapter 4 as the essential—indeed definitive—factor that finally allowed passage of a compulsory pooling act in Texas. These two cases invalidated allowable formulas with large per-well factors as unreasonable and discriminatory, even under the substantial evidence rule. As we have seen, the court was the only institution with enough distance from the political process to be able to deal this blow to small-tract drilling. The *Normanna* decision, perhaps more than any other action by the judicial branch, encouraged unitization by outlawing allowable formulas that are irreconcilable with unitized operations. However, the decision operates only prospectively. The Hawkins field, East Texas field, Yates field, and many others are untouched by the *Normanna* decision or by the compulsory pooling act which operates only on fields discovered and produced after the date of this opinion. The hapless large-tract owners in the Appling field could find no succor in the *Normanna* decision.[95]

Still, lest the supreme court's heroism in defying the enormous strength of Texas independent producers be exaggerated, it should be noted that it took the court more than a quarter of a century to gather up the courage to kill living allowables. As early as 1935, the supreme court was well aware of the confiscation of large-tract owners that would result if Rule 37 exception permits were granted to small tracts under a per-well allowable system. In sustaining the commission's authority to grant Rule 37 exception permits to small tracts under the proper circumstances, the court in *Brown v. Humble Oil & Refining Co.*,[96] stated that "in all such instances, it is the duty of the Commission to adjust the allowable, . . . so as to give to the owner of such smaller tract only his just proportion of the oil and gas. By this method each person will be entitled to recover a quantity of oil and gas substantially equivalent in amount to the recoverable oil and gas under his land."[97] Had the supreme court enforced this duty, prorationing orders which allowed small-tract producers to drain oil and gas from adjoining properties would have been prohibited at an early date. However, on motion for rehearing in this case, the court stated that it would be "erroneous" to construe this language as "a ruling that acreage must be used as the sole or controlling factor in determining . . . how much oil or gas may be taken from a tract of land."[98] The court then refused to prescribe any rule or standard to guide the commission's discretionary authority to regulate oil and gas production, except that the agency's actions must be "legal, reasonable, and not arbitrary."[99] When six justices of the U.S. Supreme Court upheld the East Texas prorationing system as rational in the *Rowan & Nichols* cases (see chapter 4 text at notes 110 to 113), it became most unlikely that the Texas judiciary would rule that per-well allocation factors were arbitrary or unreasonable. The Texas judges who sustained the 50–50 formula in the Hawkins field case were clearly influenced by the opinion of the highest court in the land that the East Texas system was reasonable.[100] Likewise, these same judges read the Texas Supreme Court's opinion on rehearing in the *Brown v. Humble Oil & Refining* case as clearly sustaining the commission's authority to issue allowable formulas based on factors other than acreage or recoverable reserves.[101] Had the Texas Supreme Court stood fast with its original opinion in *Brown,* small-tract drilling would have been rendered unprofitable at an earlier time, and compulsory pooling legislation probably would have been enacted sooner at the instigation of the small-tract producer. But with the *Rowan & Nichols* cases in 1940, the Goliath of living allowables became shielded by the U.S. Supreme Court in addition to the impenetrable political armor it wore. It would take a quarter of a century to find a hero willing to slay this formidable opponent. Thus, the courts must accept considerable blame for institutionalizing the per-well allowable system, as well as the ultimate credit for overturning it.

The elimination of large, per-well factors in allowable formulas did not eliminate all problems associated with prorationing orders. After 1961, allowables were usually set on the basis of productive surface acres. The surface acres overlying a reservoir often vary substantially in quality. Some acreage lies over thick sands, some over thin sands, and the sands will vary in porosity and permeability. The owners of high-quality tracts have sought to invalidate as unreasonable allocation formulas that are not based on acre-feet of productive sand, a better measure than surface acres of the volume of recoverable reserves underlying a tract. In *Pickens v. Railroad Commission,*[102] the court upheld a commission order allocating production from the Fairway (James Lime) field on the basis of 50 percent surface-acreage, 50 percent acre-feet of productive sand. Pickens held leases over the best part of the stratum, which varied in thickness from 15 to 115 feet. Pickens argued that he would suffer drainage under the commission's formula, and that the formula should be based wholly on acre-feet of productive sand.

The court upheld the commission order under the substantial evidence rule, even though the justices admitted that much of the evidence would have supported a different order. The justices accepted evidence that Pickens's tract was structurally advantaged and Pickens would benefit from net in-migration of oil, unlike the plaintiffs who had successfully overturned prorationing formulas in the Normanna and Port Acres fields. In addition, the thinner tracts at the edge of the fields would be the first to flood out in the water drive, and so the owners of these tracts needed a formula which would allow them to withdraw oil at a higher rate than thicker tracts in the middle of the fields. If not, their oil would be pushed updip before they could recover it. The court cited the *Rowan & Nichols* cases and the Hawkins field case, all of which had expressly repudiated the argument that proration orders had to be based on recoverable reserves in place under each tract.

The prorationing formula promulgated by the commission in the Fairway field was the same formula used to determine participation factors in a voluntary unitization agreement signed by 88 percent of the producers in this field. Pickens had, of course, refused to join the unit on these terms.[103] One can imagine the disruption to the voluntary unitization effort in this field if the commission had not adopted the same allocation formula used by the unit proponents. The court was not about to create additional chaos and delay in the negotiating process. By affirming the commission order, the court clearly indicated that the owners of structurally favored tracts need not be awarded the full benefits of their natural advantage. Updip tracts in a water-drive field need not be accorded allowables that give them more than the amount of recoverable reserves in place. In post-1961 fields, the incentive to hold out is not likely to arise from the operation of a regulatory system which favors small-tract drilling, but rather from the existence of structural advantages. The court in *Pickens* has sanctioned prorationing orders that reduce this advantage and thus discourage holdouts.

Rule 37 Orders

The analogue of the *Pickens* case in the context of judicial review of Rule 37 cases under the substantial evidence rule is the case of *Railroad Commission v. Manziel.*[104] Having held that trespass law did not govern the disposition of the case, the supreme court assessed whether the commission's grant of a Rule 37 exception location for a saltwater injection well was reasonably necessary to protect the correlative rights of the Whelans who had unitized their leases and were water flooding. The evidence showed that the Manziels were benefiting from the fieldwide migration of oil toward their tract because they had not repressured part of their land and the oil was flowing from the areas of high pressure on the Whelans' tracts to the low-pressure areas on the Manziels' tract. The court found substantial evidence to support the commission's decision that the irregular well location was necessary to protect the correlative rights of the Whelans. The data showed that the Whelans had received a disproportionately small share of the primary oil in the field as compared with their ownership of reserves. The irregularly spaced injection well would allow the Whelans to recover a disproportionately large share of the secondary oil left in the field.[105] The court approved of this result. The inequities of prorationing during primary recovery can be counterbalanced by commission orders affecting the distribution of secondary recovery oil. The court clearly considered it reasonable to prevent the Manziels from reaping the benefit of the repressuring operations next door when they did not pay any of the costs. Just as in the *Pickens* case, the court sanctioned a commission order that reduced the advantage of a structurally favored tract. Thus, both prorationing orders and well-spacing orders can be used by the commission to discourage holdouts by assuring unit participants an opportunity to recover the oil and gas in place underneath their tracts. This measure of correlative rights is fully consonant with unitization principles.

Most Rule 37 exception cases brought to the courts for review involve permits granted to drill oil wells during the field's primary production, not for saltwater injection wells used in secondary recovery. Just as the courts played an important, albeit belated, role in rationalizing prorationing formulas to accord better with the principles of unitization, so did the courts provide the only bulwark against the commission's profligate grant of Rule 37 exception wells during the era from the 1930s to the 1960s. Because fields are more easily unitized when fewer owners have vested interests in existing, but unnecessary wells, the Texas courts provided some encouragement to unitization by policing the Rule 37 exception process. Often the commission had absolutely no basis in fact or law for granting the Rule 37 exception applications.[106] When the courts overturned commission permits to drill Rule 37 wells on the basis that voluntary subdivisions had no legal right to be protected from confiscation, the commission started granting

Rule 37 permits on the basis that they were necessary to prevent waste. The "more wells, more oil" theory was rejected by the Texas courts, and the definition of waste was tightened to include only the ultimate loss of oil.[107] If existing or regularly spaced wells could recover the oil from underneath the voluntary subdivision, waste exceptions were invalid. The courts required proof of unusual reservoir conditions in order to justify a waste-exception permit.[108] The courts also refused to uphold commission orders granting Rule 37 permits to prevent fieldwide drainage which would "drill down the field" to the density of the most densely spaced area in the field.[109] Fieldwide drainage was to be prevented by a change in the field's prorationing formula, not by the drilling of unnecessary wells. That prorationing formulas were immutable, absent changed conditions, was immaterial to the court. In this regard, the court interdicted the commission's generous approach to granting Rule 37 well permits. Still, the number of Rule 37 cases actually contested and appealed to the courts was so small in comparison to the number granted without opposition, that the court's reversal of some of the permit grants did not provoke a pro-pooling attitude among small-tract owners and producers, as did the court's reversal of per-well allocation formulas in 1961.

By the early 1960s, even the commission had to acknowledge that its profligate granting of Rule 37 exception wells had the perverse effect of squeezing producers at both ends of a cost–price pinch. The drilling of hundreds of unnecessary wells both raised costs and caused overcapacity, which threatened to lower prices. Texas crude could not compete with lower-priced imports. The commission began to police itself and tighten its Rule 37 procedures and decisions.[110] The court generally sustained the commission's tighter standards.[111]

While the Texas courts thus played an important role in policing the Rule 37 orders issued by the commission, judicial activism in this area did not extend as far as desired by the owners of large tracts. On several occasions, large-tract owners argued that Rule 37 permits should be denied the small-tract owner who had been offered a fair opportunity to pool with an adjoining larger tract. The very purpose of granting Rule 37 exceptions was to assure that small-tract owners (who had not voluntarily subdivided) had an *opportunity* to recover the oil and gas underneath their tract.[112] If this opportunity existed via a pooling offer, an exception permit would not be necessary.

The Railroad Commission actually accepted the merits of this argument in its efforts to regulate the Old Ocean field, a condensate reservoir discovered in 1934. Without compulsory pooling or unitization legislation, the commission faced the nearly impossible task of trying to assure efficient recovery of gas and condensate in this field. In the late 1930s, when the owner of a 20-acre tract applied for a Rule 37 permit, the commission denied the application on the ground that the owner had been offered a fair opportunity to pool. The district court sustained the commission's denial of the permit, but the court of civil appeals in *Dailey v. Railroad Commission*[113] reversed because the commission had not made a general rule or regulation prescribing and defining pooling. The commission had not rewritten Rule 37 to conform with its new pro-pooling policy, nor did the Old Ocean field rule seem to include any mention of a change in the Rule 37 process. If the commission wanted to enforce such a far-reaching limitation on the property rights of small-tract owners, due process required that the commission first issue rules governing the change.[114]

While the court in *Dailey* implied that the commission had the authority to condition Rule 37 permits on pooling offers if proper rule-making procedures were followed, this decision made it more difficult to implement such a policy. As elected officials, the commissioners would have found it nearly suicidal to their political careers to amend Rule 37 on a statewide basis to institute this new pro-pooling policy. There is evidence that the commission did, on at least one occasion, issue field orders that conditioned Rule 37 exception permits on the applicant's first having made a *bona fide* effort to pool.[115] However, even those commentators who urged the commission to take a more active stance in favor of pooling, recognized that this sort of "pooling by indirection" presented innumerable, and perhaps insurmountable, legal problems for the agency.[116]

Nonetheless, a bolder court might have encouraged the commission's pro-pooling stance in the Old Ocean field by sustaining the commission's denial of the Rule 37 exception permit in the *Dailey* case. After all, pooling by indirection was exactly what the

Texas Supreme Court directed eleven years later, in 1961, when it invalidated allowable formulas with large per-well factors. Undeterred by the legal problems that would undoubtedly arise in administering pooling without statutory authority, the majority of justices in the Port Acres field case adopted the reasoning that an offer to pool adequately protected the correlative rights of small-tract owners. The majority wrote:

> There is some indication in the record by bill of exception that they (the small tract owners) were offered unitization that would prevent confiscation and afford to them a fair share of the mineral proceeds from the common source. . . . It is to be reemphasized that their permits were granted for the purpose of avoiding confiscation of the minerals underlying their properties and not for the purpose of enabling them to drain the minerals underlying adjoining lands to pay the cost of their operations plus profits.[117]

The time was not ripe in 1939 for such a bold stroke however.

No-Flare and No-Waste Orders

Despite the *Dailey* court's reversal of the commission's pro-pooling order, the commission was not deterred from its ultimate goal of preventing waste in the Old Ocean field. In 1947 the commission held a hearing to discuss ways to prevent waste in this condensate field. Witnesses testified that 14 million cubic feet of gas were being flared daily; that this waste could be prevented if the field were operated as a unit and the gas recycled; that ultimate oil recovery could be increased by 50 percent or 50 million barrels, and condensate recovery could be increased by 20 million barrels.[118] The purpose of this hearing was obviously to threaten the producers in the field with a shutdown order if they did not agree to operate the field cooperatively. In 1948 the Old Ocean field was voluntarily unitized.[119] Thus, the *Dailey* decision did not ultimately impair the commission's power to prevent waste, although unitization was accomplished only after fourteen years had elapsed since the field's discovery.

As we have seen, the court's working partnership with the commission to prevent waste granted the agency enormous regulatory authority. The courts coupled this power with an insistence that the commission's no-flare and no-waste orders were reasonable. The commission's first no-flare order was issued for the Seeligson field in 1947, a few short months after the Texas Supreme Court had committed itself fully to the substantial evidence rule. After interpreting the statutes as bequeathing to the commission all-inclusive powers to prevent waste, the court reminded the parties that the ultimate legality of the order would depend on whether it was reasonably supported by substantial evidence.[120] The outcome of this trial on the merits seemed a foregone conclusion. The Seeligson producers quickly found a market for their casinghead gas and the case was dismissed.

In 1949 the supreme court found that the no-flare order in the Heyser field was reasonable, even though the operators alleged that their pipeline connections would take another five months to complete, and that the shutdown would cause irreparable damage and loss to some wells.[121] One justice dissented, arguing that the order was unreasonable. His opinion illustrates just how hard-hitting and activist the commission had become:

> It is common knowledge that large scale manufacture of natural gasoline and liquefied petroleum gases from casinghead gas is a relatively recent development, and it is also common knowledge that the general availability of beneficial usage for such residue gas is a World War II development still in its infancy and still greatly impeded by the notorious war and post-war shortage of equipment and machinery related to the gas transmission business and other enterprises making large use of petroleum gas. In the Heyser field almost since its discovery some twelve years ago there has been in operation a plant manufacturing natural gasoline and liquefied petroleum gases from the raw casinghead gas—a highly valuable and beneficial operation, which is, of course, specifically approved by the conservation statutes. At the time the plant was erected such operations were generally regarded as rather forward-looking measures for the conservation of the state's resources. Until only three or four years ago there was no available use for most of the residue from that plant and the state evidently thought its flaring—even in the substantial volume of which the Commission now complains—to be a proper incident to the production of oil from the field. When a large potential gas

> purchaser moved into the area, efforts were evidently made in good faith by some of the leading producers of the field to bring about a sale of the residue, but the problem was such as naturally to require considerable negotiation and delay—particularly since the field was divided into a substantial number of different ownerships all or many of whom had to agree on a program, including the investment of very large sums of money in equipment, before the terms offered by the potential buyer could be met. The evidence shows that from the standpoint of the producers the sale of the residue gas on the terms finally agreed upon is of doubtful economic benefit, while the evident independence of the buyer about seeking the product suggests that the arrangement is at least not a highly desirable one from its standpoint. In the face of all this, however, and after the arrangement has been concluded, the equipment ordered, work started for its installation, and an evidently reliable estimate made that deliveries of the residue to the buyer would begin within the not unreasonable time of five months, the field is in effect ordered shut down until deliveries of the gas can actually begin.[122]

In the Flour Bluff field, the no-flare order was upheld as reasonable even though the evidence showed that injecting the gas back into the ground was uneconomic, that a pipeline connection would not be complete for another five months, and that the shutdown would cause a permanent loss of 275,000 barrels of oil by driving the oil into a sand formerly filled with gas, compared to an expected gain of 217,000 barrels of oil equivalent to be saved by not flaring.[123] The court found evidence suggesting that the loss of oil could be minimized by injecting gas to retard the water drive, and then blithely observed that:

> If the prevention of waste of natural resources such as gas is to await the time when direct and immediate profits can be realized from the operation, there would have been little need for the people of Texas to have amended their Constitution by declaring that the preservation and conservation of natural resources of the State are public rights and duties and directing that the Legislature pass such laws as may be appropriate thereto, . . . for private enterprise would not need the compulsion of law to conserve these resources if the practice were financially profitable.[124]

In 1953, in the Spraberry field case, the court made it clear that a no-flare order would be reasonable even if it reduced operators' profits, delayed their recovery of gas, caused problems with creditors, and even if the operators had been diligent in their efforts to make a productive use of the gas.[125] The court wrote:

> The Railroad Commission's hands should not be tied so that it could not prevent the flaring of a great amount of gas in order to recover a small amount of oil even though there were no immediate market for the gas and even though no one could be blamed for the lack of market.[126]

After this series of decisions, operators no longer litigated the commission's no-flare and no-waste orders.

Orders Under the 1949 Voluntary Unitization Act

The courts have seldom been summoned to review Railroad Commission orders issued under the authority of the 1949 voluntary unitization statute. Unit operators have a marked propensity to settle disputes with nonconsenters privately.[127] Only two cases have been found which involve any judicial interpretation of the 1949 Act. The first, *Railroad Commission v. Manziel,* has already been discussed in depth (see chapter 7 text at notes 135 to 162; and chapter 8 text at notes 37 to 40). In the second, *Staples v. Railroad Commission,*[128] the plaintiffs alleged that the commission had illegally approved a unitization agreement in the Spraberry field. Plaintiffs owned a tract of land located immediately adjacent to the eastern boundary of the unit. They alleged that their tract was within the area reasonably defined by development in the field, yet they had never been given an opportunity to enter the unit on the same yardstick basis as the other owners. The plaintiffs had appeared at the commission hearing on the unitization application and had protested the unit's approval. The commission nonetheless had approved the unit and its water-flooding operations, recognizing the plaintiffs' rights only to the extent of providing that water injection wells could not be placed within one mile of the unit boundary. The plaintiffs contended that this provision would not eliminate injury to their tract from the water-flooding operations and that, under these circumstances, the 1949 Act required the commission to deny approval of the unit because the rights of all owners—whether signers or not—were not protected. The commission answered the plaintiffs' complaint by denying all the allegations and filed a motion for summary judgment. The district court granted the motion, and the plaintiffs appealed. The court of civil appeals held that the trial court had erred in granting the summary judgment because the plaintiffs' allegations raised disputed issues of fact which required determination at a trial on the merits.[129] The issue most in dispute seemed to be whether the plaintiffs' tract was within the area reasonably defined by development. If so, the 1949 Act required that they be offered an opportunity to enter the unit on equal terms. The court remanded the case for trial, but there is no further record of the litigation on this case.

The commission order in this case does appear odd. If the plaintiffs' tract were truly outside the productive boundary of the field, then there would have been no need to limit the placement of injection wells to protect the tract. If the tract were inside the productive boundary, then the plaintiffs should have been made a fair offer. It seems that the commission was not sure whether the tract was inside or outside of the field's limits, and the commission had thus issued a compromise order. On remand, the substantial evidence rule would no doubt apply to determine the issue of whether the tract were inside or outside the area reasonably defined by development, and the commission's order would probably be upheld, whichever position the agency finally adopted.

Summary of Judicial Review of the Reasonableness of Commission Orders

Under the substantial evidence rule and APTRA, Railroad Commission orders attain a high degree of finality. Consequently, whether the orders encourage or discourage unitization and pooling, the courts have usually upheld them. For decades, the protection of the correlative rights of small-tract owners was more important to the commission than was the prevention of economic waste and the inequity to large-tract owners provoked by these formulas. For decades, prorationing formulas with large, per-well allowables were sanctioned by the courts under the substantial evidence rule. Ultimately, however, the court outlawed the per-well prorationing system and ended this cause of wasteful drilling in fields discovered after March 8, 1961.

The commission's attitude toward preventing physical waste was quite different from its complacent acceptance of the unnecessary drilling and unfair allocation of oil and gas under the per-well prorationing system. The commission's no-flare and no-waste orders were often the functional equivalent of orders to unitize. The courts upheld these orders as reasonable, despite the protests of even the most conservation-minded majors such as Humble Oil.

In the 1960s two cases involving the correlative rights of unitized landowners versus holdouts reached the supreme court for review under the substantial evidence rule. In both the *Manziel* and the *Pickens* cases, the court upheld commission orders that did not allow holdouts to benefit unduly from fieldwide migration toward their tracts that had been induced by others' repressuring operations.

Still, this antiwaste and pro-unitization partnership of the courts and the commission is not a complete substitute for a compulsory unitization statute. After Pickens lost his court battle to overturn the prorationing order in the Fairway field in February 1965, the commission—in an obvious effort to hasten the completion of a voluntary unitization agreement in this field—drastically cut the field's allowable for March 1965. In July, hearings were held to approve the proposed unit. At the hearing, Pickens continued to protest the approval of the unit on the grounds that his correlative rights were not fairly protected. The commission approved the unit despite his protests, but the unit operator was compelled to place injection wells in suboptimal locations in order to prevent Pickens from recovering more oil than he would have recovered by operating independently.[130] The ultimate lesson of the *Manziel* and *Pickens* cases is that wells which are unnecessary from a conservation perspective must be drilled to protect a unit's boundary lines, absent legislation that can force holdouts into the unit.

The Antitrust Laws

The last fount of judicial opinion which has an important effect on the unitization process in Texas is the body of case law applying antitrust statutes, both state and federal, to unitized oil and gas activities. Very few cases exist on this issue despite the great ferment on the subject in the 1930s and 1940s, when the emerging concept of unitization was alternately praised as a foremost conservation device and condemned as a monopolistic scheme of the major oil companies. While the conservation potential of unitization became an accepted fact, the belief that the antitrust laws would impede operators from cooperating to produce the output of a unitized field led many states to enact legislation which exempted from the state antitrust laws those unitization agreements which were in the public interest of increasing the ultimate recovery of oil and gas. However, the Texas 1949 voluntary unitization act gives only begrudging immunity to such agreements. The act states that in the event a court finds a conflict between the state antitrust laws and the 1949 Act and finds that the provisions of the 1949 Act are not a reasonable exception to the antitrust laws, the legislative intent is that the court declare the provisions of the 1949 Act invalid, not the antitrust laws.[131] As we have seen, the Texas antitrust laws originating in 1889 were antithetical to the very concept of unitization.[132]

In 1983 the Texas legislature enacted a wholesale revision of Texas antitrust laws,[133] modeling them closely on the federal antitrust laws, and even providing that the new Texas statutes "shall be construed in harmony with federal judicial interpretations of comparable federal antitrust statutes."[134] The new Texas Free Enterprise and Antitrust Act declares unlawful "every contract, combination, or conspiracy in restraint of trade or commerce" and declares that "it is unlawful for any person to monopolize, attempt to monopolize or conspire to monopolize any part of trade or commerce."[135] The act then expressly provides that it shall not apply to "actions required or affirmatively approved by any statute of this state or of the United States or by a regulatory agency of this state . . . duly acting under any constitutional or statutory authority vesting the agency with such power."[136] Thus, according to the terms of the 1983 antitrust act, unitization agreements approved by the Railroad Commission are not violative of the state's antitrust law. It is significant that full antitrust immunity for unitization agreements ultimately came through revision of the general antitrust laws, not through a change in the antitrust provisions found in the oil- and gas-conservation statutes.

As we have seen in chapter 4, the 1949 Act grants to the Railroad Commission the authority to approve only certain types of voluntary unitization agreements, mainly those which are necessary to effect secondary recovery operations that will increase the recovery of oil or gas. By expressly limiting antitrust immunity only to the types of agreements which meet the detailed requirements of the 1949 Act, and by expressly excluding only joint operating agreements from the 1949 Act's requirements, the Texas legislature rendered other voluntary cooperative agreements more susceptible to attack under the antitrust statutes. The passage of the 1983 antitrust act does not seem to alter this fact.[137]

Only a handful of state cases even suggest the antitrust issues that might arise from unitization, much less decide the merits of these issues.[138] One case alleged antitrust violations under state law by large-tract producers to eliminate small-tract producers from competing in the Appling field and involved unitization agreements, but was dismissed when the federal courts disposed of the same antitrust issues under the federal antitrust laws.[139] The Texas Free Enterprise and Antitrust Act of 1983 does not grant immunity from the federal antitrust laws to members of a voluntarily formed unit. Thus, the federal antitrust laws are now a greater constraint on commission-approved, unitized operations than the state laws. Also, because the Texas antitrust laws are now so similar to the federal laws, it is interesting and important to analyze the federal court's treatment of unitization issues in the context of antitrust complaints.

Very few cases exist at the federal level involving antitrust issues and unitization.[140] The federal government brought an action under the Sherman Antitrust Act against the operators of the Cotton Valley Unit in Louisiana in 1947, but the case was never brought to trial. The producers of 90 percent of the hydrocarbons in this condensate-gas field had signed a voluntary unitization agreement to recycle the gas and maximize the recovery of both liquid and gaseous hydrocarbons. The Louisiana Conservation Commission had approved the unit's pressure maintenance and cycling operations. The Justice Department did not attack the joint activity of the operators in this regard, but attacked the practice of jointly processing, refining, and reselling the products removed from the wet gas through selected trade channels at fixed prices and terms.[141] The members of the unit were charged with conspiring to exclude others from the business of extracting, processing, and marketing the hydrocarbons from the field, with the effect of eliminating competition among themselves and between themselves and other producers. The government's complaint did not clearly demark the line between legitimate conservation activities and alleged monopolistic acts, but the government did recognize that some of the concerted joint activity by the defendants was proper and necessary and in the public interest of conservation. The federal district court required that the government make a more definite statement in its complaint showing *inter alia* when and how the permissible acts of conservation ended and the alleged illegal conspiracy began. The government failed to supply the additional information and the case was dismissed.[142]

The only case that has decided that unitized activities by a group of operators actually violated the antitrust laws is *Woods Exploration and Producing Co. v. Aluminum Co. of America,*[143] the denouement between the large- and small-tract producers in the Appling gas field. The small-tract producers had bested the large-tract owners in all previous litigation: they had secured Rule 37 permits, a one-third–two-thirds prorationing formula, and additional allowables through manipulation of the gas-prorationing nomination system (see this chapter text at notes 50 to 55). The large-tract owners had litigated every possible avenue to stop the small-tract drilling that would result in massive drainage from their lands. The small-tract owners countered by filing antitrust actions against the large-tract owners in both state and federal courts.[144] In 1971 the Fifth Circuit Court of Appeals, observing that "this battle-scarred antitrust case and its antecedents, both lineal and collateral, have been in litigation without surcease, armistice, or truce since the early 1960's,"[145] held that the large-tract owners who controlled 90 percent of the land in the Appling field had violated Section 2 of the Sherman Antitrust Act by refusing to unitize or pool with the small-tract owners, refusing to transport their gas, and harassing their drilling and pipeline operations.[146] Because of these acts, the small-tract owners had incurred significant additional costs, particularly in having to build their own pipeline.[147] The small-tract owners were also awarded damages caused by the defendants' thwarting of an opportunity for the small-tract owners to participate in the construction of a proposed liquid extraction plant.[148] The court wrote:

> In essence, plaintiffs paint a picture of concerted action by defendants to restrain, hinder, or eliminate plaintiffs' extraction of gas from the common gas reservoir shared with defendants. We are not saying that pooling, unitization and joint operating agreements are in themselves maligned under the Sherman Act, but even if we consider that the Act impliedly immunizes these collective activities as benign in themselves, they cannot be the instruments of economic predatism or oppression.[149]

Thus the unitization and pooling agreements made among the large-tract owners were monopolistic because the small-tract owners were not offered an opportunity to join the unit. The Texas voluntary unitization statute requires that all owners in the field be offered an opportunity to join the unit on the same yardstick basis.[150] Of course, the small-tract owners and large-tract owners will have very different views of what constitutes a fair yardstick. The yardstick offered by the large-tract owners is unlikely to be accepted by small-tract owners, who have a per-well prorationing formula. Whether or not the commission actively enforces this provision of the 1949 Act, the very real threat of incurring antitrust liability for the refusal to deal virtually ensures that all owners in a field are offered an opportunity to join the unit. In light of the *Woods Exploration* case, no large-tract operator would risk gerrymandering unit boundaries to squeeze out owners who have not been offered a chance to join the unit. The judges on the Fifth Circuit defined the relevant market area within which monopoly power existed as the single field.[151] The court's rationale for this definition was that under the rule of capture, one landowner can drain another from the same common source of supply and can thereby destroy competition in the field. Small-tract owners *must* be bargained with under both the 1949 Act and the federal antitrust laws. This does much to explain why case law on the tort liability of a unit to nonconsenters is virtually nonexistent in Texas; the nonconsenters generally have been appeased.

The Fifth Circuit judges also held that the plaintiffs' charge that the large-tract owners had filed false nomination forecasts with the Railroad Commission to reduce the total field allowable and thereby reduce the plaintiffs' individual well allowables was actionable under the antitrust laws.[152] The district court judge had directed a summary judgment in favor of the defendants on this count, reasoning that any possible injury inflicted on the plaintiffs due to prorationing orders resulted from the action of the Railroad Commission.[153] Under the "state action" doctrine established by the U.S. Supreme Court in *Parker v. Brown,*[154] the antitrust laws apply only to private business practices, not to state action. Because state regulation of natural gas prorationing was so pervasive, including regulation of production through mandatory prorationing orders, regulation of pipeline activities under the Texas Common Purchaser Act, and regulation of gas prices by the Federal Power Commission, the district court found that the state action doctrine applied to immunize the defendants from the federal antitrust laws.[155] The Fifth Circuit Court of Appeals, however, found that the commission could not supervise or verify the gas nominations made by producers, and, therefore, the commission often accepted the nominations at face value. The commission was not the real decision maker in ascertaining market demand, and the defendants' alleged subversion of the prorationing system for anticompetitive purposes was actionable. This issue was remanded for a trial on the merits.[156] On remand, the plaintiffs dropped the claim based on false nominations,[157] probably because the small-tract owners were themselves abusing the gas nomination system by filing very high nominations.[158]

In sum, the appellate court disagreed completely with the lower court's opinion that exempted the oil and gas industry from federal antitrust strictures because of the industry's extensive regulation by the state. The appellate judges repudiated this view on the basis that antitrust immunity was never to be lightly implied, and because the Texas statutory scheme itself recognized the vitality of the antitrust laws in the oil and gas industry.[159]

In light of the Fifth Circuit's opinion in *Woods Exploration,* it is interesting to speculate about whether Railroad Commission approval of a voluntary unitization agreement could ever invoke the state action doctrine of *Parker v. Brown* and provide federal antitrust immunity to the unit. If so, this might explain why some producers are so insistent upon receiving commission approval of their cooperative projects, even to the extent of drilling unnecessary injection wells in order to qualify their project as a secondary recovery operation (see chapter 6 at notes 6 to 11).

The state regulation at issue in *Parker v. Brown* is very similar to much of the state machinery used to regulate oil and gas production in Texas. *Parker* involved a raisin-prorationing system established by the state of California to prevent the "economic waste" of raisins, that is, to raise the price of raisins. The state commission's authority to prorate raisins was conditioned on the program receiving the consent of 65 percent of the producers in each raisin zone. A raisin producer sued the prorating commission which sought to restrict his production. The Supreme Court assumed that the prorationing program would violate the Sherman Act if it were organized and effected solely through private contracts among the private producers. But, because the program derived from the legislative command of the state, the court held that federal antitrust law was not a restraint. The court wrote: "We find nothing in the language of the Sherman Act or in its history which suggests that its purpose was to restrain a state or its officers or agents from activities directed by its legislature."[160]

Texas' market-demand prorationing system has many parallels with the raisin program at issue in *Parker*. This is why the federal district court in *Woods Exploration* held that the state regulation of gas prorationing exempted the large-tract owners from federal antitrust liability. However, as we have seen, the Fifth Circuit reversed the lower court and refused to apply *Parker* because there was inadequate state control of the gas-prorationing system to justify a state action defense. The question arises: is it likely that the court would accord federal antitrust immunity to participants in a commission-approved voluntary unitization agreement?

In recent years, the Supreme Court's interpretations of the *Parker* doctrine have resulted in extraordinarily murky law.[161] In 1976 the court held that the *Parker* doctrine extended antitrust immunity only to actions *required* by the state.[162] This holding would preclude any *Parker* defense for members of a unitized operation because the Railroad Commission has no authority to require producers to unitize (although the agency's no-waste orders are often the functional equivalent of a requirement to unitize), nor does the commission even require that operators seek approval of a voluntary unitization plan. However, more recently, the Supreme Court has held that state compulsion of private action is not required in order to exempt private parties from the federal antitrust laws. In *Southern Motor Carriers Rate Conference, Inc. v. United States,*[163] the court decided that state-authorized, collective ratemaking by private groups of motor common carriers was immune from federal antitrust attack under the state action doctrine of *Parker v. Brown.*

The court used a two-prong test to determine when state regulation of private parties is shielded from federal antitrust law. First, the challenged state restraint on competition must be "clearly articulated and affirmatively expressed as state policy," and, second, the state policy must be "actively supervised by the state itself."[164] Texas' 1949 voluntary unitization act could satisfy both prongs of this two-part test. The 1949 Act does articulate an express state policy of excepting certain types of joint behavior from the antitrust laws. The 1949 Act also requires state supervision by the Railroad Commission of the unitized operations, both at the approval stage and throughout the life of the field.[165]

State regulation of unitization agreements is designed to displace the wasteful and deleterious effects of the rule of capture, and this is certainly the type of regulation which *Parker* aimed at protecting from the federal antitrust statutes. The raisin-prorationing program in *Parker* arose because of the market imperfections and instability inherent in agriculture, an area over which the federal government itself has promoted price-support programs and cooperatives.[166] The premise of state regulation of unitization agreements is to avoid the market imperfections which arise under the rule of capture where each person's pumping imposes an externality on his neighbor, and where free-rider problems are pervasive.[167] The Railroad Commission does look carefully at whether the unitization plan being approved will increase the ultimate recovery of oil and gas (see chapter 6 text at notes 19 to 21). The hearing process is open to all interested parties to present evidence for or against the unit's formation. In such a case, absent proof of any false information filed by the unit applicant to mislead the commission, it can be strongly argued that the *Parker* doctrine applies to shield the unit members from federal antitrust liability, if they secure state approval of their plan. In *Parker,* the prorationing scheme required the consent of a large majority of the private producers, but this did not negate a finding of state action. Also, the possiblity that *Parker* would allow the states to create legal havens of monopoly power favoring powerful special interest groups is much more remote with unitization agreements than with prorationing schemes, which are expressly designed to raise prices. Thus, it may be that members of a commission-approved unitization agreement are shielded from the federal as well as the state antitrust laws, notwithstanding the Fifth Circuit's opinion in the *Woods Exploration* cases. However, the *Woods Exploration* cases attest to the power of the federal antitrust laws to provide protection to independent producers in fields dominated by a few majors if commission supervision of the actions of private parties is inadequate. The dearth of antitrust cases against unit operators over the past half-century suggests that unitization agreements that treat all owners in the field fairly and that are conducted to increase the recovery of oil and gas are safe from antitrust liability under both state and federal laws.[168]

Summary of the Role of the Courts

These last two chapters have examined court decisions involving the common law and judicial review of commission orders affecting unitization. Most common law doctrines are inimical to unitization, especially the rule of capture, implied covenant law, and the tort of trespass. The Texas courts have grafted other antiunitization doctrines into Texas common law, usually unintentionally, but nonetheless with the result that voluntary efforts to unitize are made more difficult. The best example of such a grafting is the rule that the owner of the executive right to lease another's land cannot pool or unitize the interest subject to his leasing power without the consent of the nonexecutive owners. The Texas courts have refused to recognize equitable pooling or unitization by judicial decree, it being their view that the legislature, not the courts, should change Texas' fundamental rules of property. Yet within this hostile framework, the same courts have demonstrated very strong support for unitized operations that redound to the public benefit by increasing the ultimate recovery of oil and gas, when this issue has squarely presented itself. The Texas Supreme Court has ruled that secondary recovery operations approved by the commission cannot be enjoined as a trespass in a cause of action brought to invalidate a commission order. More than likely, the same public interest in secondary recovery will prevent holdouts from receiving injunctive relief or punitive damages in private suits against unit operators alleging injury to the nonconsenter's property interests. In such a lawsuit, the technical rules of trespass will probably be ignored and the reasonableness of the unit operator's actions will be considered in judging liability.

Still, the correlative rights of the nonconsenters must be protected, and many sources of law perform this function. The first is the constitutional constraint against the taking of private property. This constraint keeps the common law of tort alive. Second is the commission's statutory duty to protect correlative rights, not just as enunciated in the 1949 Act, but also as interpreted by the courts in their judicial review of agency orders generally. Third is the federal antitrust law. While the production operations of a commission-approved unit are probably exempt from the antitrust law, the antitrust law will still apply to assure that unit participants do not refuse to deal with some operators in the field who are then denied access to pipeline and processing facilities owned by the unit members. All of these sources of law explain why participation formulas in voluntary unitization agreements generally assure that the owners of structural or regulatory advantages receive the economic rents from their positions of favor. It also explains why nonconsenters seldom protest the unit's formation at commission hearings. Their correlative rights to "go it alone" have been protected.

The common law also explains why the unit's members invariably seek commission approval. Agency approval of the unit will be persuasive, although not conclusive, evidence of the reasonableness of the unit's activities in private lawsuits alleging that the unit has committed a tort. Texas' new antitrust law provides further reason to seek this approval: absolute immunity from state antitrust laws. The U.S. Supreme Court's recent applications of the *Parker* doctrine strongly suggest that the production operations of a unit will be shielded from the federal antitrust laws if the unit has secured Railroad Commission approval of the unit.

The court's greatest role in encouraging unitization has been in bequeathing the commission vast authority to prevent waste, unhindered by the prohibition in Article 6014(g) against issuing orders which require repressuring or unitization. The overthrow of prorationing formulas with large per-well factors is often cited as the court's most significant contribution to conservation in Texas, and there is no doubt of its importance. But the court's action here was belated, the ruling operates prospectively only, and the justices were essentially correcting the mistakes made by their brethren in the *Rowan & Nichols* cases and the Hawkins field case which had supported irrational and wasteful prorationing formulas for decades.

Despite the Texas courts' refusal to sanction a wholesale revision of old-style prorationing formulas in already developed fields, the commission undoubtedly has the power to revise prorationing formulas to prevent waste. A changed condition such as the need for repressuring or secondary recovery operations to prevent waste can justify a change in field rules. The fact that the new prorationing formula omits a per-well factor or fails to credit a structural advantage and thus discourages "going it alone" should not invalidate the new order. Commission orders which strongly encourage pooling and unitization are within the commission's statutory authority and will be upheld by the courts. Even if the new prorationing formula has a serious impact on the correlative rights of once-favored owners, the new order will be upheld because the public interest in conservation transcends private rights.

Yet despite the court's working partnership with the commission to promote unitization in Texas oil and gas fields, the lack of compulsory unitization legislation is still felt. In the Fairway (James Lime) field, the commission resorted to three administrative measures to push operators into unitizing the field for pressure maintenance soon after its discovery. First, the commission adopted a rule to limit drilling by permitting 160-acre spacing. This rule ran afoul of the court's policy of strictly construing pooling clauses in leases. Consequently, some of the lessees who had pooled acreage into the permitted, but not prescribed, 160-acre units, found that the pooling was invalid and that their leases had expired. The unit operator then had to negotiate with the unleased mineral interest owners of these tracts of land. The commission's second act was to promulgate a field prorationing order that used the same formula as the participation formula in the voluntary unitization agreement signed by most of the operators in the field. The court upheld this order despite the contentions of Pickens, a holdout, that it violated his correlative rights and amounted to compulsory unitization. Third, the commission drastically reduced the field's allowable to hasten the completion of the voluntary unitization agreement before it was to expire by its own terms. Still Pickens did not join the unit. He had a strategically placed tract, and the unit had to incur additional expenses to protect its boundary lines and prevent the migration of oil to Pickens' land.

Similarly, without compulsory unitization in gas fields, the commission's authority to protect correlative rights is limited by the statutory requirement that fields cannot be prorated below market demand except to prevent waste. Under this constraint, it is virtually impossible for the commission to protect the correlative rights of cyclers against those who refuse to cycle, other than to limit the field's allowable to that which can be produced without waste. In dry-gas fields, physical waste is not an issue, and the commission can protect correlative rights only by granting Rule 37 exception wells.

Both the courts and the commission have made the prevention of physical waste paramount to the protection of correlative rights. The result is to create a new-style regulatory advantage in the owners and operators who benefit from orders designed to prevent waste. Ironically, this regulatory advantage makes voluntary unitization more difficult to achieve in some fields. It is unlikely that the Fig Ridge oil field will ever be unitized now that the Texas Supreme Court has approved the drainage of 800,000 barrels of oil from Texaco to Sun Oil in this field on the basis that this drainage is an unavoidable incident of the commission's higher duty to prevent waste. Sun Oil has no incentive to accept a voluntary unitization agreement offered by Texaco. Similarly, those producers who own oil wells in a gas-cap field have little incentive to unitize with the gas-cap well owners whose wells are severely restricted in the amount of gas that can be withdrawn under statewide Rule 49(B). The oil producers have a regulatory advantage in the non-unitized status quo.

In the final balance, the court's influence on the progress of unitization in Texas must be assessed as positive. The courts have done their best to accommodate the needs of unit operators to the hostile framework of Texas' early common law and legislation. This inimical common law came into being before pooling and unitization were recognized concepts in legal, scientific, or industry circles. Once pooling and unitization became accepted conservation measures in the 1930s, antithetical legislation was enacted to assure that the Texas independent would never be forced to pool or unitize. While instances can be found where judicial activism fell short of promoting pooling and unitization to the maximum extent possible under the existing law, the courts in most cases simply could not reform the common law and the statutes without usurping the function of the legislative branch in a fundamentally undemocratic manner.

9

Statistical Assessment of the Degree of Unitization Achieved in Texas

Texas lacks a compulsory unitization statute, but the Railroad Commission and the courts have forged a legal framework that has nonetheless encouraged—sometimes through not-so-subtle coercion—the unitization of many oil and gas fields in Texas. It is impossible to estimate statistically how much additional production is directly attributable to Railroad Commission field orders designed to force operators to voluntarily agree to unitize. But an assessment of the extent to which Texas oil fields are actually unitized, especially in comparision with those of Louisiana and Oklahoma, will provide some perspective on the effectiveness of Texas' unique approach to unitization and on the ultimate question of whether a compulsory unitization statute is needed to maximize Texas resource potential.

The Extent of Unitization

In 1979 secondary recovery projects, defined as projects using injection wells to increase recovery, produced about 60 percent of Texas' total oil production for that year.[1] Not all of these secondary recovery projects involved unitized fields. In fact, of the 3,298 secondary recovery projects active in 1978 and 1979, only 821 were unitized projects.[2] If only unitized projects are analyzed, the data show that 48 percent of all of the oil produced in Texas in 1979 came from such operations.[3] This percentage is a minimum because it does not include production from fields that are completely owned by one landowner and leased to one operator, which therefore can be produced as a unit without the need to join anyone else. For example, Exxon owns the oil and gas leases on the huge King Ranch and can operate the fields within the ranch borders as a unit without unitizing. This lease totals more than 1 million acres.[4] The 48 percent figure also does not include fields like East Texas in which pressure maintenance is occurring without unitization. By comparison, Oklahoma produced 39 percent of its oil from secondary recovery projects in 1982.[5] The Oklahoma Corporation Commission does not calculate the amount of oil that comes only from unitized operations, but it is likely that a significant number of secondary projects in Oklahoma, as in Texas, are not unitized. Thus the 39 percent figure is probably an overestimate of the percentage of oil derived from unitized operations in Oklahoma, compared with a minimum of 48 percent for Texas.[6]

Thus, Texas produces a larger amount of oil from unitized areas than does Oklahoma, a neighboring state with a long history of oil production much like Texas' history, but which has had a compulsory unitization statute since 1945.[7]

Louisiana has had an even longer experience with compulsory unitization. In 1940 Louisiana enacted a statute that allowed the commissioner on his own motion to compel unitization for cycling in condensate-gas fields.[8] The 1940 Act also included provisions for compulsory pooling into drilling units in all fields.[9] In 1960 a broader, compulsory unitization act was passed, applicable to all oil and gas fields whether in primary or secondary recovery.[10] This act requires the consent of 75 percent of all working interest and royalty interest owners and requires that the entire reservoir be unitized. With this strong conservation legislation, 64 percent of Louisiana's total crude oil production in 1982 was produced from unitized fields.[11] Louisiana's wells are, on average, a third more productive than those of Texas. In 1982 the daily average production per well was 13.4 barrels in Texas, 17.9 barrels in Louisiana, and only 4.4 barrels in Oklahoma.[12] Thus, while Texas appears to have accomplished more unitization and greater efficiency than Oklahoma, notwithstanding the lack of compulsory process, Texas lags significantly behind Louisiana in both respects.

The percentage of oil produced from secondary recovery operations in Texas has steadily increased from about 25 percent in 1960 to 53 percent in 1970, to about 60 percent in 1976 through 1979.[13] While historical data are not available for the percentage of oil produced from unitized operations, this statistic undoubtedly follows the same upward trend over the years.

Appendix IV lists the number of unitization agreements approved by the commission annually from 1949 to 1978. The data show a slow but steady increase from one agreement in 1949 to twenty-two in 1958. Then in 1959, the number doubled to forty-four and continued to increase, averaging about sixty-four per year from 1960 to 1968, more than one per week. In 1969 the number dropped to forty-nine, then to thirty-eight in 1970 and has averaged about fifteen per year from 1971 to 1978. Clearly, the 1960s were the decade of unitization, caused by the cost–price squeeze which pinched producers' profits and forced more efficient practices. The decline in unitization agreements in the 1970s probably reflects the fact that only the most difficult fields to unitize remain.

Since 1949, the first year of the act authorizing Railroad Commission approval of voluntary unitization agreements, more than 1,000 units have been approved. The Railroad Commission reports 1,012 units as of 1978 (see Appendix IV); Byram's reports 1,028 as of 1982.[14] The amount of additional oil expected to be recovered by these unitizations totals almost 17 billion barrels of oil.[15] This is a whopping amount, especially when compared to Texas' yearly production of crude, which averaged about 1 billion barrels per year from 1951 to 1979. Thus, the average unitization agreement approved by the commission was expected to increase additional oil recovery by almost 17 million barrels. This average conceals vast differences among projects, however, ranging from SACROC's expected additional recovery of 720 million barrels to many unitized projects covering only a few hundred acres and expected to increase additional recovery by a few hundred thousand barrels.

Characteristics of the Unitized Projects

More detailed data for Railroad Commission Districts 3, 4, and 8 were analyzed to determine some of the characteristics of voluntarily unitized projects in Texas. These three districts were chosen because they are representative of unitization in the state as a whole. The three accounted for 41 percent of Texas oil production in 1981.[16]

Timeliness of Unitization

The most striking characteristic of the voluntary unitization process in Texas is that it occurs only many years after a field's discovery. The average time between its discovery and a field's unitization is fifteen years for District 3, nineteen years for District 4, and twenty-one years for District 8.[17] The three districts together average about eighteen years between the date of discovery and the date of unitization. It is clear that few projects are unitized early in the life of the field for pressure maintenance purposes.[18] Only 518, or 16 percent, of the 3,298 active injection projects in Texas in 1979 were for pressure maintenance.[19] The vast majority, 82 percent, were for water flooding. By contrast, in Louisiana where the 1960 compulsory unitization statute explicitly allows unitization to reduce wasteful drilling and increase recovery during primary production, 150, or 67 percent, of the 223 units formed in the first seven years after the act's passage were for primary recovery, and only 73, or 33 percent, were for secondary recovery operations.[20] To some extent, the lack of pressure maintenance in Texas was offset by the severe restrictions on producing rates imposed by market demand prorationing for many years. With fields operating at 30 percent or less of their capacity, reservoir pressure was not dissipated rapidly and additional efforts to maintain pressure may not have been required. However, the lack of pressure maintenance during primary recovery also reflects both the difficulty of negotiating unitization agreements early in the life of a field when reservoir data are not fully available, and the difficulty of negotiating agreements in fields with large per-well allowables that encourage small-tract owners to drill and produce on their own.[21] One detailed study of the negotiating time required to form units in seven oil fields in Texas shows that an average of almost six years elapsed between the time that unit negotiations began and the time the unit was formed.[22] Once an oil field approaches depletion of its primary oil, the operators share more of a common interest in cooperative, secondary recovery operations. This type of secondary recovery is relatively easy to agree to because the alternative is abandonment of the field and because much more information is available about each tract's remaining reserves.[23] Pressure maintenance projects, which are not absolutely essential to produce near-term income and which may, indeed, reduce current income, are more difficult to negotiate, even though they may reap the largest rewards in terms of increasing the longevity of the field and its operational efficiency.

Partial Versus Fieldwide Unitization

One would expect that without compulsory unitization, fields often would be only partially unitized. Landowners and lessees in advantageous positions would refuse to join the unit, and thus certain areas of the reservoir would remain outside the unit or the field would be divided into several separate units, each acting independently. The data from Districts 3, 4, and 8 show that this is often the case. In District 8, for example, the average ratio of the number of productive acres in the unitized projects to the number of productive acres in the reservoir was only 48 percent.[24] Only 43 of the 190 unitized projects in this district incorporated 95 percent or more of the productive acreage in the reservoir.[25] Many projects covered less than 20 percent of the reservoir, and were mainly low-pressure, water-flooding operations.[26] Operators often successfully can control the migration of oil across the lines of such units by the placement of boundary wells. Fieldwide unitization is not as important to these water floods as it is to pressure maintenance projects which involve the migration of fluids under higher pressures. Districts 3 and 4 show much higher percentages of fully unitized fields than District 8. In District 3, the units covered an average of 92 percent of the reservoir's productive acreage, and in District 4 the percentage was 79 percent.[27] About half of the unitizations in these two districts were 100 percent fieldwide. Even so, it is clear that a significant amount of partial unitization of reservoirs occurs in Texas.[28] It has been estimated that only 20 percent of Texas oil production comes from fieldwide units.[29] Partial unitization is not as efficient as fieldwide unitization in reducing drilling and operating costs, nor as productive in terms of increasing ultimate recovery.[30] For example, the 87,000-acre Slaughter field is divided into twenty-five unitized areas, and the field also includes twenty-eight nonunitized, secondary recovery projects.[31] Libecap and Wiggins estimate that 427 offset injection wells were drilled along subunit boundaries in the Slaughter field at a total cost of $156 million.[32] These wells prevented migration across subunits lines, but were unnecessary for production purposes.

The Unit Operators

The unitized projects in Districts 3, 4, and 8 range in size from 40 acres to 43,750 acres in the giant Spraberry unit. The average unit in Districts 3, 4, and 8 included 3,348 acres.[33]

As would be expected, the large oil companies are usually the unit operators for the larger units. There were sixty-three unitized projects in 1979 that produced more than 1 million barrels of oil each in that year.[34] These sixty-three giant units produced 381,430,000 barrels of oil in 1979.[35] This represents 80 percent of the total amount of oil produced from unitized operations in Texas in 1979, and 39 percent of the total oil produced in the state in 1979. The operators of these large units were as follows:[36]

Company name	*Number of units operated*
Amoco	14
Exxon	6
Texaco	5
Sun	5
Cities Service	4
Mobil	4
Amerada-Hess	3
Getty	3
Conoco	3
Shell	2
Arco	2
Union	2
Gulf	1
Sohio	1
Phillips	1
Chevron (SACROC)	1
Marathon	1
Quintana	1
Union Texas Pet.	1
Hunt Oil	1
Smith, R.E.	1
Cornell	1

Clearly, the major operators dominate the list, but it is significant that the last four companies listed are not major oil companies by any commonly accepted definition. Three other companies on the list—Quintana, Marathon, and Amerada–Hess—are not often ranked as majors.

The major operators were also the most frequent unit applicants, as shown by the following:[37]

Company name	*Number of unit applications filed from 1949–82*
Amoco	49
Texaco	49
Exxon	47
Sun	46
Arco	40
Mobil	35
Gulf	31
Shell	31
Cities Service	26
Conoco	19
Phillips	14
Union	14
Amerada-Hess	12
Burns, L.T. Estate	12
Chevron	9

Of the top sixteen producers in Texas, only Quintana and Marathon are missing from the list above, and they each had five unit applications. The L. T. Burns Estate is conspicuous as the only independent on the list. The 434 units represented by these fifteen companies account for 42 percent of the total 1,028 units approved in Texas from 1949 to 1982.

Still, while these fifteen companies loom large in both the number of unitization projects which they operate and the size of the projects under their control, many independent oil companies also filed unitization applications. The 1,028 unitization agreements approved from 1949 to 1982 represent 375 different applicants.[38] Eliminating the top 15 or 20 major oil companies from the list still leaves over 350 other companies that secured unitization approvals. Most of these are producers that applied only once, and judging by the unfamiliarity of their names, they are small independents.

Thus, the data bear out the hypothesis that the largest oil companies operate the largest unitized projects. Such projects require legal, financial, accounting, managerial, and engineering skills and resources that are not within the grasp of the average independent. But many independent producers have also availed themselves of the benefits of unitization, and some of the largest independents are the operators of the largest units in Texas.

Data on the Biggest Texas Fields

According to the Railroad Commission, Texas had 10,182 producing oil fields in 1980 and 8,655 gas fields. A detailed annual review of each of these fields to assess their potential for unitization would obviously impose too great an administrative burden on the commission. However, a relatively small number of fields account for most of Texas oil production. Six fields (Wasson, East Texas, Kelly–Snyder or SACROC, Slaughter, Hawkins, and Yates) accounted for 28.8 percent of the state's total crude production in 1976.[39] Ten fields produced 50 percent; 110 fields produced 71 percent. A mere 207 fields produced 78 percent of Texas crude oil in 1976.[40] The Railroad Commission's policy of pushing operators into unitization focused on fields that were sure to bring large additional amounts of oil if unitized. Thus, it is interesting to analyze the extent of unitization in the largest Texas oil fields. Of the 207 largest oil fields in the state, 117, or 56 percent, are listed as having secondary recovery operations that are either fully or partially unitized.[41] Another 43, or 21 percent (including East Texas), are in secondary recovery, but not unitized.[42] The remaining 47, or 23 percent, are not listed and are presumably still in primary production without any pressure maintenance activity.

Again, these data show both that a significant amount of unitization exists in Texas, and that a significant number of large fields are still not unitized. The real question is whether these fields need unitization to be efficiently operated.

10

Conclusions and Proposals for Reform

Does Texas Need a Compulsory Unitization Law?

Survey Evidence

The real question is not how much unitization has occurred, as documented in the last chapter, but how much more is needed. While the commission in the past has undertaken internal staff reviews of some large oil fields in Texas to ascertain whether waste was occurring,[1] it does not have a systematic policy of making annual or biennial surveys of the potential for increased efficiency of even the largest few hundred oil fields.[2] Thus, the only method left to determine whether Texas still requires a compulsory unitization bill is to ask for opinion evidence from the experts in the fields. The major oil companies almost unanimously believe that a compulsory unitization bill is needed in Texas, largely because the lack of such a statute results in less-efficient and less-productive partial unitization and because the expense and delay of negotiating voluntary agreements is so high.[3] The one major who felt that compulsory unitization was not necessary stated that the Railroad

Commission could indirectly force unitization through its control of prorationing and its authority to prevent waste.

At the hearings on H.B. 311 in 1973, it was estimated that an additional 2 billion barrels of oil would be recovered if compulsory process were available.[4] State Representative Milton Fox now believes that a compulsory unitization law would yield a half-billion barrels more oil than that which can be recovered by voluntary methods.[5] William Murray thinks that there is still some room for improvement in the conservation of oil and gas in Texas and that compulsory unitization may be needed, but to a far less extent now than was necessary during his tenure on the commission when it was "desperately" needed.[6] Neither the major oil companies nor any of the persons interviewed during the writing of this book are pressing for passage of a compulsory unitization statute. Representative Fox prefiles a bill on the subject during every legislative session, but the bills receive little attention.

All of the parties surveyed agree that almost all the large oil fields in Texas that need unitization have already been unitized.[7] However, seven of the ten major oil companies responding to the author's questionnaire stated that they knew of smaller fields in Texas that could be unitized profitably but which were not even in the process of being so.[8] The Railroad Commission staff could think of only one oil field that currently needs unitization—a newly discovered field in the heart of the city of Bryan, Texas.[9] The Bryan (Woodbine) field is overlain by hundreds of town lots; an estimated 15,000 to 30,000 royalty interest owners have shares in the reservoir. The field was discovered in 1981 by independent operators and now has more than 200 working interest owners. When the author visited the Railroad Commission in January 1983, one staff member lamented that the field would probably go from primary production to tertiary recovery within a matter of years because of the enormous overdrilling in the field. Many of the town lots qualified for Rule 37 exceptions, and with high oil and gas prices, small-tract drilling was profitable.[10] The commission obviously was aware of the field's problems, but without a major operator to push into organizing a unitization plan and with the daunting prospect of dealing with so many royalty and working interest owners, the Railroad Commission had left the field alone. Also, the commission felt that until the field was fully explored and its limits ascertained, no unitization plan was possible.

The Bryan (Woodbine) Episode

Two months after the author's visit to the commission, the following headline appeared in a Houston newspaper: "Bryan Faces Revenue Loss from Order on Oil Wells."[11] The article continued:

> Conservation-minded Texas Railroad Commission officials, concerned over the large amounts of residue gas being flared at 21 oil well sites in the Woodbine Trend formation, have handed down a no-flare order that has severely limited or halted production in some areas of Bryan.
>
> The order startled City officials, who have received about $1.4 million in bonuses and royalties in the past two years. It surprised oil companies, who say the Railroad Commission has known for two years about the flares and took no action. . . .
>
> Without pipelines to remove the gas, oil companies have been burning the gas off in flares, which are a common sight to passers-by in Bryan's residential neighborhoods and commercial areas.
>
> Jim Morrow, director of the Railroad Commission's oil and gas division, said the order was issued Friday not because the gas posed a health hazard but because of the sheer waste involved.
>
> In December alone—the last month for which the Railroad Commission has complete data—about 235 million cubic feet of gas was burned off in flares. One oil company official said the gas is worth about $3,000 per million cubic feet.

The emergency no-flare order was effective on March 8, 1983, for a period of fifteen days.[12] Fifteen wells were shut down completely, and six wells were given a very short period of time to find pipeline connections.[13] On March 21, the Railroad Commission held a hearing to determine whether the no-flare order should be continued in effect. The findings of fact and conclusions of law in the hearing examiners' report were fully adopted by the commissioners in their final order prohibiting flaring in the Bryan (Woodbine) and other nearby fields.[14]

In April 1983, one month after the emergency no-flare order, the commission held another hearing to consider revisions to the Bryan (Woodbine) field rules in order to prevent underground waste.[15] Some wells drilled high on the reservoir structure were already below their bubble point, the pressure point at which the oil begins to release gas from solution. After the hearing, the commission reduced the gas–oil ratio of the field to maintain the gas pressure in the reservoir. The new gas–oil ratio rule caused average reductions in production from the field's wells of 50 percent.[16] This new rule also served to protect the correlative rights of the owners of shut-in wells by severely limiting the allowables of those wells that had pipeline connections. The legacy of the Spraberry field lives on.

After all this attention, the operators in the field became serious about organizing a unitization effort. Getty Oil Co. spearheaded the effort and is to be unit operator under a planned unitization that includes more than 100 existing wells.[17] History does, indeed, repeat itself.

This Bryan (Woodbine) situation epitomizes all that is right and all that is wrong with Texas' approach to unitizing its oil and gas fields. The Bryan (Woodbine) field will be unitized; the gas flaring will be stopped; the gas will be used for pressure maintenance. But the unitization will take place only after excessive drilling has occurred in the field, only after underground waste has already begun, and only after the Railroad Commission has arm-twisted the operators into preventing waste. The commission waited more than two years to take action in the Bryan (Woodbine) field, unable, and perhaps somewhat unwilling, to regulate the field's early development. Even with the availability of a compulsory pooling statute and a prorationing formula based on acreage rather than per-well factors, the East Texas legacy of Rule 37 exception permits to small tracts and the lack of a compulsory unitization statute created unnecessary and costly investment in oil wells and an inefficient recovery of oil in the early stages.[18] While pressure maintenance and secondary recovery will probably succeed in ultimately recovering most of the field's potential, this success will come only at a higher than necessary cost.

Delay and high costs—these are two characteristics of the unitization process in Texas. A third is the significant degree of partial unitization. This partial unitization results in additional unnecessary drilling as the unit seeks to protect its boundaries from drainage to adjacent tracts, and as the nonjoined owners on the neighboring tracts similarly seek competitive advantage in well placement. Even with lease line wells, partial unitization may also result in an unfair distribution of the field's output, as the non-joining tracts benefit from the repressuring operations of the unit next door, especially during pressure maintenance operations in primary recovery when the high-pressure fluids and gases are more difficult to control.

A fourth consequence of the Texas approach to unitization is that those with a regulatory advantage under the existing prorationing and well-spacing system are able to capture large rents from the unit. This transfer of wealth means that the expected increase in revenues from the unitized operations must be high enough to pay these rents and still reward the other participants in the unit. The expected increase in revenues must also be large enough to cover the increased administrative costs to the unit of negotiating to secure unanimous consent, rather than the consent of only a supra-majority of the interest owners in the field, and to cover the legal risks under trespass and implied covenant law of proceeding without unanimity.

In addition, the lack of more complete unitization of oil and gas fields in Texas increases the administrative burden on the Railroad Commission to resolve intrapool squabbles between the owners inside the unit and those outside, especially as the reservoir's performance changes through time. Disputes over well placement, allowable rates, injection rates, and so forth consume a considerable amount of the commission's time.

Two Other Episodes

This failure to achieve the margin of excellence between the good and the best in Texas is illustrated by two further examples, both of which represent just one of many similar episodes in the administration of Texas' conservation statutes. Any observer of the Railroad Commission's day-to-day activities will recognize these examples as familiar happenings.

The first example illustrates the small losses that occur with partial unitization of a water-flooding operation. In January 1971, Mobil Oil requested commission approval to drill a water-injection well 163 feet from a tract owned by the Otis Chalk

Estate.[19] This estate refused to participate in the water flood. Mobil Oil's spokesperson, C. K. Adam, stated that several operators' meetings had been held to discuss unitization, but that it had been impossible to reach agreement. Adam testified that even with the injection well only 163 feet from the lease line, about 10,000 barrels of recoverable oil would be swept from Mobil's tract to the Otis Chalk Estate. Because of the estate's desire not to join the water flood, Mobil would have to expend $42,000 in carried drilling and operating costs over the next ten to fifteen years. This additional cost would make the water flood uneconomical at an earlier time than if the Otis Chalk Estate had participated, and the earlier abandonment would result in a minimum loss of additional recovery of 6,000 barrels of oil. This 6,000 barrels of ultimate loss was in addition to the 10,000 barrels that Mobil would lose but the estate would gain. Adam stated that Mobil was willing to bear all of these costs, but any additional losses due to the proposed well being placed further from the lease line would cause Mobil to reevaluate whether to proceed. The spokesperson for the Otis Chalk Estate was, of course, arguing that the injection well should be placed 330 feet from the lease line, the regular spacing distance.

The possible loss of 6,000 barrels of oil is not, in itself, significant. But when this type of thing occurs hundreds of times, the small losses accumulate into a significant amount. To this sum must be added the administrative costs to both the industry and the commission of conducting hearing upon hearing on the placement of individual wells and of litigating some of the ultimate decisions in court, as in the *Railroad Commission v. Manziel* case. More important, were the field unitized, the well being sought by Mobil probably would not even need to be drilled.

Fred Young, the only person among the many interviewed who opposes compulsory unitization legislation, argues that the use of compulsory process would increase the length of time and cost of securing commission approval of a unitization plan and that this would ultimately delay unitization more than Texas' current approach of relying on voluntary agreements.[20] Undoubtedly, contested issues would arise more often at the commission level as some owners and operators are forced into units unwillingly. But balanced against this factor is the reduced time necessary to negotiate and form a unit were compulsory process available. Under

Texas' current system, unitization hearings are short and uncontested because the unit proponents have spent many months, and often many years, securing near-unanimity. When unanimity is impossible to obtain, partial unitization often occurs, but this brings with it additional administrative, legal, and operating costs to the unit, as well as an increased administrative burden on the commission to resolve conflicts between those inside and those outside the unit. With wells costing from a half-million to a million dollars each, a unitization agreement that can save the cost of drilling just one well probably justifies any additional administrative costs incurred by both the operators and the commission in securing unitization through contested hearings under compulsory process.

The second example illustrates the continuing costs to society of administering the non-unitized East Texas field. For many years the field was prorated at market-demand factors below 50 percent. In March 1972, the oncoming energy crisis led the commission to raise the field's market-demand factor (MDF) to 86 percent. During eighteen months of operation at this new rate, large schisms arose between the producers in the field as to the field's proper regulation. In August 1973, the Railroad Commission held many days of hearings concerning the field's operation.[21] The principal matter under discussion was a proposal by one operator, General American, to divide the field into a northern section that would operate at a 100 percent MDF and a southern section that would operate at a 50 percent MDF. General American testified that as many as 50 million barrels of oil would be lost to recovery if the field were allowed to continue to produce at the 86 percent rate.[22]

Opposing the proposal were Exxon Co., USA (Humble Oil's successor), Shell Oil, and Mobil Oil. Exxon proposed that the entire East Texas field be allowed to produce at 100 percent MDF, without any severance into northern and southern sections. Both Exxon and Mobil testified that the field's MER was 400,000 barrels per day, not the 210,000 barrels at which it was currently producing, and that the energy crisis justified an increase in the field's total allowable to a 100 percent MDF.[23]

While the battle of words that ensued throughout the month of testimony was couched in terms of the objective of preventing waste in the field, particularly to stop a decline in pressure on the eastern edge of the field, the real fight was clearly over the allocation of the millions of barrels of remaining reserves in the field. With the rise in the market-demand factor, those operators with the best wells—those very wells that had been so invidiously discriminated against in favor of the marginal wells for so many decades after the *Rowan & Nichols* decisions (see chapter 4 text at notes 97 to 117)—gained allowables relative to the operators of marginal wells that were incapable of producing at higher rates. Exxon owned many of the better wells; independent operators owned many of the stripper wells. Thus, the spokesperson for the Texas Independent Producers and Royalty Owners Association (TIPRO) urged the commission to reduce the field's MDF immediately and substantially. TIPRO prefaced its testimony with the following speech:

> Throughout the history of the East Texas field, but particularly since 1966, we have witnessed a series of proposals calling for various types of adjustments in the production patterns and practices in this field. Almost without exception these have involved significant shifts in equity positions of the interested operators and mineral interest holders.
>
> In almost all cases TIPRO officials have concluded it would be inappropriate for an Association attempting to represent independent producers and royalty owners throughout the state to take a position for or against such specific proposals, particularly since most tended in our judgment to promote interests of one or more equity holders at the expense of others. Rarely has it appeared clear to us that the proposed changes are solely motivated by a desire to advance conservation in the field without doing violence to the basic concepts of correlative rights.
>
> The proposal under consideration today would seem to bear far more upon the fundamental issue of conservation than have many earlier proposals.
>
> Clearly, the time has come to make some changes in production practices and patterns in the East Texas field.[24]

For years, TIPRO and independent producers had successfully opposed almost all changes in the operation of the East Texas field, including more liberal on-lease allowable transfers that were adopted statewide (see chapter 5 text at notes 115 to 121). That

TIPRO now proposed so drastic a change in the field's producing rate is proof of the enormous stakes at issue.

Questions of distributive equity aside, the East Texas hearings illuminate many of the existing sources of waste in the field which define the margin between the good and the best.

First, the regulation of gas–oil ratios in the field imposes significant testing and administrative costs on operators and on the commission because of the large number of existing wells. Many of these wells are not necessary for efficient production and, were the field unitized, significant savings in operating costs would be possible.[25]

Second, the various rules which award bonus allowables to operators who reinject salt water in this field have the effect of encouraging the production of water. As the water edge advances, the water must be transported over increasingly lengthy distances from the producing wells to the injection wells, and then must travel this lengthier distance through the reservoir rock to reach the producing wells again. A more stringent water–oil ratio rule or rescission of the bonus-allowable credits would greatly discourage water production and decrease the costs of producing and reinjecting more than a half-million barrels of salt water every day. At the 1973 hearings, Exxon argued that the solution to any possible decline in pressure on the east side was not to reduce the production of oil, but to reduce the production of water. However, recognizing that any drastic change in the water–oil ratio or bonus-allowable rules would have serious consequences to operators of wells producing large amounts of water, Exxon proposed that the commission retain all the existing rules and add another. Exxon recommended that the commission permit the off-lease transfer of allowables from high-water wells to wells with little water production. The existing lease-allowable system permitted on-lease transfers, but not off-lease transfers. Of course, in a non-unitized field, transfer of an oil allowable from one landowner's tract to another would require that the oil company pay double royalties on the transferred oil. Both the landowner from whose well the allowable was transferred and the landowner from whose well the oil is actually produced would have to be paid royalties on the transferred allowable. To encourage off-lease transfers, Exxon proposed that the commission assign an additional allowable to compensate operators for the double royalties and for the administrative costs of making the transfer.[26] With the off-lease transfer-allowable system, Exxon argued that all producers could share in the recommended increased market-demand factor of 100 percent, yet the costs of saltwater production and reinjection would be decreased.

Six weeks after these hearings concluded, the Arab oil embargo destroyed TIPRO's chances of convincing the commission to reduce the field's MDF. But Exxon also did not prevail. The MDF remains at 86 percent today, and off-lease transfers from high-water to low-water wells are not allowed.

In 1978 the staff of the Railroad Commission undertook another three-week review of the East Texas field. The oil companies again appeared with various proposals either to split the field into two parts or to increase its allowable. This time the commission staff independently recommended that the commissioners eliminate the bonus-allowable rule. According to the staff, the rule had become an incentive to produce water rather than oil, and the higher price for crude oil in 1978 made high-water wells profitable to produce even without the bonus allowable. The staff report concluded that the rule was being abused and that it "results in the production of excessive amounts of water without benefit to the reservoir and at the expense of the other interest owners in the field."[27]

This staff recommendation touched off a flurry of letters and petitions to the commission from owners and operators who claimed that elimination of the bonus-allowable rule would be "economic suicide" for many wells and for local school districts and cities dependent on the field's tax revenues. A newspaper report on the furor over the bonus-allowable rule concluded that "the commission is virtually certain to weigh such alleged abuses of the rule against the economic impact on cities in the four-county area."[28] The bonus-allowable rule remains in effect today.

The most disturbing aspect of the East Texas situation is that as long as the field is not unitized, the testimony from operators about the field's proper rate of production—the single, most important issue from society's standpoint—will often be a masquerade for operators' attempts to secure competitive advantage. The Railroad Commission continues to prorate the East Texas field at an 86 percent MDF today on the basis that this figure is the field's

MER. Is it? The public would have a much greater degree of confidence in this conclusion were the field unitized so that political considerations concerning which group of citizens should receive the economic rents from the field would be totally irrelevant to the commissioners' decision making. Indeed, were the field unitized, the question of division of the economic rents of the field over time as production practices changed would be irrelevant to every owner in the field. The participation formula in the unitization agreement would govern. The unit operator would be free to maximize profits by shutting in poor wells and transferring their allowables to good wells without paying double royalties. The commission's regulatory apparatus for the field, including the bonus-allowable rules and the lease-allowable transfer rules, could be dispensed with.

The final consequence of Texas' failure to unitize more fully is that correlative rights in many oil and gas fields cannot be fully protected. The commission's use of its strong conservation powers achieves good results in preventing waste, but these good marks often come at the expense of fairness in the allocation of a field's production. It is this inescapable conflict between correlative rights and conservation in non-unitized fields that creates the possibility that political influence will intrude into the commission's decision making, especially if the issue is one which pits majors against independents. Also, because the commission is charged with a duty to protect correlative rights, any agency order that does result in a significant amount of net uncompensated drainage from one owner to another is open to legal challenge and will often result in lengthy litigation at the agency level and in the courts. This litigation is another social cost.

During the no-flare campaign of the late 1940s, the agency's quandary over the inescapable conflict between its two duties was starkly depicted by Commissioner Murray:

> The Texas Railroad Commission has tried to formulate regulations which will help preserve equities under such cases, but so far the best we have been able to devise are admittedly far from adequate. . . . (I)t is seemingly impossible to accomplish equity by such regulations. Consequently, in many fields conservation measures have halted pending efforts to negotiate unitization agreements.[29]

While the Texas Supreme Court has given the commission considerable freedom to relegate correlative rights in order to further the paramount public interest in conservation, those operators favored by the commission's waste-prevention orders now have a regulatory advantage in the status quo, and this makes voluntary unitization agreements even more difficult to accomplish in many fields (see chapter 8 text at notes 68 to 70 and page 313).

The entire structure of the natural gas prorationing system in Texas defies fairness in the allocation of production from nonunitized gas fields. While the waste of gas is not an issue in most nonassociated gas fields, the drilling of Rule 37 exception wells as a device to protect correlative rights hauntingly recalls the East Texas legacy (see chapter 8 text at notes 66 to 70).[30]

Yet despite all these deleterious consequences of not having compulsory process to facilitate unitization—the long delays, higher than necessary costs, partial unitizations, inequitable allocations, and greater administrative burdens on the commission—the important point is that much unitization does, in the end, take place. Almost half of Texas' oil production comes from unitized fields. That the unitization is accompanied by excessive costs, especially in overdrilling, is not and never has been a serious sin to Texans. The drilling of unnecessary wells brings many virtues: booming local economies and communities, employment opportunities, statewide tax revenues, income growth, and pride. With properly controlled production rates, the unneeded wells are not themselves a cause of physical waste of oil, except in the last stages of a field's depletion, when the higher cost of operating a field may result in its abandonment at an earlier date than would occur if the profit margin between costs and revenues were wider. With controlled rates of production which assure long-lived fields, the discounted present value of the additional oil or gas which might be produced thirty years later at the margin is a remote loss compared to the immediate benefits of drilling now and creating income for rig owners, service companies, field hands, and local communities. Similarly, the hundreds of thousands of worker-hours spent in preparing for and debating the proper operation of the East Texas field every few years provide payroll checks for many engineers, economists, lawyers, and accountants living in both rural and metropolitan areas of Texas.

Thus, the fact that the commission and the courts cannot prevent unnecessary drilling under the existing conservation legislation is not likely to bring reform of the Texas conservation framework. The unnecessary well that Quintana drilled in order to qualify its pseudo-secondary recovery project under the 1949 Act will not inspire a public outcry as did the gas flaring after World War II. Nor will the holding in *Amoco Production Co. v. Alexander*[31] that dictates that lessees must drill Rule 37 replacement wells to protect their lessors against fieldwide drainage, even though these wells are unnecessary to the ultimate recovery of the field's oil and gas. The many recent gas wells drilled by independents on small tracts and the Rule 37 exception wells granted to large-tract operators to protect them from drainage by these same small-tract producers are all sources of wealth to the state of Texas. The excessive drilling in old fields like East Texas and in new ones like the Bryan (Woodbine) spreads the underground wealth to more landowners and to more members of the Texas community. Economists can properly deride Texas' conservation framework, but they should not be surprised if their criticisms do not persuade Texas legislators to reform the system.

Delays in the unitization process are not so sinful either to a Texas politician. If oil is produced inefficiently at first, so much the better for the ultimate success of secondary and enhanced oil-recovery operations. The higher costs of this two-stage process are just as beneficial to the Texas economy as the drilling of unnecessary wells during primary recovery and, in the end, the reservoirs are swept clean of their recoverable oil and gas. Of course, this higher-cost production ultimately results in decreased profits to producers. But when Texas producers are caught in a cost–price squeeze, as in the 1960s and the current oil field recession of the 1980s, they will decide in their own self-interest that unitization is a good idea in order to reduce costs. Voluntary unitization will often result, with time-split participation formulas which give small-tract owners and producers the economic rents of their position of regulatory advantage during primary production.

The greatest evil of Texas' system of voluntary unitization is that partial unitization is so prevalent. A less-than-fieldwide unit may be unable to protect its boundaries completely by a line of offset wells. If a significant number of holdouts refuse to join and the unit cannot prevent migration of oil and gas to the holdout's tracts, the economic feasibility of the unit operation is seriously threatened and the unit may not be formed.[32] However, when the problems of profitable obstructionism, information uncertainties, or ideological opposition are so severe as to threaten defeat of the unit and the waste of a substantial amount of oil and gas, the Railroad Commission is willing and able to arm-twist operators into cooperating through show-cause hearings, no-flare orders, reduced gas–oil ratios, and lower field MERs. Little oil is ultimately lost to recovery. The Texas system works quite well indeed from a Texan's perspective.

Summary of the Effect on Unitization of the Interrelationships Between the Legislature, the Commission, and the Courts

This inquiry into the legal institutions affecting the course of unitization in Texas was prompted by the noteworthy, indeed amazing, fact that Texas—the largest oil- and gas-producing state in the nation—is the only state without a compulsory unitization law. Two hypotheses flowed from this fact, one as remarkable as the other. Either Texas has achieved the efficient recovery of oil and gas by voluntary methods, a feat that no other state has been able to accomplish. Or, Texas has not achieved the efficient recovery of oil and gas, and yet, has not been compelled by the federal government to pass a compulsory unitization law that would demonstrably increase the ultimate recovery of these critical resources to the nation during an energy crisis that has been described as the moral equivalent of war and a decade of enormous federal intervention into all other aspects of the oil and gas industry.

This book has analyzed the evidence required to ascertain which of these hypotheses—remarkable as each is—best mirrors the truth. The conclusion regarding the first hypothesis is that Texas has achieved a remarkable degree of unitization without a compulsory unitization statute. Yet this unitization has not been accomplished by purely voluntary methods. Much of the unitization has been achieved by the coercive power of a state agency. The commission's no-flare and no-waste orders are the functional equivalent of compulsory unitization. Other sources of coercion have also been documented: the PAW's allocation of scarce materials only to pooled or unitized tracts during World War II; and the force of federal and state antipollution laws. Moreover, while much unitization has occurred, it takes a particular pattern in Texas that still results in inefficient production practices—namely, excessive well drilling.

As to the second hypothesis, it is true that Texas has escaped the preemptive hand of the federal government. However, the premise for requiring Texas to adopt a compulsory unitization statute is that such a law would demonstrably increase the ultimate recovery of oil and gas in Texas. Were a federal congressional investigating committee to pursue any allegations of physical waste of oil and gas in Texas reservoirs, the evidence would be difficult to find. The Railroad Commission can point to a long history of dedication to the prevention of waste and to the unitization of Texas oil and gas fields that predates the energy crisis of the 1970s. Texas already has a compulsory pooling statute that aims at preventing the drilling of unnecessary wells which might impose inflationary costs on society (although it applies only to fields discovered after March 8, 1961). And the Railroad Commission in the postembargo era has been actively committed to pressing for the margin of excellence in Texas fields. One by one, Texas' giant old oil fields, which had resisted unitization for so long, were vanquished in the 1970s by the commission's coaxing and coercing and by the organizational efforts of the major oil companies.

Undoubtedly, the investigating committee would find ample evidence of economic waste in Texas' excessive well drilling, but Congress has itself promoted and fostered such drilling for decades through federal tax preferences, such as the percentage-depletion allowance and the expensing of intangible drilling costs. Something in the national political psyche favors well drilling and its employment-related effects, especially on independent operators, almost as much as Texans do. Independent producers have considerable political influence at the national level as evidenced by a long and successful history of securing special legislative and regulatory advantages under the tax laws and price-control laws.[33] Economic efficiency has seldom been the lodestar of federal energy policy.

Nonetheless, assuming that this hypothetical congressional investigating committee concerned itself with economic waste, the committee would probably recognize that the economic waste in Texas' oil fields ultimately results in the physical waste of oil and gas because fields are abandoned at an earlier time. Assume further that this committee discovers some physical waste in some of Texas' smaller oil fields that have not received Railroad Commission attention. The question still remains whether a compulsory unitization law would prevent this waste. Oklahoma has had a compulsory unitization law since 1945, but it is by no means clear that its fields are operated more efficiently than those in Texas (see chapter 9 text at notes 1 to 13). The compulsory unitization laws of most states do not authorize the state's conservation agency to unitize fields on its own motion.[34] Most statutes require that 75 to 85 percent of the owners first reach agreement before presenting the unit plan to the agency for approval.[35] Only a few states authorize the formation of exploratory units or units whose purpose is to save drilling expenses.[36] Most compulsory unitization statutes focus on the prevention of physical, not economic, waste. Few statutes relieve the unit operator from continued regulation by the state conservation agency. The unitization process in states with compulsory laws often is marked by long delays and still results in less-than-fieldwide units and the attendant problems of trespass, protection of lease lines, and disputes over the allocation of the field's output.[37] Were the burden of proof on the federal investigating committee to show that Texas was not meeting its national responsibilities through its own unique framework of conservation legislation, administrative orders, and court decisions, it is unlikely that this burden could be met.

Thus, the reason that the federal government in the 1970s did not override the raw political power of the Texas independent at the state level and the self-interest of Texas citizens in their own state's economic health was because it was not clear that any significant, preventable physical waste of oil and gas existed in Texas that a compulsory unitization bill like that proposed in H.B. 311 or existing in other states would prevent.[38] In an imperfect world, the good is often synonymous with the best. World Wars I and II and the 1956 Suez Canal crisis had not pushed the federal lawmakers to pass a national compulsory unitization law, and the Arab oil embargoes of 1973 and 1978 were no more effective than the previous emergencies in eliciting this sort of federal legislative response.[39]

Other factors also explain congressional inaction during the energy crises of the past decade. The federal energy agencies, which were created to regulate the oil and gas industry, needed cooperation—not confrontation—with the producing states. The expertise, knowledge, and experience of the state regulatory officials were essential to effective regulation of any sort. To accuse any state commissioners or governors of derelictions in their duties would be a step backward in resolving many of the pressing problems at hand.[40] Many of these state officials had been warning federal policymakers for years that excessive reliance on foreign imports of crude oil, and inadequate price incentives to explore for domestic oil and gas, risked the nation's economic and defense security. From 1973 to 1977, the congressional docket was overloaded with energy legislation involving issues of transcendent national concern, such as the distribution of gasoline and heating oil supplies in times of shortages; the imposition of price controls on a large and complex, multilayered industry; conversion of public utilities to coal; and conflicts with environmental legislation such as the Clean Air Act. Texas' lack of a compulsory unitization law was dwarfed by higher-priority concerns involving resource allocation decisions on a far more massive scale.[41] Only in 1976 did a congressional agency, the Office of Technology Assessment, finally begin to study the effect on federal energy policies of the lack of compulsory unitization process in Texas. By the time the agency issued its report indicating that the lack of compulsory unitization in Texas could significantly inhibit enhanced oil recovery projects in that state,[42] the Hawkins, Conroe, Webster, and Yates fields were all either unitized or in the process of being unitized.

In the end, then, the very success of the commission and the courts in thwarting and offsetting the anti-unitization fervor of the Texas legislature explains why Texas does not now have a compulsory unitization act and why one is unlikely to be passed either by the state or by preemptive federal power.[43] Simply put, the good is the enemy of the best. This is a very great irony indeed. Had the Railroad Commission not pursued conservation with such vigor, and had the courts not been so willing to accept the commission's waste-prevention orders and to relegate correlative rights to the primary goal of conservation, then it seems likely that a compulsory unitization bill would have been enacted in Texas, either with or without federal persuasion behind it. Had the Railroad Commission obeyed the intent and spirit of Article 6014(g), and had the courts invalidated any orders which did not; had the commission refused to approve pseudo-secondary recovery projects and thus defeated the formation of certain units; had the Texas courts invalidated commission orders that authorized trespassory invasions of salt water in secondary recovery, and interpreted implied covenant law to make lessees strictly liable for any injury to hold-out lessors; and had the courts invalidated the bonus-allowable rule in water-drive fields and the volumetric displacement rule in gas-cap fields because they violated correlative rights, then many fewer fields would be unitized or operated efficiently today. Waste would be evident and inescapable. The need for a compulsory unitization statute would be incontrovertible. But such waste is difficult to find. The Texas independent can now tout as the strongest reason to oppose enactment of a compulsory unitization bill the fact that it is simply not needed.

The haunting question then arises: if the courts had not been so accommodating toward the Railroad Commission's orders and toward unit operators whose secondary recovery activities were involved in litigation, would a compulsory unitization bill have been enacted as a necessity, bequeathing Texas the best possible conservation framework rather than merely a good one? After all, it was the Texas Supreme Court's invalidation of a commission prorationing order in the Normanna field, and the court's suggestion that sandfracing was an actionable trespass that finally propelled passage of the Mineral Interest Pooling Act. Similarly, the origins of pressure maintenance in the East Texas field arose from a finding that tort liability did exist for saltwater pollution, not from a judicial posture that excused such pollution from the traditional norms of tort law. Had Justice Frankfurter and his brethren refused to accept the obvious imperfections in the regulation of the East Texas field, a more rational, less wasteful, and more equitable solution would have had to be found. Compulsory unitization would had to have been at least considered; and Texas might have adopted a regulatory framework that resolved the conflicts between conservation, correlative rights, and the common law in a better, more expeditious, and more direct fashion than the framework which exists today.

Administrative Agencies and Democratic Values

In the end, then, perhaps the strongest argument for passage of a compulsory unitization law is not based on the increased recovery that it might bring, but on the merits of having a legal framework that is visible, open, and solidly based—not invisible, hidden, and precariously perched. Texas has accomplished a substantial degree of unitization only by virtue of the coercive powers of the state's regulatory agency. This *is* compulsory unitization. It should be recognized as such and codified to remove its secretive and legally tenuous status. In his study of the Railroad Commission's policies and politics, David F. Prindle labels the commission's greatest failure as its nonpolicy toward unitization, its refusal to consider that unitization is within its jurisdiction. Prindle writes:

> Since the 1940s, observers of the petroleum industry who desired its long-run best interests have been advocating that some authority in Texas compel the unitization of its fields. The implementation of this policy would be fairly simple: the Commission would manipulate well spacing and allowable rules so that it was to the advantage of operators to unitize. . . .
>
> But the mere whisper of the phrase "compulsory unitization" around many independent producers or Texas landowners is sufficient to provoke opposition of a most violent and vocal sort. Although there has been some sentiment on the Commission in favor of imposing unitization, the opposition has been sufficiently intense to prevent Commissioners from even considering it as a potential course of action. They have instead, as a group, denied that they have the authority to deal with the question and referred it to the state legislature. But the legislature has on numerous occasions demonstrated that it is no more capable than the Commission of braving the political opposition created by the possibility of compulsory unitization.
>
> Against the billions of barrels of oil that the Commission has preserved with conservation rules and prorationing must be set the millions of barrels that have been lost to future generations because of nonunitized production. Although most of the criticisms of the Commission's policies seem unfounded or overharsh, therefore, there is legitimate reason to condemn its most important nonpolicy.[44]

That a scholar of the Railroad Commission could label the agency's attitude toward unitization as a "nonpolicy" is a testament to the invisibility of the legal framework used to achieve unitization in Texas. True, no commissioner has ever spoken out publicly in favor of a compulsory unitization law for oil fields. But the commission does have a policy of effecting unitization. By necessity, the policy is unspoken and unwritten: since 1931 the law has clearly stated that the waste-prevention powers of the commission are not to be used to require repressuring or unitization of an oil and gas pool.

Prindle also mentions the encouragement to voluntary unitization provided by the 1949 Act, and the success of this statute in finally unitizing the Yates field in 1976.[45] In fact, the 1949 Act played an insignificant role in forming the Yates unit compared to the commission's manipulation of the field's MER, and the 1949 Act is, by its own terms, a begrudging and half-hearted encouragement to cooperative agreements. In the unitization of oil and gas fields in Texas, things are not what they seem to be.

In our tripartite system of government, the legislature is to make the law, the administrative branch to enforce it, and the judicial branch to interpret it. In Texas, the legislature has made laws which are inimical to unitization, and the commission and the courts have largely ignored and bypassed them. The Railroad Commission's policy of effecting unitization has unquestionably served the public interest. The problem is that in so doing, the commission has acted either as a lawmaker or a lawbreaker, neither of which fits the model of a democratic society of checks and balances among the separate branches of government. If the agency were expressly delegated the statutory authority to coerce unwilling owners to unitize, the law would be visible, the agency's legal authority would be firmly rooted in substance rather than form, and academicians, political scientists, and economists would no longer condemn the commission for nonpolicies that cannot be seen and thus are not known.

Time and time again the Railroad Commission has been criticized for not exercising more fully its administrative powers to impose greater rationality on the regulatory system, to encourage more pooling, to decrease the favored status of marginal wells under the Marginal Well Act, and to promote unitization. Undoubtedly, if the commission's traditional tools fail to work as well in the future as they have in the past, the commission will be urged to use its discretionary powers to find new methods of overriding the legislative process, and criticized if it does not. This seems a strange way indeed to run a democracy.

A Final Episode: The Boonsville Cases

The risk of switching from reliance on the commission and the courts to effect unitization, and forcing a showdown with the Texas legislature to compel it to pass a compulsory unitization law is, of course, that unitization will lose the fight. Then no legal mechanisms would remain to achieve unitization in an indirect fashion. One final example epitomizes both the risks and the rewards of forcing the law out into the open, and provides a dramatic illustration of the interrelationships of the three branches of lawmaking in the determination of conservation policy in Texas.

The final episode involves prorationing and the efficient recovery of gas in the Boonsville field. Texas conservation statutes long have prohibited the downhole commingling of oil or gas from different strata. Operators who drill wells that pass through several different strata must install a separate string of casing to produce from each stratum.[46] Without separate casing, gas from a high-pressure zone could move through the well bore and enter a zone with lower pressure, causing a loss of recoverable hydrocarbons. However, in some fields, production by downhole commingling will not cause waste; indeed, it can be an efficient way of producing gas in lenticular fields that consist of many thin layers of sands, called "stringers," which cannot justify the costs of separate well completions and individual casings. Consequently, the commission allows exceptions to the rule against downhole commingling upon a showing that commingling would prevent waste and protect correlative rights.

In 1957 the commission consolidated a large number of separate strata into the Boonsville (Bend Conglomerate Gas) field and treated all the zones as one common source of supply for proration purposes. On August 1, 1975, the commission suspended prorationing in the Boonsville field because of the high demand for natural gas during the energy shortages prevalent at that time. Without prorationing, every well could produce at capacity, and it became profitable to drill on small tracts, which heretofore had limited allowables under the field's acreage formula. Almost three years later, the large-tract operator in the field that had requested the initial suspension asked the commission to reinstate prorationing because the wells which had been drilled on the small tracts were recovering more than their fair share of gas. On July 31, 1978, the commission reimposed prorationing in the Boonsville field on a 100 percent acreage basis. Several owners of wells on the small tracts, calling themselves the Gage group, appealed the commission's proration order, claiming that the commission did not have the statutory authority to prorate production from separate sources of supply. The Texas Supreme Court agreed with the Gage group that the oil and gas statutes only authorized the commission to prorate oil and gas produced from a common source of supply, and that each stringer in the Boonsville field was a separate source of supply.[47] The opinion thus prevented the commission from prorating wells with commingled production. Actually, the Texas Supreme Court had ruled on this very issue only two years earlier, holding in 1977 that the commission had no power to prorate production from separate zones that had been allowed to produce through downhole commingling.[48] The commission had issued its 1978 Boonsville order in direct and flagrant violation of this recent opinion from the highest judicial authority in the state.

Now faced with a second supreme court opinion prohibiting this type of prorationing, the commission began to deny all requests for commingling in order to maintain the integrity of the statewide prorationing system. This meant that oil and gas in certain lenticular reservoirs would not be recoverable.[49] The Texas legislature needed to reform the law if the state was to maximize its recovery of oil and gas.

In 1979 Senate Bill 257 was introduced into the legislature. It authorized prorationing from multiple accumulations of oil and gas upon a commission finding that the multiple strata needed to be operated as a common reservoir to prevent waste. However, the political influence of the independent operators was so great that the legislature refused to pass the bill in this form.[50] As finally passed, the bill did nothing more than grant the commission the authority to permit downhole commingling; it did not allow prorationing of commingled production.

Incredibly, braving two recent Texas Supreme Court opinions and the plain language and legislative history of S.B. 257, the commission called another hearing on the prorationing of the Boonsville field. Claiming that it had the authority to prorate Boonsville's commingled gas production, the commission entered an order on October 20, 1980, reinstating prorationing on a 100 percent acreage basis. The Gage group of operators again appealed this order, and the district court held that the commission had no statutory authority to enter such an order. The commission appealed the holding, but with virtually no chance of success in the higher courts.[51] The commission's lawbreaking had come to an end.

The only solution was true legislative reform. In 1981 the Texas legislators finally granted the commission the authority to prorate production from commingled zones as if they were a single pool.[52] However, the political influence of the independent producers was so strong that several provisions were inserted into the law to restrict the commission's prorationing authority over commingled production. One provision, for example, requires that the commission base its prorationing formula in commingled fields on at least two factors, thus foreclosing a formula based on 100 percent acreage, the bane of the small-tract operator.[53]

On September 1, 1982, the commission issued a final order in the Boonsville field, prorationing the field according to a 95 percent acreage and 5 percent deliverability formula.[54]

If the political power of the independent producers is strong enough both to delay passage of meaningful reform for four years and to secure special treatment in the final law, it is by no means certain that a Texas Supreme Court opinion which invalidated a no-waste order or which enjoined a repressuring operation as a trespass would force the passage of a compulsory unitization bill. Given this political environment, the Texas Supreme Court's opinions restricting the commission's statutory authority in the two commingling cases become the target of criticism: the court could have held that the artificial communication among zones created by commingled production techniques resulted in the field's becoming a common source of supply.[55] This interpretation would have resulted in granting the commission greater authority to prorate commingled production than the 1981 bill that was finally enacted. If this had been done, the statutes would no longer appear to mean what they say, but such an approach avoids the risk that the Texas lawmakers will not choose conservation as a more important goal than the allocation of a reservoir's resources for other politically popular purposes.

Of course, the Railroad Commission is itself a political institution. The commissioners are elected officials, and many of their decisions in the East Texas field—even in the rather recent past—are not above suspicion as having been influenced by political considerations rather than conservation (see this chapter text at notes 21 to 28). To let the system continue to exist in its present form is to premise conservation on the personal convictions and political courage of only two persons (a vote by two of the three commissioners suffices to promulgate an order). The commissioners' slap in the face to the Boonsville independents in the guise of the 5 percent deliverability factor shows that these officials are willing to brave the wrath of the Texas independent, as do the recent shutdown orders in the Bryan (Woodbine) field. These episodes raise the level of public trust and confidence in the commission's decision making and control over other fields, including East Texas. Without exception, every person interviewed during the writing of this book expressed the opinion that the current Railroad Commission is dedicated to the prevention of waste. Even Michel Halbouty, still angry that the agency did not grant his requested order requiring cycling in the Normanna field in the 1960s, is certain that the commission today would take steps to assure that the cycling was instituted.[56]

The very fact that the commissioners are elected officials also serves to legitimatize their active role as lawmakers. While the judiciary is not a strong check on the commission's decision making, the commissioners are ultimately accountable to the electorate. A check and balance does exist to preserve the democratic process against administrative despotism, lawlessness, or irrational decision making not in the public interest. Even when the commission has unquestionably acted outside its statutory authority, this lawlessness does not justify reform of the system or a change in the traditional principles of administrative law. Richard B. Stewart argues that when an agency and all interests affected by an agency decision are content with an unlawful policy, a court should not grant liberalized standing to allow a stranger, or a plaintiff with a strong ideological interest in law enforcement, to challenge the agency action. He writes: "Permitting a stranger to thwart such a mutually advantageous agreement would stifle creative compromise with the dead hand of past legislatures. . . . The abstract concern for vindicating the bare words of statutes seems too attenuated a justification for disturbing mutually satisfactory arrangements, struck by all the relevant public and private interests."[57] If none of the parties affected by the commission's illegal orders (such as the no-waste orders that expressly required repressuring in oil fields, or the orders that approved unitization agreements not meeting the 1949 Act's definition of secondary recovery) cared to litigate the commission's lack of authority, the orders should stand. Thus, even considering the democratic values of maintaining a system of tripartite government, Texas' unique approach to unitization stands up well.

So we are back to where we began: that the Texas method of effecting unitization works well. It is a close call indeed whether Texas should pass a compulsory unitization bill.

Conclusion: Texas Should Have a Strong Compulsory Unitization Law

The problem is, however, that the commission's bag of tools is no longer as useful as it once was. With most fields at 100 percent MERs, the commission cannot promise increased "carrots" to operators as a reward for unitization without injuring the field. Discovery allowables, capacity allowables for water flooding, and increases in a field's MER no longer can function as incentives to channel private investment decisions into desirable types of activity. And in times of oil and gas shortages, shutdown orders and large reductions in a particular field's production rate may do serious harm to users of the oil and gas. Without slack in the system, substitute supplies from other fields may not be available to provide energy to homes and industries affected by the shutdown or slowdown order. The Bryan (Woodbine) episode took place during a time of recession and reduced demand for oil and gas in Texas. Would the commission have issued the shutdown and reduced gas–oil ratio orders during the throes of an embargo-induced energy crisis? Also, drastic reductions in a field's allowable, such as occurred in the Fairway (James Lime) field, penalize both those operators and royalty owners who have agreed to unitize and those who are holding out. A compulsory unitization law would give the commission a finer and more equitable instrument to use in fulfilling its conservation duties than the bludgeon of shutdown orders.

Also, some doubt now exists whether the commission can prorate nonwasteful oil wells below market demand. In 1953 the commission prorated the production from nonwasteful oil wells in the Spraberry field, and the court seemingly approved this result as necessary to protect the correlative rights of the owners of shut-in wells. Yet a few years later, the court refused to allow the commission to prorate gas wells below market demand in order to protect correlative rights (see chapter 8 text at notes 50 to 65). Thus, the commission's use of no-waste orders in the future may be subject to greater legal challenge. The owners of prorated but nonwasteful wells may challenge the orders on the basis that they exceed the authority of the commission. If the commission does not prorate the nonwasteful wells, then the owners of shut-in wells may challenge the shut-in order on the basis that it does not adequately protect the correlative rights of all the owners in the field. Only the most difficult fields remain to be unitized in Texas, including smaller fields that are more likely to be owned predominately by independent producers. These producers are apt to challenge the commission's authority to prevent waste much more readily than the majors.

We have seen that the Railroad Commission's powers to prevent waste through regulation of gas–oil and water–oil ratios, gas-cap withdrawals, MER prorationing, and well spacing and placement assure that good conservation techniques are practiced. These general conservation powers do not assure the best, however. In this new era of energy scarcity with market-demand factors at 100 percent, the margin between the good and the best is all important, even if it is small. It is not enough that Texas' large oil and gas fields have now been unitized or, if not unitized, that they, like the East Texas field, are being operated with careful attention to MERs and pressure maintenance. In these large fields, incomplete unitization causes small losses on the margin. The higher costs of operating a field that is overdrilled because of delays in unitizing or because of partial unitization can result in abandonment of the field at an earlier date than would occur if the profit margin between costs and revenues were wider. Similarly, while the failure to unitize and repressure small-to-medium fields does not result in the physical loss of large amounts of oil and gas since these fields produce only a small percentage of Texas' oil, still, small amounts of physical waste are at issue. For the nearly half-century of the oil industry's history that was characterized by enormous overcapacity, the prevention of such small amounts of waste was not worth pursuing. This half-century ended in 1973.

Texas has the opportunity to pass a strong compulsory unitization law, and in the end, the reasons for so doing are compelling, notwithstanding the efficacy of the current methods of achieving unitization. A strong compulsory unitization act could increase the margin of oil and gas recovered in Texas oil reservoirs. Such an act could effect earlier unitization, reduced drilling and operating costs, and more equitable allocation of a pool's reserves. Even if the additional recovery amounted to no more than a 5 percent increase, this difference is significant in the energy-supply situation confronting the nation now. Just as important, the unitization law would be open and obvious, and the public's trust in the commission's decision making about a field's operating practices would be untinged with doubts about political motivations arising from the inevitable conflict between conservation and correlative rights in non-unitized fields. The Texas legislature has modernized other oil and gas laws in response to the energy crisis, including laws which grant the power of eminent domain to firms who want to use depleted underground reservoirs for oil- and gas-storage purposes.[58] The Texas legislature should take a leadership role among states and pass a strong compulsory unitization law, which would assure that Texas conservation laws and practices are the finest in the nation.

Proposed Reforms in the Absence of Compulsory Unitization Legislation

The history of compulsory unitization proposals in Texas does not augur well for passage of such a bill in any form. Texas has developed a unique approach to unitization that works quite well, and the passage of a compulsory unitization law that is no better than those which exist in many states is not a *sine qua non* for conservation of Texas oil and gas reserves. Nonetheless, some reforms in Texas' current laws, short of enacting compulsory unitization, would seem to be called for. Some of these reforms may be as difficult to achieve as passage of a compulsory unitization law, but they are nonetheless avenues worth pursuing.

Article 6014(g)

The first reform is to eliminate from the definitions of waste that part of Article 6014(g), now appearing in Section 85.046(a)(7) of the Natural Resources Code, which states that "it is not the intent of this Act to require repressuring of an oil pool or that the separately owned properties in any pool be unitized under one management, control or ownership." As we have seen, the spirit and intent of the statute are not honored, yet its existence creates doubt as to the authority of the Railroad Commission to issue waste-prevention orders, which have the inescapable and sometimes clearly intended effect of requiring unitization. This provision performed an understandable role in the events surrounding the enactment of market demand prorationing in 1931 and 1932. The fear of major oil companies' monopolization of Texas' oil and gas fields was not baseless at that time. The Railroad Commission was a weak, inept institution hardly able to inspire the public trust that it could regulate the oil fields in the public interest. In particular, Humble Oil and Refining's business strategy of block leasing and promotion of unitization seemed like propaganda for its broader aim of securing a dominant position in Texas' oil fields. But those days have long passed. The provision against repressuring and unitization now stands largely as a monument to the small-tract owner's right to secure the economic rents from a position of regulatory advantage bequeathed by the Marginal Well Act, per-well prorationing, and liberal Rule 37 exceptions. By removing this part of Article 6014(g) from the statute books, the legislature would perform the symbolic gesture of legitimizing those Railroad Commission orders issued in the past, which by their very terms required repressuring and which often required unitization in substance, though not in form. It is because of these orders that the legislature was spared the necessity of making the hard political choice of passing a compulsory unitization law in order to prevent large-scale waste in Texas oil and gas fields.[59] It is right and fitting that the lawmakers acknowledge their indebtedness to the commission for shouldering the unitization burden which they themselves refuse to carry by eliminating this part of Article 6014(g) from the statutes.

While the abolition of this anti-unitization provision is recommended, it may not be worthwhile to pursue legislative reform only for the sake of symbolism. Although Article 6014(g) clearly constrained the commission in its decision not to issue a forced cycling order in the Port Acres field, there is no evidence that the provision currently affects the commission's decision making, and the Texas Supreme Court seems consciously bent on ignoring it. Thus, even though there is no public policy justification for retaining the statute, it may be more useful to focus legislative reform on other provisions in the law which do constrain unitization in Texas in more substantial ways.

The 1949 Voluntary Unitization Act

As we have seen, the 1949 Act authorizing the Railroad Commission to approve voluntary unitization agreements is a restrictive, begrudging statute. The act was passed in order to grant antitrust immunity to cooperative agreements, but in the end, the statute asserts the primacy of antitrust law over voluntary unitization plans. The act only grants the commission the authority to approve certain types of voluntary agreements, thus increasing the risk that agreements which do not meet the strictures of the act will be found to violate public policy and the antitrust laws. The reasons for many of the restrictive provisions in the 1949 Act parallel the reasons why Texas does not yet have a compulsory unitization act. By 1949, the East Texas legacy was firmly vested in the Texas independent producer who had quite selfish motives for wanting to assure that the voluntary unitization act was limited in purpose and effect. The Texas attorney general's comments on the 1947 version of the voluntary unitization bill reinflamed fears of condoning monopoly. While some of the restrictions in the revised 1949 Act may have reflected a valid public interest in restricting the grant of antitrust immunity to an industry as large and powerful as the oil industry, much of the furor over the proposed 1947 bill seems to have been motivated by the political strength of the independents rather than by a real concern about monopoly. The passage of the Texas Free Enterprise and Antitrust Act of 1983 heralds a new respect for unitization agreements which have commission approval by exempting these agreements unequivocally from the reach of the state's antitrust laws. However, cooperative plans which fall outside the purview of the commission's authority, as defined in the 1949 Act, do not receive the benefit of this reform in the antitrust laws.

Given the many serious shortcomings of the 1949 Act in promoting unitization and the possible political problems involved in reforming the act, it is worthwhile to consider a recommendation that the act be discarded altogether and not replaced with any other. Under Texas' new antitrust laws, the "rule of reason" will be applied to analyze whether voluntary unitization agreements violate the antitrust laws.[60] If the unit members can show that their agreement was designed to effect an increase in the ultimate recovery of oil and gas, and that all owners in the field were given a fair and reasonable opportunity to join the unit, it is most unlikely that a state court would find the unitization agreement to violate the antitrust law, even if the unit had not secured Railroad Commission approval of the cooperative plan. Moreover, a significant degree of public control over the unit would still exist through the commission's regulation of well placements and producing rates in the field. The federal courts have never interpreted the federal antitrust laws to preclude a cooperative agreement that added to the supply of oil and gas and offered all owners in the field a fair opportunity to recover their oil and gas. Thus, having no act whatsoever governing voluntary unitization agreements may be better than having the existing 1949 Act.

However, the courts have considered the commission's approval of secondary recovery projects under the 1949 Act as significant in other contexts. First, the secondary recovery activities of an approved unit are not subject to an injunction in a suit brought by an adjoining, nonsigned landowner to invalidate a commission order pertaining to the unit's location of wells. In addition, Railroad

Commission approval of the unitization agreement probably insulates the unit from injunctions, from strict liability in tort, and from liability for punitive damages in a lawsuit brought by a private party against the unit operator for injury to adjoining property (see chapter 7 text at notes 135 to 155 and chapter 8 text at notes 37 to 40). The courts in both Texas and other states have used legislative enactments that support secondary recovery and unitized operations to justify modifications of traditional common law rules in order to accommodate unitized activities. Thus, a statute enacted to encourage and promote voluntary unitization agreements is valuable to unit participants aside from any antitrust benefits to be secured therefrom.

A simple one-paragraph act, like the following, would suffice:

> An agreement by two or more persons owning, claiming, or controlling production, leases, royalties or other interests in the same oil or gas field, or in what appears from geological data to be the same oil or gas field, is authorized and may be performed and shall not be held or construed to violate any of the statutes of this state relating to trusts, monopolies, or contracts and combinations in restraint of trade, if the agreement is approved by the Railroad Commission, after notice and hearing, and upon a finding by the Commission that the agreement is in the public interest, that it protects correlative rights, and that it is reasonably necessary to increase ultimate recovery, or to prevent waste, including the wasteful drilling and operation of unnecessary wells, or to protect correlative rights.[61]

Such an act imposes no restrictions on the type of agreements that can be approved—whether for primary recovery, secondary recovery, tertiary recovery, or *in-situ* combustion—as long as the agreement prevents waste and protects correlative rights. The act also grants to the commission the authority to approve agreements which protect correlative rights, independent of any showing that waste would be prevented. This allows the commission to approve, *inter alia,* unitization agreements in nonassociated gas fields, which are made to facilitate equity in the distribution of the gas or to avoid some of the administrative burdens of the state's gas-prorationing rules, but where physical waste is not a factor in forming the unit.[62]

If replacement of the 1949 Act by a single paragraph is too radical a turn from the past, then the framework of the 1949 Act can be retained but with the following revisions and amendments:

a. Section 101.011 should be lengthened to include a third category of pooled units subject to Railroad Commission approval, which category consists of "pooled units reasonably necessary to increase the ultimate recovery of oil or gas, or to prevent waste, including the wasteful drilling and operation of unnecessary wells, or to protect correlative rights."[63] Alternatively, this proposed new category can simply displace the two existing types of pooled units which are subject to commission approval. This reform will obviate the need for operators to drill unnecessary injection wells in order to qualify their projects for commission approval. By expressly providing that the elimination of wasteful costs due to unnecessary wells is a legitimate purpose for entering into voluntary unitization agreements, the proposed language will end the speculation which surrounds this issue in Texas. Also, under the proposed revision, the protection of correlative rights is an independent basis for approving a unitization plan.

To further expand the scope of Section 101.011, the often-used word "necessary" should be replaced with the phrase "reasonably necessary," and the section should provide that the act applies to the property "in the same oil field or gas field, or in what appears from geological or other data to be in the same oil or gas field." This change would allow approval of voluntary units formed in the exploration stage.

b. Section 101.013 should be revised to make it clear that voluntary unitization agreements may still be legal and effective even if the Railroad Commission has not approved them.[64] The unit would simply not be accorded the benefits of state antitrust immunity. With the revisions in Section 101.011 suggested above, no unitization agreement should fall outside the Railroad Commission's approval authority, but some operators may want to avoid the administrative costs of securing commission approval, especially for small water-flooding projects, and they should be able to do so.

Throughout Section 101.013, the words "reasonably necessary" should replace the absolutism of "necessary." Section 101.013(a)(2) should be omitted as redundant, or else limited to a finding that the agreement is "in the public interest." Section 101.013(a)(4) should simply require a finding that the proposed operation is reasonably expected to be economic. This section was taken from Oklahoma's compulsory unitization act and only makes sense in the context of coerced unitization. The section could be eliminated altogether, but because the commission often indirectly coerces operators into unitizing, it is probably prudent to retain a finding of expected profitability so that the commission does not either consciously or unwittingly, force producers into an uneconomic business enterprise.

Section 101.013(a)(5) is a senseless invitation to delay the unitization hearing process and should be eliminated.

Section 101.013(a)(6) should be deleted because Section 101.013(a)(3) provides sufficient protection of correlative rights. If retained, the section should be redrafted to apply to areas "that have reasonably been defined by exploration or development," rather than by development alone, thus allowing the formation of exploratory units.

c. Consideration should be given to amending Section 101.017(c) which now prohibits authorizing agreements which provide for cooperative marketing of gas.[65] The efficient development and production of gas may be enhanced by cooperative marketing agreements which allow producers to bargain more effectively with large pipeline purchasers.

d. If the legislature refuses to revise Sections 101.011 and 101.013 in order to expand the type of agreements which the commission can approve, a provision should be added to the act stating that failure to submit a cooperative agreement to the commission for approval shall not imply or constitute evidence that the agreement or operation under it violates the antitrust law.

Other Legislative Proposals

In addition to, or instead of, revising the current conservation statutes, Texas lawmakers could pass other types of legislation designed to encourage voluntary unitization. For example, a statute could provide that any person complaining of injury to his property due to unit operations which have been approved by the Railroad Commission shall be limited to an action in damages.[66] Such a statute would protect the unit from injunctions. The courts can reach this result without legislative support, but the public policy reasons for so doing are more evident when the legislature has enacted such a statute.

In its study of legal obstacles to enhanced oil recovery, the Office of Technology Assessment suggested that unitization could be encouraged by enacting a statute that exempted producers from liability for any damages caused by commission-approved operations which did not involve negligence on the part of the unit operator.[67] The constitutionality of such a statute can certainly be questioned. But if the statute affords procedural due process to all affected parties through adequate notice and hearing provisions, and if nonliability is conditioned on a finding that the injured party received a fair and reasonable offer to join the unit, the statute might be upheld. Such a law would certainly reduce the incentive to hold out. A modified version of such a statute could provide that damages to any person who refuses a fair and reasonable offer to join an approved unit and who is subsequently injured by the unit's activities shall be limited to the decreased value of the primary oil and gas reserves in place underneath his tract. This measure would significantly reduce the incentive to hold out by precluding the recovery of damages based on the economic rents flowing from a position of regulatory or structural advantage. If such far-reaching limits on a unit's liability are legislated, it might be advisable to require by law that the unit first receive the voluntary consent of at least 75 percent of the working interest and royalty interest owners in the field. Of course, the passage of a compulsory unitization law would seem a simpler, more direct, and better established method of accomplishing unitization than these proposals to limit a unit's liability.

Other legislative proposals could focus on providing carrots rather than sticks to encourage unitization. With most fields producing at 100 percent capacity, the Railroad Commission no longer has the freedom to play with bonus allowables and increased MERs in order to spur cooperative agreements. Two economists have proposed that Texas legislators enact tax benefits to leaseholders who agree to unitize a field before drilling begins.[68] Of course, if exploratory units are to be encouraged, the 1949 Act must be revised to allow the Railroad Commission to approve these units.

In the event that the Texas legislature fails to act on any of the proposals advanced above, the duty falls on the Railroad Commission to use its considerable powers to encourage unitization further. The commission has the power, as yet unused, to change existing prorationing formulas in fields where waste is occurring.

Whether or not the law is reformed, the commission should actively investigate on its own motion, the conservation practices used in the larger non-unitized or partially unitized fields that produce most of Texas oil. A revival of Commissioner Murray's carefully sequenced field hearings on waste, along with greater publicity about the commission's dedication to "permissive unitization," would probably spark more voluntary efforts among operators. Such renewed vigilence would also reduce the "surprise," such as occurred in the Bryan (Woodbine) field, when unanticipated shutdown orders and reduced gas–oil ratio rules upset established purchasing relationships.

Epilogue

In October 1983, on the tenth anniversary of the Arab oil embargo and almost fifty years from the date of Doherty's letter to President Coolidge warning of the need to conserve oil and gas so that the oil industry and government officials would not be blamed for bankruptcy in the nation's oil and gas reserves, Commissioner Wallace, chairman of the Texas Railroad Commission, addressed a strikingly similar letter to President Reagan. After warning of the severely debilitating effects of another embargo and the absence of any excess producing capacity in the domestic oil and gas industry, Commissioner Wallace criticized the federal government for its inability to recognize the seriousness of the situation and for the "national disgrace" of failing to remove the regulatory shackles from natural gas. He then called upon President Reagan to appoint a bipartisan national energy conference composed of congressional leaders, federal energy officials, and representatives of energy-producing states, the consuming public, the bigger oil companies, and the independent producers to outline the steps necessary to secure America's energy future. The letter ended: "A national consensus must be formulated and decisive action must be taken. Our children are entitled to a future based on promise and opportunity, not regret and recrimination."[69]

To many Americans, Chairman Wallace's letter is simply another call for higher prices to consumers and more wealth to Texans. Commissioner Wallace's message might be received more hospitably were he to address his own governor and legislature and his constituents in the state's oil and gas industry with the need to effect every possible efficiency in the production of oil and gas in Texas. The Texas legislature has the power to remove the shackles to fuller implementation of unitization in Texas oil and gas fields and achieve the margin of excellence in the conservation of these precious resources. It is also within Commissioner Wallace's powers to initiate a study, drawing on the enormous expertise and prestige of the Railroad Commission staff, to assess the need for compulsory unitization legislation in Texas and to so advise the Texas lawmakers. Such a study need not await a bipartisan national energy conference. Indeed, Commissioner Wallace could champion a compulsory unitization bill as part of his efforts to secure America's energy future. At the same time, the commissioner can point with great pride to the commission's record of achieving a very substantial degree of unitization in Texas already. This monumental achievement by the commission and the Texas courts in the face of sustained legislative antipathy to the very concept of unitization, should not be used to mask the need for legal reforms that could add a small but critical increase to the nation's recovery of domestic oil and gas. Nor should the passage of a compulsory unitization statute in Texas signal any lack of admiration for the unique legal framework developed in Texas to achieve so much unitization with so little legislative support.

Appendix I.

Texas Natural Resources Code, Section 85.046 (Vernon Supp. 1982)

Sec. 85.046. Waste

(a) The term "waste," among other things, specifically includes:

(1) operation of any oil well or wells with an inefficient gas–oil ratio and the commission may determine and prescribe by order the permitted gas–oil ratio for the operation of oil wells;

(2) drowning with water a stratum or part of a stratum that is capable of producing oil or gas or both in paying quantities;

(3) underground waste or loss, however caused and whether or not the cause of the underground waste or loss is defined in this section;

(4) permitting any natural gas well to burn wastefully;

(5) creation of unnecessary fire hazards;

(6) physical waste or loss incident to or resulting from drilling, equipping, locating, spacing, or operating a well or wells in a manner that reduces or tends to reduce the total ultimate recovery of oil or gas from any pool;

(7) waste or loss incident to or resulting from the unnecessary, inefficient, excessive, or improper use of the reservoir energy, including the gas energy or water drive, in any well or pool; however, it is not the intent of this section or the provisions of this chapter that were formerly a part of Chapter 26, Acts of the 42nd Legislature, 1st Called Session, 1931, as amended, to require repressuring of an oil pool or to require that the separately owned properties in any pool be unitized under one management, control, or ownership;

(8) surface waste or surface loss, including the temporary or permanent storage of oil or the placing of any product of oil in open pits or earthen storage, and other forms of surface waste or surface loss including unnecessary or excessive surface losses, or destruction without beneficial use, either of oil or gas;

(9) escape of gas into the open air in excess of the amount necessary in the efficient drilling or operation of the well from a well producing both oil and gas;

(10) production of oil in excess of transportation or market facilities or reasonable market demand, and the commission may determine when excess production exists or is imminent and ascertain the reasonable market demand; and

(11) surface or subsurface waste of hydrocarbons, including the physical or economic waste or loss of hydrocarbons in the creation, operation, maintenance, or abandonment of an underground hydrocarbon storage facility.

(b) Notwithstanding the provisions contained in this section or elsewhere in this code or in other statutes or laws, the commission may permit production by commingling oil or gas or oil and gas from multiple stratigraphic or lenticular accumulations of oil or gas or oil and gas where the commission, after notice and hearing, has found that producing oil or gas or oil and gas in a commingled state will prevent waste, promote conservation, or protect correlative rights.

Appendix II.
Texas Natural Resources Code, Sections 101.001–.018

Subchapter A. General Provisions

101.001 Definition

In this chapter, "commission" means the Railroad Commission of Texas.

101.002 Existing Agreement Rights

None of the provisions in this chapter restrict any of the rights that a person now may have to make and enter into unitization and pooling agreements.

101.003 Applicability

None of the provisions in this chapter impair the power of the commission to prevent waste under the oil and gas conservation laws of the state except as provided in Section 101.004 of this code or repeal, modify, or impair any of the provisions of Sections 85.002 through 85.003, 85.041 through 85.055, 85.056 through 85.064, 85.125, 85.201 through 85.207, 85.241 through 85.243, 85.249 through 85.252, or 85.381 through 85.385 of this code or Subchapters E and J of Chapter 85 of this code, relating to oil and gas conservation.

101.004 Conflict With Antitrust Acts

(a) Agreements and operations under agreements which are in accordance with the provisions in this chapter, being necessary to prevent waste and conserve the natural resources of this state, shall not be construed to be in violation of the provisions of Chapter 15, Business & Commerce Code, as amended.

(b) If a court finds a conflict between the provisions in this chapter and Chapter 15, Business & Commerce Code, as amended, the provisions in this chapter are intended as a reasonable exception to that law, necessary for the public interests stated in Subsection (a) of this section.

(c) If a court finds that a conflict exists between the provisions in this chapter and Chapter 15, Business & Commerce Code, as amended, and finds that the provisions in this chapter are not a reasonable exception to said Chapter 15, it is the intent of the legislature that the provisions in this chapter, or any conflicting portion of them, shall be declared invalid rather than declaring Chapter 15, Business & Commerce Code, as amended, or any portion of it, invalid.

Subchapter B. Cooperative Agreements in Secondary Recovery Operations

101.011 Authorized Agreements for Separately Owned Properties

Subject to the approval of the commission, as provided in this chapter, persons owning or controlling production, leases, royalties, or other interests in separate property in the same oil field, gas field, or oil and gas field may voluntarily enter into and perform agreements for either or both of the following purposes:

(1) to establish pooled units, necessary to effect secondary recovery operations for oil or gas, including those known as cycling, recycling, repressuring, water flooding, and pressure maintenance and to establish and operate cooperative facilities necessary for the secondary recovery operations;

(2) to establish pooled units and cooperative facilities necessary for the conservation and use of gas, including those for extracting and separating the hydrocarbons from the natural gas or casinghead gas and returning the dry gas to a formation underlying any land or leases committed to the agreement.

101.012 Persons Bound by Agreements

Agreements for pooled units and cooperative facilities do not bind a landowner, royalty owner, lessor, lessee, overriding royalty owner, or any other person who does not execute them. The agreements bind only the persons who execute them, their heirs, successors, assigns, and legal representatives. No person shall be compelled or required to enter into such an agreement.

101.013 Commission Approval

(a) Agreements for pooled units and cooperative facilities are not legal or effective until the commission finds, after application, notice, and hearing:

(1) that the agreement is necessary to accomplish the purposes specified in Section 101.011 of this code;

(2) that it is in the interest of the public welfare as being reasonably necessary to prevent waste and to promote the conservation of oil or gas or both;

(3) that the rights of the owners of all the interests in the field, whether signers of the unit agreement or not, would be protected under its operation;

(4) that the estimated additional cost, if any, of conducting the operation will not exceed the value of additional oil and gas so recovered, by or on behalf of the several persons affected, including royalty owners, owners of overriding royalties, oil and gas payments, carried interests, lien claimants, and others as well as the lessees;

(5) that other available or existing methods or facilities for secondary recovery operations or for the conservation and utilization of gas in the particular area or field concerned or for both are inadequate for the purposes; and

(6) that the area covered by the unit agreement contains only that part of the field that has reasonably been defined by development, and that the owners of interests in the oil and gas under each tract of land in the area reasonably defined by development are given an opportunity to enter into the unit on the same yardstick basis as the owners of interests in the oil and gas under the other tracts in the unit.

(b) A finding by the commission that the area described in the unit agreement is insufficient or covers more acreage than is necessary to accomplish the purposes of this chapter is grounds for the disapproval of the agreement.

101.014 Jointly Owned Properties

None of the provisions in this chapter shall be construed to require the approval of the commission of voluntary agreements for the joint development and operation of jointly owned property.

101.015 Commission Regulation

An agreement executed under the provisions of this chapter is subject to any valid order or rule of the commission relating to location, spacing, proration, conservation, or other matters within the authority of the commission, whether adopted prior to or subsequent to the execution of the agreement.

101.016 Permissible Provisions

(a) An agreement authorized by this chapter may provide for the location and spacing of input wells and for the extension of leases covering any part of land committed to the unit as long as operations for drilling or reworking are conducted on the unit or as long as production of oil or gas in paying quantities is had from any part of the land or leases committed to the unit. However, no agreement may relieve an operator from the obligation to develop reasonably the land and leases as a whole committed to the unit.

(b) An agreement authorized by this chapter may provide that the dry gas after extraction of hydrocarbons may be returned to a formation underlying any land or leases committed to the agreement and may provide that no royalties are required to be paid on the gas so returned.

101.017 Prohibited Provisions

(a) No agreement authorized by this chapter may attempt to contain the field rules for the area or field, or provide for or limit the amount of production of oil or gas from the unit properties, those provisions being solely the province of the commission.

(b) No agreement authorized by this chapter may provide directly or indirectly for the cooperative refining of crude petroleum, distillate, condensate, or gas, or any by-product of crude petroleum, distillate, condensate, or gas. The extraction of liquid hydrocarbons from gas, and the separation of the liquid hydrocarbons into propanes, butanes, ethanes, distillate, condensate, and natural gasoline, without any additional processing of any of them, is not considered to be refining.

(c) No agreement authorized by this chapter may provide for the cooperative marketing of crude petroleum, condensate, distillate, or gas, or any by-products of them.

101.018 Effect of Approval Outside of Unit

The approval of an agreement authorized by this chapter shall not of itself be construed as a finding that operations of a different kind or character in the portion of the field outside of the unit are wasteful or not in the interest of conservation.

Appendix III.
Rule 37 Applications, 1940–81

Year	*Rule 37 applications approved*	*Rule 37 applications denied*
1981	2,725	7
1980	3,031	3
1979	2,392	10
1978	2,042	0
1977	2,493	5
1976	2,197	8
1975	2,369	11
1974	2,039	11
1973	2,010	13
1972	1,616	24
1971	1,617	13
1970	1,562	18
1969	1,503	13
1968	1,594	20
1967	1,616	33
1966	1,640	13
1965	1,644	26
1964	1,688	48
1963	1,517	24
1962	1,193	35

Appendix III. (*continued*)

Year	*Rule 37 applications approved*	*Rule 37 applications denied*
1961	1,509	39
1960	1,295	55
1959	1,256	64
1958	1,256	43
1957	1,447	28
1956	1,444	19
1955	2,315	57
1954	1,343	57
1953	1,308	32
1952	1,273	24
1951	1,731	40
1950	1,942	50
1949	1,706	22
1948	1,537	33
1947	1,357	41
1946	1,033	62
1945	519	2
1944	408	2
1943	321	4
1942	693	31
1941	2,314	485
1940	3,276	40

Appendix IV.

Number of Unitization Agreements Approved by the Railroad Commission

Year	*Total*
1949	1
1950	11
1951	19
1952	15
1953	14
1954	13
1955	19
1956	23
1957	22
1958	22
1959	44
1960	64
1961	57
1962	65
1963	56
1964	75
1965	74
1966	59
1967	70
1968	59
1969	49

Appendix IV. (*continued*)

Year	*Total*
1970	38
1971	20
1972	23
1973	22
1974	26
1975	12
1976	12
1977	13
1978	15
TOTAL	1,012

Notes

Chapter 1

1. As quoted in Robert E. Hardwicke, *Antitrust Laws v. Unit Operation of Oil or Gas Pools* (rev. ed., Dallas, Tex., Society of Petroleum Engineers of the American Institute of Mining and Metallurgical Engineers, 1961) p. 186.

2. In October 1973, the Arab oil-producing countries ordered an embargo of crude oil shipments to the United States in response to the U.S. pro-Israeli stance in the Arab–Israeli war, and OPEC quadrupled the posted price of crude oil. In late 1978 Iranian oil exports were suspended because of political unrest in that country, and long gasoline lines appeared throughout the United States in 1979, accompanied by further large increases in the price of imported crude. See James M. Griffin and David J. Teece, eds., *OPEC Behavior and World Oil Prices* (Edison, N.J., Allen and Unwin, 1982) pp. 4–13; Robert Stobaugh and Daniel Yergin, eds., *Energy Future* (New York, Random House, 1979) pp. 3–56; and "The Great Energy Mess," *Time,* July 2, 1979, pp. 14–27.

By the time the 1973 embargo was lifted in March 1974, unemployment had increased by 300,000 to 500,000 persons, national income had fallen by $10 to $20 billion, the United States was paying $2 billion per month for imported energy, which in 1973 had cost $500 million, and higher energy prices had driven the inflation rate to 14 percent. These effects ushered in a recession that lasted throughout 1974. See Benjamin S. Cooper, "U.S. Policies and Politics," in Daniel N. Lapedes, ed., *McGraw-Hill Encyclopedia of Energy* (New York, McGraw-Hill, 1976) pp. 41–42. This cycle was largely repeated in 1979 and 1980.

3. The Ninety-third Congress enacted sixteen major pieces of energy-related legislation between November 1973 and January 1975, including the Emergency

Petroleum Allocation Act, rights of way for the Trans-Alaskan pipeline, a national speed limit of 55 miles per hour, and increased funding for energy research and development, especially solar and geothermal projects. See Cooper, "U.S. Policies," pp. 39–44. The pace did not slow under subsequent congresses and presidents. See, for example, *Energy Initiatives of the 95th Congress,* Committee Print 96-10, Senate Committee on Energy and Natural Resources, 96 Cong., 1 sess. (1979).

The Federal Trade Commission launched an antitrust complaint against the eight largest oil companies in July 1973, alleging almost every possible form of monopolistic act by the majors. United States v. Exxon Corp., FTC Docket No. 8934, filed July 18, 1973. The Senate and House held months of hearings on legislative proposals for both horizontal and vertical divestiture of the oil majors, for establishing a federally owned oil and gas corporation to compete with the majors, and for regulating the oil companies as public utilities. Several states enacted laws prohibiting petroleum producers and refiners from operating retail stations within the state. See, for example, Howard R. Williams, Richard C. Maxwell, and Charles J. Meyers, *Oil and Gas: Cases and Materials* (4th ed., Mineola, N.Y., The Foundation Press, 1979) pp. 176–211.

4. At this time, Henry Doherty was president of H. L. Doherty and Co., fiscal agent for the Cities Service Companies, and a director of the American Petroleum Institute (API). Under his direction, much early scientific research into the nature of oil and gas reservoirs was conducted, especially on the concept of gas drives as reservoir energy necessary to propel oil to the surface. Based on this research, Doherty became the first, and probably the most ardent spokesperson ever, for compulsory unitization of oil and gas fields. Few members of the oil and gas fraternity other than the American Institute of Mining and Metallurgical Engineers accepted Doherty's early advocacy of compulsory unitization. The API treated him as a pariah in 1924. Hence, Doherty presented his views directly to President Coolidge rather than through established oil industry lobbying channels. See Hardwicke, *Antitrust Laws,* pp. 1–13 and 256–258.

5. For example, Tennessee, Wyoming, and West Virginia passed compulsory unitization laws between 1971 and 1972 when natural gas shortages were appearing. Many other states with existent, but weak, compulsory unitization statutes amended their laws significantly in the early 1970s in order to broaden and strengthen their reach. See David W. Eckman, "Statutory Fieldwide Oil and Gas Units: A Review for Future Agreements," *Natural Resources Lawyer* vol. 6, no. 3 (1973) pp. 339–387; and Comment, "Compulsory Unitization in Florida: A New Emphasis on the Energy Crisis?" *University of Florida Law Review* vol. 27, no. 1 (1974) pp. 196–223.

6. California—another large and established producing state—has very limited compulsory unitization laws, applicable only to townsites and areas where land subsidence is occurring as a result of oil and gas withdrawals. California Public Resources Code secs. 3321–3342 and 3630–3690 (West 1972). Illinois and Kentucky are the only other producing states without a compulsory unitization law, but together they account for less than 1 percent of U.S. crude oil production

while Texas accounts for 31 percent. See U.S. Department of Commerce, Bureau of the Census, *Statistical Abstract of the U.S. 1982–83* (Washington, D.C., GPO, 1983) p. 730.

7. The General Accounting Office has warned that the United States imported relatively more oil from OPEC members in 1982 than was the case prior to the 1973 embargo (41 percent versus 37 percent), and that shortages could result in major disruptions similar to those in 1973 and 1979. See *Energy Management* no. 500, August 18, 1982 (Chicago, Ill., Commerce Clearinghouse) p. 4. The Congressional Research Service forecasts massive unemployment and recession if access to Persian Gulf oil is cut off. See *Energy Management* no. 562, October 25, 1983 (Chicago, Ill., Commerce Clearinghouse) p. 1.

8. The price of a barrel of domestic crude oil rose from $3.18 in 1970 to $31.84 in 1981. During this same period, the price of natural gas rose from 17 cents per MCF (thousand cubic feet) to $1.85 per MCF. Railroad Commission of Texas, *1981 Annual Report of the Oil and Gas Division* (Austin, Tex., 1981) p. 16 [hereafter cited as *RRC 1981 Annual Report*].

9. Amoco Production Co. v. Alexander, 622 S.W.2d 563 (Tex. 1981).

10. See, for example, Stuart E. Buckley, ed., *Petroleum Conservation* (Dallas, Tex., American Institute of Mining and Metallurgical Engineers, 1951); Interstate Oil Compact Commission, *Oil and Gas Production: An Introductory Guide to Production Techniques and Conservation Methods* (Norman, University of Oklahoma Press, 1951); Leroy H. Hines, *Unitization of Federal Lands* (Denver, Colo., F. H. Gower, 1953); Stephen L. McDonald, *Petroleum Conservation in the United States: An Economic Analysis* (Baltimore, Md., Johns Hopkins University Press for Resources for the Future, 1971); and Mid-Continent Oil and Gas Association, *Handbook on Unitization of Oil Pools* (Saint Louis, Mo., Blackwell Wielandy, 1930).

11. The Railroad Commission was created in 1891 to regulate railroad rates, and in 1917 it was given additional authority to regulate oil pipelines. In 1919 Texas passed a strong conservation law to prevent waste and to regulate oil and gas production. The Railroad Commission naturally was given the additional duties of enforcing this act. Soon oil and gas regulation dominated its operations, but the agency's name has never been changed to more accurately reflect its activities. The commission's decision-making power is in the hands of three commissioners, who are elected to staggered six-year terms in statewide elections.

12. The economic analysis used in this book draws primarily on McDonald, *Petroleum Conservation.*

13. National Petroleum Council, *Enhanced Oil Recovery: An Analysis of the Potential for Enhanced Oil Recovery from Known Fields in the United States—1976 to 2000* (Washington, D.C., National Petroleum Council, 1976) p. 11.

14. It is possible to argue that price supports for domestic oil are efficient because the market price of foreign oil does not fully reflect the risk and cost of supply disruptions that can cause major unemployment and recession in the

United States. Government interference in the marketplace is often justified to correct externalities that arise when the market price of a good does not fully reflect its cost or value to society as a whole. However, economic analysis still may show that the goal of national security can be achieved more efficiently (that is, at less cost) with an alternative policy such as tariffs on foreign oil or the purchase of standby strategic reserves rather than with price supports on domestic oil. See, for example, Douglas R. Bohi and Milton Russell, *Limiting Oil Imports: An Economic History and Analysis* (Baltimore, Md., The Johns Hopkins University Press for Resources for the Future, 1978) pp. 247–330.

Chapter 2

1. One explanation for the long delay and ultimate enactment of only very limited compulsory unitization laws in California is that its reservoirs are physically so unhomogeneous that formulas for allocating production reserves and expenses from the unit area are difficult to negotiate and promulgate. The early and active promotion of unitization in the midcontinent states is partly attributed to their homogeneous reservoirs that allow simple allocation formulas, such as one based on surface acreage. See Winfield S. Payne, Jr., "The Engineering Phases of Unit Plans," *UCLA Law Review* vol. 5, no. 3 (1958) p. 405; and Mid-Continent Oil and Gas Association, *Handbook on Unitization of Oil Pools* (Saint Louis, Mo., Blackwell Wielandy, 1930).

2. Most of the discussion that follows on the engineering aspects of oil and gas production derives from Stuart E. Buckley, ed., *Petroleum Conservation* (Dallas, Tex., American Institute of Mining and Metallurgical Engineers, 1951). This excellent volume, sponsored by the Henry L. Doherty Memorial Fund, pays tribute to Doherty's early and brash championing of compulsory unitization.

3. Ibid., p. 130.

4. The force of gravity can be considered as a fourth type of displacement mechanism. Gravity separates the fluids within the reservoir: the lighter gas on top, oil in the middle, and water at the bottom. During the last stages of depletion, gravity drainage of oil from the upper part of the reservoir to the lower part may be the primary agent of recovery (ibid., pp. 122 and 221).

5. American Petroleum Institute, *Primer of Oil and Gas Production* (Dallas, Tex., American Petroleum Institute, 1976) p. 11.

6. Buckley, *Petroleum Conservation,* p. 139.

7. Ibid., pp. 149–170 and 212–216.

8. Ibid., pp. 208–209. For example, when gas remains dissolved in the reservoir oil, the viscosity of the oil is reduced and this increases the ability of the oil to flow more easily. These beneficial effects of pressure maintenance can be very significant in deep reservoirs containing oil saturated with dissolved gas.

9. Ibid., pp. 151–163. However, the MER concept is still criticized by economists for its failure to include economic criteria. A more detailed discussion of economic efficiency in oil and gas conservation appears in the text of this chapter at notes 32 to 36.

10. Buckley, *Petroleum Conservation,* pp. 157–158. In some dissolved gas reservoirs, no segregation of fluids is possible, and therefore no reduction in the producing rate would increase oil recovery; the reservoir does not have an MER. Nonetheless, secondary recovery or pressure maintenance operations may increase oil-recovery rates.

11. Ibid., pp. 196–237; and Shofner Smith, "The Engineering Aspects of Pressure Maintenance and Secondary Recovery Operations," *Rocky Mountain Mineral Law Institute,* vol. 6 (Albany, N.Y., Matthew Bender, 1961).

12. About 85 percent of the oil wells in the United States in 1982 were stripper wells averaging less than 10 barrels per day. See "Stripper Well Production Higher in Texas, U.S.," *Houston Chronicle,* May 30, 1983. Stripper wells have always been common in the United States. In 1950, 311,000 (or 69 percent) of the 450,000 producing oil wells in the United States were stripper wells, averaging about 3 barrels of production per day. See Buckley, *Petroleum Conservation,* p. 182.

13. Buckley, *Petroleum Conservation,* pp. 187–188.

14. Examples of successful gas-injection projects appear in Rafael Sandrea and Ralph Nielson, *Dynamics of Petroleum Reservoirs Under Gas Injection* (Houston, Tex., Gulf Publishing, 1974) pp. 3–15.

15. Buckley, *Petroleum Conservation,* pp. 184–186, 193, 210–211, and 229–234. For example, improperly abandoned wells or irregular encroachment of water can trap oil in the formation so as to render it unrecoverable. The premature loss of dissolved gas shrinks the oil in the reservoir, increases its viscosity, and dissipates the fuel that could be used for secondary recovery.

16. This is the definition of EOR used by the National Petroleum Council (NPC), an industry trade association which advises the federal government on oil and gas issues. See National Petroleum Council, *Enhanced Oil Recovery: An Analysis of the Potential for Enhanced Oil Recovery From Known Fields in the United States—1976 to 2000* (Washington, D.C., National Petroleum Council, 1976) p. 3. The term *tertiary recovery* is sometimes used synonymously with EOR, especially in the federal legislation governing crude oil pricing and the windfall profits tax. Tertiary recovery implies a third-stage process following secondary recovery. Yet tertiary recovery is often used instead of, not after, secondary recovery.

17. Ibid., pp. 13–17.

18. Ibid., p. 51.

19. Ibid., pp. 5–7. This projection assumes a crude oil price of $25 per barrel.

20. Buckley, *Petroleum Conservation,* pp. 176–177 and 243.

21. Ibid., pp. 177–181 and 244–247. One advantage of operating a condensate reservoir with a water drive is that the water maintains the reservoir pressure,

allowing producers to process and market the gas immediately rather than recycling it.

22. Ben Daviss, "Enhanced Gas Technique Looks Profitable," *Gulf Coast Oil Reporter,* December 1982, pp. 9–10. Because 90 percent of the natural gas in a reservoir is commonly produced by simple expansion, little gas remains to be recovered in a second stage. However, certain gas fields may have been watered out at relatively high pressure, leaving 15 to 50 percent of the gas in the reservoir. Forty-two such fields have been targeted in Texas.

23. National Petroleum Council, *Enhanced Oil Recovery,* p. 17. In 1967 the federal government experimentally fractured gas formations with underground nuclear explosions. See Ben R. Howell, "Project Gasbuggy—Legal Problems," in *Rocky Mountain Mineral Law Institute,* vol. 14 (Albany, N.Y., Matthew Bender, 1968).

24. Estimates of recoverable gas from tight sands range from 25 to 600 trillion cubic feet. The latter would be a thirty-year supply for the United States at current rates of use. See Roscoe C. Born, "Tight Sands Gas," *Barrons,* April 14, 1980.

25. These words are from the opinion of the Pennsylvania Supreme Court in one of the first cases to define the rule of capture as the common law governing the rights of separate owners to oil and gas in a common reservoir. Barnard v. Monongahela Natural Gas Co., 216 Pa. 362, 65 A. 801 (1907). The rule was established at a time when very little was known about the reservoir sciences, and the opinion reflects a state of knowledge that precluded an accurate determination of how much of the oil and gas flowing from a well on one owner's tract had probably originated from someone else's land. The justices borrowed the rule of capture from the early common law of fugitive wild animals; the person who captured the animal owned it, even though the animal may have originated on another's land.

The definition of correlative rights under the common law is not a simple matter. Initially, the term referred to an owner's right under the rule of capture to drain oil and gas from others without liability. Subsequently, the concept of correlative rights was expanded to include the right to be protected against other owners' wasteful or negligent operations that injured the common source of supply. One eminent scholar has defined common law correlative rights to include "duties not to take an undue proportion of the oil and gas." See Walter L. Summers, *The Law of Oil and Gas,* vol. 1 (2d ed., Kansas City, Mo., Vernon Law Book, 1954) sec. 63. However, not all scholars or courts recognize this aspect of correlative rights. See, for example, Eugene Kuntz, "Correlative Rights of Parties Owning Interests in a Common Source of Supply of Oil or Gas," in *Institute on Oil and Gas Law and Taxation,* vol. 17 (Albany, N.Y., Matthew Bender, 1966); and Elliff v. Texon Drilling Co., 146 Tex. 575, 210 S.W.2d 558 (1948). Most states now have defined correlative rights in their oil- and gas-conservation statutes. The statutory definitions generally accord each owner of a common reservoir the opportunity to produce a fair share of the oil and gas in the reservoir. See

R. O. Kellam, "A Century of Correlative Rights," *Baylor Law Review* vol. 12, no. 1 (1960) pp. 1–42.

26. Prorationing is the restriction of production by a state regulatory commission. The prorationing may be done solely to prevent the physical waste of oil, in which case the commission determines the MER for the field and issues an order limiting production to this rate. Many states developed prorationing systems that went beyond preventing physical waste. When the productive capacity of the state's fields under MER restraints was so great that the oil produced would glut the market and result in serious price declines and losses on producers' investments, the states authorized the regulatory commissions to prorate oil and gas to meet reasonable market demand, that is, to stabilize the market. Market demand prorationing is discussed in detail in chapter 3.

27. Most states passed compulsory pooling laws at a much earlier date than compulsory unitization laws. Between 1935 and 1945, ten states enacted pooling statutes; by 1945, only Oklahoma had passed a compulsory unitization law applicable to both oil and gas fields. (Louisiana, Florida, and Georgia had compulsory unitization statutes, but they were applicable only to cycling operations in condensate-gas fields at this time.) By 1955, twenty-five states had compulsory pooling acts, but only five states had joined Oklahoma in passing compulsory unitization acts for both oil and gas fields. See J. Frederick Lawson, "Recent Developments in Pooling and Unitization," in *Institute on Oil and Gas Law and Taxation,* vol. 23 (Albany, N.Y., Matthew Bender, 1972) pp. 210–213.

28. A more detailed statistical example of this situation appears in Buckley, *Petroleum Conservation,* pp. 237–242.

29. The state's traditional "police power" to regulate for the health and welfare of its citizens encompasses the power to regulate oil and gas production in the public interest of preventing waste and protecting correlative rights. However, such regulation must not violate the superior law of the state and federal constitutions. In the last thirty years or so, the courts have interpreted the state's police power in an expansive way, which allows state regulations to have serious financial effects on privately owned interests without violating the taking clause of the constitutions. However, in the 1930s when much of the oil- and gas-conservation legislation was passed by the states, the taking clause posed a substantial legal barrier to state orders that significantly decreased the value of private property. See, for example, Pennsylvania Coal Co. v. Mahon, 260 U.S. 393 (1922); Thompson v. Consolidated Gas Utilities Corp., 300 U.S. 55 (1937); and Laurence H. Tribe, *American Constitutional Law* (Mineola, N.Y., The Foundation Press, 1978) pp. 1–11, 444–446, and 456–465.

Because of constitutional constraints, the courts will recognize a state's duty to protect correlative rights even if the state's conservation legislation makes no mention of such a duty. In Pattie v. Oil and Gas Conservation Comm'n, 402 P.2d 596 (Mont. 1965), the Montana commission denied that it had the statutory authority to protect correlative rights because the state's laws were silent on this issue. The Montana Supreme Court held that Montana's laws impliedly granted the commission this authority by implication because otherwise the conservation acts would have to be held unconstitutional as a deprivation of property without due process.

30. In Texas, for example, a gas-cap well can produce only that quantity of gas which is the volumetric equivalent in reservoir displacement of the gas and oil produced from the oil well in the reservoir that withdraws the maximum amount of gas. Texas Administrative Code tit. 16, sec. 3.49(B) (1982), commonly called Rule 49(B).

31. Buckley, *Petroleum Conservation,* pp. 158, 165, 175–176, and 271. The recovery of oil in some dissolved gas fields can be significantly enhanced by segregation of fluids through gravity drainage and gas buoyancy. In these fields, unitization will facilitate optimal well placement and controlled rates of production.

If the water in a water-drive field underlies all parts of the oil stratum evenly, MER prorationing, regular spacing, and pooling may be adequate to prevent waste and protect correlative rights without the need for unitization. Drainage between tracts depends to a great extent on the shape of the reservoir, the slope of the formation, the continuity of permeability, and the area underlaid by water (ibid., p. 238).

32. Most of this discussion on economic efficiency derives from Stephen L. McDonald, *Petroleum Conservation in the United States: An Economic Analysis* (Baltimore, Md., The Johns Hopkins University Press for Resources for the Future, 1971) pp. 59–110.

33. Ibid., pp. 93–110. McDonald defines the *short run* as the period of time in which decisions are made about the rate of production from existing developed reservoirs. The *intermediate run* is a period long enough to permit development of already discovered reservoirs. The *long run* is a period that allows time to explore and develop prospects. McDonald's analysis also shows that the price of crude oil is more stable under unitized operations than under the rule of capture.

34. Ibid., pp. 150–198. For example, MER prorationing creates significant conservation benefits by correcting much of the waste induced by the rule of capture. However, MER prorationing is based on technical and engineering studies and fails to include economic criteria that would increase its usefulness in determining whether a reservoir is being operated efficiently. Economists would incorporate the economic cost of time, and expected changes in the cost of capital and the price of crude oil into the determination of MERs. Only if these economic criteria are used is it possible to determine whether it is more efficient to produce 20 million barrels over twenty years or 22 million barrels over thirty years. The economist's MER is defined as that rate of output which maximizes the present discounted value of the stream of net future receipts from oil production in the reservoir, not just the physical recovery of oil.

35. Ibid., pp. 198–209.

36. See, for example, Mid-Continent Oil and Gas Association, *Handbook on Unitization of Oil Pools,* pp. 44–55 (ultimate recovery of oil in Cromwell field

would be doubled with unitization); and Interstate Oil Compact Commission, *Oil and Gas Production: An Introductory Guide to Production Techniques and Conservation Methods* (Norman, University of Oklahoma Press, 1951) pp. 68–81 (unitization of Shuler–Jones field increased ultimate recovery by 50 percent from 32 to 50 million barrels).

37. For example, the owners of oil tracts in a gas-cap field have a strong incentive to negotiate with the owners of gas-cap tracts. In return for shutting in or never drilling the gas-cap wells, the gas owners will be offered a fair share of the production proceeds from the wells on the oil tracts. The oil-well owners will recover more oil at less cost because the gas-cap pressure is maintained. The gas-cap owners will not be any worse off; indeed, the unitization agreement may well accord them more revenue than they would receive operating under Rule 49(B) (see note 30 in this chapter) and at substantially less cost because they will not have to drill wells in order to recover income.

38. Buckley seems to have coined this term (Buckley, *Petroleum Conservation,* pp. 289–290). Economists often describe this situation as the problem of the "free rider." Free riders and holdouts long have plagued the provision of public goods (that is, goods owned in common) and have contributed much to the law and economics literature of externalities. See, for example, Alan E. Friedman, "The Economics of the Common Pool: Property Rights in Exhaustible Resources," *UCLA Law Review* vol. 18, no. 5 (1971) pp. 855–887; and Garrett Hardin, "The Tragedy of the Commons," *Science* vol. 162, no. 3859 (1968) pp. 1243–1248.

The most common justification for eminent domain laws is the holdout problem. One scholar would grant condemnation powers over real estate to private entities upon a finding that a degree of monopoly exists in the land to be condemned and that the use of eminent domain to take the land (in exchange for just compensation) would increase property values of the surrounding land and benefit large numbers of persons in a nonexclusionary and nondiscriminatory manner. Monopoly is defined as control of the only property on which a project can reasonably be carried out. See Lawrence Berger, "The Public Use Requirement in Eminent Domain," *Oregon Law Review* vol. 57, no. 2 (1978) pp. 203–246. This same analysis of monopoly power can be applied to tracts of land located on parts of the reservoir that are critical to efficient operations. Compulsory unitization laws are, in essence, condemnation statutes: the unwilling holdouts are forced into the unit at a price that returns a fair, but not monopolistic, share of the unit's proceeds.

39. About $2 million and two years were expended in obtaining and presenting information to determine proper well spacing in the Aneth field in Utah in 1960. See Frank W. Cole, *Well Spacing in the Aneth Reservoir* (Norman, University of Oklahoma Press, 1962).

40. Prior to drilling, estimates of the value of oil reserves in a formation have a range of error from 10 percent to 100 percent. As the field is developed, the range of error falls. See John M. Campbell, *Oil Property Evaluation* (Englewood Cliffs, N.J., Prentice-Hall, 1960) p. 149. Once 10 percent of the original oil in place in the reservoir has been produced, it is usually possible to ascertain the

physical characteristics of the formation in order to permit proper planning and efficient control of reservoir performance.

The distinct role that imperfect information plays in thwarting the formation of fieldwide units, apart from any holdout strategy of profitable obstructionism, is analyzed in Steven N. Wiggins and Gary D. Libecap, "Oil Field Unitization: Contractual Failure in the Presence of Imperfect Information" (draft, College Station, Texas A & M University, Department of Economics, March 1984).

41. A working interest owner is typically the operator under an oil and gas lease, that is, the lessee who has acquired the right to exploit the minerals. The working interest must pay for the costs of exploration and development. A royalty interest owner is entitled to a share of production free of any costs of production. Thus, if a landowner leases Blackacre to X Oil Company in return for a one-eighth royalty, X Oil Company acquires the working interest and must pay for all the costs of drilling and production. The landowner receives one-eighth of the revenues from production cost-free, but has no operating rights. If several different parties have acquired concurrent working interests in the same acreage, one party is often selected as the operator whose task is to develop the acreage for the joint account of all the parties. In this event, the other parties are called nonoperating working interest owners.

42. Maurice H. Merrill, "Compulsory Oil and Gas Unitization: Effect on Overriding Royalty Obligations," *Michigan Law Review* vol. 62, no. 3 (1964) pp. 407–408. The time value of money may differ significantly among the common owners of the reservoir. A risk-averse, ninety-year-old royalty interest owner may be quite unimpressed with the prospect that recovery rates in the long run are expected to double under unitization if the immediate effect is a small reduction in royalties. Artificial persons like business corporations may be less risk averse and have much lower discount rates on future income than mortal beings. If the present value of the oil and gas in the reservoir is greater with unitization in the opinion of many market participants, one would expect economically rational, royalty interest owners with high discount rates to sell their royalty shares to persons who have lower discount rates at a mutually satisfactory price. Perhaps royalty owners are less apt always to behave in an economically rational way than more sophisticated market participants, such as working interest owners.

43. This listing summarizes obstacles from the point of view of an individual royalty interest or working interest owner facing a decision of whether to join a unit. The list does not include legislative policy objectives that militate against the adoption of efficient conservation practices, such as the deliberate encouragement of unnecessary well drilling to stimulate local economies or the deliberate provision of excess producing capacity in domestic oil fields for national security reasons. These policy considerations (discussed in chapter 4) are important to an understanding of the legislative reluctance to pass compulsory unitization statutes.

44. One attorney experienced in the negotiation of many unitization agreements related that even employees of large, pro-unitization companies have personal interests that conflict with unitization. Division production superintendents

in charge of large tracts of land are often reluctant to encourage unitization that would turn operational control over to another company. See R. M. Williams, "The Negotiation and Preparation of Unitization Agreements," *Institute on Oil and Gas Law and Taxation,* vol. 1 (Albany, N.Y., Matthew Bender, 1949) p. 76.

45. Buckley, ed., *Petroleum Conservation,* p. 290; and Harold Garvin, "The Effect of Field Unit Operation Upon the Royalty Interest," *Oklahoma Bar Journal* vol. 1 (1950) p. 1793.

46. For example, the unitization of the Kelly–Snyder field in Texas involved almost 2,000 royalty interest and working interest owners. See Robert E. Sullivan, ed., *Conservation of Oil and Gas: A Legal History, 1948–1958* (Chicago, Ill., American Bar Association, 1960) p. 227. The 1980s revival of the Ranger oil field in West Texas, which went from boom to bust between 1917 and 1920, involved contacting at least 5,000 owners, some of whom owned less than one ten-thousandth of an acre. Sun Oil's team of landmen spent more than $750,000 in title examination work on the 15,000-acre field. See John Bloom, "Old Oil: The Ranger Revival," *Texas Monthly* (February 1981) pp. 113–116 and 194–198. In some old Texas fields, mineral interests were bought up by promoters, and fractional shares were sold as speculative ventures to hundreds of investors scattered across the nation and abroad. See A. Allen King, "Pooling and Unitization of Oil and Gas Leases," *Michigan Law Review* vol. 46, no. 3 (1948) p. 327.

47. Robert E. Hardwicke, *Antitrust Laws v. Unit Operation of Oil or Gas Pools* (rev. ed., Dallas, Tex., Society of Petroleum Engineers of the American Institute of Mining and Metallurgical Engineers, 1961) p. 43.

48. Twenty-seven states have antitrust exemptions for voluntary unitization agreements. See Howard R. Williams and Charles J. Meyers, *Oil and Gas Law* vol. 6 (Albany, N.Y., Matthew Bender, 1980) sec. 911. The federal government has not granted immunity from the federal antitrust statutes to unitization agreements, however.

49. Hardwicke, *Antitrust Laws,* p. 154.

50. National Petroleum Council, *Enhanced Oil Recovery,* p. 65.

51. Texas 2000 Commission, *Texas: Past and Future: A Survey* (Austin, Tex., Office of the Governor, 1982) p. 80 [hereafter cited as Texas 2000 Report].

52. Ibid.

53. All the data in this paragraph are found in Railroad Commission of Texas, *Secondary and Enhanced Recovery Operations in Texas to 1980: Bulletin 80* (Austin, Tex. 1980) pp. II–V [hereafter referred to as *Bulletin 80*].

54. Texas 2000 Report, p. 84.

55. Governor's Energy Advisory Council, *Executive Summaries of Project Reports of the Council* (Austin, Tex., 1975) p. 69. The National Petroleum Council estimates that by the year 2000, Texas could contribute about one-third of all the EOR oil projected to be produced in the United States. See National Petroleum Council, *Enhanced Oil Recovery,* pp. 46–48.

Chapter 3

1. The changing attitude of the industry toward unitization is well chronicled in Robert E. Hardwicke, *Antitrust Laws v. Unit Operation of Oil or Gas Pools* (rev. ed., Dallas, Tex., Society of Petroleum Engineers of the American Institute of Mining and Metallurgical Engineers, 1961) pp. 14–102, 162–170, and 262. Doherty's letter spurred President Coolidge to form the Federal Oil Conservation Board to investigate waste in the petroleum industry. The American Petroleum Institute (API), a trade association of the leading oil companies in the United States, appointed a committee of eleven members to prepare a report for the new federal board. The API's 1925 report refused to admit that waste in the production of petroleum was anything but "negligible." The report was completely silent on the issue of unitization, but vociferous on the need for the "exercise of initiative, liberty of action, the play of competition and the free operation of the law of Supply and Demand." The report by the Committee of Eleven was scathingly attacked by Doherty and ridiculed in the industry press by no less a personage than Archy, Don Marquis's redoubtable cockroach, for its blind denial of any wasteful practices in the industry or of any need for improvements (ibid., pp. 24–28, and app., items 3 and 4). In 1929 the API finally endorsed unitization, but only by cooperative agreement (ibid., p. 48). Unitization was even more controversial among independent producers who were not typically members of any of the above listed organizations.

2. A chronology of the passage of compulsory unitization laws appears in J. Frederick Lawson, "Recent Developments in Pooling and Unitization," in *Institute on Oil and Gas Law and Taxation,* vol. 23 (Albany, N.Y., Matthew Bender, 1972) p. 213.

3. 1931 Tex. Gen. Laws, ch. 26, sec. 1, p. 48, now appearing at Texas Natural Resources Code Ann. sec. 85.046(a)(7) (Vernon Supp. 1982).

4. Carl C. Rister, *Oil! Titan of the Southwest* (Norman, University of Oklahoma Press, 1949) p. 315.

5. James E. Nugent, "The History, Purpose, and Organization of the Railroad Commission," in *Oil and Gas: Texas Railroad Commission Rules and Regulations* (Austin, State Bar of Texas, 1982) p. A-18.

6. House Journal of Texas, 42nd Leg., 1st Called Sess., p. 14 (1931).

7. The transcripts appear in the House Journal of Texas, 42nd Leg., 1st Called Sess., pp. 14–684 (1931) [hereafter cited as 1931 H.J.]. Also see, the Senate Journal of Texas, 42nd Leg., 1st Called Sess., pp. 30–556 (1931) [hereafter cited as 1931 S.J.].

8. 1919 Tex. Gen. Laws, ch. 155, pp. 285–288.

9. 1929 Tex. Gen. Laws, ch. 313, sec. 2, p. 695. The Texas legislature met in 1929 to revise the state's original conservation law, because the 1925 codification of the Texas civil statutes inadvertently had omitted the Railroad Commission's enforcement provisions and because certain sections of the original act needed

clarification. See Senate Journal of Texas, 41st Leg., Reg. Sess., p. 903 (1929), for a letter from the Railroad Commission to the legislature explaining the need for amending the statutes.

By this time, the large Yates oil field and several West Texas gas fields had been discovered. The Railroad Commission had issued its first prorationing order in 1927 for the Yates field. This order encountered little opposition because it had been requested by the operators of the field who had agreed to develop the field cooperatively (see notes 54 and 97 in this chapter). Yet the possibility of widespread, market demand prorationing was already foreseeable, and the 1929 Act—while ostensibly involving only minor amendments to the 1919 Act—added the proviso that waste was not to include economic waste. The background of the 1929 Act is further discussed in the text of this chapter at notes 95 to 96. Even without this proviso, enormous litigation over the Railroad Commission's subsequent prorationing orders, particularly in East Texas, was likely, because the types of waste prohibited in the 1919 and 1929 Acts clearly encompassed only physical waste. Nonetheless, the 1929 proviso would make it even more difficult for the courts to uphold the Railroad Commission's earliest prorationing orders. See, for example, MacMillan v. Railroad Comm'n, 51 F.2d 400 at 405 (W.D. Tex. 1931), *rev'd per curiam and dismissed,* 287 U.S. 576 (1932).

10. 1931 H.J., pp. 130–131, 140; 1931 S.J., p. 166.

11. 1931 H.J., pp. 419, 437, 471–472, and 491.

12. The Standard Oil Co. trust had been busted only twenty years before the 1931 hearings. See Standard Oil Co. of N.J. v. United States, 221 U.S. 1 (1911). To many Americans, John D. Rockefeller and the Standard Oil Co. were the very essence of corporate monopoly. See Ida M. Tarbell, *The History of the Standard Oil Company* (New York, McClure, Phillips, 1904). One witness at the 1931 hearings defined an independent as a "company in which the Rockefellers own no stock." See 1931 H.J., p. 164. This was an accepted definition. The 1911 Court decree had separated the Standard Oil Co. of New Jersey (SONJ) into numerous smaller, vertically integrated oil companies, but each of these still had considerable market power in its geographic area. Indeed three spin-offs (now known as Exxon, Mobil, and Standard Oil of California) were to become three of the infamous "Seven Sisters" who would dominate much of the world oil trade in succeeding years. See Anthony Sampson, *The Seven Sisters* (New York, Bantam, 1976). At the 1931 hearings, Dan Harrison, an independent producer, refreshed the legislators' memories about the recent efforts of Shell and the Standard Oil Co. of New Jersey to carve up the world oil market, by inserting into the record several newspaper articles about the oil magnates' mysterious meetings at Achnacarry Castle and about Shell's public campaign to secure orderly petroleum markets worldwide through industry cooperation. Prorationing in Texas was viewed as an extension of this global campaign. See 1931 H.J., pp. 571–581 and 601–602; and 1931 S.J., pp. 470–480 and 556–559 (statistics showing the domination of refining capacity and crude production by SONJ's subsidiaries or spin-offs). The history of SONJ's activities in Texas, particularly through its affiliate, Humble Oil and Refining Co., is detailed in the text of this chapter at notes 73 to 108.

13. See, for example, the testimony of Tom Hunter and Ed Mayer, both independent producers, in 1931 H.J., pp. 124–125 and 276. The major oil companies were questioned in detail about the price differential between crude oil and refined lubricants. The latter price had not declined, and this fact supported the contention that vertical integration was being used anticompetitively (1931 H.J., pp. 432–433, 460, and 465–469). Other testimony elicited data that the Texas Co. had made $9.6 million in profits from its pipeline operations in 1930 (1931 H.J., p. 240), and Humble Oil had made more than $1 million a month in pipeline profits in 1930 (1931 H.J., p. 476). The size of these profits was especially grating to those lawmakers who represented wheat and cotton farmers suffering through the Great Depression. See, for example, 1931 H.J., pp. 503–504, and 620; 1931 S.J., pp. 188–189, 242, and 278. Both oil companies clearly stood to make enormous inventory profits on stored oil which had been bought at depressed prices, if prorationing was enforced and succeeded in raising the price. See 1931 S.J., p. 141.

The legislators were also very interested in the timing of changes in the majors' crude price postings. The major oil companies were the largest purchasers of crude and used a system of posted prices in each oil field to announce to producers the price at which they would buy crude. The majors had dropped the posted price from 67 cents to 35 cents a barrel shortly before the special session was called. Some independents viewed this price cut as evidence of the majors' attempt to punish the independents for not supporting prorationing laws.

14. The Texas Co. imported 2 million barrels per year of Mexican crude and held large oil concessions in Venezuela. See 1931 H.J., p. 264.

One accepted definition of a major oil company, proffered to the legislators by an independent, was a company "that had passed the stage of financial troubles." See 1931 H.J., p. 161. The testimony of R. C. Holmes, the president of the Texas Co., clearly showed that the large companies could take a long-run view and sacrifice immediate drilling and production returns. See 1931 H.J., p. 255.

15. 1931 H.J., p. 314.

16. See, for example, 1931 H.J., pp. 159, 348–356 and 397.

17. R. D. Parker, the chief supervisor of oil and gas at the Railroad Commission, admitted this (1931 H.J., p. 75), as did William S. Farish, the president of Humble Oil (1931 H.J., pp. 495–496). Parker's statutory duties included being a "pipeline expert," but he knew very little about pipelines (ibid., pp. 86–90 and 105). The lack of enforcement of the Common Purchaser Act was explained to some extent by the fact that the former governor had vetoed all appropriations to enforce it (1931 S.J., p. 295).

18. The argument that the majors "stole" leases at "thieves' prices" was made by virtually every independent producer or landowner who testified. See, for example, 1931 H.J., pp. 107–108, 119, 127, and 162.

19. One independent testified that the geologists for Humble Oil had publicly admitted that they "should bow their heads in humiliation" at the thought of East Texas (1931 S.J., p. 302).

20. 1931 H.J., p. 458.

21. The data showed that the largest twenty companies owned 74 percent of the domestic refining capacity (1931 H.J., p. 314). The travails of Joe Danciger and Capt. E. H. Eddleman, two independent refiners, were chronicled at the hearings (1931 H.J., pp. 400 and 516). The Patent Club restricted its membership to those oil companies that had made important inventions in refining. The major oil companies' patented refining technology was so superior to the small refiners' simple skimming plants that the extinction of the latter was clearly foreseeable (1931 H.J., pp. 278–280 and 523–528; and 1931 S.J., pp. 268–272).

22. Those who testified for market demand prorationing included: Robert R. Penn, a wealthy established independent who served as president of the Central Prorationing Committee (CPC); William S. Farish, president of Humble Oil and Refining; R. C. Holmes, president of the Texas Co.; G. S. Rollin, vice president of Shell Oil Co.; William R. Boyd, vice president of the API; and E. V. Foran, a petroleum engineer and consultant to the CPC. Those opposing market demand prorationing included small, independent producers and refiners like Dan Harrison, president of Harrison Oil Co.; Harry Pennington, president of the San Antonio Independent Petroleum Association; and J. M. West, owner of the South Texas Petroleum Co. Yet the lines were not always easy to draw between the groups. Gulf Oil Co., a large producer and refiner, opposed market demand prorationing because Gulf's code of business principles opposed any government intervention into the private enterprise system (1931 S.J., pp. 334–347). H. F. Sinclair of the Sinclair Oil Co. also seemed philosophically opposed to placing additional power in the hands of a state agency (1931 H.J., 537–539). Joe Danciger, an independent producer, violently opposed market demand prorationing, but he had his own small pipeline and refinery and did not represent the condition of most independent producers (1931 H.J., pp. 402–404). E. A. Landreth, president of the Landreth Production Corp., an independent, supported market demand prorationing and served on the West Texas Division of the CPC, but he did not seem to have any interest in the East Texas field and thus did not represent East Texas producers. The legislators had a difficult time sorting out the titles and viewpoints of the various associations of independents to determine whether they supported or opposed market demand prorationing (1931 S.J., pp. 429–430 and 481–483). This developed into a "battle of the telegrams," with various groups trying to convince the lawmakers that they best represented the voice of the independent in Texas (1931 S.J., pp. 539–544 and 599–623). Some independents seemed to support market demand prorationing in principle, if fairly enforced by the Railroad Commission, but opposed it as currently practiced. See, for example, 1931 H.J., p. 274 (testimony of Ed. R. Mayer, a director of the Independent Petroleum Association); and 1931 H.J., p. 124 (testimony of Tom Hunter, an independent producer).

23. See, for example, 1931 H.J., p. 230 (testimony of the Texas Co.); pp. 379–380 (testimony of the API); and pp. 413–434 (testimony of Humble Oil). Charles E. Bowles, the statistician for the Independent Petroleum Association, sought to counter all the majors' statistics about supply and demand with

a publication entitled "The Fallacy of 'Overproduction' in 1929" (ibid., pp. 508–510).

24. See, for example, 1931 H.J., pp. 173–174. Those operators who filed injunction suits in court against the Railroad Commission's orders were immune from penalties and therefore produced oil without restriction (1931 H.J., p. 71). Operators with land bordering these wide-open wells were then allowed by the commission to produce without limit in order to protect their land from drainage (ibid., p. 76). The Railroad Commission's list of those who violated its prorationing orders included only independents (ibid., pp. 69–70). Joe Danciger, one of the "outlaws," had produced more than a million barrels of oil from a 63-acre tract in a matter of months (1931 S.J., p. 538).

25. See, for example, 1931 H.J., pp. 200–201 (testimony of Robert R. Penn); ibid., pp. 413–419 and 437 (testimony of William S. Farish); and 1931 S.J., pp. 46–66 (testimony of E. V. Foran).

26. See, for example, 1931 H.J., pp. 173–174 (testimony of Robert R. Penn that Texas would lose at least $200 million in tax revenues due to the low price of crude); 1931 S.J., pp. 170–171 (testimony of R. C. Holmes, president of the Texas Co., that non-Texans would generally bear 87 percent of the increased cost of refined petroleum products due to higher-cost crude, but Texans would get 100 percent of the benefits).

27. For example, with remarkable candor, the Texas Co.'s R. C. Holmes described an April meeting at the API's New York offices in which the major producers in the East Texas field agreed to try to purchase enough crude to raise its price (1931 H.J., p. 256). That the effort was unsuccessful was not as telling as its having been made. A prior API proposal to divide the U.S. market into regions and control the production rates in each to achieve industry stability had been condemned by the attorney general of the United States as a violation of the antitrust laws (1931 S.J., p. 126). Holmes did not advance the cause of prorationing when he opined that the economic depression in the United States was largely a result of the antitrust laws, which "encouraged and supported overactivity in every line of endeavor" and advocated the repeal of the Sherman and Clayton Antitrust Acts (1931 H.J., pp. 231 and 250).

Also, the chief umpire in the East Texas field testified that Humble, Gulf, and the Texas Co. each had pipelines adjacent to the East Texas field but refused to build lines into the field, and that this forced independent producers to seek outside markets at low prices (1931 H.J., p. 224). The railroad commissioners as much as admitted that the majors set the price of oil (1931 H.J., pp. 55–56). Also see text of this chapter at notes 52 to 55.

28. 1931 H.J., p. 327. The chairman of the Railroad Commission, C. V. Terrell, also stated that the East Texas field's total allowable of 250,000 barrels was "largely based on the outlet, the demand." See 1931 H.J., p. 47; and also 1931 H.J., pp. 653–654 (testimony of Commissioner Terrell on the agency's use of its existing waste-prevention powers); and 1931 H.J., p. 174 (testimony of Robert R. Penn, president of the CPC).

29. MacMillan v. Railroad Comm'n, 51 F.2d 400 (W.D. Tex. 1931), *rev'd per curiam and dismissed,* 287 U.S. 576 (1932). A few months prior to the *MacMillan* court decision, a Texas trial court had upheld prorationing as a valid waste-prevention measure in the Panhandle field. Joe Danciger had brought suit in this Texas court in 1930, alleging that the Railroad Commission had no authority to prorate on the basis of market demand (the 1929 Act expressly denied authority to prevent economic waste) and also because his operations involved no physical waste. Undoubtedly, this ruling by the Texas court, which upheld a prorationing order, influenced MacMillan to bring his suit attacking the East Texas order in a federal court. Danciger's case was on appeal in the state courts during the 1931 special session. The Texas Court of Civil Appeals did not uphold the validity of the Panhandle prorationing order until March 23, 1932. See Danciger Oil & Refining Co. v. Railroad Comm'n, 49 S.W.2d 837 (Tex. Civ. App. 1932), *rev'd on other grounds and dismissed,* 122 Tex. 243, 56 S.W.2d 1075 (1933). Even had the *Danciger* opinion been before the legislators at the 1931 session, it is questionable whether a market demand prorationing law would have been enacted in light of the testimony adduced at the 1931 hearings. Also, the Panhandle situation was significantly different from the East Texas situation. See text of this chapter at notes 148 to 165. Thus the *MacMillan* opinion invalidating the East Texas order was not necessarily in conflict with the *Danciger* opinion upholding the Panhandle order. But see Danciger Oil & Refining Co. v. Smith, 4 F. Supp. 236 (N.D. Tex. 1933), *rev'd per curiam and vacated as moot,* 290 U.S. 599 (1933).

30. MacMillan v. Railroad Comm'n, 51 F.2d 400, 402 (W.D. Tex. 1931), *rev'd per curiam and dismissed,* 287 U.S. 576 (1932). The hearings simply confirmed the court's finding.

31. 1931 H.J., pp. 638–639. This message was something of a flipflop for the governor since he had called the special session in response to hundreds of telegrams, letters, and petitions that had poured into his office urging the adoption of market demand prorationing (1931 H.J., p. 14). This flood of telegrams seems to have been at least partly orchestrated by Humble Oil, Governor Sterling's former employer.

32. 1931 H.J., pp. 47–48, 83–84, 113, 207–222, and 331–334. The Central Prorationing Committee (also called the Central Advisory Committee) was appointed by the operators. This committee submitted nominations of purchasing companies' demands for crude oil to the Railroad Commission and supplied the commission with almost all its statistical data (ibid., pp. 171–173). Commissioner Neff described the existing prorationing system: "That we [the Railroad Commission] will pass these orders prepared by and suggested by the representatives of the companies, and then these same companies, they select an advisory committee in the various fields, and that committee selects an umpire, paid for and employed by the companies, and then we, very delicately and sanctimoniously send out our orders . . . , and he [the umpire] goes out in the field and proceeds to allocate to the different producers out there what they are allowed. . ." (1931 H.J., p. 333). The majors' incentive to nominate fictitious well potentials in order to achieve higher allowables had not escaped the eye of the independents or of the legislators (1931 H.J., pp. 107–108 and 397).

33. 1931 H.J., pp. 94–98 and 215–217.

34. 1931 H.J., pp. 52–53, 219–222, and 655–656. Fifty-five lives had been lost in the East Texas drilling. The chief umpire for East Texas estimated that 30 percent of the wells in the field were drilled improperly. One witness described how a woman with four children had been "burned to a crisp" because of a "busted second-hand pipe" (ibid., p. 521).

35. 1931 H.J., p. 104; and 1931 S.J., p. 288.

36. 1931 H.J., pp. 96–97.

37. 1931 H.J., pp. 84, 110–111, 327–328, and 621–624; and 1931 S.J. pp. 303–304.

38. 1931 H.J., pp. 54–56, 59, 61–63, and 343. By disavowing any knowledge about market prices, the commissioners left the distinct impression that the majors controlled the price.

39. 1931 H.J., p. 337. This accusation by Commissioner Neff so angered Commissioner Terrell that he insisted on returning to the hearing to defend his honor and that of the commission (ibid., pp. 644–662). The disharmony among the commissioners prompted some senators to urge the passage of a resolution that "the three Railroad Commissioners should immediately divorce from their minds these animosities and enter into a wholehearted and cooperative effort to the end that the commission may function properly, . . ." or else resign (1931 S.J., pp. 565–566).

40. 1931 H.J., pp. 670–679.

41. 1931 H.J., pp. 665–670.

42. In all fairness to the commission, it was underfunded and understaffed because the former governor had cut its appropriations (1931 H.J., p. 188). The larger oil companies, with their financial, legal, and technical resources, moved to fill the resulting vacuum. The legislators at the special session seriously considered creating a new Conservation Committee to regulate the oil and gas industry, leaving the Railroad Commission only its original function of regulating the railroads. After eleven days of testimony, which had painted a rather dismal picture of the Railroad Commission's competence, Governor Sterling sent the House and Senate a letter endorsing a new commission that would be appointive rather than elective. On this issue, Governor Sterling echoed the advice of Governor Hogg, who had successfully campaigned in 1890 for the establishment of the Railroad Commission. Hogg wanted the commission to be appointive from its start in 1891 in the belief that an appointive committee would be less subject to vote-buying by the powerful railroad interests. Also, the credentials of appointed commissioners could be better screened. Governor Hogg lost his bid for an appointive Railroad Commission when he angered the Farmers Alliance by refusing to appoint an agrarian to one of the three seats. See Robert C. Cotner, *James Stephen Hogg: A Biography* (Austin, University of Texas Press, 1959) pp. 220–249.

43. See, for example, 1931 H.J., p. 53.

44. See, for example, 1931 S.J., pp. 384, 426, and 504–505. The first proration order had been based on well potentials; it was then changed to the "unit plan" of allocating production to 20-acre units (ibid., pp. 384, 426, and 504).

45. The East Texas field was the first large field discovered that had a water drive. E. V. Foran, the petroleum engineer, testified about the technical difficulties of determining the proper rate of production for such a field (1931 S.J., pp. 49–63). Two other expert geologists did not agree with Foran's technical presentation about the need for slow rates of production (1931 S.J., pp. 403–405 and 424). The independents viewed all such testimony as pure theory and speculation. See, for example, 1931 S.J., pp. 258–259 and 286.

46. 1931 H.J., pp. 209–210 and 223–224.

47. The proponents of unitization included Robert R. Penn, the chairman of the CPC (1931 H.J., pp. 192–197); R. C. Holmes, the president of the Texas Co. (ibid., pp. 234–235); William R. Boyd, vice president of the API (ibid., p. 381); William S. Farish, president of Humble Oil (ibid., pp. 419–420); H. F. Sinclair, chairman of Sinclair Oil (ibid., pp. 537–539); E. V. Foran, a petroleum engineer and consultant to the CPC (1931 S.J., pp. 63–76); and G. S. Rollin, the vice president of the Dutch Shell Corp. (1931 H.J., p. 328). Sinclair, who opposed prorationing because it would give sweeping power to the Railroad Commission to exercise arbitrary judgments over the marketplace, viewed unit operations as the only answer to overdrilling and waste (1931 H.J., pp. 537–539). The other proponents listed above supported the adoption of market demand prorationing as well. No proponent of unitization favored a compulsory unitization law, however (1931 H.J., pp. 199 and 381).

48. 1931 H.J., p. 419.

49. 1931 H.J., p. 196 (Robert R. Penn); 1931 S.J., p. 328 (testimony of G. S. Rollin, vice president of Shell, that the Van field was unitized to reduce costs, not to conserve oil).

50. 1931 H.J., pp. 117–118; and 1931 S.J., p. 260.

51. 1931 H.J., pp. 419, and 480–481.

52. 1931 H.J., pp. 480–481.

53. 1931 S.J., p. 50.

54. The proration formula adopted at Yates was based on 100-acre units and average well potentials. See 1931 H.J., pp. 169–170, 418, and 480.

55. 1931 H.J., pp. 423–424. One independent testified that Humble initially proposed to build a pipeline to East Texas only if the operators unitized the field. When this concept was rejected, Humble proposed prorationing (1931 S.J., pp. 300–301). The large number of operators in the East Texas field—Humble's data showed 625 operators as of July 8, 1931 (1931 H.J., p. 457)—and its easier access to transportation by railroad and other pipelines doomed Humble's efforts

to secure prorationing in this field, although such a move had succeeded in the Yates field.

56. 1931 H.J., pp. 413–504.

57. See note 1 in this chapter. Robert R. Penn, the independent producer who most ardently supported prorationing at the special session, was chairman of the API's Production Division at that time. He organized a pro-unitization symposium on "A New Conception of Oil Production" for the API's November 1931 annual meeting, at which the API's Board of Directors formally announced the "new" conception that Doherty had propounded in 1924. See Hardwicke, *Antitrust Laws,* pp. 87–92. Because the API's wholehearted acceptance of unitization in 1931 came only after its failure to secure antitrust immunity for a worldwide system of quotas on crude oil, it seemed that the API wanted to use unitization as an indirect prorationing device to raise prices.

58. 1931 H.J., pp. 194–196, and 480. William S. Farish had urged that the antitrust laws be modified to allow cooperative agreements as early as 1926. Farish had tried to convince the Texas legislature to pass a "Permissive Bill" in 1929 providing antitrust immunity for voluntary unitization and prorationing agreements. This effort seems to have been partly responsible for the proviso prohibiting economic waste in the 1929 amendments to Texas' conservation law. See 1931 H.J., pp. 578 and 596–599; and text of this chapter at notes 94 to 96.

59. 1931 H.J., pp. 192–193. At an API meeting a month earlier, Penn had proposed that the independents in East Texas form a large corporation to negotiate a unitization agreement with the majors in the field. See Hardwicke, *Antitrust Laws,* p. 87, n. 92.

60. Harry Pennington, an articulate independent, made it quite clear to the legislators that passage of a prorationing bill amounted to passage of a unitization bill. See 1931 S.J., p. 282. Another independent testified: "Unitization means simply every producer in their field coming into a common pool and turning the management over to one company for organization. I don't think the people will ever agree to turn over 100 miles of oil fields to one big company, . . .for when they do that they have kissed their property good-bye forever" (ibid., p. 259).

61. 1931 Tex. Gen. Laws, ch. 26, sec. 1, p. 48:

> Neither natural gas nor crude petroleum shall be produced, transported, stored, or used in such manner or under such conditions as to constitute waste; provided, however, this shall not be construed to mean economic waste, and the Commission shall not have power to attempt by order, or otherwise, directly or indirectly, to limit the production of oil to equal the existing market demand for oil; and that power is expressly withheld from the Commission.

62. 1931 Tex. Gen. Laws, ch. 26, sec. 1, p. 48.

63. 1931 H.J., p. 457. Of these 586 operators, 334 had not yet developed their acreage, so 252 operators owning 10 percent of the acreage produced 49 percent of the oil. The data also showed that the average well of the nineteen majors produced 421 barrels per day versus 288 barrels per day for the other operators

(ibid., p. 459). This suggests that the majors' operations were more efficient, probably because their wells were more widely spaced, and could be profitable even at lower prices.

64. 1931 H.J., pp. 175, 343, 392 and 437; and 1931 S.J., pp. 68 and 216–220. The House version of H.B. 25 had included a section that sought to protect correlative rights under conditions of oversupply by allowing an operator of a common source of supply to take therefrom "only such proportion of all crude oil. . .as the potential production of the well or wells of any operator bears to the total potential production of such common source of supply, having due regard to the acreage drained by each well" (1931 S.J., pp. 650–651). The attorney general was asked to review the constitutionality of this section and he declared it invalid on two grounds: (1) that the state of Texas' police power extended only to regulating oil and gas production to prevent physical waste; and (2) that the power to determine fair and equitable shares of an oil pool was a purely judicial question for the courts, not an administrative agency. See Texas Attorney General Opinion, August 7, 1931, as cited in 1931 S.J., pp. 704–708. Dispute over the Railroad Commission's authority to protect correlative rights independent of physical waste prevention was to plague the administration of oil and gas production in Texas for years to come. See this chapter notes 148 to 165 and chapter 8 text at notes 41 to 45.

Ultimately, the 1931 Act did include a definition of waste based on inequitable withdrawals from a common pool, but the Railroad Commission could invoke this definition only to prevent discrimination between producers with unequal access to regulated carriers. See 1931 Tex. Gen. Laws, ch. 26, sec. 1, pp. 48–49. This definition was eliminated in the 1932 revision of the 1931 Act. See 1932 Tex. Gen. Laws, ch. 2, pp. 3–10.

65. 1931 S.J., pp. 313–315 and 431–432; and 1931 H.J., pp. 582 and 627.

66. 1931 Tex. Gen. Laws, ch. 26, secs. 1, 7, and 15, p. 46. For example, the act gave the Railroad Commission express authority over gas–oil ratios, and defined waste broadly as "physical waste incident to or resulting from drilling, equipping, locating, spacing, or operating wells as to reduce, or tend to reduce, the ultimate recovery of crude petroleum oil or natural gas from any well or pool." The act required that the Railroad Commission apportion any decrease in production to prevent physical waste "justly and equitably."

67. Ibid., secs. 18, 22, and 23. To protect against further pipeline abuses, the legislature also passed a stronger Common Purchaser Act requiring that oil and natural gas pipelines purchase without discrimination in favor of their own production or in favor of one producer against another in the same field, and without unjust and unreasonable discrimination between fields. See 1931 Tex. Gen. Laws, ch. 28, pp. 58–66.

68. The chairman of the Board of Regents of the University of Texas was especially troubled by the huge amounts of gas being flared on university-owned lands (1931 S.J., pp. 191–200). The majors and independents alike may have worried that in their zeal for prohibiting physical waste, the lawmakers would

prevent gas flaring without regard to the profitability of building facilities to process and market gas or to use it in repressuring. See 1931 H.J., pp. 334, 338–339, 490–491, 652, and 661; and 1931 S.J., pp. 377–378. This seems an inadequate explanation for the statute's wording, however. The existing commission was most unlikely to ever order producers to invest in unprofitable facilities, nor was the legislature interested in such a policy. Repressuring did carry the taint that it generally required unitization (1931 H.J., p. 80), but the statute already prohibited the commission from requiring unitization, so the repressuring phrase was not necessary for this purpose.

69. See, for example, J. Howard Marshall and Norman L. Meyers, "Legal Planning of Petroleum Production," *Yale Law Journal* vol. 41, no. 1 (1931) pp. 33–68; and J. Howard Marshall and Norman L. Meyers, "Legal Planning of Petroleum Production: Two Years of Proration," *Yale Law Journal* vol. 42, no. 5 (1933) pp. 702–746. Robert Hardwicke's account of the events in East Texas is often used as the definitive history of the oil conservation laws in Texas in the 1930s. See Robert E. Hardwicke, "Legal Histories of Conservation of Oil and Gas in Texas as to Oil," in *Legal History of Conservation of Oil and Gas* (Chicago, Ill., American Bar Association, 1939) pp. 214–268 [hereafter cited as Hardwicke, "Legal History of Oil"]. Unfortunately, Hardwicke's active role in support of prorationing (he was the attorney for the Central Prorationing Committee and represented the Railroad Commission as special counsel in almost all the prorationing cases) leads to a somewhat distorted account of the 1931–32 events in this work. Hardwicke clearly viewed the failure to enact market demand prorationing in 1931 as due to anticonservationists who did not want to prevent physical waste. This view does not explain the enactment of the provision in the 1931 Act that expressly authorized the commission to prorate production to prevent waste and conserve oil and gas. Hardwicke also wrote that the several federal district court holdings which invalidated commission prorationing orders were unsupported on the facts and an unwarranted substitution of the court's judgment for that of the commission's experts. Hardwicke's account fails to acknowledge the extraordinary lack of expertise at the commission; the legislators' rational fear of monopolistic practices in the oil fields; and the testimony of the commissioners themselves that their prorationing orders were primarily designed to raise prices.

Because Hardwicke focused almost exclusively on the courts' and commission's orders rather than on the extant legislative policies which the judiciary and the agency were bound to interpret, his account virtually ignores the early foundations of Texas unitization laws and policy. Other scholars have treated Hardwicke's legal history of the East Texas field as definitive, and this reliance has unfortunately compounded the bias against judicial review of Railroad Commission orders. See, for example, Kenneth Culp Davis, "Judicial Emasculation of Administrative Action and Oil Proration: Another View," *Texas Law Review* vol. 19, no. 1 (1940) pp. 39–44.

70. Northcutt Ely, *Oil Conservation Through Interstate Agreement* (Washington, D.C., Government Printing Office, 1933) pp. 17–18; Anthony Sampson, *The*

Seven Sisters, pp. 70–103; and Donald N. Zillman and Laurence H. Lattman, *Energy Law* (Indianola, Ind., Foundation Press, 1982) pp. 815–819.

71. Harold F. Williamson, Ralph L. Andreano, Arnold R. Daum, and Gilbert C. Klose, *The American Petroleum Industry: The Age of Energy 1899 to 1959* (Westport, Conn., Greenwood Press, 1963) pp. 350–362 and 570–602. This thorough history of the early petroleum industry documents many other parts of the testimony heard at the 1931 session. The oil pipeline business had always been monopolistic because the large capital investment required to build a pipeline created barriers to entry. In 1906 Congress passed the Hepburn Act which made interstate pipelines common carriers. In 1914 the Supreme Court upheld the constitutionality of this act, reciting the need to control the economic power of the pipelines. The Court wrote:

> Availing itself of its monopoly of the means of transportation the Standard Oil Company refused through its subordinates to carry any oil unless the same was sold to it or to them and through them to it on terms more or less dictated by itself. In this way it made itself master of the fields without the necessity of owning them and carried. . .a great subject of international commerce coming from many owners but, by the duress of which the Standard Oil Company was master, carrying it all as its own.

The Pipeline Cases, 234 U.S. 548, 561 (1914). Also see, Melvin G. DeChazeau and Alfred E. Kahn, *Integration and Competition in the Petroleum Industry* (Philadelphia, Pa., Porcupine Press, 1959) pp. 512–514.

A recent study confirms the existence of continuing pipeline monopsony power in crude oil-purchasing markets. In Texas, the four largest, ownership-connected pipeline firms received 96.52 percent of all the crude oil received by interstate pipelines in Texas in 1977. The weighted average four-firm concentration ratio for the seventeen state markets examined was 95.90 percent. See John A. Hansen, *U.S. Oil Pipeline Markets* (Cambridge, Mass., MIT Press, 1983) p. 49. Hansen writes (p. 116): "In the absence of effective regulation, this market power would probably allow pipelines to either drive independent producers out of business or to force sales of crude at prices that would enable pipelines to capture a portion of the rent that would have accrued to producers in a competitive market." Even in markets where water transportation was available to most refineries, this shipping alternative did not promote competition because pipelines were still necessary to move domestic oil from the fields to the nearest waterway.

72. According to David F. Prindle, in *Petroleum Politics and the Texas Railroad Commission* (Austin, University of Texas Press, 1981), pp. 32–34, by 1930, the post of railroad commissioner was "largely a sinecure for over-the-hill Texas politicians." Of the three commissioners in power during the prorationing debate, two were former state legislators and one was a former governor.

73. Henrietta M. Larson and Kenneth W. Porter, *History of Humble Oil and Refining Company: A Study in Industrial Growth* (Salem, N.H., Ayer, 1959) pp. 264–326, 390–425, and 446–487.

74. Ibid., p. 456.

75. Ibid., pp. 10–20. Ironically, Ross Sterling, a founder and first president of Humble, had organized the Texas Oil Producers and Landowners Association in 1915. This association was formed to oppose the "Texas Company" bill, a bill which would allow producing companies to receive a single business charter to function as vertically integrated refiners and marketers. Independent producers such as Sterling and Farish had had unfortunate experiences with the majors who owned pipelines and who had superior bargaining power over the nonintegrated producer. In 1915 Farish scathingly denounced the big companies for their monopolistic practices, and in 1916 he headed the Gulf Coast Oil Producers Association, another group of independents. In 1917 this association won support of a bill constituting pipelines as common carriers under the control of the Railroad Commission. With this legal safeguard in place, the independents were no longer so hostile to a Texas Company bill, and such an act was also passed in 1917. When Humble Oil and Refining Co. incorporated in 1917, it availed itself of this newly passed law, and the company's charter authorized it to perform all the functions of a vertically integrated company (ibid., pp. 39–55).

76. Oil first was produced commercially in Texas in 1894 with the discovery of the Corsicana field in East Central Texas. This field was developed with capital from Standard Oil of New Jersey, but the field was not of national importance (ibid., pp. 11–15).

77. In 1906 Standard Oil controlled 72 percent of the crude oil in the Appalachian region, 95 percent in Indiana, 100 percent in Illinois, 45 percent in the mid-continent region, and only 10 percent in the Gulf Coast. See Williamson and coauthors, *The American Petroleum Industry*, p. 7.

78. 1889 Tex. Gen. Laws, ch. 117, sec. 1, p. 141.

79. Cotner, *James Stephen Hogg*, pp. 105–220.

80. The early Texas cases against the Standard Oil trust are discussed in Bruce Bringhurst, *Antitrust and the Oil Monopoly: The Standard Oil Cases, 1890–1911* (Westport, Conn., Greenwood Press, 1979) pp. 40–67. Before Spindletop, Texas imported almost all of its petroleum products from Appalachia and the Midwest. The Waters–Pierce Oil Co. controlled 90 percent of the Texas market and maintained its position through many anticompetitive practices, especially rebates. When Waters–Pierce first lost its permit to do business in Texas in 1898, Henry Pierce, the head of the company, dissolved it and formed a new company, not even bothering to change its name. Pierce concealed the fact that Standard Oil owned 60 percent of the stock, and with the use of political influence secured by secretly lending money to a leading Texas politician, he obtained a new permit to operate in Texas. The second antitrust case resulted in the forfeiture of Waters–Pierce Co.'s charter in 1907. In 1909 three other large, Standard Oil-owned companies—the Navarro Refining Co., the Corsicana Petroleum Co., and the Security Oil Co.—were similarly expelled from Texas. The early excesses of Standard Oil in other states from 1890 to 1911 and the many other state antitrust suits against the trust are also well described by Bringhurst.

81. Larson and Porter, *History of Humble Oil*, pp. 74–75.

82. State v. Humble Oil & Refining Co., 263 S.W.2d 319 (Tex. Civ. App. 1924).

83. Larson and Porter, *History of Humble Oil*, pp. 138, 158, and 165.

84. Ibid., pp. 168–169. Humble's pipeline earned $22 million in profits in 1931 alone (ibid., at app. tab. XI). This was far more than the estimate of $12 million in profits made by one witness at the 1931 hearings. See note 13 in this chapter.

85. Larson and Porter, *History of Humble Oil*, p. 171. In recognition of his new role as a major purchaser, Farish resigned from the Gulf Coast Producers Association in 1919.

86. Ibid., pp. 184–185.

87. Ibid., pp. 248–257. Farish was a member of the API's Committee of Eleven and opposed any government regulation of the oil industry, either state or federal, in 1925. See note 1 in this chapter.

88. Ibid., pp. 257–262, 279–287, and 434–445.

89. Ibid., p. 265.

90. Ibid., pp. 267–279 and 295.

91. Ibid., pp. 285–287, 293, and 295.

92. Ibid., pp. 303–306. Bills to this effect were introduced at both the 1930 and 1931 legislative sessions. See House Journal of Texas, 41st Leg., 4th Called Sess., p. 50 (1930); and House Journal of Texas, 42d Leg., 1st Called Sess., p. 750 (1931).

93. Larson and Porter, *History of Humble Oil*, pp. 309 and 315–316. Actually the Van field unitization agreement went into effect after the discovery well was drilled, but before it was generally known. Humble drew up the agreement and took the risk that it would escape antitrust prosecution.

94. Ibid., pp. 307–309.

95. Ibid., pp. 312–314.

96. Ibid., pp. 314–315; and see note 9 in this chapter. The railroad commissioners addressed a letter to the legislators in 1929 urging them to enact a bill which would give the commission stronger powers to prevent physical waste. After referring to the "extensive discussion in the public press" of H.B. 388, the commissioners' letter noted that "the Commission desires to make it plain that it has not suggested any other legislation tending...to enlarge its powers over matters heretofore placed within its jurisdiction." See Senate Journal of Texas, 41st Leg., Reg. Sess., pp. 903–904 (1929). This last quotation seems to be an effort by the commissioners to dissociate themselves from Humble's attempts to include in H.B. 388 authorization of voluntary prorationing or unitization agreements by the commission. At the 1931 hearings, Dan Harrison, an independent producer who vehemently opposed prorationing, testified that Farish's attempt to secure antitrust immunity for cooperative agreements in 1929 resulted in the

legislative proviso prohibiting the commission from controlling economic waste. See 1931 H.J., pp. 578 and 596–599.

97. Larson and Porter, *History of Humble Oil,* pp. 316–332. For example, Humble had the only pipeline in the vicinity of the Yates field and offered to extend its line if the producers would agree to ratable sharing of the outlet. Thus pressured, the operators agreed on a prorationing plan and appealed to the Texas attorney general for approval of its legality, but he refused to approve or disapprove it (ibid., p. 317). The field was developed cooperatively, and production costs were less than 5 cents per barrel.

98. Humble's attitude was neatly summarized by Farish as follows (ibid., p. 325):

> We are interested in conservation; we are interested in proration or other forms of cooperative development and production, whether voluntary or compulsory; we are interested in unit operation; we are interested in producing our oil at the lowest cost under the best engineering practices. . . . Intelligent controlled production may influence prices, . . . but as such controlled production is going to be dependent . . . upon some political body in any state, it is reasonable to assume that fair prices will prevail and that the power to control can never be abused because public opinion will prevent [abuse].

99. Ibid., p. 323. Humble's biography states that Farish's public declaration that no excess crude production existed six weeks before the price cut was true at that time.

100. Ibid., pp. 323–325.

101. Ibid., pp. 324–325.

102. Ibid., p. 325. This price cut led to another antitrust suit against Humble and other majors (ibid., p. 465).

103. Ibid., p. 397.

104. Ibid., pp. 398–399. Humble's large land purchases in East Texas were only part of its overall strategy of acquiring large tracts. Between 1925 and 1940, Humble increased its share of U.S. reserves from 0.7 percent to 14 percent. So quick was Humble in East Texas that between the first showing of oil in Joiner's well and its actual completion, Humble leased 12,000 acres at an average cost of about $20 per acre. Later, Humble bought leases for more than $1,000 per acre. Doubtless this prompted the testimony of many witnesses at the 1931 hearings that Humble had "stolen" many of the leases. The most spectacular of Humble's block leases was the million-acre King Ranch lease in Southwest Texas, negotiated in 1933 (ibid., pp. 405–406).

105. Ibid., pp. 449–452.

106. Ibid., p. 453.

107. Humble's own account of the hearings appears in ibid., pp. 455–459. The fury and outrage that accompanied Humble's price cuts and its charges to producers for storage fees on the oil that it purchased and could not use were so severe

that Humble's biographers conclude, rather wistfully, that "the better course from a public-relations viewpoint might have been to purchase all the oil from the beginning and take the full loss rather than to arouse such animosity" (ibid., p. 517).

108. Crude oil production from East Texas totaled 106 million barrels in 1931, 120 million barrels in 1932, and 172 million barrels in 1933. Of this total, 3 million barrels in 1931, 25 million barrels in 1932, and 35 million barrels in 1933 were "hot" oil, that is, oil produced in violation of prorationing orders. See Williamson and coauthors, *The American Petroleum Industry,* p. 545.

109. In United States v. Socony-Vacuum Oil Co., 310 U.S. 150 (1940), the U.S. Supreme Court held that a group of major oil companies had violated the Sherman Antitrust Act by combining and conspiring to artificially raise the tank car price of gasoline in the spot markets in the East Texas and Mid-continent fields. The major oil companies were found to have combined and agreed to buy gasoline from independent refiners in the East Texas field who had no storage or marketing facilities and who therefore would dump gasoline on the market at distress prices, causing price wars and economic losses to the majors' own investments in refining and marketing facilities.

The genesis of the problem was, of course, the huge surplus of crude oil pouring out of the East Texas field. Each major oil company involved in the agreement selected an independent refiner as a "dancing partner," and assumed responsibility for purchasing its distress supply. The major oil companies under the antitrust indictment (who included Shell, Continental Oil, Sinclair, Phillips, and Socony-Vacuum; Socony-Vacuum had been the Standard Oil Co. of New York, and later was renamed Mobil) posed as a defense that their plan was designed to eliminate the competitive evils stemming from the rule of capture and the failure to control the production of hot oil in the East Texas field. The court held that price-fixing agreements are unlawful *per se,* and that the attempt to eliminate ruinous and harmful competition could not be a legal justification for a price-fixing conspiracy. The antitrust laws do not permit an inquiry into the reasonableness of price-fixing agreements. The majors also argued as a defense that their efforts had been in vain and that they had not been able to affect market prices for gasoline. The court considered this fact irrelevant to the issue of the defendants' guilt. The majors then argued as a defense that the federal government knew of, condoned, and encouraged the gasoline buying program as part of the National Industrial Recovery Act (NIRA). The court responded that, even though the buying program may have been consistent with the general objectives and ends of the NIRA, after the NIRA expired in June 1935, the price-fixing combination lacked Congressional sanction and was *per se* illegal. In sum, the majors could not replace free market forces, imperfect as they were in achieving economic efficiency, with private agreements to affect prices without violating the antitrust laws.

110. Larson and Porter, *History of Humble Oil,* pp. 323–324, 457, 461, 474, and 481.

111. The independents successfully blocked a bill allowing business charters for vertically integrated companies until they secured a bill conferring common carrier status on oil pipelines in 1919. See note 75 in this chapter.

112. The independents also secured the Marginal Well Act in April 1931 (discussed in text of this chapter at notes 136 to 145) which exempted many wells from prorationing even before the special session was called to discuss market demand prorationing in the summer of 1931.

113. The shutdown of the wells under martial law gave the commission time to hold new hearings on waste in East Texas. On September 2, 1931, a new proration order was promulgated, holding the East Texas field to 450,000 barrels per day of production, allocated on a flat, per-well basis with no consideration of acreage or well potentials. The National Guard enforced these orders until February 18, 1932, when a federal district court decided that the governor's declaration of martial law was invalid because no real threat of violence existed. Constantin v. Smith, 57 F.2d 227 (E.D. Tex. 1932) *aff'd,* 287 U.S. 378 (1932). The Railroad Commission promulgated a new proration order on February 25, 1932, limiting the field to 325,000 barrels per day, or 75 barrels per day per well. The commission attempted to enforce this order with the aid of National Guardsmen acting as ordinary peace officers, but many operators again flouted the orders. On October 24, 1932, a federal district court enjoined the commission's new order on two grounds: (1) the evidence showed that the order was based on market demand, a factor forbidden by the 1931 Act; and (2) the flat, per-well allocation was unreasonable and confiscatory because it did not distinguish good from poor wells. People's Petroleum Producers v. Smith, 1 F. Supp. 361 (E.D. Tex. 1932). This court decision led Governor Sterling to call the November 1932 special session. The events of this era are chronicled in Larson and Porter, *History of Humble Oil,* pp. 447–487; Rister, *Oil!,* pp. 317–322; and Hardwicke, "Legal History of Oil," pp. 214–268.

114. The Senate held hearings on the proposed Market Demand Bill from November 4 to 9, 1932. The transcripts appear in Senate Journal of Texas, 42d Leg., 4th Called Sess., pp. 2–5 and 46–208 (1932) [hereafter cited as 1932 S.J.].

115. 1932 Tex. Gen. Laws, 4th Called Sess., ch. 2, p. 3.

116. Rister, *Oil!,* pp. 320–321.

117. See, for example, 1932 S.J., pp. 206–209 (testimony of Charles Roeser and Carl Estes). Commissioner Terrell also praised the virtues of raising the price of oil and shipping it out of state to non-Texans who would bear most of the price burden (ibid., p. 48). Danciger and Gulf Oil continued to oppose market demand prorationing, despite its proven record in raising prices (1932 S.J., pp. 200–202; and 152–180). Mr. Harrell of the Associated Gasoline Consumers naturally opposed it (ibid., pp. 107–111 and 129–152).

118. The influence of Commissioner Thompson cannot be underestimated. His testimony was masterful (1932 S.J., pp. 71–106), especially when compared to that of Commissioner Terrell, who again admitted that his support for the latest proration order in East Texas was based partly on the need to raise the price

of oil rather than on physical waste alone (ibid., pp. 62–64). Clearly, this was forbidden by the 1931 Act. Thompson even placated Senator DeBerry, an ardent foe of market demand prorationing, with the following notion about the East Texas oil: "It (the oil) was put there... by the God Almighty to be a benefit to mankind through the years and we are charged with the duty of conserving the resources of the State and looking after the interests of the greatest number" (ibid., p. 95). Thompson also indicated how actively he would use his power to promote the state's interest. If major oil companies imported oil into Texas adding to the oversupply, Thompson would "work on the pool in which they are interested," that is, cut their Texas production down (ibid., p. 86). Thompson would serve on the Railroad Commission for almost thirty-three years. His enormous influence and political skills are further described in Prindle, *Petroleum Politics,* pp. 35–40.

119. President Roosevelt would be inaugurated in January 1933. Thompson clearly foresaw and supported the New Deal legislation which would be passed (1932 S.J., pp. 88 and 103). Many others foresaw it and condemned it. The 1932 hearings were far more ideological than the 1931 hearings. A Rotary Club speech was submitted into the record criticizing the alarming degree of government aid and control of U.S. private industry as Russian-inspired (ibid., pp. 23–30); a law professor viewed prorationing as the beginnings of collectivism (ibid., pp. 112–120); and Gulf Oil Co. reiterated its 1931 stand against prorationing, this time linking the proposed bill to communism and socialism (ibid., pp. 152–181). Despite the ideological tone of the hearings, no witness ever mentioned unitization, probably because the 1931 Act already protected operators against it.

120. 1932 S.J., pp. 71–72 and 99. Actually it was not until December 1932—a month *after* the special session ended—that the Railroad Commission conducted scientific tests to ascertain the pressure characteristics of the East Texas field. See Prindle, *Petroleum Politics,* p. 29. During the 1932 hearings, Gulf Oil called on the Railroad Commission to study the East Texas field so that a prorationing order based on physical waste could be entered and upheld by the courts, precluding the need for market demand prorationing (1932 S.J., p. 154). Clearly the Railroad Commission still did not have a firm scientific basis for selecting the maximum total allowable for East Texas. Prorationing based on market demand was easier to implement administratively than prorationing to prevent physical waste. See 1932 S.J., p. 196 (testimony by Chief Supervisor Parker of the commission that the recent 10 percent reduction in allowables of Gulf Coast wells was arbitrarily ordered without advance engineering data). Despite this evidence that the Railroad Commission did not have a well-supported conservation basis for prorationing oil, Commissioner Thompson refused to admit that the orders had anything to do with price or market demand, and he did so credibly (1932 S.J., pp. 75, 78, and 98).

121. Champlin Refining Co. v. Oklahoma Corp. Comm'n, 286 U.S. 210 (1932) (upholding Oklahoma's market demand prorationing statute as a reasonable method of preventing waste). See also Danciger Oil & Refining Co v. Railroad Comm'n, 49 S.W.2d 837 (Tex. Civ. App. 1932), *rev'd on other grounds*

and dismissed, 122 Tex. 243, 56 S.W.2d 1075 (1933) (upholding Railroad Commission's proration order in the Panhandle field as reasonably necessary to prevent physical waste under the Texas 1929 conservation act).

122. Larson and Porter, *History of Humble Oil,* p. 469.

123. Ibid., pp. 464–465. Farish had learned that "one of the easiest ways to defeat a thing is for us to ask for it" (ibid., p. 465). Also, Humble's pricing policy in 1932 had provoked further acrimony with independent producers. Farish continued to call for low-cost production with correspondingly low prices for consumers. The president of Sun Oil urged Farish to join him in raising crude prices in 1932 in order to bolster the independents' support for market demand prorationing. Humble refused to do so. Farish often expressed the position that only low prices would bring producers to accept regulation. This position, sound as it was, made many enemies (ibid., pp. 472–474).

124. 1932 Tex. Gen. Laws, 4th Called Sess., ch. 2, sec. 13, p. 9, now appearing at Texas Natural Resource Code Ann. sec. 85.002 (Vernon 1978).

125. Ibid., at sec. 14. The need for periodic reenactment of the 1932 Act (generally every two years) was finally ended in 1941. See Blakely M. Murphy, ed., *Conservation of Oil and Gas: A Legal History, 1948* (Chicago, Ill., American Bar Association, 1949) pp. 450–451 [hereafter cited as Murphy, *Legal History, 1948*].

126. 1932 Tex. Gen. Laws, 4th Called Sess., ch. 2, sec. 6-A, p. 7, now appearing at Texas Natural Resources Code Ann. sec. 85.056 (Vernon 1978). The senators from the cotton districts had tried to secure an amendment to the bill that prohibited the Railroad Commission from restricting the production of crude oil when the retail price of gasoline exceeded the price per pound of spot middling cotton (1932 S.J., p. 22).

127. 1932 Tex. Gen. Laws, 4th Called Sess., ch. 2, sec. 1, now appearing at Texas Natural Resources Code Ann. sec. 85.046 (Vernon Supp. 1982). This section of the 1932 Act also strengthened and further expanded the commission's power to prevent physical waste. See App. I of this book which contains the full version of Section 85.046 as it now exists.

128. See text of this chapter at notes 63 to 66. Similarly, Montana's conservation law, enacted in 1953, deliberately omitted any mention of correlative rights. The independent producers in Montana feared that if the conservation agency had power over correlative rights, the independents would be forced to share their discoveries with the majors. Note, "Application of the Doctrine of Correlative Rights by the State Conservation Agency in the Absence of Express Statutory Authorization," *Montana Law Review* vol. 28, no. 2 (1967) p. 216. Despite this express exclusion of authority over correlative rights, the Montana Supreme Court held that the conservation agency had the implied power and duty to consider and protect correlative rights. By judicially implying such authority, Montana's 1953 conservation act was held to be constitutional. Ironically, the plaintiff in the case was an independent oil company that was being drained. See Pattie v. Oil & Gas Conservation Comm'n, 145 Mont. 531, 402 P.2d 596 (1965).

129. According to the Texas attorney general, one effect of the commission's lack of authority to prorate oil production for any purpose other than to prevent waste was that the agency had no power to require an operator to underproduce his wells in order to make up previous production above the legal allowable. See Texas Attorney General Opinion No. 2932 (1933).

130. 1932 Tex. Gen. Laws, 4th Called Sess., ch. 2, sec. 4, p. 7, now appearing at Texas Natural Resources Code Ann. sec. 85.053 (Vernon 1978). Many state conservation laws are more explicit in recognizing and defining correlative rights. See R. O. Kellam, "A Century of Correlative Rights," *Baylor Law Review* vol. 12, no. 1 (1960) pp. 30–35; and Jed B. Maebius, Jr., "Statutory Guidelines for Determining 'Fair Share'," *St. Mary's Law Journal* vol. 2, no. 1 (1970) pp. 63–80. The Interstate Oil Compact Commission's model conservation act defines the protection of correlative rights as action by a state agency which affords "a reasonable opportunity to each Person entitled thereto to recover or receive the Oil or Gas in his tract or tracts or the equivalent thereof, without being required to drill unnecessary wells or to incur unnecessary expense." See Stephen L. McDonald, *Petroleum Conservation in the United States: An Economic Analysis* (Baltimore, Md., The Johns Hopkins University Press for Resources for the Future, 1971) p. 244.

131. 1932 S.J., p. 188; and also see text of this chapter at notes 62 to 63.

132. 1932 S.J., pp. 106–107. Thompson also testified that the bill gave the commission the power to consider the cost of drilling in setting allowables (ibid., p. 100).

133. 1932 Tex. Gen. Laws, 4th Called Sess., ch. 2, sec. 6, p. 7, now appearing at Texas Natural Resources Code Ann. sec. 85.054 (Vernon 1978).

134. Hardwicke, "Legal History of Oil," pp. 240–243. When the price of crude oil reached 10 cents a barrel, the commission tried to coax operators into shutting down by promising them that if they refrained from producing any oil at all "until oil reaches the price it ought to bring we will allow you to make up this production upon proper application to the Commission for back allowable permit." See Flannery v. State, 85 S.W.2d 1052 at 1054 (Tex. Civ. App. 1935). The commission's wildly varying, total allowables for East Texas show that it still did not fully understand the field's MER.

135. See discussion in text of this chapter at note 132. Parker, the Railroad Commission's chief supervisor, testified that the per-well formula was adopted to protect small producers (1932 S.J., pp. 187–188).

136. 1931 Tex. Gen. Laws, Reg. Sess., ch. 58, p. 92, now appearing in Texas Natural Resources Code Ann. secs. 85.121–.125 (Vernon 1978). The Railroad Commission's prior prorationing orders in fields other than East Texas had not inspired a push to protect marginal wells. The commission had previously issued prorationing orders for the Hendricks field (dated April 4, 1928), for the Panhandle field (dated October 10, 1929), and a statewide order (dated August 14, 1930). This last order was designed to reduce statewide production by the relatively small sum of 50,000 barrels per day out of the previous year's total of 800,000 barrels per day. See Nugent, "The Railroad Commission," pp. A-17–A-18.

137. 1 F. Supp. 361 (E.D. Tex. 1932).

138. Texas Attorney General Opinion No. 2916 (1933).

139. Ibid.

140. Ibid.

141. Ibid.

142. 1933 Tex. Gen. Laws, Reg. Sess., ch. 97, pp. 215–216.

143. A. W. Walker, Jr., "The Problem of the Small Tract Under Spacing Regulation," *Texas Law Review,* Proceedings of the 57th Annual Session of the Texas Bar Association (1938) p. 165. By 1938, there were about 25,000 wells in the East Texas field. The Railroad Commission had imposed a 10-acre spacing rule on the field, but so many exceptions to the spacing rule had been granted that the average well density was one well to 5 acres. The statewide spacing rule, commonly called Rule 37, was adopted as early as 1919. However, the commission and courts interpreted the rule as requiring exceptions for small tracts that had been subdivided prior to the discovery of oil or gas in the vicinity. If these small tracts were not granted well permits, the oil and gas underneath them would be drained away by adjoining large tracts, in the absence of pooling. For a full discussion of the relationship between Rule 37 exception wells, prorationing, and the Marginal Well Act, especially in East Texas, see ibid. at 157–169; and also ch. 4 text at notes 97 to 124.

144. Some writers have accepted the view that the Marginal Well Act was passed primarily to prevent physical waste and that without the act, marginal wells (also called stripper wells) that were prorated would no longer be profitable and would be abandoned. Because the small amount of oil remaining in the ground would not justify reopening the well or drilling another well later when market demand or crude prices rose, this oil would be permanently lost. See Erich W. Zimmermann, *Conservation in the Production of Petroleum* (Northford, Conn., Elliot's Books, 1957) pp. 81–82; and Prindle, *Petroleum Politics,* pp. 199–200. Economists have unanimously considered the Marginal Well Act as creating rather than preventing waste because the act encourages high-cost production at the expense of low-cost production. See McDonald, *Petroleum Conservation,* p. 186; and Wallace F. Lovejoy and Paul T. Homan, *Economic Aspects of Oil Conservation Regulation* (Baltimore, Md., Johns Hopkins University Press for Resources for the Future, 1967) pp. 185–195. One economist calculated that shutting down stripper wells in Texas would cause producers to lose $1.12 billion in capital value, but the gain to the owners of non-stripper wells would total $2.41 billion, resulting in a net gain to Texas producers of $1.29 billion. See Lovejoy and Homan, *Economic Aspects,* p. 190.

145. Amazon Petroleum Corp. v. Railroad Comm'n, 5 F. Supp. 633 (E.D. Tex. 1934), *rev'd on other grounds sub nom.* Panama Refining Co. v. Ryan, 293 U.S. 388 (1935). From April 22, 1933 until the fall of 1933, the total allowable for the East Texas field averaged about 700,000 barrels per day, the price of East Texas crude fell to 10 cents per barrel in the summer of 1933, and reservoir pressures

declined. Gradually the commission reduced the field's total to about 450,000 barrels per day. The commission was aided in its effort to control the field by the passage of the National Industrial Recovery Act (NIRA) in the summer of 1933. Under this act, a Code of Fair Competition for the Petroleum Industry was promulgated on September 13, 1933, and federal agencies undertook to regulate all aspects of production, transportation, refining, and marketing of oil in the United States. In early 1935 the Supreme Court held that certain sections of the NIRA were unconstitutional because they delegated legislative power to the executive branch. Panama Refining Co. v. Ryan, 293 U.S. 388 (1935). On this basis, the Supreme Court reversed the lower court opinion that had upheld the East Texas order. Shortly thereafter, Congress passed the Connally Act which prohibited the interstate transportation of "hot oil" produced in violation of state prorationing orders. Under the Connally Act, a Federal Tender Board controlled movements of oil from the East Texas field. Also in 1935, Congress ratified the Interstate Oil Compact which allowed member producing states to coordinate their crude oil demand and supply forecasts. These federal acts, in conjunction with state prorationing laws, finally brought some order to the East Texas field. See generally, Hardwicke, "Legal History of Oil," pp. 240–254; and Williamson and coauthors, *The American Petroleum Industry*, pp. 545–551. However, even in late 1935, an investigating committee of the Texas Senate found much to criticize in the commission's administration of prorationing. Hot oil was still being run; the Oil and Gas Division still employed mostly "inexperienced, incompetent men"; the agency's records were hopelessly inaccurate; the commission was "victimized" by "designing operators" who filed false statements; the commission was not even trying to enforce the spacing rule; and a "vicious" and "reprehensible" practice of offering increased potentials for sale was discovered. See House Journal of Texas, 44th Leg., 1st Called Sess., pp. 108–113 (1935). Litigation about the East Texas prorationing formula, which continued to be based on well potentials, also did not end. Ultimately, this issue came to the U.S. Supreme Court which upheld the commission's order. Railroad Comm'n v. Rowan and Nichols Oil Co., 310 U.S. 573 (1940), *modified,* 311 U.S. 614 (1940), *aff'd, on rehearing,* 311 U.S. 570 (1941), is discussed in ch. 4 text at notes 110 to 113.

146. See, generally, Prindle, *Petroleum Politics,* pp. 44–55. In other fields, the Railroad Commission commonly allocated allowables by allotting one-half of the allowable to well potentials and one-half to acreage, a formula which still greatly favored the small tract. Small tracts also received the benefit of the full Marginal Well Act allowance, no matter how small the tract. See Walker, "The Problem of the Small Tract," pp. 166–167. By 1938, the Marginal Well Act controlled the allocation of 78 percent of the state's total allowable, and the proration laws controlled only 22 percent. See Northcutt Ely, "The Conservation of Oil," *Harvard Law Review* vol. 51, no. 7 (1938) pp. 1226–1227.

147. Ely, "The Conservation of Oil," pp. 1226–1233.

148. 1935 Tex. Gen. Laws, 44th Leg., Reg. Sess., ch. 120, sec. 21, p. 318 (often referred to as House Bill 266).

149. These facts are found in Maurice Cheek, "Legal Histories of Conservation of Oil and Gas in Texas as to Gas," in *Legal History of Conservation of Oil and Gas* (Chicago, Ill., American Bar Association, 1939) pp. 269–270 [hereafter cited as Cheek, "Legal History of Gas"]. See also Walter L. Summers, *The Law of Oil and Gas,* vol. 1A (2 ed., Kansas City, Mo., Vernon Law Book, 1954), sec. 97, on the history of gas prorationing in Texas generally.

150. 1899 Tex. Gen. Laws, 26th Leg., Reg. Sess., ch. 49, sec. 3, p. 68. The 1899 Act did not apply to oil wells. Thus, oil well producers could, and did, strip the casinghead gas of its liquid gasoline content and then vent the residue (containing 95 percent of the heat value of the gas) into the air.

151. 1931 Tex. Gen. Laws, 42nd Leg., 1st Called Sess., ch. 26, sec. 2, p. 49 (the Anti-Market Demand Act). The legislators' focus at the 1931 special session was on East Texas oil, but the waste of natural gas in the Panhandle was often discussed (see note 68 in this chapter) as was the need for antitrust immunity for cooperative development among producers. See text at notes 92 and 96 in this chapter and 1931 S.J., p. 188 (message from the governor about the need for cooperative marketing of natural gasoline).

152. The Common Purchaser Act was first passed in 1930 and provided that a purchaser of oil who was a common carrier must purchase oil ratably with no discrimination. See 1930 Tex. Gen. Laws, 5th Spec. Sess., ch. 36, sec. 7, p. 173 (now in Texas Natural Resources Code Ann., sec. 111.052 (Vernon 1978)). The legislators at the 1931 special session amended the act to include gas. See 1931 Tex. Gen. Laws, 1st Spec. Sess., ch. 28, secs. 8, 8-A, p. 62 (now in Texas Natural Resources Code Ann. secs. 111.081–.083 (Vernon 1978)).

153. Texoma Natural Gas Co. v. Railroad Comm'n, 59 F.2d 750 (W.D. Tex., 1932) *aff'd, on other grounds sub nom.* Thompson v. Consolidated Gas Utilities Corp., 300 U.S. 55 (1937); Texoma Natural Gas Co. v. Terrell, 2 F. Supp. 168 (W.D. Tex. 1932). In 1931 the Texas attorney general had advised that no law could be valid in Texas which curtailed a person's production solely on the theory that he was obtaining more than a fair share of the reservoir. Curtailments could be authorized only to prevent waste. Texas Attorney General Opinion, Aug. 7, 1931, as cited in 1931 S.J., p. 704.

154. 1932 Tex. Gen. Laws, 42d Leg., 4th Called Sess., ch. 2, sec. 4, p. 6:

> Whenever the full production, from wells producing gas only, from any common source of supply of natural gas in this State is in excess of the reasonable market demand, the Railroad Commission shall inquire into the production and reasonable market demand therefor and shall determine the allowable production from such common source of supply, which shall be the reasonable market demand which can be produced without waste, and the Commission shall allocate, distribute or apportion the allowable production from such common source of supply among the various producers on a reasonable basis, and shall limit the production of each producer to the amount allocated or apportioned to such producer.

155. Cheek, "Legal History of Gas," p. 276.

156. Canadian River Gas Co. v. Terrell, 4 F. Supp. 222 (W.D. Tex. 1933). The federal court's interpretation of the Texas prorationing law seems wrong in light of the newly passed law cited in note 154 in this chapter.

157. Often the applicants for stripping permits did not truly desire to engage in stripping, but hoped to use the permits as a bargaining chip against the pipeline companies: if the pipelines did not accept the permit holder's gas, then large-scale stripping operations would begin and drain gas away from the pipelines. The pipeline owners did not give in, and the stripping operations started up. See Cheek, "Legal History of Gas," pp. 277–278.

158. Ibid., p. 280. The courts had despaired of solving the enormous waste of gas or its equitable distribution. See Sneed v. Phillips Petroleum Co., 76 F.2d 785 (5th Cir. 1935).

159. 1935 Tex. Gen. Laws, 44th Leg., Reg. Sess., ch. 120, pp. 318–327, now appearing in Texas Natural Resources Code Ann. secs. 86.001–.225 (Vernon 1978). Many of the waste definitions for gas parallel those for oil in the 1932 Market Demand Act, but without any proviso against required repressuring or unitization.

160. Ibid., sec. 1, pp. 318–319; now appearing at Texas Natural Resources Code Ann. sec. 86.001 (Vernon 1978).

161. Ibid., sec. 10, pp. 322–323, now appearing in Texas Natural Resources Code Ann. sec. 86.081 (Vernon 1978). Section 86.083 of this code also states that "the commission shall exercise its authority to adjust correlative rights and opportunities of each owner of gas in a common reservoir whenever the potential capacity to produce all the gas wells in a common reservoir is in excess of the reasonable market demand."

162. Ibid., sec. 21. The wording of this section does not expressly provide antitrust immunity for cooperative development, but this was clearly the purpose of requiring the attorney general's approval of the agreements.

163. John R. Stockton, Richard C. Henshaw, and Robert W. Graves, *Economics of Natural Gas in Texas* (Austin, The University of Texas Press, 1952) p. 230. These factors also result in gas being sold under long-term contracts rather than by the posted-price system of short-term sales used for oil. The large interconnected network of oil pipelines and its alternate transportation modes made market demand prorationing of oil on a statewide basis relatively easy to implement. The Railroad Commission seldom had to resort to the Common Purchaser Act to force pipeline purchasers to take oil ratably; proration orders performed this function. See Dee J. Kelly, "Gas Proration and Ratable Taking in Texas," *Texas Bar Journal* vol. 19, no. 11 (1956) pp. 763–764 and 795–798.

164. Most Railroad Commission orders prorating gas-well gas are made to adjust correlative rights under Section 10(b) of the 1935 Act, quoted in the text of this chapter at note 161, rather than to prevent waste under Section 10(a). See John W. Stayton, "Proration of Gas," *Institute on Oil and Gas Law and Taxation,* vol. 14 (Albany, N.Y., Matthew Bender, 1963) p. 9.

165. Because the independent producers had no vested advantage in the Panhandle field as they did in East Texas, much less dispute existed about the fairest allocation formula to use in setting allowables. The legislators in 1932 had been unable to agree to any more specific guide to oil prorationing than that it be on "a reasonable basis." See text of this chapter at notes 130 to 132. The 1935 Act required that the commission consider the size of the tract, its daily producing capacity compared with the reservoir's aggregate capacity, and the area efficiently drained by the well as determined by formation pressure, permeability, porosity, and the well's structural position. See 1935 Tex. Gen. Laws, 44th Leg., Reg. Sess., ch. 120, sec. 13, pp. 323–324, now appearing in Texas Natural Resources Code Ann. sec. 86.089 (Vernon 1978).

166. The 1931 legislature had considered, but did not pass, a bill authorizing cooperative development (1931 H.J., pp. 750–751).

167. The different legislative treatment of gas fields versus oil fields continued in 1945, when the legislature passed bills authorizing guardians to enter into pooling or unitization agreements covering their ward's rights to gas and liquid condensate produced from gas reservoirs (1945 Tex. Gen. Laws, ch. 80, sec. 1, p. 117); and authorizing the commissioner of the General Land Office to execute pooling agreements for state royalty interests in natural gas, expressly excluding authority over crude oil or casinghead gas (1945 Tex. Gen. Laws, ch. 309, secs. 1–3, p. 507).

168. The 1935 Act did not contain a provision asserting the primacy of antitrust law as did Section 13 of the 1932 Act, quoted in the text of this chapter at note 124.

169. Texas Panhandle Gas Co. v. Thompson, 12 F. Supp. 462 (W.D. Tex. 1935); Consolidated Gas Utilities Co. v. Thompson, 14 F. Supp. 318 (W.D. Tex. 1936), *aff'd on other grounds,* 300 U.S. 55 (1937).

170. By contrast, the Railroad Commission had found that the pipeline purchasers had withdrawn more than twice the amount of gas per acre than the nonintegrated producers and, therefore, the property of the latter group was being taken. See Hardwicke, "Legal History of Oil," p. 283, note 38.

171. Thompson v. Consolidated Gas Utilities Corp., 300 U.S. 55 (1937).

172. Ibid., pp. 75–76.

173. For a description of the mechanics of the gas prorationing system which developed in Texas and its many unresolved legal problems, see Stayton, "Proration of Gas," pp. 1–69. It was not until 1945 that the Railroad Commission's statutory authority to prorate natural gas to protect correlative rights was confirmed in Corzelius v. Harrell, 143 Tex. 509, 186 S.W.2d 961 (1945). The lengthy and bitter litigation over gas prorationing in the Appling field, discussed in chapter 8 at notes 50 to 61, shows the failure of the gas prorationing system to protect correlative rights.

174. To give some relief to the nonintegrated sweet-gas producers in the absence of prorationing or ratable take orders, the legislature amended the law in

1941 to authorize the use of sweet gas in carbon black plants in the Panhandle. See Murphy, *Legal History, 1948*, pp. 454–455.

175. Landowners were also often unhappy about the pipeline's private prorationing. For example, in Biskamp v. General Crude Oil Co., 452 S.W.2d 515 (Tex. Civ. App. 1970), a landowner who had leased to an operator in return for a one-eighth royalty sued the operator for failing to produce a fair share of gas from her wells. The operator took equal amounts from the wells which it produced, but the landowner's well had a substantially larger open-flow potential and she alleged that the wells should have been prorated on the basis of well potentials. For a highly critical account of the recent Railroad Commission's failure to establish an equitable statewide gas prorationing system, see Prindle, *Petroleum Politics*, pp. 97–107.

176. See John R. Hays, Jr., "Gas Pipelines, NGPA Filings, and Public Utilities," *Oil and Gas: Texas Railroad Commission Rules and Regulations* (Austin, State Bar of Texas, 1982) pp. K-10–K-14.

Chapter 4

1. 1949 Tex. Gen. Laws, Reg. Sess., ch. 259, pp. 477–483, now codified at Texas Natural Resources Code Ann. secs. 101.011–101.052 (Vernon 1978).

2. Blakely M. Murphy, ed., *Conservation of Oil and Gas: A Legal History, 1948* (Chicago, Ill., American Bar Association, 1949) pp. 460–461 [hereafter cited as Murphy, *Legal History, 1948*].

3. Robert E. Hardwicke, *Antitrust Laws v. Unit Operation of Oil or Gas Pools* (rev. ed., Dallas, Tex., Society of Petroleum Engineers of the American Institute of Mining and Metallurgical Engineers, 1961) pp. 104–107.

4. During the 1940s, condensate production from gas pools had become so large that many bills were introduced to prorate this production in a manner similar to oil pools. See Murphy, *Legal History, 1948*, pp. 457–458. A bill authorizing guardians to enter into unitization agreements in gas fields, particularly condensate-gas fields, was passed in 1945. See 1945 Tex. Gen. Laws, ch. 80, sec. 1, p. 117. In 1946 a bill requiring compulsory cycling if necessary to increase the ultimate recovery of oil or gas was introduced into the Texas legislature, but it failed to pass. See Henrietta M. Larson and Kenneth W. Porter, *History of Humble Oil and Refining Company: A Study in Industrial Growth* (Salem, N.H., Ayer, 1959) p. 650.

5. Murphy, *Legal History, 1948*, p. 470.

6. The status of secondary recovery and repressuring projects in Texas in 1948 is well described in American Petroleum Institute, *Secondary Recovery of Oil in the United States* (2 ed., Dallas, Tex., American Petroleum Institute, 1950) pp. 17–18 and 561–627. Many of the dissolved gas fields in North Central Texas had been

inefficiently produced in their primary stage and were the object of gas- and air-injection operations in 1948. The West Pampa Repressuring Association had been formed in the Texas Panhandle and was the largest, cooperative secondary recovery program in the United States, containing 11,880 productive acres and 1,000 wells owned by twenty-five operators (ibid., pp. 598–613). Cooperative projects differ from unitization in that each operator retains control over his own tract, but agrees to develop it under a joint plan (ibid., pp. 90–93). Gas injection for pressure maintenance existed in all areas of Texas. Water flooding was uncommon in Texas at this time, although the reinjection of salt water into the East Texas field for pressure maintenance and to avoid surface pollution was well under way (ibid., pp. 582–591).

7. Hardwicke, *Antitrust Laws,* pp. 134–137 and 211–228. The *Cotton Valley* case was eventually dismissed without being tried on its merits (ibid., pp. 329–330). All cases involving unitization and the antitrust laws are discussed in chapter 8 at notes 131 to 168.

8. Eugene V. Rostow, *A National Policy for the Oil Industry* (Northford, Conn., Elliot's Books, 1948) pp. 34–42.

9. See, for example, the speech by Hines Baker, president of Humble Oil and Refining Co., stating that lack of a statute in Texas authorizing voluntary unitization of oil pools was greatly hindering gas recycling and pressure maintenance operations in oil pools with dissolved gas and gas-cap drives, as quoted in John R. Stockton, Richard C. Henshaw, and Robert W. Graves, *Economics of Natural Gas in Texas* (Austin, The University of Texas, 1952) pp. 237–238.

10. The 1889 antitrust act is quoted in ch. 3 at note 78. Many writers had discounted the need for antitrust immunity for unitization agreements in the belief that such agreements would never be held to violate the antitrust laws because their very purpose was to increase oil and gas production, rather than restrain it. As early as 1929, an American Bar Association study found little cause for antitrust anxiety. See Hardwicke, *Antitrust Laws,* pp. 55–58 and 119–134. The Legal Committee of the Interstate Oil Compact Commission (IOCC) reported in 1947 that fear of incurring antitrust liability was often not real, but was used by operators as a poker chip to secure a more favorable unitization agreement. However, the IOCC report specifically excluded expressing an opinion on the validity of unitization agreements under Texas' particular antitrust laws. To remove antitrust fears, whether real or imaginary, the IOCC report recommended that oil-producing states pass laws specifically authorizing unitization agreements. See American Petroleum Institute, *Secondary Recovery,* p. 65.

11. The Supreme Court had enunciated the "state action" doctrine in 1942 in the case of *Parker v. Brown,* 317 U.S. 341 (1942). This case held that the federal antitrust laws were directed only at acts of private parties, not at state action. If the Railroad Commission had regulatory control over voluntary unitization agreements, producers might use this control to argue a "state action" immunity from any possible federal antitrust suits. The state action defense is discussed further in chapter 8 at notes 152 to 168.

12. As of January 1, 1948, nine states had already passed bills authorizing compulsory unit operations, although most of these acts limited compulsory process to pools containing state-owned lands or to condensate pools requiring cycling. Texas' nearest neighbors were especially advanced: Oklahoma had a comprehensive unitization statute and Louisiana's law authorized its commission to require cycling in condensate fields on its own initiative. See American Petroleum Institute, *Secondary Recovery*, pp. 46–62. Data on unitized operations formed from 1931 to 1948 show the limitations of voluntary unitization: of forty-seven unitized projects studied, ten did not unitize the royalty interest owners, five had only partially unitized royalty interests, and twelve avoided the problems of unitizing royalty owners by selecting pools already under one ownership. Many of the units included only part of the field. The first large unitization agreement occurred in 1931 in the Kettleman Hills pool in California, owned largely by the federal government, and the second large project included a large block of land under the government-controlled Osage Indian Reservation in Oklahoma (ibid., pp. 72–79).

13. Robert E. Hardwicke, "Legal Histories of Oil and Gas in Texas as to Oil," in *Legal History of Conservation of Oil and Gas* (Chicago, Ill., American Bar Association, 1939) p. 266. Hardwicke was so distressed by the chaos in East Texas from 1931 to 1934 that he even pondered government ownership of the oil fields as an alternative to state regulation (ibid., pp. 267–268).

14. Ibid., p. 266.

15. The Cole Committee was a House Subcommittee of the Committee on Interstate and Foreign Commerce which held lengthy hearings in 1940 to discuss proposed legislation for extensive federal regulation of the oil industry. Hardwicke's testimony at the hearings is reprinted in Robert E. Hardwicke, "Oil Conservation: Statutes, Administration, and Court Review," *Mississippi Law Journal* vol. 13, no. 3 (1941) pp. 381–416.

16. Ibid., p. 407. Hardwicke's testimony seems inconsistent with his strong support of prorationing. If petroleum science had advanced enough to provide rational prorationing formulas, it is not clear why unitization formulas would be considered arbitrary and unreasonable.

17. Ibid., p. 408.

18. Robert E. Hardwicke, "Unitization Statutes: Voluntary Action or Compulsion," *Rocky Mountain Law Review* vol. 24, no. 1 (1951) pp. 38 and 42. By 1951, Hardwicke's views against compulsory process had softened though. See Hardwicke, *Antitrust Laws*, pp. 161–165. Other observers of this time did not see a determined commitment to unitization on the part of leading executives of oil and gas companies. See Dean Terrill, "Unit Agreements and Unitized Operations: A Review of Their Past and Some Speculations as to Their Future," *Institute on Oil and Gas Law and Taxation* vol. 1 (Albany, N.Y., Matthew Bender, 1949) pp. 17–20.

19. Leslie Moses, "Some Legal and Economic Aspects of Unit Operations of Oil Fields," *Texas Law Review* vol. 21, no. 6 (1943) p. 769. However, one Texas scholar was skeptical of the ability of voluntary processes to achieve unitization. See H. P. Pressler, Jr., "Legal Problems Involved in Cycling Gas in Gas Fields," *Texas Law Review* vol. 24, no. 1 (1945) pp. 32–33. In contrast to Hardwicke and Moses, many non-Texas writers had long advocated compulsory process. See, for example, Northcutt Ely, "The Conservation of Oil," *Harvard Law Review* vol. 51, no. 7 (1938) pp. 1209–1244; W. P. Z. German, "Compulsory Unit Operation of Oil Pools," *California Law Review* vol. 20, no. 2 (1932) pp. 111–131; and John C. Jacobs, "Unit Operation of Oil and Gas Fields," *Yale Law Journal* vol. 57, no. 7 (1948) pp. 1207–1228.

20. This disapproval by Texas scholars of compulsory process for unitization did not extend to pooling. See, for example, Robert E. Hardwicke, "The Rule of Capture and Its Implications as Applied to Oil and Gas," *Texas Law Review* vol. 13, no. 4 (1935) pp. 391–422; and A. W. Walker, Jr., "The Problem of the Small Tract Under Spacing Regulations," *Texas Law Review,* Proceedings of the 57th Annual Session of the Texas Bar Association (1938) pp. 157–169. Of course, the enormous overdrilling of wells in East Texas that had occurred in the absence of pooling was difficult to ignore.

21. See ch. 3 text at notes 3 to 107. Evidence of this prevailing belief appears in a review of Rostow's book by a Texas attorney who criticized Rostow's call for a federal compulsory unitization law, because if enacted, the law would "result in the speedy elimination of all independent producers and in the creation of monopoly." See Jack Blalock, "Review of E. Rostow's *A National Policy for the Oil Industry,*" *Texas Law Review* vol. 27, no. 1 (1948) p. 128.

22. The IOCC's 1947 model form appears in American Petroleum Institute, *Secondary Recovery,* p. 65. Alabama, Arkansas, California, Georgia, Indiana, Mississippi, and North Carolina had adopted statutes similar to the IOCC's model (ibid., pp. 47–59). Section 21 of Texas' 1935 act is similar to the IOCC's model and is quoted in full in ch. 3 text at note 162.

23. Texas Attorney General Opinion No. V-97 (1947) p. 11.

24. Ibid., pp. 2–10.

25. Ibid., p. 7.

26. The Texas attorney general may have been influenced by the U.S. attorney general's press release on the *Cotton Valley* suit which disavowed any intention of attacking joint activity by the defendants which was "necessary or essential" to the conservation of resources. The press release is reprinted in Hardwicke, *Antitrust Laws,* pp. 224–225.

27. Texas Attorney General Opinion No. V-97 (1947) p. 9. This statement by the attorney general followed a long history of antipathy to authorizing the commission to prevent economic waste. See ch. 3 at notes 8 to 11.

28. Texas Attorney General Opinion No. V-97-A (1947).

29. Murphy, *Legal History, 1948,* pp. 460–461.

30. 1949 Tex. Gen. Laws, 51st Leg., Reg. Sess., ch. 259, p. 483. The act was codified in Texas Revised Civil Statutes Ann. art. 6008b (Vernon 1972), and now appears virtually unchanged in Texas Natural Resources Code Ann. secs. 101.001–101.052 (Vernon 1978). Appendix II contains the full text of these sections. For convenience, the statute will be referred to as the 1949 Act throughout this book, but the section numbers used will be from the current Texas Natural Resources Code.

31. Texas Natural Resources Code Ann. sec. 101.011 (Vernon 1978). The act never uses the word "unitization" and unfortunately uses the term "pooled units." As discussed in this chapter at notes 44 to 45, the act does not authorize Railroad Commission approval of pooling into drilling units.

32. The federal statutes and regulations governing unitization of federal lands are discussed in Leroy H. Hines, *Unitization of Federal Lands* (Denver, Colo., F. H. Gower, 1953). By 1948, more than half of all oil produced on the public domain came from unitized leases. See Murphy, *Legal History, 1948,* p. 612.

33. The federal laws do not force the owners of state or private lands to join in the unit, but most states have compulsory unitization laws which can be used to this effect.

34. Ibid., pp. 16–19 and 30–32. Wide spacing rules and compulsory pooling can avoid much unnecessary drilling, but not all. For example, early unitization of a gas-cap field can avoid the costs of drilling any wells at all into the gas cap.

35. Stephen L. McDonald, *Petroleum Conservation in the United States: An Economic Analysis* (Baltimore, Md., The Johns Hopkins University Press for Resources for the Future, 1971) pp. 208–209.

36. See, for example, the Washington and Illinois statutes reprinted in American Petroleum Institute, *Secondary Recovery,* pp. 50–51 and 61.

37. The Benton field in Louisiana was unitized in 1945 after just three producing wells and two dry holes had been drilled and before the limits of the pool were known. The agreement provided for drilling test wells and for retroactive adjustments in participation formulas as data became available. See M. Darwin Kirk, "Content of Royalty Owners' and Operators' Unitization Agreements," *Institute on Oil and Gas Law and Taxation,* vol. 3 (Albany, N.Y., Matthew Bender, 1952) pp. 54–58. Other examples of early unitization plans on private lands appear in R. M. Williams, "The Negotiation and Preparation of Unitization Agreements," *Institute on Oil and Gas Law and Taxation* vol. 1 (Albany, N.Y., Matthew Bender, 1949) pp. 65–70. Humble Oil had advocated exploratory units in 1929 (see ch. 3 text at notes 92–93).

38. See 1945 Okla. Sess. Laws, sec. 5, p. 162, now appearing in slightly modified form at Okla. Statutes tit. 52, sec. 287.4 (1951). This statute requires that "only so much of a common source of supply, as has reasonably been defined by actual drilling operations, may be so included within the unit area."

39. The evils of premature unitization from the royalty owners' viewpoint are vividly described in Harold Garvin, "The Effect of Field Unit Operation Upon the Royalty Interest," *Oklahoma Bar Journal* vol. 21 (1950) pp. 1794–1798.

40. In 1982, independent producers drilled 88 percent of all the wells drilled in the United States. Independents drilled 87 percent of all the wells drilled in Texas in 1982. See "Producing Oil Wells Reach Record Level," *World Oil,* Feb. 15, 1983, p. 121. The Market Demand Prorationing Act of 1932 contained a specific provision that nothing in the act "shall be construed to grant the commission any authority to restrict or in any manner limit the drilling of wells to explore for oil or gas, or both, in territory that is not known to produce either oil or gas." 1932 Tex. Gen. Laws, 42d Leg., 4th Called Sess., ch. 2, sec. 2, p. 5, now appearing at Texas Natural Resources Code Ann. sec. 85.057 (Vernon 1978). The 1932 Act also prohibited the commission from prorationing on the basis of market demand the production of oil from any newly explored field until the total production of the field equaled 10,000 barrels of oil a day. This section now appears at Texas Natural Resources Code Ann. sec. 85.048. This restriction on limiting the allowables of exploratory wells is obviously designed to allow explorers to recoup their expenses more quickly.

41. The Cotton Valley unit operators were prepared to argue and prove this point. See Burns H. Errebo, "Unit Operation at Cotton Valley: An Alleged Violation of the Sherman Act," *Tulane Law Review* vol. 24, no. 1 (1949) pp. 79–83.

42. See ch. 3 text at notes 174 to 176. McDonald is especially persuasive on the possible benefits to be secured from cooperative marketing, including the benefit of increased competition. See McDonald, *Petroleum Conservation,* p. 209.

43. American Petroleum Institute, *Secondary Recovery,* p. 65.

44. Ibid., p. 63. The API and IOCC defined other relevant terms as follows: Pressure maintenance is a primary or secondary recovery operation conducted to control reservoir pressure decline; recycling is a continuous reinjection of produced gas; cycling is an operation which displaces condensate-bearing gas from a gas zone by injection of dry gas; repressuring is the introduction of a gas or liquid into a formation to increase reservoir pressure.

45. The IOCC's 1947 report on secondary recovery urged states to encourage cooperative operations which minimized unnecessary expense and thus increased profits to producers (ibid., p. 63). Of course, pooling small tracts into properly sized drilling or prorationing units is the most common device used to avoid unnecessary drilling and operating costs and to protect correlative rights. The 1947 unitization proposal in Texas, which died in the conference committee, expressly stated that no provisions of the act should be construed "as requiring Railroad Commission approval of voluntary agreements between operators of tracts embraced within an established proration or drilling unit to integrate their interests and operate their tracts as a unit." See 1947 House Journal of Texas, 50th Leg., p. 1971. The 1949 Act omitted this provision, but the restrictions of Sections 101.011 and 101.013(a) do not authorize the commission to approve pooling agreements. Pooling agreements are generally made to reduce drilling costs and protect correlative rights during primary production, not to increase the recovery of oil and gas through secondary recovery operations. Also, the acreage in a

pooled drilling unit would undoubtedly be found insufficient to accomplish the act's purpose of increasing the production of oil and gas through secondary recovery or gas-injection operations, and so a pooling agreement would be disapproved under Section 101.013(b) as well. Moreover, it is doubtful that fear of the antitrust laws caused operators to refuse to make pooling agreements. Pooling agreements were not common in 1949, mainly because exceptions to the spacing rule were easy to acquire and because per-well allowables discouraged pooling.

46. This is Rule 49(b), formerly known as Rule 6(b), Texas Administrative Code tit. 16, sec. 3.49(b) (1982).

47. For this reason, the rule has been severely criticized as confiscatory. Comment, "Proration in Texas: Conservation or Confiscation?" *Southwestern Law Journal* vol. 11, no. 2 (1957) pp. 198–200.

48. American Petroleum Institute, *Secondary Recovery*, p. 65.

49. Murphy, *Legal History, 1948,* p. 460. The lengthy revised version of the 1947 proposal also would probably have allowed Railroad Commission authorization of this unitization agreement. See 1947 House Journal of Texas, 50th Leg., p. 1970.

50. *In-situ* combustion relies largely on heat. In some instances, only one well bore is used for *in-situ* combustion. Steam is injected into the one well for several days, and later the oil is produced from the same well. See National Petroleum Council, *Enhanced Oil Recovery: An Analysis of the Potential for Enhanced Oil Recovery from Known Fields in the United States—1976 to 2000* (National Petroleum Council, 1976) pp. 13–14. Miscible displacement and chemical flooding require the use of injection wells and so can be classified as secondary recovery operations, but the carbon dioxide and chemicals injected do not in themselves affect reservoir pressure. They are injected to effect hydrocarbon vaporization, to lower interfacial tension, or to increase sweep efficiency. An accompanying water flood supplies the sweeping drive from the injection wells to the producing wells. When the chemicals are added to the water used in the water drive, the operation might easily be termed water flooding under Section 101.011. But sometimes the chemicals are injected separately to form "slug zones" to be pushed by a subsequent water flood (ibid., pp. 13–17).

51. Also, the act does not read "including, but not limited to," nor does the list of permitted operations end with "or any other form of joint effort" as used in other states' statutes. Thus the list could be interpreted as exclusive. The original 1947 proposal authorized cooperative plans "including, but not limited to, secondary recovery operations." See Murphy, *Legal History, 1948,* p. 460.

52. 231 P.2d 997 (Okla. 1951).

53. Ibid., p. 1001. The plaintiff's complaint read:

> We submit that any compulsory unitization law, competing with the general conservation law whose real purpose is exactly the same, should not be given constitutional sanction under the police power unless it specifically provides that the Commission finds that any plan of unitization approved thereunder will

accomplish the conservation of oil and gas with substantially greater results than is being accomplished under the general conservation law still in full force and effect.

54. 1945 Okla. Sess. Laws, sec. 4, p. 162, now appearing at Oklahoma Statutes tit. 52, sec. 287.3 (1951).

55. As of January 1, 1948, a considerable number of water-flooding operations in Texas had failed. See American Petroleum Institute, *Secondary Recovery*, pp. 17 and 561–581.

56. The Cotton Valley unit's cycling operation was the first of its kind. Also see the discussion of the SACROC unit in ch. 5 at notes 56 to 60.

57. Royalty owners are also protected by implied covenant law against negligent and unreasonable operations by their lessees which injure the reservoir. See Maurice Merrill, *Covenants Implied in Oil and Gas Leases* (2 ed., Saint Louis, Mo., Thomas Law Book, 1940) at sections 72–83.

58. In 1962 two major producers in the East Texas field proposed that the Railroad Commission shut down about seven-eighths of the wells in the field and allow the field to produce at the same rate from the remaining wells. In 1965 major operators in the field proposed another plan to the Railroad Commission to reduce the number of wells by half. The Railroad Commission rejected both proposals. See Wallace Lovejoy and Paul T. Homan, *Economic Aspects of Oil Conservation Regulation* (Baltimore, Md., Johns Hopkins University Press for Resources for the Future, 1967) p. 121.

59. Because the East Texas field is prorationed largely on a per-well basis, few operators would voluntarily abandon any wells to reduce operating costs until the economic break-even point was reached. Operators with only one well on a small tract would, of course, refuse to abandon any profitable wells unless they could secure their fair share of production from other wells in the unitized area.

60. One might wonder why the oil operators would desire this unitized project only to protect the correlative rights of the gas owners and to save them the costs of drilling gas wells. However, if the operation of Rule 49(b) as applied to this reservoir resulted in large, uncompensated drainage of the gas owners' gas, all parties might want to avoid the costs of administrative hearings and court battles on the validity of the commission's orders in light of their confiscatory effect. Also, some of the oil operators may have interests in the gas zone as well, and may want to save the costs of drilling gas wells without suffering drainage to other gas-cap owners.

61. Texas Attorney General Opinion Nos. V-97 (1947), V-97-A (1947), discussed in this chapter at note 27.

62. 1940 La. Acts 157, sec. 4(b).

63. W. J. McAnelly, Jr., "A Review of Poolwide Unitization Under Act 441 of 1960," in George W. Hardy III, ed., *Institute on Mineral Law*, vol. 15 (Baton Rouge, Louisiana State University Press, 1968) p. 10.

64. See Morris G. Gray and Oscar E. Swan, "Fieldwide Unitization in Wyoming," *Land and Water Law Review* vol.7, no.2 (1972) pp. 438–439; Charles J. Meyers and Howard R. Williams, "Petroleum Conservation in Ohio," *Ohio State Law Journal* vol. 26, no. 4 (1965) p. 615. One writer asserts that unitizing a field to save drilling costs does not benefit non-cost-bearing royalty interest owners and so should not be approved. See R. F. Bryant, Jr., "The Negotiation and Execution of a Voluntary Unit Agreement," *Institute on Oil and Gas Law and Taxation,* vol. 3 (Albany, N.Y., Matthew Bender, 1952) p. 188. Few unitization laws mention the avoidance of unnecessary drilling as a purpose of unitization, although most pooling statutes do.

65. The desirability of unitizing to protect correlative rights and avoid the regulatory morass of ratable take orders is discussed in chapter 3 at notes 174 to 176.

66. See ch. 2 at notes 35 to 36. Also, Section 101.003 of the 1949 Act states that none of the act's provisions repeal, modify, or impair any of the provisions of the oil- and gas-conservation laws, such as the Marginal Well Act.

67. See Railroad Comm'n Docket No. 8A-76,908, In re: Application of Union Oil Co. of California to Add Injection Wells in Its North Riley Unit and to Continue to Use Ogallala Water for Injection (1981). Union Oil's application in this matter was opposed by the Southwest Soil and Water Protection Association which asserted that the Railroad Commission had the authority to prohibit the wasteful use of fresh water for secondary recovery. The commission disavowed such authority. The Ogallala, a nonrecharging aquifer, is virtually the only source of water in the High Plains of Texas. Under Texas law, the mineral estate owner can use as much of the surface estate, including the fresh water, as is reasonably necessary to produce the minerals. Thus, in Sun Oil Co. v. Whitaker, 483 S.W.2d 808 (Tex. 1972), the Texas Supreme Court allowed Sun Oil to use 4,200,000 barrels of Ogallala fresh water for secondary recovery, even though this use would shorten the life of Whitaker's farm by at least eight years.

68. Texas Water Code sec. 27.051 (Vernon Supp. 1984).

69. Maurice H. Merrill, "Unitization Problems: the Position of the Lessor," *Oklahoma Law Review* vol. 1, no. 2 (1948) pp. 123–124; and Garvin, "The Effect of Field Unit Operation Upon the Royalty Interest."

70. American Petroleum Institute, *Secondary Recovery,* p. 65.

71. Of course, Texas lawmakers had their own 1932 Market Demand Act's forceful declaration of the primacy of antitrust laws to use as a model. See ch. 3 at note 124. Section 101.003 of the 1949 Act expressly affirms the continuing vitality of this 1932 antitrust provision, which now appears in Section 85.002 of the Texas Natural Resources Code Ann. (Vernon 1978).

72. In 1983 the Texas legislature revised the state's antitrust laws. The effect of this wholesale revision on antitrust immunity for voluntary unitization agreements is discussed in chapter 8 at notes 136 to 168.

73. See note 6 in this chapter. Several pre-1949 Act unitization agreements are listed in Raymond M. Myers, *The Law of Pooling and Unitization,* vol. 2 (Albany, N.Y., Banks, 1967) sec. 15.01.

74. Myers, *The Law of Pooling and Unitization,* vol. 1, sec. 7.01; and Hardwicke, *Antitrust Laws,* pp. 152–154.

75. See, for example, New York Environmental Conservation Law sec. 23.0701 (McKinney 1973); and Oregon Laws sec. 520.230 (1981). Of course, the 1949 Act may not have passed containing such a provision.

76. One other currently existing section of the 1949 Act, not discussed above, authorizes the commissioner of the General Land Office, on behalf of the state of Texas, to execute contracts committing state royalty interests or leases to unitization agreements. See 1949 Tex. Gen. Laws, 51st Leg., ch. 259, p. 477, sec. 2, now appearing at Texas Natural Resources Code Ann. secs. 101.051–101.052 (Vernon 1978). About 9 percent of Texas oil and gas production derives from state-owned lands. See Dan S. Boyd, "Legal Aspects of State-Owned Oil and Gas Energy Resources," Texas Governor's Energy Advisory Council, Project No. L/R-5 (Austin, Tex., 1974) p. 1.

77. In 1953 Texas passed another act authorizing cooperative facilities for the conservation and utilization of gas. See 1953 Tex. Gen. Laws, 53d Leg., ch. 117, p. 407, now codified at Texas Natural Resources Code Ann. secs. 103.041–103.046 (Vernon 1978). This act duplicates many sections of the 1949 Act, adding new mortar only by granting the commission authority to approve agreements for the construction and operation of cooperative gas-processing facilities upon a finding that the facilities "are in the interest of conservation and that secondary recovery operations are *not* feasible or necessary" [italics added]. The 1949 Act expressly allows agreements to establish cooperative facilities for extracting and separating the hydrocarbons from natural gas or casinghead gas and returning the dry gas to the reservoir. The 1953 Act expressly allows agreements to establish cooperative facilities for extracting and separating hydrocarbons from gas or casinghead gas, even though the dry gas is not returned to the reservoir. Evidently the restrictions of the 1949 Act led gas producers in noncondensate-gas fields to sell their gas to pipelines which transported it to other states for processing and removal of liquid hydrocarbons. The antitrust problems arising from joint ownership of gas processing plants are discussed in Myers, *The Law of Pooling and Unitization,* sec. 12.05(4).

78. Walter J. Mead, "Petroleum: An Unregulated Industry?" in Robert J. Kalter and William A. Vogely, eds., *Energy Supply and Government Policy* (Ithaca, N.Y., Cornell University Press, 1976) p. 141.

79. Wallace F. Lovejoy, "Oil Conservation, Producing Capacity and National Security," *Natural Resources Journal* vol. 10, no. 1 (1970) pp. 65-71 [hereafter cited as Lovejoy, "Producing Capacity"]; and Lovejoy and Homan, *Economic Aspects,* pp.102–107.

80. Lovejoy, "Producing Capacity," pp. 66–67; and Lovejoy and Homan, *Economic Aspects,* pp. 115–122.

81. For example, under Texas' 1947 yardstick allowable formula, an operator with 160 acres to develop could get an allowable of 1,296 barrels daily if the Railroad Commission approved 10-acre spacing, and only 444 barrels daily on 40-acre spacing. See Lovejoy, "Producing Capacity," p. 66.

82. Lovejoy and Homan, *Economic Aspects,* pp. 99–102; and McDonald, *Petroleum Conservation,* pp. 165–166.

83. Mead, "Petroleum: An Unregulated Industry?" p. 141. A well's yardstick allowable is determined by a formula adopted by the Railroad Commission which uses two factors: depth of the well and the amount of acreage assigned to the well under the field's spacing pattern.

84. Lovejoy, "Producing Capacity," pp. 77–78. Lovejoy estimates the cost of maintaining this idle capacity at almost $1.5 billion a year (ibid., pp. 83–95).

85. Alfred E. Kahn, "The Combined Effects of Prorationing, the Depletion Allowance and Import Quotas on the Cost of Producing Crude Oil in the United States," *Natural Resources Journal* vol. 10, no. 1 (1970) p. 53–61.

86. Edward W. Erickson, "Crude Oil Prices, Drilling Incentives and the Supply of New Discoveries," *Natural Resources Journal* vol. 10, no. 1 (1970) p. 33.

87. John D. LaRue, "The Rising Cost of New U.S. Oil Reserves 1959–1975," in *Exploration and Economics of the Petroleum Industry,* vol. 14 (Albany, N.Y., Matthew Bender, 1976) pp. 202 and 226.

88. Erickson, "Crude Oil Prices," p. 31.

89. In 1957 the Suez Canal was closed to tankers bringing crude oil to western markets and this resulted in a crude oil price increase of 25 to 35 cents per barrel in Texas. This price increase led to the antitrust indictment of twenty-nine oil companies for conspiring to control prices. See Lovejoy, "Producing Capacity," p. 70. Similar antitrust fervor accompanied the crude oil price increases in the 1970s.

90. See generally, Melvin G. DeChazeau and Alfred E. Kahn, *Integration and Competition in the Petroleum Industry* (Philadelphia, Pa., Porcupine Press, 1959); Lovejoy and Homan, *Economic Aspects;* Erickson, "Crude Oil Prices"; Mead, "Petroleum: An Unregulated Industry?"; Stephen L. McDonald, "Taxation System and Market Distortion," in Robert J. Kalter and William A. Vogely, eds., *Energy Supply and Government Policy* (Ithaca, N.Y., Cornell University Press, 1976); and Note, "The Antitrust-Federal Tax Conflict in the Petroleum Conservation System," *Stanford Law Review* vol. 21, no. 2 (1969) pp. 316–338.

The domestic oil industry's lobbying efforts succeeded in establishing a Mandatory Oil Import Program (MOIP) in 1959 that restricted the entry of lower-priced imports into the United States. The combined effects of MOIP and state prorationing were estimated to cost consumers more than $4 billion a year in 1961. See Mead, "Petroleum: An Unregulated Industry?" pp. 148–150. Federal tax provisions such as percentage depletion and expensing of intangible drilling costs subsidized the oil industry and induced much overinvestment in drilling. See McDonald, "Taxation System," pp. 26–44; and Kahn, "The Combined Effects of Prorationing." The low tax rate on drilling and production activities influenced vertically integrated companies to acquire their own reserves and to keep the price of crude oil high, even though this resulted in less profitable refining and marketing operations. Independent refiners without crude oil operations then found it difficult to compete against the vertically integrated refiner. Dechazeau and Kahn develop this argument in detail in their book.

91. In 1964 the IOCC undertook a large study of state conservation laws in an attempt to thwart criticism and possible federal intervention into the industry. The study urged states without adequate unitization laws to consider enactment of such. See Governor's Special Study Committee of the Interstate Oil Compact Commission, *A Study of Conservation of Oil and Gas in the United States 1964* (Oklahoma City, Okla., Interstate Oil Compact Commission, December 11, 1964) [hereafter cited as *IOCC Study of Conservation 1964*]; and Jack M. Campbell, "Conservation and the States," *National Institute for Petroleum Landmen*, vol. 7 (Albany, N.Y., Matthew Bender, 1966) p. 703. See also Lovejoy and Homan, *Economic Aspects;* McDonald, *Petroleum Conservation;* Frank J. Allen, "An Argument for Enforced Unit Development of Oil and Gas Reservoirs in Utah," *Utah Law Review* vol. 7, no. 2 (1960) pp. 197–207; Stephen L. McDonald, "Unit Operation of Oil Reservoirs as an Instrument of Conservation," *Notre Dame Lawyer* vol. 49, no. 2 (1973) pp. 305–316; Blakely M. Murphy, "The Unit Operation of Oil and Gas Fields," *Notre Dame Lawyer* vol. 27, no. 3 (1952) pp. 405–422, and *Notre Dame Lawyer* vol. 28, no. 1 (1952) pp. 73–126; and Raymond M. Myers, "The Necessity of Unitization," *Mississippi Law Journal* vol. 33, no. 1 (1961) pp. 1–13.

92. From 1948 to 1962, the annual production of oil from unitized projects in the United States increased from less than 50 million barrels to almost 400 million barrels. See *IOCC Study of Conservation 1964,* p. 67. These numbers reflect production from compulsorily formed units as well as voluntary ones.

93. See David W. Eckman, "Statutory Fieldwide Oil and Gas Units: A Review for Future Agreements," *Natural Resources Lawyer* vol. 6, no. 3 (1973) pp. 339–387. Articles discussing some of the newly passed laws are, in chronological order: Thomas M. Winfiele, "New Legislation Relating to the Conservation Department," in Carlos E. Lazarus, ed., *Institute on Mineral Law,* vol. 8 (Baton Rouge, Louisiana State University Press, 1961); Arthur B. Custy and Sam D. Knowlton, II, "Compulsory Fieldwide Unitization Comes to Mississippi," *Mississippi Law Journal* vol. 36, no. 2 (1965) pp. 123–141; McAnelly, " A Review of Poolwide Unitization"; Ernest E. Smith, "The Kansas Unitization Statute: Part I," *University of Kansas Law Review* vol. 16, no. 4 (1968) pp. 567–583, and "The Kansas Unitization Statute: Part II," *University of Kansas Law Review* vol. 17, no. 1 (1968) pp. 133–146; Gray and Swan, "Fieldwide Unitization in Wyoming"; Robert G. Rogers and E. Spivey Gault, "Mississippi Fieldwide Unitization," *Mississippi Law Journal* vol. 44, no. 1 (1973) pp. 185–207; Comment, "Conservation of Oil and Gas in Tennessee," *Tennessee Law Review* vol. 41, no. 2 (1974) pp. 323–338.

94. These regulatory changes are discussed in chapter 5. The Texas legislature did enact a compulsory pooling act in 1965, but by this time the Railroad Commission was accomplishing *de facto* compulsory pooling through the administration of the prorationing regulations. See text of this chapter at notes 194 to 214.

95. Donoghue, "Report of TIPRO Advisory Committee to Interstate Oil Compact Commission," *Oil and Gas Compact Bulletin* (December) 1956, p. 73, as quoted in Comment, "Prospects for Compulsory Fieldwide Unitization in Texas," *Texas Law Review* vol. 44, no. 3 (1966) p. 524.

96. See ch. 2 text at notes 28 to 30. Much debate exists as to whether unitization formulas must or should give credit for structural advantage in order to be fair and equitable. Kuntz argues that structural advantage must be credited as an important factor in order to protect the correlative rights of the fortunately located landowners. See Eugene Kuntz, "Correlative Rights of Parties Owning Interests in a Common Source of Supply of Oil or Gas," *Institute on Oil and Gas Law and Taxation,* vol. 17 (Albany, N.Y., Matthew Bender, 1966) pp. 242–243. McDonald refutes this position. See McDonald, *Petroleum Conservation,* p. 244. Oklahoma's compulsory unitization statute specifically directs the commission to consider "location on structure" in determining each tract's equitable share of unit production, but other states do not give statutory recognition to this factor. For a comparison of the various states' standards for determining a tract's fair share of production under prorationing, pooling, and unitization statutes, see R. O. Kellam, "A Century of Correlative Rights," *Baylor Law Review* vol. 12, no. 1 (1960) pp. 30–39; and Jed B. Maebius, Jr., "Statutory Guidelines for Determining 'Fair Share'[6]," *St. Mary's Law Journal* vol. 2, no. 1 (1970) pp. 63–80.

97. Texas was the first state to enact a well-spacing rule. See James E. Nugent, "The History, Purpose, and Organization of the Railroad Commission," *Oil and Gas: Texas Railroad Commission Rules & Regulations* (Austin, State Bar of Texas, 1982) p. A-16. Rule 37 now appears in Texas Administrative Code tit. 16, sec. 3.37 (1982). It has been amended over the years to establish increasingly larger spacing units and currently provides for 40-acre units in oil fields.

In 1919 well-spacing rules were considered important largely to reduce fire hazards and minimize water intrusion into oil strata, not to prevent the economic waste of unnecessary drilling. See Railroad Comm'n v. Bass, 10 S.W.2d 586 (Tex. Civ. App. 1928). The benefits of well spacing as a method of reducing drilling and operating costs seem to have become recognized a decade later when flush production from large, new fields in Oklahoma and Texas reduced crude oil prices and profits.

98. In the 1930s, the Railroad Commission was nonetheless predisposed to grant Rule 37 exception requests to these subdivisions as well, in violation of its own rules. See Gulf Land Co. v. Atlantic Refining Co., 134 Tex. 59, 131 S.W.2d 73 (1939); and Gulf Land Co. v. Atlantic Refining Co., 113 F.2d 902 (5th Cir. 1940), which ultimately upheld a Railroad Commission order granting a permit for a seventh well on a 6.88-acre tract. Waste exceptions were also freely granted in the early years. See Elton M. Hyder, "Some Difficulties in the Application of

the Exceptions to the Spacing Rule in Texas," *Texas Law Review* vol. 27, no. 4 (1949) pp. 494–501. Not surprisingly, the administration of Rule 37 exceptions spawned enormous litigation. By 1952, the state and federal courts had decided about 185 reported cases involving Rule 37. Robert E. Hardwicke, "Oil-Well Spacing Regulations and Protection of Property Rights in Texas," *Texas Law Review* vol. 31, no. 2 (1952) p. 104. The rule became increasingly complex. See Hyder, "Some Difficulties"; and Charles J. Meyers, "Common Ownership and Control in Spacing Cases," *Texas Law Review* vol. 31, no. 1 (1952) pp. 19–35. Indeed, the formulation of equitable unitization agreements might have taken less time and legal talent than that involved in defending, opposing, and explaining Rule 37 exception requests. One commentator has noted that the complexity and uncertainty of Rule 37 renders its fair administration difficult and therefore does not adequately protect those who have invested large sums of money in oil and gas properties. See Note, "Oil and Gas—Rule 37 Exception," *Texas Law Review* vol. 30, no. 1 (1951) p. 138.

99. By 1938, the Marginal Well Act controlled the allocation of 78 percent of the state's total allowable, and the burden of market demand prorationing thus fell on the wells producing the remaining 22 percent of the oil. See Ely, *The Conservation of Oil*, pp. 1226–1227. In 1963, 15 percent of Texas total crude production was exempt from prorationing because of the Marginal Well Act. At the same time, nonexempt wells were prorated to 28 percent of their capacity. See Lovejoy and Homan, *Economic Aspects*, pp. 156–157.

100. In Railroad Comm'n v. Sample, 405 S.W.2d 338 (Tex. 1966), the defendant Sample had overproduced his wells by falsely reclassifying them from "top allowable" to "high marginal" wells. Production from Sample's wells increased from 49 barrels per day to 150 barrels per day because of their new exempted status. In Stewart v. Humble Oil & Refining Co., 377 S.W.2d 830 (Tex. 1964), an operator who had intentionally deviated his well in an attempt to drain more oil from a richer part of the East Texas field, was allowed to keep the deviated bottomhole locations (which did not cross lease lines), because the evidence showed that had he drilled straight down, his well would have been a marginal well, and he then would have been able to drain even more oil from the adjacent tracts. Therefore, Humble, the adjacent lessee in this case, could not prove injury from the slant-well drilling and the Railroad Commission's orders granting Rule 37 exceptions to Stewart's irregular bottomhole locations were upheld. A list of common methods used to manipulate production in order to obtain marginal well status appears in Robert E. Hardwicke and M. K. Woodward, "Fair Share and the Small Tract in Texas," *Texas Law Review* vol. 41, no. 1 (1962) pp. 90–91.

101. The commission's position on the East Texas field is described in Rowan & Nichols Oil Co. v. Railroad Comm'n, 28 F.Supp. 131 at 135 (W.D. Tex. 1939).

102. In fields without any small tracts, regularly spaced drilling units were established and the prorationing formula used was a simple "yardstick" based on the depth of the well and the size of the spacing unit. In fields with small and irregularly shaped tracts, the commission used schedules with specific per-well

factors. In 1948, 17 percent of all fields in Texas were prorated on a 100 percent per-well basis; 35 percent were prorated using a formula which allocated 50 percent of the field's allowable on an acreage basis and 50 percent on a per-well basis; 26 percent of all fields were prorated on the basis of a 75 percent acreage factor and a 25 percent per-well factor. Only 7 percent of all fields were allocated on acreage alone. See Erich W. Zimmerman, *Conservation in the Production of Petroleum* (Northford, Conn., Elliot's Books, 1957) p. 333.

103. For example, in the Hawkins field, the 50 percent per-well and 50 percent surface acreage formula resulted in the drainage of 30 million barrels of oil from large tracts to the densely drilled townsite. See Railroad Comm'n v. Humble Oil and Refining Co., 193 S.W.2d 824 (Tex. Civ. App. 1946), *aff'd,* 331 U.S. 791 (1947), *aff'd on rehearing,* 332 U.S. 786 (1948). In an effort to provide some protection to large-tract owners in East Texas from drainage by the thousands of small-tract wells, the Railroad Commission established the "eight times area" rule. This allowed a producer to drill wells on his tract equal in number to the average well density existing in the surrounding area eight times the size of his tract. Thus, even large tracts in East Texas were drilled more closely than the spacing rule of one well to 10 acres. The eight-times-area rule obviously results in a chain reaction of "drilling down the field" to the density of the most closely drilled tract, completely negating the well-spacing rule. The commission recognized this and, in fits and starts, attempted to contain the chain reaction. First, the Railroad Commission refused to extend the eight-times-area rule to fields which were prorationed on a basis other than 100 percent per well. See Kraker v. Railroad Comm'n, 188 S.W. 2d 912 (Tex. Civ. App. 1945). Second, the commission refused to grant Rule 37 exceptions to prevent confiscation due to fieldwide migration of oil and gas rather than to localized drainage. Miller v. Railroad Comm'n, 185 S.W.2d 223 (Tex. Civ. App. 1945); and Woolley v. Railroad Comm'n, 242 S.W.2d 811 (Tex. Civ. App. 1951). This refusal required some prodding from the courts. See Byrd v. Shell Oil Co., 178 S.W.2d 573 (Tex. Civ. App. 1944). A Rule 37 applicant's recourse against fieldwide drainage was to seek a change in the field's general spacing or prorationing orders. This gave scant comfort to applicants because prorationing orders were almost impossible to overturn. See text of this chapter at notes 203 to 204. The failure of the commission to protect tracts from regional drainage is sharply criticized in Hyder, "Some Difficulties," pp. 508–512; and Note, *Texas Law Review* vol. 23 (1944). The Railroad Commission was simply in an untenable situation: damned if it did grant Rule 37 permits which drilled down the field, and damned if it did not and failed to protect correlative rights.

104. Walker, "The Problem of the Small Tract," pp. 164–165. The commission has also been severely criticized by scholars who felt that the agency did not do enough to prevent the fraudulent classification of wells as marginal or to issue regulations that defined the criteria for testing a well's maximum capacity more tightly, especially in the East Texas field. See Prindle, *Petroleum Politics,* pp. 81–92; and Hardwicke and Woodward, "Fair Share and the Small Tract," pp. 90–92. The agency's limited budget partly explains its failure to police the thousands of marginal wells in Texas more effectively.

105. As to nonmarginal wells drilled on small tracts as Rule 37 exceptions, the Railroad Commission was similarly urged to prorate their production so as to reduce the small tract's drainage of oil and gas from other tracts. See W. L. Summers, "Does the Regulation of Oil Production Require the Denial of Due Process and the Equal Protection of the Laws?" *Texas Law Review* vol. 19, no. 1 (1940) p. 16; and Note, *Texas Law Review* vol. 13, p. 129. While surface acreage prorationing would achieve this end, the authors do not suggest how the owner of a small tract which is not profitable to drill and produce under surface acreage prorationing would then be protected against drainage to wells on adjoining tracts. Absent compulsory pooling, the only compromise solution would necessitate a well-by-well analysis to ascertain the lowest possible allowable which would still permit the small-tract owner to operate profitably.

106. Robert E. Hardwicke, "Oil-Well Spacing Regulations," pp. 120–122. Walker also urged the Railroad Commission to pool tracts under its general power to control well spacing and prevent waste, unrelated to the Rule 37 exception process. See Walker, "The Problem of the Small Tract," pp. 168–169.

107. So obvious was this solution that other states started to enact compulsory pooling legislation at an early date. In 1935 Oklahoma and New Mexico did so, and in 1936 Louisiana followed. See Walker, "The Problem of the Small Tract," p. 167. By 1952, sixteen states had enacted such laws; at this time, only seven states had compulsory unitization statutes, four of which were for cycling operations only. See Ely, "The Conservation of Oil," pp. 1169 and 1171–1173. Ironically, Texas was the first state to pass a well-spacing rule and the last state to enact compulsory pooling.

108. The Marginal Well Act states that marginal wells include any wells that fit into the specified depth and production brackets, without any other condition. Texas Natural Resources Code Ann. sec. 85.122 (Vernon 1978). In 1940 the Texas attorney general declared that the act prohibited the commission from limiting the production of a marginal well below the amount stated in the act. Texas Attorney General Opinion No. 0-1845 (1940). A legislative amendment would be necessary to exempt from prorationing only those wells which would be in fact damaged or prematurely abandoned if curtailed. See Comment, "Proration in Texas," pp. 207–208. The political success of such a proposed amendment would doubtless equal that of a compulsory unitization bill for all the same reasons.

109. Walker interpreted this declaration of intent to apply only to fieldwide unitization, not to pooling, and asserted that the commission's general powers to control well spacing were not limited by this provision so as to prohibit the commission from requiring pooling. See Walker, "The Problem of the Small Tract," p. 168. In light of the legislative history of this phrase (discussed in chapter 3 text at notes 3 to 112), and the absence of compulsory pooling authority, Walker's view seems improbable.

Hardwicke argues that a Railroad Commission rule that denied a Rule 37 exception permit to a small-tract applicant who refused to pool would not be an order compelling pooling. It would simply pool by "indirection." See Hardwicke, "Oil Well Spacing Regulations," pp. 122 and 124–125. Hardwicke's semantics are unconvincing. In addition, it is not clear that pooling would always be induced. Under his scheme, the Railroad Commission would deny a Rule 37 permit to a small-tract owner who was made a fair and reasonable offer to pool by the owner of an adjoining larger tract. This would induce the small-tract owner to pool. But what about the small-tract owner who offered to pool on reasonable terms with an adjoining larger-tract owner and was refused? He would be granted a Rule 37 exception permit under Hardwicke's plan, but if allowable formulas give no advantage to small tracts, as Hardwicke simultaneously urges, it might not be profitable for the small-tract owner to drill, and his oil would be confiscated. Hardwicke offers no solution to this other than passage of a compulsory pooling statute. Indeed, in a footnote, he views his own proposal that the Railroad Commission indirectly pursue pooling as impractical. See ibid., p. 121, n. 43.

110. Railroad Comm'n v. Rowan & Nichols Oil Co., 310 U.S. 573 (1940), *modified,* 311 U.S. 614 (1940), *aff'd on rehearing,* 311 U.S. 570 (1941). Three justices dissented and would have upheld the opinions of the two lower federal courts that had invalidated the East Texas order as unreasonable and confiscatory. The Supreme Court's decision is praised in Kenneth Culp Davis, "Judicial Emasculation of Administrative Action and Oil Proration: Another View," *Texas Law Review* vol. 19, no. 1 (1940) pp. 29–58, and condemned in Summers, "The Regulation of Oil Production." Humble Oil and Refining Co. also brought suit seeking to invalidate the East Texas order. Humble claimed that it owned 14 percent of the reserves in East Texas, but was allowed to produce only 9 percent of the field's allowable. On August 11, 1939, a federal district court granted Humble's plea for an injunction against the order. This ruling allowed Humble to produce 5,000 additional barrels per day in East Texas. The Railroad Commission appealed both the *Humble* case and the *Rowan & Nichols* case to the U.S. Supreme Court. When this court decided in favor of the commission in the *Rowan & Nichols* case, it also reversed Humble's victory in the lower courts. See Railroad Comm'n v. Humble Oil & Refining Co., 311 U.S. 578 (1941).

The timing of Humble's crude oil price cuts during this litigation caused another public furor. On August 11, 1939, the same day that Humble had won the injunction which allowed it to produce 5,000 more barrels a day from East Texas, Humble cut the price of crude oil by as much as 32 cents a barrel. Many interpreted this price cut as Humble's chortling over its ability to produce more oil in East Texas and thus being able to afford the price cuts. Jerry Sadler, a railroad commissioner, warned that unless Humble either cut the retail price of gasoline by five cents a gallon or restored the crude price, the Standard Oil Co. of New Jersey would stand convicted of the intent to rule or ruin the oil industry, to drive every independent out of business, and to deprive the state of Texas of funds for the care of its blind and aged. The Texas attorney general began a probe to see if Humble was violating the antitrust laws. On August 14, the commission ordered a fifteen-day shutdown of all Texas fields. The shutdown was ostensibly to investigate whether waste was occurring, but the real objective was presumably to decrease the supply of crude oil. Five other states joined Texas in the shutdown, which soon had the intended effect of reducing inventories of crude oil stocks. On August 28, Humble rescinded its price cut retroactively and the commission ended the shutdown. Commissioner Thompson then proclaimed that Texas had shown Standard Oil that its "fair-haired child" could not trample the rights of Texas producers under its "golden boot heels." See Larson and Porter, *History of Humble Oil,* pp. 498–499 and 531–533.

111. Railroad Comm'n v. Rowan & Nichols Oil Co., 310 U.S. 573 at 582 (1941).

112. 311 U.S. 570 at 575.

113. 311 U.S. 570 at 576–577. In Justice Frankfurter's words:

> The real answer to any claims of inequity or to any need of adjustment to shifting circumstances is the continuing supervisory power of the expert commission. In any event, a state's interest in the conservation and exploitation of a primary natural resource is not to be achieved through assumption by the federal courts of powers plainly outside their province and no less plainly beyond their special competence.

Justice Frankfurter was a relative newcomer to the Supreme Court in 1941, having been carefully selected by President Roosevelt as a justice who was likely to uphold the New Deal legislation which bequeathed enormous powers to administrative agencies to regulate private business and increase social welfare. In 1938, when Frankfurter was a law professor at Harvard, he enunciated his predisposition to uphold the decisions of administrative agencies in the foreword of an issue of the *Yale Law Journal* devoted entirely to the revolution in administrative law wrought by the 1930s. See Felix Frankfurter, "Foreword," *Yale Law Journal* vol. 47, no. 4 (1938) pp. 515–518b.

114. Brown v. Humble Oil & Refining Co., 126 Tex. 296, 83 S.W.2d 935, 944 (1935), *rehearing denied,* 87 S.W.2d 1069 (1935). Humble's euphoria in winning this case was somewhat diminished by the supreme court's language in its subsequent opinion denying the motion for rehearing. This opinion disavowed any construction of the language quoted in the text above so as to require prorationing on a surface acreage basis. Still, Humble considered that it had won a major legal victory which could constrain the commission's profligate grants of Rule 37 exception wells. See Larson and Porter, *History of Humble Oil,* pp. 493–495. Humble did, however, lose the battle to prevent this specific well based on a Rule 37 exception from being drilled. On remand, the permit for this well was granted by the commission and upheld by the court on the incredible basis that substantial evidence existed to show that the well was necessary to prevent waste, that is, to recover all the oil under the 1 1/2 acre tract at issue. Humble Oil & Refining Co. v. Railroad Comm'n, 112 S.W.2d 222 (Tex. Civ. App. 1937).

115. Marrs v. Railroad Comm'n, 142 Tex. 293, 177 S.W.2d 941 (1944). The formula found to be unreasonable and discriminatory in this case was based 50 percent on a per-well factor and 50 percent on acreage potentials. The plaintiff's portion of the field contained 87 percent of the recoverable reserves, but under prorationing received only 53 percent of the allowables.

116. Railroad Comm'n v. Humble Oil & Refining Co., 193 S.W.2d 824 (Tex. Civ. App. 1946), *aff'd,* 331 U.S. 791 (1947), *aff'd on rehearing,* 332 U.S. 786

(1948), commonly called the Hawkins field case. Under the formula, a well on one-tenth of an acre obtained an allowable of 46 barrels of oil a day, whereas a well on a tract 200 times as large, drilled in accordance with the 20-acre spacing rule, was awarded an allowable of only twice as much, or 92 barrels a day.

117. 193 S.W.2d 832.

118. 193 S.W.2d at 832–833. The court stated that the 20-acre spacing rule saved Humble $10 million in costs compared to drilling on a 10-acre spacing pattern. The court clearly viewed Humble's loss of 5 percent of the field's reserves as being offset by these cost savings. Humble owned 76.5 percent of the recoverable reserves in the field but would be able to produce only 71.4 percent of the reserves under the prorationing formula.

119. Jim C. Langdon, "The Influence of Court Decisions Upon Railroad Commission Policy in Rule 37 Cases and the Allocation of Allowables to the Small Tract Well," *Texas Bar Journal* vol. 27, no. 3 (1964) pp. 149–150 and 177–185.

120. Kraker v. Railroad Comm'n, 188 S.W.2d 912 (Tex. Civ. App. 1945).

121. Trapp v. Shell Oil Co., 145 Tex. 323, 198 S.W.2d 424 (1946). The substantial evidence rule was first used by the Texas Supreme Court in 1934, but was inconsistently applied until 1946. The 1944 case of Marrs v. Railroad Comm'n, 142 Tex. 293, 177 S.W.2d 941 (1944), discussed in this chapter at note 115, which invalidated a per-well prorationing order as confiscatory, was decided before the adoption of the substantial evidence rule. Query whether the court would have reversed the commission's order under the new standard of review. The effect of the substantial evidence rule on Railroad Commission decisions is discussed further in the text in chapter 8 at notes 80 to 92; and in Hyder, "Some Difficulties," pp. 487–494; A. W. Walker, Jr., "Developments in Oil and Gas Law During World War II," *Texas Law Review* vol. 25, no. 1 (1946) pp. 5–6; and A. W. Walker, Jr., "The Application of the Substantial Evidence Rule in Appeals from Orders of the Railroad Commission," *Texas Law Review* vol. 32, no. 6 (1954) pp. 640–659.

122. Hardwicke, "Oil Well Spacing Regulations," pp. 118–120. Standard Oil Co. of Texas, a large-tract owner in the Yates field, had applied to the commission in September 1945 to amend the field's prorationing formula from one based on the average well potentials of all wells in 50-acre units, to one based wholly on reserves and oil in place. Despite the Hawkins field ruling in 1946, Standard continued to press its case in court after the commission refused to change the formula. Standard Oil alleged that it had 15.5 percent of the field's acreage and 17 percent of recoverable reserves, but would recover only 11.4 percent of the oil under the potential formula. The court refused to invalidate the commission's order, citing the Hawkins field case extensively as precedent. Standard Oil Co. of Texas v. Railroad Comm'n, 215 S.W.2d 633 (Tex. Civ. App. 1948).

The cost of litigating a prorationing case is high. Expert testimony is required as to the geological and producing conditions in the entire field. In one such case, the statement of facts consisted of more than 2,000 pages with several hundred pages of engineering exhibits. Railroad Comm'n v. Marrs, 161 S.W.2d 1037 at 1039 (Tex. Civ. App. 1942), *rev'd,* Marrs v. Railroad Comm'n, 142 Tex. 297, 177 S.W.2d 941 (1944). By contrast, Rule 37 exception cases generally involve only the issue of local drainage, and are thus less costly to litigate.

123. Professor Summers prophesied that the effect of the U.S. Supreme Court's holding in the *Rowan & Nichols* cases would be to destroy the oil industry's and landowners' support of state conservation legislation and its enforcement through administrative agencies see Summers, "The Regulation of Oil Production," p. 27:

> If this decision of the supreme court means what it appears to mean, that is, that the taking of private property without due process has become an administrative and not a judicial question, it is hardly reasonable to expect that in the future the oil industry will lend its wholehearted support to the enforcement of the policy of conservation of oil and gas.

That the Railroad Commission was not discredited in the following years perhaps reflects recognition by the oil industry that the failure to protect property rights was a failure of legislative, not administrative, decision making.

The changing standard of judicial review of agency decision making exemplified in the *Rowan & Nichols* cases also profoundly affected scholarly attitudes and analysis of the Railroad Commission's orders. Before 1940, scholars discussed the intractability of the small-tract problem in Texas and called for compulsory pooling legislation. See Hardwicke, "The Rule of Capture"; and Walker, "The Problem of the Small Tract." Scholars also criticized the legislature more generally for failing to pass laws that recognized a legal right against uncompensated drainage. See A. W. Walker, Jr., "Property Rights in Oil and Gas and Their Effect Upon Police Regulation of Production," *Texas Law Review* vol. 16, no. 3 (1938) pp. 370–381; and W. L. Summers, "Legal Rights Against Drainage," *Texas Law Review* vol. 18, no. 1 (1939) pp. 27–47. With the *Rowan & Nichols* decision, the focus of academic writing shifted to the Railroad Commission's administrative decision making, not legislative reform. In effect, the Railroad Commission was faulted for failing to implement administratively that which the legislature had refused to enact. Thus, Summers criticized the East Texas prorationing order on the basis that the Railroad Commission had alternatives: pooling, surface acreage prorationing, and denial of Rule 37 exceptions. The first alternative was outside the commission's statutory authority and the other two would result in equally impermissible confiscation of small tracts. Conversely, Davis's praise for the *Rowan & Nichols* decision was largely based on the Railroad Commission's lack of regulatory alternatives. See Davis, "Judicial Emasculation." Both approaches ignore the legislative framework which prevented any satisfactory solution to prorationing.

124. The federal district court opinion in *Rowan & Nichols* specifically questioned the constitutionality of the Marginal Well Act because marginal wells confiscated oil from better wells and also caused waste by bringing up large amounts of water which lowered the pressure of the water drive. See 28 F. Supp.

131, 135, 137 (W.D. Tex. 1939). The district court also clearly envisioned pooling legislation as a solution to the Railroad Commission's quandary (ibid., p. 136).

Of course, any radical change in the prorationing system might cause existing wells to become unprofitable. The lower federal courts in *Rowan & Nichols* refused to apply the doctrine of laches to protect producers who had invested thousands of dollars in reliance on existing prorationing orders. Compulsory pooling would have allowed small-tract owners to secure some share of the East Texas oil, but perhaps not enough in all cases to pay for the costs of the wells already drilled. This underscores the importance of securing legislative and regulatory reform before large-scale investments are made in reliance on an imperfect system.

125. Mancur Olson, Jr., *The Logic of Collective Action* (Cambridge, Mass., Harvard University Press, 1965). As the number of members in the group increases, voluntary agreement is less likely to succeed because each member has a proportionately smaller stake in the outcome of group action. Thus, voluntary unitization would be unlikely in a field owned by a large number of independents without a dominant major. In this instance, the cost of negotiating a voluntary agreement might exceed the benefit to any single operator. Compulsory unitization might be welcomed by all the lessees in this field, including the independents.

126. It is not immediately obvious that crude prices would be lower under unitization than under market demand prorationing. Market demand prorationing forcibly restricts the quantity supplied by producers, and this implies that the price under prorationing is held too high. However, because unrestrained extraction under the rule of capture results in too great a supply on the market and a corresponding too low price, a rise in price induced by market demand prorationing may be more efficient than the price under unrestrained production (see ch. 2 text at notes 32 to 36). Theoretically the switch from market demand prorationing to unitized operation could result in producers reducing their current production rates to the private and socially optimal rate. The resulting higher price of crude oil would be more efficient than the lower price under market demand prorationing. See McDonald, *Petroleum Conservation,* pp. 186–187. However, most economists and operators seemed to believe that the changeover to unitized operations would result in lower prices during the 1960s.

127. McDonald, *Petroleum Conservation,* pp.242–248. Under McDonald's ideal approach to unitization, all reservoirs would be unitized a reasonable time after discovery. As a practical matter, McDonald urged the adoption of phased unitization, beginning with new reservoirs and then moving to older fields over the next ten years in order to minimize price declines and abandonment of marginal wells. McDonald also suggested that the federal government further the goal of unitization by slowing down the leasing of new federal lands and by purchasing developed reservoirs to use as national defense storage reserves.

128. Between 1955 and 1971, the four largest oil producers increased their share of U.S. production from 19 percent to 31 percent, and the eight largest increased their share from 31 percent to 50 percent. Ford Foundation, Energy

Policy Project Staff, *Time to Choose: America's Energy Future, Final Report* (Cambridge, Mass., Ballinger, 1974).

129. This factor played an important role in the independent producer's opposition to compulsory unitization in California. Reputedly, no large oil company in Oklahoma had ever lost an argument before the commission at a unitization hearing because of the number of expert witnesses at the majors' command. See Comment, "Conservation and Price Fixing in the California Petroleum Industry," *Southern California Law Review* vol. 29, no. 4 (1956) pp. 478–479.

130. See Garvin, "The Effect of Field Unit Operation Upon the Royalty Interest."

131. McDonald, *Petroleum Conservation,* pp. 208–209.

132. Alvin C. Askew, "Texas Energy Policy," *Baylor Law Review* vol. 29, no. 4 (1977) pp. 643–667.

133. Prindle has documented the independent producers' financial contributions to Railroad Commission elections. Generally the independents have supplied more than 60 percent of all the contributions to the winning commissioners. See David F. Prindle, *Petroleum Politics and the Texas Railroad Commission* (Austin, University of Texas Press, 1981) pp. 163–180. There is no reason to believe that the independents' organization and financial strength is not used in elections of other state officials and lawmakers. Royalty interest owners are generally not as well organized but they are numerous, totaling 650,000 persons in Texas in 1983 (ibid., p. 126).

Compulsory unitization has many attributes of a "special interest" issue that generates large gains for a small number of constituents while imposing small individual costs on a large number of other voters. Voters with a small personal stake in the issue will not vote either for or against a politician on this matter, especially if the issue is a relatively complex one to understand. On the other hand, the special interest group that is vitally affected by the issue will help politicians who favor their group and oppose those who do not. Thus the special interest issue is likely to be decided in favor of the special interest group. See James M. Buchanan and Gordon Tullock, *Calculus of Consent: Logical Foundations of Constitutional Democracy* (Ann Arbor, University of Michigan Press, 1962) pp. 283–295.

134. In 1978, about 88 percent of all the Texas firms engaged in oil and gas production and 85 percent of Texas' oil and gas exploration services firms employed less than 20 persons. Texas 2000 Commission, *Texas: Past and Future, A Survey* (Austin, Tex., Office of the Governor, 1982) pp. 29–30 [hereafter cited as *Texas 2000 Report*].

135. Oil and gas and mining employment grew from 1.6 percent of Texas' total employment in 1930 to 3.0 percent in 1960 (ibid., p. 24).

136. James M. Whittier, "Compulsory Pooling and Unitization: Diehard Kansas," *University of Kansas Law Review* vol. 15, no. 3 (1967) pp. 318–319.

137. Prindle, *Petroleum Politics,* pp. 125–129. At a 1938 meeting of the IOCC, the chairman of the Railroad Commission stated that wells allowed as Rule 37 exceptions were granted in part to increase employment and add to the state's taxable properties. See Hardwicke, "Legal History of Oil" p. 256, n. 52. In 1962 the Railroad Commission refused to order the shutdown of many of the unnecessary wells in East Texas, partly on the basis that it would depress the local economy dependent on the field's payrolls. See Lovejoy and Homan, *Economic Aspects,* p. 121.

138. Prindle, *Petroleum Politics,* pp. 123–141 and 145–151. See also the testimony of Commissioner Thompson at the 1940 hearings of the Temporary National Economic Committee that the price-maintenance effect of prorationing was necessary in order to protect stripper wells from being squeezed out of the picture, as quoted in Kenneth Culp Davis and York Y. Willbern, "Administrative Control of Oil Production in Texas," *Texas Law Review* vol. 22, no. 2 (1944) p. 155; and note 110 in this chapter on the commission's response to Humble's 1939 price cut.

139. William Murray, former Texas railroad commissioner, telephone interview with author, February 4, 1983. See also Prindle, *Petroleum Politics,* pp. 139–140, describing the commission's most important nondecision as its refusal to consider compulsory unitization. In contrast, Louisiana's conservation commissioner actively sponsored a compulsory unitization bill. See Winfiele, "New Legislation Relating to the Conservation Department," p. 19. In 1946, the Railroad Commission did support a bill for compulsory cycling in gas fields (see ch. 5 text at note 18).

140. Prindle, *Petroleum Politics,* pp. 63–68, 81–91.

141. William J. Murray, Jr., "Engineering Aspects of Unit Operation," *Institute on Oil and Gas Law and Taxation,* vol. 3 (Albany, N.Y., Matthew Bender, 1952) pp. 2–3.

142. Ibid., pp. 13–14.

143. Fred Young served in the capacity of general counsel to the Oil and Gas Division of the Railroad Commission from 1948 to 1978. He cannot recall any commissioner publicly supporting compulsory unitization during this time (letter to the author from Fred Young, Feb. 8, 1983).

144. See, for example, Ernest O. Thompson, "The Texas Market Demand Statute on Oil and Gas and Its Application," *Texas Law Review* vol. 39, no. 2 (1960) pp. 146–149. In 1953 Commissioner Thompson requested the eleven leading importers of crude to report on their planned imports for the year. In response to the report, Thompson stated: "It seems that Texas crude is being supplanted by foreign crude by some companies. This works a hardship on Texas producers. The average Texas oil well produces 20 barrels a day. The average well in the Persian gulf produces 5,000 barrels a day. . . . How can a Texas 20-barrel-well compete with a 5,000 barrel well?" (quoted in Zimmerman, *Conservation,* pp. 380–381).

145. Federal inaction is discussed further in chapter 10 text at notes 33 to 43. Clearly some federal lawmakers perceived the excess capacity in the oil fields induced by the state conservation regimes as in the national security interest. In 1952 the Defense Department had specifically requested that the oil industry keep a reserve capacity of 15 percent of annual consumption. See Zimmerman, *Conservation*, p. 70, note 30.

Economic sectionalism between states in the supply of energy is common. For example, when Montana's coal industry began to boom during the energy crisis of the 1970s, Montana passed a severance tax of as much as 30 percent on the price of coal. Montana exported 90 percent of its coal, so the tax burden fell primarily on citizens of other states. The U.S. Supreme Court upheld the severance tax against the challenge that it violated the commerce clause and the supremacy clause of the U.S. Constitution. Even though the state tax may have had the effect of frustrating the national energy policy of encouraging coal use, the state law was valid absent more express congressional intent to preempt all state legislation that adversely affected coal use. See Commonwealth Edison Co. v. Montana, 453 U.S. 609 (1981) (three justices dissenting).

Of course, Texas' choice of an inefficient conservation regime would encourage firms to explore and produce in other states that permitted more efficient practices. However, it is difficult to quantify this loss of income to Texas. Other states also had inefficient regulatory structures. See McDonald, *Petroleum Conservation*, pp. 152–182.

146. McAnelly, "A Review of Poolwide Unitization," p. 7.

147. Rogers and Gault, "Mississippi Fieldwide Unitization," p. 187.

148. For example, Arkansas passed its compulsory unitization statute in 1951 after the Arkansas Supreme Court's decision in Dobson v. Arkansas Oil and Gas Comm'n, 218 Ark. 160, 235 S.W.2d 33 (1951), holding that the commission had no authority to force a royalty owner into a unit plan merely because 96 percent of the other royalty owners and 100 percent of the operators had voluntarily agreed to the unit which would avert a substantial waste of mineral resources. The profitable obstructionism of a Nebraskan oil operator who refused to join a large unit doubtless influenced the Nebraskan legislature to pass a compulsory unitization statute. See Baumgartner v. Gulf Oil Corp., 184 Neb. 384, 168 N.W.2d 510 (1969), *cert. denied,* 397 U.S. 913 (1970), for this history. California passed a narrow compulsory unitization statute under emergency conditions when the City of Long Beach started sinking into the ocean because of land subsidence from massive withdrawals of oil and gas underneath the city. See Robert E. Sullivan, ed., *Conservation of Oil and Gas: A Legal History, 1948–1958* (Chicago, Ill., American Bar Association, 1960) pp. 48–52. Wyoming's inertia to compulsory unitization was overcome in 1971 with the discovery of the Powder River Basin area which required early unitization to prevent substantial waste. See Gray and Swan, "Fieldwide Unitization in Wyoming," pp. 435–436.

149. J. Frederick Lawson, "Recent Developments in Pooling and Unitization," *Institute on Oil and Gas Law and Taxation,* vol. 23 (Albany, N.Y., Matthew Bender, 1972) pp. 152–153. Oklahoma's statute enacted in 1945 requires 65 percent voluntary approval. Later statutes in Colorado, Montana, North Dakota,

Utah, Mississippi, and Maine require 80 percent to 85 percent voluntary approval.

150. Lovejoy, "Producing Capacity," p. 63, n. 3.

151. See statement of C. John Miller, president, Independent Petroleum Association of America, in *Profitability of Domestic Energy Company Operations, 1974,* Hearings on Excess Profits Tax Legislation Before the Senate Finance Committee, 93rd Cong., 2nd sess. (1974) p. 45.

152. See, for example, David Gross, "Unitization, The Fair Advantage," *National Institute for Petroleum Landmen,* vol. 8 (Albany, N.Y., Matthew Bender, 1967) pp. 97–98.

153. Mead, "Petroleum: An Unregulated Industry?" p. 141.

154. Ibid., p. 144.

155. *Oil Regulation Report,* vol. 37, February 18, 1969 (Austin, Texas State House Reporter, Inc.) pp. 6–11; *Oil Regulation Report,* vol. 37, May 8, 1969, p. 7. A Senate committee did hold hearings on the 1971 proposal, but opponents of the bill flooded it with amendments. See *Oil Regulation Report,* vol. 39, February 23, 1971, pp. 5–6; also, March 30, 1971, p. 9, and May 5, 1971, p. 1.

156. *Oil Regulation Report,* vol. 39, February 19, 1971, p. 5.

157. *Oil Regulation Report,* vol. 40, September 15, 1972, p. 3.

158. *Oil Regulation Report,* vol. 40, March 5, 1973, p. 3; March 13, 1973, p. 2.

159. Photocopy of Texas House Bill 311, 63rd Leg., Reg. Sess. (1973) with amendments, received from Legislative Reference Service, Austin, Texas [hereafter cited as H.B. 311].

160. Ibid., sec. 2(11).

161. The focus on secondary recovery is also evident in the requirement that the applicant for the compulsory unitization order submit an "engineering and geological report showing a plan to increase the ultimate recovery for the common reservoir by unitized operations including, but not limited to, proposed injection wells, proposed injection fluids, and estimate of increased recovery of oil or gas or both. . . ." See H.B. 311, sec. 3(a)(4).

162. See text of this chapter at notes 58 and 59 for a discussion of this issue and its importance under the 1949 Act. See also Commissioner Murray's view of this issue in text of this chapter at note 141.

163. H.B. 311, secs. 4(a)(7), 5.

164. H.B. 311, sec. 5(a) & (b).

165. H.B. 311, secs. 5(c), 5(f).

166. H.B. 311, sec. 5(j).

167. In the end, however, H.B. 311 required that the commission select as unit operator the working interest owner best qualified in terms of "financial responsibility, adequate organization and personnel," a requirement that would preclude most independent producers who do not possess large engineering,

legal, and accounting staffs. See H.B. 311, sec. 5(a). Nonetheless, under H.B. 311, the commission, not the working interest owners, selected the unit operator, and this assured selection by an unbiased tribunal.

168. H.B. 311, sec. 5(q) (added as a floor amendment on April 12, 1973).

169. H.B. 311, secs. 5(h) and 5(i). The withdrawal provisions did not prevent a working interest owner from selling and assigning its interest to any purchaser or assignee, nor did they create a preferential right to purchase by the other unit participants.

170. H.B. 311, sec. 5(e).

171. By contrast, the 1969 compulsory unitization bill which TIPRO refused to support defined a separate tract's fair share of unit production as the value of the tract for oil and gas purposes "taking into account acreage, the quantity of oil and gas recoverable therefrom, location on the structure, its probable productivity of oil and gas in the absence of unit operations, the burden of operation to which the tract will or is likely to be subject, or so many of said factors or such other pertinent engineering, geological, or operating factors, as may be reasonably susceptible of determination." See *Oil Regulation Report,* vol. 37, February 18, 1969, pp. 6 and 9.

172. H.B. 311, sec. 5(e). The bill expressly allowed a two-part participation formula: the first part to allocate shares in the remaining primary reserves, and then, after an agreed period of time or a certain volume of production had been reached, the second part of the formula became effective to allocate the reserves producible by secondary recovery. The status quo was also retained by Section 11 of the bill which provided that if the commission ordered unitization of a reservoir that embraced a previously unitized area, production from the established unit must be allocated among the tracts within it as provided by the voluntary agreement or previous order which authorized the unit. Section 15 provided that H.B. 311 was supplemental to other unitization laws, thus retaining the 1949 voluntary unitization act.

173. H.B. 311, sec. 6. The required percentage of royalty interest owners had to be secured within a period of six months from the date on which the unitization order was made by the commission, although the commission could extend this period for an additional year if good cause was shown. Section 7 of the bill also required that notice of the unitization hearing be mailed to all royalty owners who had not approved the plan of unitization.

174. H.B. 311, sec. 10(a).

175. H.B. 311, sec. 3(c). Under this section, the unit's proponents could also request such information from those who opposed the unit.

176. H.B. 311, sec. 8.

177. *Oil Regulation Report,* vol. 40, March 7, 1973, pp. 1–2 (statement by Forrest Hoaglund, East Texas Division Manager of Humble Oil and Refining Co.); *Oil Regulation Report,* vol. 40, March 13, 1973, p. 3 (statement by Charles Matthews, chairman of Texas Engineers for Conservation-Unitization).

178. *Oil Regulation Report,* vol. 40, March 7, 1973, p. 3 (statement by Herbert Grubb, Office of Information Services, Office of the Governor).

179. *Oil Regulation Report,* vol. 40, March 13, 1973, pp. 2–3 (statement of A. W. Rutter, Jr., representing TIPRO); ibid., March 15, 1973, p. 2; and ibid., March 23, 1973 (statement of George Mitchell, president of TIPRO).

180. A recently adopted Senate rule allowed such requests. The Senate debate on the bill had already been delayed by a bomb threat, but Senator McKnight's letter requesting the typed transcripts was described as the real bombshell of the day. Before the bomb scare, the Senate heard hyperbolic denunciations of the bill by independent producers who likened it to "giving three-fourths of a jury the right to condemn a man to death" and as a scheme by the major oil companies to "get back what the Arabs are taking away from them by taking it away from us." See *Oil Regulation Report,* vol. 41, May 3, 1973, pp. 1–3.

181. The May 19, 1973, vote to suspend was 17 yea, 13 nay (one present not voting) (Legislative Reference Service, Austin, Texas).

182. By contrast, opposition and defeat of California's proposed compulsory unitization bill in 1956 was understandably based on the bill's failure to safeguard the public interest from the possible monopolistic effects of unitization. The California bill vested almost total control of unitized operations in the unit operators rather than in a state agency. See Comment, "Conservation and Price Fixing in the California Petroleum Industry."

183. *Oil Regulation Report,* vol. 41, March 7, 1973, p. 3 (testimony of Robert Payne, who stated that a change of 1 percent in the allocation formula for East Texas meant $70 million to Humble). Also see *Oil Regulation Report,* vol. 41, March 7, 1973, p.4; ibid., March 13, 1973, pp. 1–2; ibid., March 14, 1973, pp. 4–5 (efforts by opponents to raise the 75 percent voluntary approval required to 85 percent, and to prohibit the bill from altering allocation formulas in effect in fields for ten years or more); and ibid., March 20, 1973, pp. 1–3.

184. *Oil Regulation Report,* vol. 41, March 13, 1973, p. 3 (statement by F. Hoaglund of Humble Oil); ibid., April 13, 1973, p. 3 (proposed amendment to exempt East Texas from the bill was offered but failed after Rep. Dave Finney stated that the East Texas field would not be unitized anyway).

185. *Oil Regulation Report,* vol. 41, April 13, 1973, p. 3 (testimony of Representative Williamson of East Texas).

186. Section 14 of H.B. 311 provided absolute antitrust immunity to unitization agreements made pursuant to the bill. Attorney General Hill stated that Texas could validly exempt the unitization plans permitted by H.B. 311 from the state's antitrust laws, although he suggested that the legislature consider adopting the language of the 1949 Act (see text of this chapter at note 70) which gave only conditional immunity. See Texas Attorney General LA-41 (Letter Advisory) (1973) pp. 113–114 (copy in author's file). The attorney general also advised that H.B. 311 was not an unconstitutional taking of private property for public use because the bill provided that the nonsigning, working interest owner could assign or sell his or her interest to the unit and thereby be relieved of unitized

operations. In the attorney general's opinion, the bill's procedures for establishing adequate compensation for the sale prevented a "taking in the literal sense of the word although the practical effects could in some instances be the same." This less-than-wholehearted blessing of the bill clearly reflected the attorney general's own political instincts. One opponent testified that if the bill passed, an amendment should be added to "change the name of Texas to Exxon" so everyone would understand the full intent of the bill, calling it "the same bill major oil companies have tried to get through for 30 years." See *Oil Regulation Report,* vol. 41, April 12, 1973, p. 9.

187. The attorney for the city of Kilgore and the Kilgore school district testified against the bill, as did a director of the Oil, Chemical, and Atomic Workers Union. Robert Payne, an attorney and independent producer, testified that seven out of eight wells in the East Texas field would be plugged by the bill and thousands of workers unemployed. *Oil Regulation Report,* vol. 41, March 15, 1973, pp. 2–3.

188. State Representative Milton Fox, who strongly supported the bill, estimates that about half the royalty interest owners in the state opposed the bill (telephone interview with the author January 27, 1983).

189. App. III of this book contains data on Rule 37 permits. The large increase in Rule 37 permits in 1973 versus 1972 reflects the jump in crude oil prices at this time. Between 1957 and 1969, an average of only 1,474 Rule 37 exception requests were granted per year (data from the Railroad Commission files).

190. State Representative Milton Fox continues to file a compulsory bill at every legislative session, but without any expectation of attention to the issue (telephone interview with author January 27, 1983).

191. *Oil Regulation Report,* vol. 41, December 14, 1973.

192. Governor's Energy Advisory Council, *Texas Energy Policy* (Austin, Tex., 1977) p. 14. The 1978 update of this document echoed this view. See Texas Energy Advisory Council, *Texas Energy Policy: 1978 Update* (Austin, Tex., 1978) p. 13. See also Joe Ventura, "Energy Policy Development in Texas," *Baylor Law Review* vol. 29, no. 4 (1977) pp. 821–846; and Askew, "Texas Energy Policy." In 1979 energy and natural resources policy development was consolidated in the Texas Energy and Natural Resources Advisory Council (TENRAC). TENRAC chose not to study the unitization issue (letter to the author from T. J. Taylor, director of Policy Analysis Division of TENRAC, dated May 18, 1982; see also Texas Sunset Advisory Commission, *Final Report on Energy Regulatory Agencies* (Austin, Tex., 1983) pp. 73–106). The 1980 election of Texas' first Republican governor in more that a century did not change this policy toward unitization, despite Governor Clements' strong interest in efficiency and long-term planning and policy development to ensure Texas' continued economic growth. Clements established the Texas 2000 Commission to examine various approaches for meeting statewide problems through the year 2000 in nine critical areas, including energy. The report is completely silent on the issue of unitization of oil and gas fields (*Texas 2000 Report*).

193. Allegations that producers were deliberately producing at less than the maximum efficient rate (MER) in order to speculate on higher prices led to passage of a section in the Energy Policy and Conservation Act of 1976 which empowers the President of the United States to require production from domestic wells at their MERs in times of severe energy supply disruptions. The Federal Power Commission also asserted authority over production rates of gas wells. See Dan A. Bruce, "'Maximum Efficient Rate'—Its Use and Misuse in Production Regulation," *Natural Resources Lawyer* vol. 9, no. 3 (1976) pp. 441–443 and 451–452.

Commissioner Langdon's reaction to the federal assertion of power over MERs was as follows: "The Railroad Commission is not interested in defying the federal government. If the President tells us we can produce more oil, and if he knows what the MER is, . . . we will review the reservoir situation, and if we conclude that we can produce more oil without waste, we will do it. But if we conclude from our engineering evidence that it would result in damage to the reservoirs . . . I think we would make a court require us to do it." See *Oil Regulation Report,* vol. 41, November 13, 1973.

The federal government's failure to preempt state conservation laws and force states to adopt strong compulsory unitization laws during the energy crisis is discussed in ch. 10 text at notes 34 to 43.

194. The failure of the legal system to protect the correlative rights of royalty interest owners and lessees of voluntary subdivisions had not moved the legislature to pass a compulsory pooling statute. In Japhet v. McRae, 276 S.W. 669 (Tex. Comm'n App. 1925, opinion adopted), the court held that the royalties from a tract of land leased as a whole were not to be shared upon the subsequent conveyance of part of the tract to a new owner. In this postlease conveyance situation, only the owner of that part of the tract containing the producing wells was entitled to receive royalties. Without compulsory pooling, the owners of other parts of the original tract would suffer drainage without remedy. Their plight did not result in legislative reform, perhaps because royalty owners were unorganized politically, and perhaps also because the court's decision was made at such an early date that most informed landowners could thereafter draft provisions to avoid this result. See William O. Huie, "Apportionment of Oil and Gas Royalties," *Harvard Law Review* vol. 78, no. 6 (1965) pp. 1113–1145.

In Ryan Consolidated Petroleum Corp. v. Pickens, 155 Tex. 221, 285 S.W.2d 201 (1955), the Texas Supreme Court refused to allow the lessee of a small tract who was not entitled to a Rule 37 exception permit (because the tract had been subdivided by lease after oil and gas were discovered in the vicinity) to share in the proceeds of an adjacent well that was indisputedly draining the lessee. This inequitable situation did not lead to legislative reform via a compulsory pooling statute. The situation at hand did not pit a large-tract owner against a small-tract owner, but involved the division of production between two small tracts, only one of which could receive a Rule 37 permit. The Railroad Commission had established an informal, but well-known rule to determine which small tract received the well permit: the first lessee got the well. If a second operator was careless enough to take a lease on the adjoining subdivided tract, neither the court nor the legislature felt compelled to protect him or her from bad judgment.

195. Well spacing helps to prevent physical waste by reducing risks of fire and blowouts and by retarding the channeling of water and dissipation of gas, in addition to preventing the economic waste of unnecessary drilling and operating expenses. Pooling allows well-spacing rules to withstand constitutional attack by protecting the correlative rights of all owners in the spacing unit and thus indirectly serves to prevent waste. See Hunter Co. v. McHugh, 202 La. 97, 11 So.2d 495 (1942) (upholding compulsory pooling order as a waste prevention measure). Of course, in the long term, lower operating costs in fields with widely spaced wells will prolong the field's productive life and increase ultimate recovery, just as unitization does.

196. Atlantic Refining Co. v. Railroad Comm'n, 162 Tex. 274, 346 S.W.2d 801 (1961) [hereafter cited as the *Normanna* case].

197. Halbouty v. Railroad Comm'n, 163 Tex. 417, 357 S.W.2d 364 (1962), *cert. denied,* 371 U.S. 888 (1962) [hereafter cited as the *Port Acres* case]. The evidence showed that 22 small tracts with 1.77 percent of the producing sand would recover 20.46 percent of the field's total gas and condensate potential (ibid., p. 371). Many attorneys seem to have considered the *Normanna* case to be an isolated special situation (ibid., p. 376).

198. The total cost of a well in the Port Acres field, including drilling, operating, and royalty expenses was $361,674 (ibid., p. 371). Under the one-third, two-thirds formula, the well would produce a total income of $407,501, leaving a none-too-fat profit margin of $45,827.

199. The majors and large established independents started a campaign against small-tract drilling in the late 1950s. See Prindle, *Petroleum Politics,* pp. 73–78.

200. For example, Peter Henderson Oil Co., one of the plaintiffs in the Port Acres field, had pooled several hundred small tracts into three large units. See 357 S.W.2d 369.

201. Relations between the plaintiffs and defendants in the *Normanna* case were openly hostile. The plaintiffs had sought to intimidate drilling contractors from drilling on the small tract. See 346 S.W.2d 816; and Atlantic Refining Co. v. Bright & Schiff, 321 S.W.2d 167 (Tex. Civ. App. 1959).

202. 357 S.W.2d at 367–368.

203. 346 S.W.2d 811. The Hawkins and Yates cases are discussed in the text of this chapter at notes 117 to 122.

204. 380 S.W.2d 599 (Tex. 1964). The civil appeals court had not found any unreasonable delay by plaintiffs in attacking the prorationing order and had invalidated it under the *Normanna* precedent. Railroad Comm'n v. Aluminum Co. of America, 368 S.W.2d 818 (Tex. Civ. App. 1963).

A timely attack on the 50–50 formula used in oil fields resulted in judicial invalidation of this formula for new fields. Railroad Comm'n v. Shell Oil Co., 380 S.W.2d 556 (Tex. 1964).

205. Atlantic Refining Co. v. Railroad Comm'n, 330 S.W.2d 494 (Tex. Civ. App. 1959); and Halbouty v. Darsey, 326 S.W.2d 528 (Tex. Civ. App. 1959). In

the first case, Atlantic Refining Co. had offered to pool the small tract and give appellees their *pro rata* share of gas without cost to them.

206. Halbouty v. Railroad Comm'n, 357 S.W.2d 364, 367–368.

207. Atlantic Refining Co. v. Railroad Comm'n, 346 S.W.2d 801 at 814–823; Halbouty v. Railroad Comm'n, 357 S.W.2d 364 at 376–383. The dissent also found substantial evidence to support the one-third per-well factor because it compensated operators for the high risks of drilling. 346 S.W.2d 823.

208. The *Normanna* proration order is quoted in the *Port Acres* case, 357 S.W.2d 364 at 377, nn. 1–4. The supreme court in the *Port Acres* case implied approval of a special allowable system which would induce large-tract owners to pool with small tracts. 357 S.W.2d 364 at 376.

209. The pooling act now appears at Texas Natural Resources Code Ann. secs. 102.001–102.112 (Vernon 1978).

210. Hardwicke proposed that the Railroad Commission deny Rule 37 exception requests to tract owners who refused to pool. See text of this chapter at notes 106 to 107. Denial of a special allowable to those who refuse to pool is not as coercive a pooling device as Hardwicke's proposal. In some instances the small-tract owner might still find it profitable to drill without a special allowable as long as he can obtain a Rule 37 exception.

211. The commissioners viewed the 1946 Hawkins field case as a judicial mandate to include large per-well factors in prorationing formulas and thus blamed the courts for creating the prorationing mess. But the *Hawkins* court was only affirming an already established Railroad Commission policy of including large per-well factors.

212. Langdon, "The Influence of Court Decisions Upon Railroad Commission Policy," p. 185.

213. For example, the constitutionality of granting special allowables that permitted uncompensated drainage was unclear, as was the status of royalty owners who refused to pool, despite their lessee's willingness to pool. Also, the special allowable system could not induce pooling in a *Ryan v. Pickens* situation (see note 194 in this chapter) where a small-tract owner had no right to a Rule 37 permit, or in a situation where the small-tract lessee was unable to secure financing for his or her own well. See Dick H. Gregg, "Dialogue on the Problems of Large Tract and Small Tract Operators and Royalty Owners," *South Texas Law Journal* vol. 7, no. 2 (1963) pp. 85–99; and Hardwicke and Woodward, "Fair Share and the Small Tract in Texas."

214. Texas Natural Resources Code Ann. secs. 102.001–102.112 (Vernon 1978). A compulsory pooling bill drafted by CEDOT (the Committee for Equitable Development of Texas Oil and Gas Resources), an organization of about 85 medium- to large-sized companies and independent operators, was introduced into the Texas House during the 1963 legislative session, but was never brought to a vote and lacked widespread support among independents. TIPRO drafted a bill in 1964 to present to the legislature in the 1965 session. See Charles E. Wallace,

"The Proposed Compulsory Pooling Bill," *Baylor Law Review* vol.16, no. 3 (1964) pp. 234–235.

215. Because the General Land Office feared that the state administrative boards which lease state-owned lands were insufficiently staffed with geologists and other expert advisors who could assess the benefits and costs of entering into pooling agreements, state lands were excluded from the scope of the Mineral Interest Pooling Act. However, with the consent of or at the instance of the commissioner of the General Land Office, or any board or agency having jurisdiction over state-owned land, such land may be pooled under the provisions of Section 102.004 of the act. Thus, pooling proposals involving state land are judged by both the General Land Office and the Railroad Commission.

216. For a discussion of similar provisions in the context of joint operating agreements, see Lee Jones, Jr., "Problems Presented by Joint Ownership of Oil, Gas, and Other Minerals," *Texas Law Review* vol. 32, no. 6 (1954) pp. 723–727.

217. In Railroad Comm'n v. Coleman, 460 S.W.2d 404 (Tex. 1970), Coleman leased his interest in a 240-acre tract to Boyd. Thereafter, Coleman sold 183 acres of the tract to Ashford. Boyd drilled a well on Coleman's tract, on an 80-acre proration unit consisting of 35 of Coleman's acres and 45 of Ashford's acres. Under Texas common law, Ashford received no royalties from this well. Ashford applied for a compulsory pooling order and the Railroad Commission granted the application. The supreme court overturned the Railroad Commission order, holding that only owners who have drilled or propose to drill could apply for a pooling order under the terms of the act. Royalty owners have no right to drill or propose drilling. While the court recognized that the common law rule bred great inequity, the legislative intent of the pooling act was "to save owners and lessees of small tracts from the devastating effect of those [the *Normanna* and *Port Acres*] decisions," not to aid royalty interest owners (ibid., p. 408).

218. A forced pooling order usually takes a minimum of six months to secure. Rule 37 exception permits can usually be obtained in less than a week. Staff of the Oil and Gas Division of the Railroad Commission, personal interviews with the author, January 13–14, 1983.

219. In 1917, 1930, and 1931, the Texas legislature also sought to provide equal access to pipelines by passing laws that conferred common carrier status on oil and gas pipelines and that authorized the commission to require ratable, nondiscriminatory purchases by pipeline owners.

220. See, for example, Ely, "The Conservation of Oil" p. 1166 (per-well allowable system did not result from conscious design of commission, but was imposed on them by the courts in the Hawkins field case and the Marginal Well Act); Davis, "Judicial Emasculation," p. 37 (ineffective regulation due to interference and ignorance of inexpert judges); Hardwicke, "Oil-Well Spacing," p. 116 (many mistakes made in East Texas); and Zimmerman, *Conservation*, p. 71 (excess capacity developed in oil fields because of unpredictable discoveries and uncontrollable forces).

221. See also Gary D. Libecap and Steven N. Wiggins, "Contractual Responses to the Common Pool: Prorationing of Crude Oil Production," *American Economic Review* vol. 74, no. 1 (1984) pp. 87–97.

222. Zimmerman, *Conservation*, pp. 275–279 and 281–325. Zimmerman does admit that there are some "weaknesses and unresolved problems" in the oil- and gas-conservation framework, such as inequitable allocation formulas, excessive well drilling, and the lack of compulsory pooling and unitization (ibid., pp. 326–349), but he considers the achievements to overshadow the weaknesses. Zimmerman concludes that compulsory unitization should be treated "as a last resort" (ibid., p. 348) and that the existing prorationing system has the advantage of safeguarding "a maximum of individual freedom" with the "least possible interference with private enterprise and initiative" (ibid., p. 345). Zimmerman's failure to explain and reluctance to criticize the lack of compulsory pooling and unitization in Texas seriously mars his work.

223. Section 85.056 of the Texas Natural Resources Code states that "the commission shall take into consideration and protect the rights and interests of the purchasing and consuming public in oil and all its products." Arguably, this section delegates broad power to the commission to control oil and gas prices and regulate competition in the industry, but the legislative history of this section shows that it was inserted into the 1932 Market Demand Act merely to assure that the commission did not use market demand prorationing to raise oil prices too much. See ch. 3 text at note 126. The commission has never claimed to have any statutory authority to regulate prices or competition, nor have the courts ever recognized any such power in the commission. In contrast, federal lawmakers delegated to the Federal Power Commission (now the Federal Energy Regulatory Commission) the express power to regulate natural gas prices "in the public interest," and this was held to require that the federal agency specifically assess the effect of its rate orders on the state of competition in the gas industry, particularly its effect on independent gas producers. See Southern Louisiana Area Rate Cases v. Federal Power Comm'n, 428 F.2d 407 at 441–442 (5th Cir. 1970).

224. In Woods Exploration and Producing Co. v. Aluminum Co. of America, 438 F.2d 1286 (5th Cir. 1971), the court adopted the rationale that the individual pool is the relevant market area for defining monopoly power because of the rule of capture. This case is fully discussed in chapter 8 text at notes 143 to 168.

The existence of local monopoly power by an integrated company that owns the only pipeline into a field will have an effect primarily on the division of the economic rents from the field between the producers and the pipeline company. In the absence of statewide or regional monopoly power over pipelines or crude oil production, vertically integrated majors will not be able to influence the price of crude oil to the ultimate consumer. Nonetheless, the antitrust laws are not restricted to protecting firms only against anticompetitive acts which affect consumer prices. Also, the public interest may include considerations of equity in the fair division of the economic rents of a field, regardless of the presence or absence of an effect on consumer prices or efficiency.

Chapter 5

1. Most of these laws prohibiting the waste of oil and gas are now codified in chapters 85 and 86 of the Texas Natural Resources Code Annotated. Appendix I of this book contains section 85.046, the statute defining waste in the oil conservation laws.

2. Article 6014(g) now appears in the Texas Natural Resources Code Ann. sec. 85.046(7) (Vernon Supp. 1982).

3. Field rules are published in *Rules and Regulations: Texas Railroad Commission* (Austin, Tex., R. W. Byram, 1958—) [hereafter cited as *Byram's Rules*]. Statewide orders are published in the Texas Administrative Code tit. 16, secs. 3.1–3.71 (1982). Many statewide orders are still commonly referred to by the numbering adopted in 1919 when the Railroad Commission promulgated thirty-eight statewide rules. For example, the statewide spacing rule is still called Rule 37.

4. Texas Natural Resources Code Ann. sec. 86.181 (Vernon 1978) requires that gas from a gas well be used for (1) light or fuel, (2) chemical manufacture, (3) pressure maintenance or cycling, or (4) extraction of natural gasoline when the residue is returned to the reservoir.

5. David F. Prindle, *Petroleum Politics and the Texas Railroad Commission* (Austin, University of Texas Press, 1981) p. 60.

6. Clymore Production Co. v. Thompson, 11 F. Supp. 791 (W.D. Tex. 1935); 13 F. Supp. 469 (W.D. Tex. 1936).

7. R. W. Byram & Co., *Texas Oil and Gas Handbook* (Austin, Tex., R. W. Byram, 1982) p. 144.

8. Prindle, *Petroleum Politics,* pp. 60–61.

9. Railroad Comm'n Order No. 20-550, Jan. 18, 1939.

10. Blakely M. Murphy, ed., *Conservation of Oil and Gas: A Legal History, 1948* (Chicago, Ill., American Bar Association, 1949) p. 473 [hereafter cited as Murphy, *Legal History, 1948*].

11. Kenneth Culp Davis and York Y. Willbern, "Administrative Control of Oil Production in Texas," *Texas Law Review* vol. 22, no. 2 (1944) pp. 149–193.

12. Ibid., p. 193.

13. Robert E. Hardwicke, *Antitrust Laws v. Unit Operation of Oil or Gas Pools* (rev. ed., Dallas, Tex., Society of Petroleum Engineers of the American Institute of Mining and Metallurgical Engineers, 1961) pp. 105–106.

14. This is Order No. 3-6475 of July 14, 1944, as cited in Murphy, *Legal History, 1948,* p. 472. See also *Byram's Rules,* vol. 2, sec. IV, p. 909.

15. Murphy, *Legal History, 1948,* p. 472.

16. Order No. 4-6,636, effective August 28, 1944, in *Byram's Rules,* vol. 7, sec. VII, pp. 6–8.

17. Order No. 4-7,889, effective July 16, 1945, in *Byram's Rules,* vol. 7, sec. VII, pp. 10–14.

18. Henrietta M. Larson and Kenneth W. Porter, *History of Humble Oil and Refining Company: A Study in Industrial Growth* (Salem, N.H., Ayer, 1959) p. 650.

19. John R. Stockton, Richard C. Henshaw, and Robert W. Graves, *Economics of Natural Gas in Texas* (Austin, The University of Texas, 1952) p. 235. From 1936 to 1941, about 50 percent of the casinghead gas produced was flared.

20. Prindle, *Petroleum Politics,* pp. 64–65; and note 18 in this chapter.

21. Davis and Willbern, "Administrative Control of Oil Production in Texas," pp. 149–193.

22. Prindle, *Petroleum Politics,* p. 64. Commissioner Beauford Jester was elected governor of Texas and he appointed Murray to fill his unexpired term. As governor, Jester spoke approvingly of unitization, and a bill authorizing voluntary unitization agreements almost passed in 1947. See Hardwicke, *Antitrust Laws,* p. 149.

23. This fact is noted in Railroad Comm'n v. Shell Oil Co., 146 Tex. 286, 206 S.W.2d 235 at 243 (1947).

24. Ibid., p. 240.

25. Ibid., p. 237.

26. The case was dismissed on March 15, 1948. See American Petroleum Institute, *Secondary Recovery of Oil in the United States* (2 ed., Dallas, Tex., American Petroleum Institute, 1950) p. 60.

27. Murphy, *Legal History, 1948,* pp. 472–473. The attorney general refused to pass on the question of whether the casinghead gas facilities violated the antitrust laws. No suit alleging a violation was ever brought.

28. James E. Nugent, "The History, Purpose, and Organization of the Railroad Commission," in *Oil and Gas: Texas Railroad Commission Rules and Regulations* (Austin, State Bar of Texas, 1982) p. A-28.

29. Railroad Comm'n v. Sterling Oil & Refining Co., 147 Tex. 547, 218 S.W.2d 415, 419 (1949).

30. Commissioner Thompson's testimony on the agency's willingness to grant exceptions to such wells reflected the same liberality that dispensed Rule 37 exception wells. Thompson admitted, "Our whole fault all along is being too liberal in our administration" (ibid., p. 421).

31. Ibid., pp. 426–428. Stockton has also remarked that the commission's campaign against flaring often required investments in casinghead gas plants which realized only very small returns initially. See Stockton, Henshaw, and Graves, *Economics of Natural Gas,* p. 233.

32. Railroad Comm'n v. Flour Bluff Oil Corp., 219 S.W.2d 506 (Tex. Civ. App. 1949).

33. Ibid., p. 508. Using the economist's definition of waste, the flaring of casinghead gas would not be wasteful if the cost of avoiding the flaring was greater than the value of the flared gas. The Flour Bluff case seems to evidence a commission preoccupation with maximizing the physical recovery of oil rather than its efficient recovery. However, within a few years, most cycling plants were profitable. The commission was careful to issue no-flare orders only when they believed cycling to be economic. See text of this chapter at notes 101 to 105.

34. Railroad Comm'n v. Flour Bluff Oil Corp., 219 S.W.2d 506 (Tex. Civ. Appl. 1949) p. 509. It seems odd that Humble attacked these waste-prevention orders. Under Farish's leadership, Humble had been at the forefront of the conservation movement to require repressuring. See ch. 3 text at notes 87 to 92. Hines Baker, the president of Humble in 1948–49, explained that the only real difference between the commission and the company on the issue of preventing waste was one of timing. Humble felt that it was investing huge sums of money in casinghead gas and cycling plants as quickly as humanly possible in the post-war era. See Larson and Porter, *History of Humble Oil,* pp. 651–653.

35. Larson and Porter, *History of Humble Oil,* p. 652.

36. These facts are found in Railroad Comm'n v. Rowan Oil Co., 152 Tex. 439, 259 S.W.2d 173, 175 (1953). The Spraberry trend is about 55 miles long and 20 miles wide. A detailed account of the field's history appears in Nelson Jones, "The Spraberry Decision," *Texas Law Review* vol. 32, no. 6 (1954) pp. 730–739.

37. Railroad Comm'n v. Rowan Oil Co., 152 Tex. 439, 259 S.W.2d 173, 176 (1953). The court cited Texas Revised Civil Statutes Ann. art. 6008, secs. 10, 11 (Vernon Supp. 1962), now codified at Texas Natural Resources Code Ann. secs. 86.081–86.083 (Vernon 1978) and (Vernon Supp. 1982), quoted in the text of chapter 3 at note 161.

38. Indeed, in the Seeligson field case, the court held that Article 6008 pertained primarily to gas wells and relied only on Articles 6014 (now Section 85.046) and 6029 (now Section 85.202) to sustain the commission's shutdown order in this oil field. Railroad Comm'n v. Shell Oil Co., 146 Tex. 286, 206 S.W.2d 235, 239 (1947).

39. 259 S.W.2d 177.

40. Jones, "The Spraberry Decision" p. 738. The financial hardships to both the flarers and the nonflarers occasioned by the Spraberry orders are described in ibid., pp. 731–732 and 739. Some producers even lost their leases. See Haby v. Stanolind Oil & Gas Co., 228 F.2d 298 (5th Cir. 1955).

41. See ch. 3 text at notes 169 to 173. Of course, although the oil prorationing statutes do not specifically authorize prorationing to protect correlative rights, the commission has a duty to consider the impact of its orders on correlative rights so as to avoid the confiscation of property which might render its orders unconstitutional.

42. Texas Natural Resources Code Ann. sec. 85.054 (Vernon 1978).

43. Not surprisingly, several legislative attempts were made from 1947 onward to eliminate the substantial evidence rule and require courts to review commission orders under stricter tests. See Murphy, *Legal History, 1948,* pp. 458–459. In 1959 a bill requiring that disputed facts be resolved by the preponderance of evidence rule was passed, but vetoed by the governor. See Robert E. Sullivan, ed., *Conservation of Oil and Gas: A Legal History, 1948–1958* (Chicago, Ill., American Bar Association, 1960) p. 225 [hereafter cited as Sullivan, *Legal History 1958*].

44. Stockton, Henshaw, and Graves, *Economics of Natural Gas,* p. 235. Today, less than one-half of one percent of all the gas produced in Texas is flared. Railroad Commission of Texas, *1981 Annual Report of the Oil and Gas Division* (Austin, Tex. 1981) p. 31.

45. For example, the Heyser field is not yet unitized; the Spraberry field was not unitized until the 1960s. Railroad Commission of Texas, *Secondary and Enhanced Recovery Operations in Texas to 1980: Bulletin 80* (Austin, Tex., 1980) pp. 10, 100, and 148 [hereafter cited as RRC, *Bulletin 80*]. The Conroe field was one of the sixteen fields in the 1948 order. Its gas was marketed; the field was not unitized until 1977 (see text of this chapter at notes 66 to 68). The Seeligson field operators did seek the attorney general's approval of a cooperative agreement in 1947, however.

46. Hardwicke, *Antitrust Laws,* p. 146.

47. RRC, *Bulletin 80,* p. 20.

48. Letter from Fred Young, general counsel to the Oil and Gas Division of the Railroad Commission from 1948 to 1978, to the author, dated February 8, 1983.

49. RRC, *Bulletin 80,* p. 160.

50. Hardwicke, *Antitrust Laws,* p. 124. The orders no longer appear in *Byram's Rules* because they were rescinded when the fields were unitized. Hardwicke describes the Fort Chadbourne order as requiring that the wells be shut in on February 1, 1952, unless the operators proved that a repressuring program had been implemented or was not feasible.

51. William J. Murray, Jr., "Engineering Aspects of Unit Operations," *Institute on Oil and Gas Law and Taxation,* vol. 3 (Albany, N.Y., Matthew Bender, 1952) pp. 17–18.

52. RRC, *Bulletin 80,* pp. 94, 98.

53. William Murray, former Texas railroad commissioner, telephone interview with author, February 4, 1983.

54. Ibid.

55. Ibid.

56. Paul Hull, "SACROC—An Engineering Conservation Triumph," *Interstate Oil Compact Commission Bulletin* vol. 12, no. 1 (1970) p. 30. The Kelly–Snyder field is located in Scurry County and produces from the Canyon Reef reservoirs;

hence the acronym SACROC—the Scurry Area Canyon Reef Operators' Committee.

57. Ibid., p. 31.

58. William Murray, former Texas railroad commissioner, telephone interview with author, February 4, 1983.

59. Sullivan, *Legal History 1958,* p. 227.

60. The original water-flooding operation succeeded in recovering 44 percent of the original oil in place—about 700 million barrels more than would have been recovered without SACROC. The unit now is injecting carbon dioxide into the field to recover another 150 million barrels of oil at a cost of about $175 million. See Hull, "SACROC," p. 33.

61. William Murray, former Texas railroad commissioner, telephone interview with author, February 4, 1983.

62. Ibid.

63. The commission uses two types of hearings to bring regulatory problems to the attention of operators. The "commission called" hearing is a "trial balloon" which gives parties an opportunity to testify about possible regulatory changes initiated by the commission. It threatens no sanctions against producers. The "show cause" hearing requires operators to appear and show cause why an action proposed by the agency should *not* be taken. The agency already has gathered information supporting its proposed action. This hearing threatens sanctions unless the commission is satisfied that the problem has been solved. See Jim C. Langdon, "Rules of Regulatory Bodies: Texas," *National Institute for Petroleum Landmen,* vol. 7 (Albany, N.Y., Matthew Bender, 1966) pp. 159 and 173–174.

64. David Gross, "Unitization: The Fair Advantage," *National Institute for Petroleum Landmen,* vol. 8 (Albany, N.Y., Matthew Bender, 1967) pp. 67, 98.

65. Wallace F. Lovejoy and Paul T. Homan, *Economic Aspects of Oil Conservation Regulation* (Baltimore, Md., Johns Hopkins University Press for Resources for the Future, 1967) pp. 81–82. Also see Comment, "Prospects for Compulsory Fieldwide Unitization in Texas," *Texas Law Review* vol. 44, no. 3 (1966) pp. 529–530.

66. RRC, *Bulletin 80,* p. 42.

67. George Singletary, technical examiner, Texas Railroad Commission, personal interview with author, January 13, 1983.

68. Order No. 3–68,046 in *Byram's Rules,* sec. IV, p. 1129.

69. The large-tract owners in the Yates field unsuccessfully had sought to change the old prorationing formula as early as 1945. See Standard Oil Co. of Texas v. Railroad Comm'n, 215 S.W.2d 633 (Tex. Civ. App. 1948).

70. In 1967 Marathon Oil Co., the largest operator in the Yates field, had applied for an increase in the MER. This request was opposed by some operators who wanted a revision in the proration formula before the field's allowable was

raised. In 1969 applications to revise the proration formula were filed. In 1971 six months of hearings were held on these applications. For the next five years, the Railroad Commission refused to act further on the prorationing applications. When the energy crisis hit in 1973 and crude oil prices started to soar, a larger MER became even more desirable to producers. Under the federal pricing rules, Marathon anticipated an average price increase of $3 per barrel if the field were unitized and the MER increased. The lure of these profits finally broke the operators' stalemate over a fair allocation formula for the field's output. See *Secondary Recovery Railroad Commission Application Summaries 1976* (Austin, Texas State House Reporter, Inc., 1976) pp. 175–178. For further discussion of the Yates field unitization, see ch. 6 text at notes 40 to 49.

71. RRC, *Bulletin 80,* p. 158. The MER was raised from 50,000 barrels per day to 100,000 barrels per day. See *Secondary Recovery Railroad Comm'n Summaries 1976,* p. 182. Similarly, the commission increased the MER of the Tom O'Connor (5800) field from 15,000 barrels per day to 18,000 barrels per day when it was unitized in June 1976 (ibid., p. 16).

72. Hull, "SACROC" p. 31. Also see ch. 6 text at note 24 on the commission's incentive allowables for SACROC's recent tertiary recovery project.

73. Many wells involved in water flooding in nearly depleted fields would qualify as marginal wells under the Marginal Well Act and would be exempt from prorationing in any case. Capacity water-flood allowables are authorized by statewide Rule 48, Texas Administrative Code tit. 16, sec. 3.48 (1982), but are granted only after hearings for each individual field.

74. See Lovejoy and Homan, *Economic Aspects,* pp. 200-201; and Wallace F. Lovejoy, "Oil Conservation, Producing Capacity, and National Security," *Natural Resources Journal* vol. 10, no. 1 (1970) p. 68 [hereafter cited as Lovejoy, "Producing Capacity"]. Many of the capacity water-flood fields operated above the break-even point, but it would have imposed too great an administrative burden on the commission to approve only that amount of incentive allowable which would cover the costs of water flooding in hundreds of small fields.

75. Stephen L. McDonald, *Petroleum Conservation in the United States: An Economic Analysis* (Baltimore, Md., The Johns Hopkins University Press for Resources for the Future, 1971) pp. 185–186.

76. As a rule of thumb, the commission considered fields that were 75 percent depleted as secondary recovery projects eligible for capacity allowables. Repressuring in less-depleted fields was considered pressure maintenance. See Brooks Peden, *The Texas Oil and Gas Hearing Aid: Evidence, Procedure, and Practice for Oil and Gas Hearings of the Railroad Commission of Texas* (Austin, R. W. Byram & Co., 1961) pp. 55 and 59–60 [hereafter cited as Peden, *Oil and Gas Hearing Aid*]. Peden was a hearing examiner for the Railroad Commission for many years and wrote this book while employed at the agency. The book was not, however, officially sponsored by the commission. It is useful as a reflection of one experienced examiner's views of the workings of the agency and the industry it serves.

77. Ibid., pp. 23, 41, and 64. Statewide Rule 49, Texas Administrative Code tit. 16, sec. 3.49 (1982) establishes a gas–oil ratio of 2,000 cubic feet of gas to each barrel of oil. Oil wells with higher ratios are restricted from producing their full allowable. For example, if the limiting ratio for the field is 2,000 cubic feet per 1 barrel of oil, and the maximum allowable for that field is 60 barrels per well, then a well can produce no more than 120,000 cubic feet of gas daily without penalty. If a well in such a field produced with a ratio of 3,000 per 1, its oil allowable would be reduced to the point where it produced only 120,000 cubic feet of gas. Instead of its 60-barrel allowable it would be granted only 40 barrels. This penalty can be avoided by applying for a net gas–oil ratio which deducts reinjected gas from the gas–oil ratio.

78. Limitations on transfers were commonly imposed to protect correlative rights, such as a rule that the allowable for any single well could not exceed twice the top allowable of any well in the field, and restrictions on transfers to wells close to an offset operator's lease lines. See Peden, *Oil and Gas Hearing Aid,* pp. 46–53.

79. Unless otherwise noted, this account of the East Texas saltwater repressuring program is taken from Staff of the East Texas Salt Water Disposal Co., *Salt Water Disposal: East Texas Oil Field* (2 ed., Austin, Petroleum Extension Service, University of Texas, 1958) pp. 1–20 and 113–116.

80. Five of the defendant oil companies appealed the judgment against them, without success. Goldsmith & Powell v. State, 159 S.W.2d 534 (Tex. Civ. App. 1942).

81. Sullivan, *Legal History 1958,* pp. 477–478. In 1973 Exxon proposed expansion of this off-lease transfer allowable system, but without success (see ch. 10 text at notes 25 to 26).

82. Sullivan, *Legal History 1958,* p. 476.

83. This happy state of affairs may not last forever. The same type of bonus-allowable rule was the subject of litigation in Texaco, Inc. v. Railroad Comm'n, 583 S.W.2d 307 (Tex. 1979). In 1954 three operators in the Fig Ridge field entered into a cooperative pressure maintenance project for which the Railroad Commission granted bonus allowables. In 1973, after almost twenty years of operation, the injection wells along the western edge had created an area of high pressure, and this was causing oil to be pushed toward the east. Texaco was losing oil to Sun Oil. Texaco applied to the Railroad Commission for a suspension of the bonus-allowable rule on the basis that it violated Texaco's correlative rights by causing net uncompensated drainage. The commission refused to suspend the rule after finding that an additional 800,000 barrels of oil would be produced as a result of the rule. The Texas Supreme Court affirmed the agency's decision. Similar litigation in the East Texas field may eventually occur. In 1978 the commission staff recommended elimination of the bonus-allowable rule, but the commission did not act on this staff recommendation. See the ch. 10 text at notes 27 to 28.

Net gas–oil ratio rules can also affect correlative rights. In 1982 the Railroad Commission held a hearing to consider suspending statewide and fieldwide net gas–oil ratios and the increased allowables associated with the rules. The hearing examiners found no persuasive evidence to indicate that the ratios were harmful to correlative rights and found that physical waste might be caused by the suspension. The commissioners accepted the examiners' recommendation to take no action, but also decided to keep the record open to study the proposal further. See Railroad Commission Legal Memorandum re: Hearing to Consider Suspension of Net Gas–Oil Ratios, September 27, 1982 (copy in author's files); and "TRC Denies Suspension of Net Gas–Oil Ratio," *Drill Bit* (January) 1983, p. 13.

84. Article 6014(g), now Texas Natural Resources Code Ann. sec. 85.046(7) (Vernon Supp. 1982).

85. William Murray, former Texas railroad commissioner, telephone interview with author, February 4, 1983.

86. Some commentators have urged conservation commissions in states without compulsory unitization laws to issue compulsory orders under their broad waste-prevention authority. See, for example, James M. Whittier, "Compulsory Pooling and Unitization: Die-Hard Kansas," *University of Kansas Law Review* vol. 15, no. 3 (1967) pp. 307 and 321. The premise that courts might uphold such orders despite the lack of a compulsory unitization law is based on court decisions upholding agency orders which fixed minimum prices for natural gas even though no statutes authorized this power over prices. For example, the Kansas Supreme Court held that the Kansas commission had the power to fix minimum gas prices because the conservation statutes were "broad enough to authorize the Commission to make any rule or order to prohibit waste of natural gas. . . ." Kansas–Nebraska Natural Gas Co. v. State Corp. Comm'n, 169 Kan. 722 at 732, 222 P.2d 704 at 712 (1950); and see also, Cities Services Gas Co. v. Peerless Oil & Gas Co., 340 U.S. 179 (1950). Because of Article 6014(g), the Railroad Commission could not even attempt to assert an implied right to issue compulsory unitization orders under its waste-prevention authority, even though the Texas courts in the no-flare cases interpreted the agency's powers to prevent waste very broadly. In fact, the Railroad Commission has refused to issue orders setting minimum prices for natural gas, stating that it lacked both specific statutory authority and implied authority. See Sullivan, *Legal History 1958*, p. 240. Commissioner Culberson's opinion (ibid., p. 240, n. 92) in the order refusing to grant the petition to establish a minimum price for gas illustrates his reason for dissenting in the Lake Creek order which required pressure maintenance:

> As has been pointed out heretofore, increased field prices would have the effect of making more conservation practices profitable to the producer, and would therefore tend to reduce the waste of such gas and its dedication to inferior uses in competition with other and more plentiful fuels . . . [T]he lack of a fair price will shorten the period of time in which the consumers of Texas and the industry of this State will be enabled to enjoy the benefits and convenience of natural gas and allow the continued dissipation of this great natural resource to the detriment of the economy, both present and future, of the State of Texas. Being a strict constructionist, however much I feel that Texas economy is being plundered of

> its economical birthright, I cannot agree that by the most liberal application of the theory of "implication," can it be said that this Commission has the authority to fix prices, as contended by the petitioners herein.

It is especially difficult to find implied authority to issue minimum-price or compulsory unitization orders when the legislature has often debated and consistently failed to pass bills authorizing such powers. Texas lawmakers considered and defeated both types of proposed bills on many occasions. See Murphy, *Legal History, 1948,* p. 456, n.19; and Sullivan, *Legal History 1958,* p. 224. The courts in other states have consistently refused to uphold compulsory unitization orders by conservation agencies in the absence of express legislative authority. See, for example, Dobson v. Arkansas Oil & Gas Comm'n, 218 Ark. 160, 235 S.W.2d 33 (1950); and Union Pacific Railroad v. Oil & Gas Conservation Comm'n, 131 Colo. 528, 284 P.2d 242 (1955). Even if Article 6014(g) did not exist, the lack of express authority to issue compulsory unitization orders probably would have led the commission to deny such power.

87. See, for example, Northcutt Ely, "The Conservation of Oil," *Harvard Law Review* vol. 51, no. 7 (1938) p. 1236, n. 94.

88. A. W. Walker, Jr., "The Problem of the Small Tract Under Spacing Regulation," *Texas Law Review,* Proceedings of the 57th Annual Session of the Texas Bar Association (1938), p. 168.

89. Hardwicke suggests this interpretation. See Robert E. Hardwicke, "Oil-Well Spacing Regulations and Protection of Property Rights in Texas," *Texas Law Review* vol. 31, no. 2 (1952) p. 123; and Robert E. Hardwicke, "The Rule of Capture and Its Implications as Applied to Oil and Gas," *Texas Law Review* vol. 13, no. 4 (1935) p. 421 (arguing that Article 6014(g) was apparently enacted "for the purpose of protecting some operators from the bugaboo of compulsory unit operations for the entire field").

The commission attempted to force one lessee of a small tract to share production from his soon-to-be-drilled well with an adjoining small-tract operator by granting a Rule 37 exception permit conditioned on locating the well exactly on the line between the two operators' adjacent lots. However, it then had second thoughts about the propriety of this order, and on rehearing, granted the well permit to the original applicant without condition. See Ryan Consolidated Petroleum Corp. v. Pickens, 155 Tex. 221, 285 S.W.2d 201 at 211–212 (1955).

90. For example, Article 6014(c), now appearing at Texas Natural Resources Code Ann. sec. 85.046(3) (Vernon Supp. 1982), was enacted in 1932, subsequent to Article 6014(g). Article 6014(c) prohibits "underground waste or loss however caused and whether or not defined in other subdivisions thereof." This definition of waste does not have the proviso against requiring repressuring or unitization. See App. I.

91. William Murray was not yet on the commission. Commissioner Thompson had pledged to protect the interests of the independent producers. See ch. 3 text at notes 118 and 132. The third commissioner, Beauford Jester, was less dedicated to this pledge than Culberson or Thompson, but a commission order requires the votes of two commissioners in order to pass.

92. In 1940 Louisiana passed a statute authorizing compulsory unitization of condensate fields for cycling. In 1945 Florida and Georgia followed suit, and Oklahoma passed its broad compulsory unitization law. See David W. Eckman, "Statutory Fieldwide Oil and Gas Units: A Review for Future Agreements," *Natural Resources Lawyer* vol. 6, no. 3 (1973) p. 382.

93. The order was issued in 1951 and appears in Sullivan, *Legal History 1958,* p. 230, n. 66. The circumstances underlying this order are not discussed, however. According to Brooks Peden, a commission hearing examiner for many years, some operators liked to have the Railroad Commission issue a compulsory order which conformed to the terms of their private voluntary agreement as a "safeguard against possible conflict with lease terms and as a hedge against unexpected extensions of the field." See Peden, *Oil and Gas Hearing Aid,* p. 62. Peden states that the commission would issue a compulsory cycling or repressuring order as a conservation measure if no question of correlative rights were involved. Presumably, this situation exists when a reservoir is completely unitized by voluntary agreement. This may explain the wording of the Shafter Lake order. It seems unlikely that Commissioner Culberson would have signed a compulsory order if the repressuring had been protested by any of the operators, although his vote would not have been necessary if the other two commissioners agreed to the order. If the commission required repressuring over the protests of some of the lessees or lessors who would benefit financially by going it alone, the lack of court challenge to the order is somewhat puzzling in the absence of any war-inspired push for conservation. The 1951 order in the Shafter Lake field was in effect until May 4, 1959, when the water flooding became ineffective. Railroad Comm'n Order No. 8–40,575, *Byram's Rules,* sec. V, p. 320.

It should be noted that the operators' belief that a compulsory order avoids conflicts with lease terms is erroneous in that the commission has no authority to modify lease terms by compulsory process. See Texas Natural Resources Code Ann. sec. 101.012 (Vernon 1978). Perhaps lessees use this argument to persuade landowners to sign unitization agreements.

94. Halbouty v. Railroad Comm'n, 163 Tex. 417, 357 S.W.2d 364 (1962), *cert. denied,* 371 U.S 888 (1962), discussed in the text of chapter 4 at notes 196 to 212.

95. 357 S.W.2d 364 at 367–368.

96. Brief of Appellant Pan American Petroleum Corp. before the Texas Supreme Court in Halbouty v. Railroad Comm'n, Docket No. A-8200 (1961). (Copy in University of Houston Law Center's Library.)

97. Ibid., app. pp. 99–102.

98. Ibid., p. 33.

99. Ibid., p. 29.

100. The commission's statement about Article 6014(g) arose in the context of a lawsuit seeking to force the commission to issue a compulsory order rather than a lawsuit seeking to invalidate a promulgated order. Had the latter case been brought to court, the commission presumably would have sung a different tune. (Both Culberson and Thompson were on the commission when it issued the

pressure maintenance, no-flare, and no-waste orders in the 1940s and 1950s, as well as during the Port Acres episode.) The Port Acres plaintiffs had backed the commission into a corner. The commission was seeking to uphold the reasonableness of the per-well prorationing infrastructure that it had built up over the years. The commission's only rationale for granting allowable formulas which so favored small tracts was that it lacked compulsory pooling authority and, therefore, per-well factors were necessary to protect the correlative rights of small-tract owners. The commission could not simultaneously argue that Article 6014(g) prevented the agency from pooling small tracts but allowed it to issue compulsory repressuring orders. Given this ticklish position, the commission might have been pushed reluctantly into making the statement quoted in the text above. Whether willing or unwilling, the statement accurately reflects the precarious legal status of some of the agency's orders.

101. The *Port Acres* case is itself a good example of form over substance. After the supreme court struck down the one-third–two-thirds prorationing formula, the Railroad Commission and the small-tract owners who were aligned with it, argued on motion for rehearing that the effect of this decision was to compel pooling of small tracts and that this was forbidden by the legislature. Motion for Rehearing of H. L. Dillon *et al.* and of the Railroad Commission in Halbouty v. Railroad Comm'n, Docket No. A–8200, pp. 4–7 and 10–14 (1961). Halbouty and the other plaintiffs rejoined that the court's decision had nothing to do with pooling, either voluntary or compulsory. See answer of Pan American Petroleum Corp. to Appellees' Motion for Rehearing, Docket No. A–8200, pp. 5–6. The supreme court denied the motion for rehearing. Only one dissenting justice was willing to recognize that, substantively, the court's holding was "the opening wedge to compulsory pooling by judicial decree." See Halbouty v. Railroad Comm'n, 357 S.W.2d 364, 376 (1962).

Even in 1932, the federal courts were impressed with the focus on waste prevention in the 1931 Anti-Market Demand Act that had created Article 6014(g). In People's Petroleum Producers, Inc. v. Sterling, 60 F.2d 1041 at 1046 (E.D. Tex. 1932), the court wrote:

> The new statute, though it saves against the requirement of integration and unitization under one management and control, recognizes waste as brought about by contribution, introduces for the first time into the statutory prohibition against waste the idea that operations are to be looked at as a whole. It condemns equally and makes equally preventable individual commission of waste, or the commission of waste by contributing to a common result.

This quote offers some basis for judicial acceptance of form over substance in the no-waste orders. Also see ch. 8 text at notes 7 to 13.

102. Peden, *Oil and Gas Hearing Aid,* p. 36.

103. See ch. 3 text at notes 3 to 30 and 156 to 157; and text of this chapter at notes 123 to 126. It should be noted that Halbouty was himself a small producer at the time that he requested the cycling order and elimination of the one-third–two-thirds prorationing formula.

104. William Murray, former Texas railroad commissioner, telephone interview with author, July 19, 1983.

105. Ibid.

106. Michel T. Halbouty, chairman of the board and chief executive officer, Michel T. Halbouty Energy Company, telephone interview with author, July 7, 1983.

107. In 1962 fifty-five oil fields were spaced at 40 acres, thirty-eight fields at 80 acres, and six fields at 160 acres or more. Only five fields received 20-acre spacing. By comparison, in 1950, thirty fields were spaced at 20 acres or less, twenty-four fields at 40 acres, and none were spaced larger than 40 acres. See Governors' Special Study Committee of the Interstate Oil Compact Commission, *A Study of Conservation of Oil and Gas in the United States 1964* (Oklahoma City, Okla., Interstate Oil Compact Commission, Dec. 11, 1964) p. 59.

108. Texas Administrative Code tit. 16, secs. 3.37–3.38 (1982).

109. Texas Administrative Code tit. 16, sec. 3.43 (1982).

110. Texas Administrative Code tit. 16, sec. 3.45 (1982).

111. Both the 1947 and 1965 yardsticks appear in Texas Administrative Code tit. 16, sec. 3.45 (1982). A comparison of the number of barrels of allowable accorded an oil well drilled to 8,100 feet at various spacings illustrates the radical change:

Year	*10 acres*	*20 acres*	*40 acres*	*80 acres*	*160 acres*
1947 yardstick	103	113	133	—	—
1965 yardstick	34	68	133	215	380

Under the 1947 yardstick, two wells drilled on 20-acre spacing would receive an allowable of 226 barrels per day (113 × 2) versus 133 barrels for one well on 40 acres. Under the 1965 yardstick, two wells on 20-acre spacing would receive 136 barrels (68 × 2), only 3 barrels more than would be received by drilling one well on 40 acres. It is unlikely that this 3-barrel difference would cover the additional cost of drilling a second well.

The 1965 yardstick—while a vast improvement toward wider well spacing—is nonetheless inefficient compared to the use of MER prorationing. Yardstick allowables are based on depth and acreage and have no systematic relationship with the recoverable reserves in a reservoir. For example, the 1965 yardstick allowables for wells in deep but thin reservoirs may greatly exceed the field's MER. For extended analyses of these issues, see McDonald, *Petroleum Conservation,* pp. 167–188; and Lovejoy and Homan, *Economic Aspects,* pp. 141–154 and 167–173.

112. William Murray, former Texas railroad commissioner, telephone interview with author, February 4, 1983.

113. Texas Administrative Code tit. 16, sec. 3.42 (1982).

114. Also, if spacing rules are not adopted until the field's discovery allowable ends, the first ten wells may prematurely determine the field's spacing pattern.

The discovery allowable yardstick is based on depth only, with no acreage factor, and this creates an incentive to space at the minimum size of 40 acres. An operator can request temporary field rules requiring larger spacing after the first well is drilled, however. Texas Administrative Code tit. 16, sec. 3.43 (1982). The commission can also initiate a hearing on its own motion to establish temporary field rules under its authority to prevent waste.

115. Texas Administrative Code tit. 16, sec. 3.52(D) (1982).

116. For example, assume a water-drive field with the oil being pushed from west to east so that western wells drown out first. The lease-allowable transfer rule would let a small operator of a 20-acre lease with two wells on it shut in his poorer well and transfer its allowable to the second, better well located more to the east. A large operator with a 1,000-acre lease and 50 wells in this field could benefit even more from shutting in many of the western wells and transferring allowables a much further distance to his easternmost wells.

117. Lovejoy, "Producing Capacity," pp. 82–83.

118. Comment, "The Lease Allowable System: New Method of Regulating Oil Production in Texas," *Texas Law Review* vol. 47, no. 4 (1969) pp. 658 and 665.

119. The 1968 statewide lease-allowable order includes several provisions to protect correlative rights. Wells nearer to the lease line than a regular location may produce no more than their normal allowables, and wells located at the minimum regular distances from lease lines may produce no more than twice their top allowables, unless waivers are obtained from operators offsetting the well.

120. Comment, "The Lease Allowable System," pp. 672–673.

121. Efforts by larger operators to change the East Texas field rules to reduce the number of wells in operation had failed in 1962 and 1965 as well. See Lovejoy and Homan, *Economic Aspects,* p. 121. The ferment over the lease-allowable system in East Texas did result in the revocation on July 1, 1968, of the March 16, 1943, order which had allowed operators to apply an administratively determined pressure decline curve to estimate the amount of allowable which could be transferred (see text of this chapter at notes 80 to 81). The Railroad Commission heard substantial evidence from operators that the rule was being abused, that operators with wells that could no longer produce oil were nonetheless transferring allowables from these wells to other wells capable of producing, thereby securing an unfair share of the field's allowables. Railroad Comm'n Order No. 6-57,901B.

122. Texas Administrative Code tit. 16, sec. 3.8 (1982).

123. Texas Water Code Ann. sec. 21.261 (Vernon 1972). The Texas Water Quality Board has primary responsibility for all other water pollution control in Texas. In 1962 the board tried to assert control over oil field pits. The legislature amended the pollution laws in 1965 to give the Railroad Commission exclusive authority over pollution from oil and gas operations. See Comment, "Control of Oil Production Pollution," *Texas Law Review* vol. 48, no. 6 (1970) pp. 1086 and 1092.

124. Comment, "Control of Oil Production Pollution," pp. 1095–1097 and 1116–1126. The commission's concern for marginal well operators has also been criticized in the context of the agency's failure to police the fraudulent classification of wells as marginal and to tighten the regulatory standards for classifying wells as marginal. See Prindle, *Petroleum Politics*, pp. 81–92.

125. Jim C. Langdon, "The Railroad Commission Looks at Pollution," *South Texas Law Journal* vol. 8, no. 3 (1966) pp. 179 and 186–187. This entire speech bears reading as an example of the commission's general philosophy of regulating the industry by exhortation rather than direct confrontation.

126. In Commissioner Langdon's words, the commission forced antipollution measures "inch-by-inch." See Langdon, "Rules of Regulatory Bodies—Texas," p. 179. For a comparison of other state's saltwater disposal efforts, see McDonald, *Petroleum Conservation*, pp. 145–147. Arkansas passed a statute in 1957 authorizing compulsory saltwater disposal units if fieldwide unitization agreements already existed.

Chapter 6

1. Much of the information used in this section comes from personal interviews by the author with the staff of the Oil and Gas Division of the Railroad Commission on January 13–14, 1983 in Austin, Texas [hereafter cited as Railroad Commission interviews]. Additional information was received from questionnaires sent to the sixteen largest oil producers in Texas in 1983 (copies in author's file). Ten questionnaires were completed and returned. These ten companies' answers are probably representative of the position of the majors on the practice of voluntary unitization in Texas. The data used from these questionnaires are hereafter cited as Company Questionnaires. TIPRO, the largest and most active of the independent producers' organizations, did not respond to the author's queries on their membership's views concerning unitization.

2. Considerable confusion still exists on this basic issue in the legal literature. Some commentators view the 1949 Act as allowing, but not requiring, commission approval of all unitizations. See, for example, Robert E. Hardwicke, *Antitrust Laws v. Unit Operation of Oil or Gas Pools* (rev. ed., Dallas, Tex., Society of Petroleum Engineers of the American Mining and Metallurgical Engineers, 1961) pp. 330–331; Raymond M. Myers, *The Law of Pooling and Unitization*, vol. 1 (Albany, N.Y., Banks, 1967) p. 240; David T. Searls, "Antitrust and Other Statutory Restrictions of Unit Operations," *Institute on Oil and Gas Law and Taxation*, vol. 3 (Albany, N.Y., Matthew Bender, 1952) p. 92.

Others view the approval as mandatory. See, for example, Roscoe Walker, Jr., "Problems Incident to the Acquisition, Use and Disposal of Repressuring Substances Used in Secondary Recovery Operations," *Rocky Mountain Mineral Law Institute*, vol. 6 (Albany, N.Y., Matthew Bender, 1961) p. 278; and Howard R.

Williams and Charles J. Meyers, *Oil and Gas Law*, vol. 6 (Albany, N.Y., Matthew Bender, 1980) sec. 934. Williams and Meyers acknowledge some uncertainty as to their conclusion, but then argue that "[i]n view of the substantial risks involved, it is not to be expected that parties to a unitization agreement will seek to proceed . . . without first obtaining the consent of the regulatory commission." The authors do not elucidate the nature of these risks. The authors do not seem to view the risk of antitrust liability as substantial (ibid., sec. 911). Tort liability from injury to nonconsenters is a substantial risk, but obtaining regulatory approval does not remove it. See ch. 7 text at notes 125 to 160. The Railroad Commission itself sees little to be gained from its approval. See text of this chapter at note 8.

3. Company Questionnaires.

4. See, for example, Railroad Commission of Texas, *Secondary and Enhanced Recovery Operations in Texas to 1980: Bulletin 80* (Austin, Tex., 1980) pp. 450 and 452; and Railroad Comm'n Docket No. 8–78,610, Dec. 13, 1982 (final order approving the unitization agreement for tertiary recovery in the South Fault Block Unit of the Andector, Ellenburger field). The Crude Oil Windfall Profits Tax Act of 1980 grants special benefits to "qualified tertiary recovery projects." Many operators use this terminology in applying for commission approval of their unitization agreements under the 1949 Act because they simultaneously seek commission certification as a qualified tertiary recovery project under the tax act.

5. Robert E. Sullivan, ed., *Conservation of Oil and Gas: A Legal History, 1948–1958* (Chicago, Ill., American Bar Association, 1960) p. 229 [hereafter cited as Sullivan, *Legal History, 1958*]; Brooks Peden, *The Texas Oil and Gas Hearing Aid: Evidence, Procedure and Practice for Oil and Gas Hearings of the Railroad Commission of Texas* (Austin, Tex., R. W. Byram, 1961) p. 75.

6. Railroad Commission Legal Memorandum re: Quintana Petroleum Corporation Unitization, January 21, 1980 (copy in author's files).

7. Ibid.

8. The legal examiner's memorandum states: "Finally, it is unclear what Quintana gains by receiving Commission approval of its unitization program. Quintana is not limited by statute from pursuing a voluntary unitization, but Commission approval will not compel those who do not voluntarily join the unit. Thus, there appears to be nothing to gain with Commission approval."

9. According to Quintana, Exxon refused to participate in the project unless Railroad Commission approval was secured. Exxon is very careful about any issue which involves potential antitrust problems. Quintana's own reasons for seeking approval derived more from the demands of its own royalty interest owners that the company obtain this approval than from any antitrust fears. (T. B. Cochran, III, attorney for Quintana, personal interview with author, January 28, 1983.) Of the ten companies surveyed, three stated that they sought Railroad Commission approval of unitization agreements primarily because landowners would sign more readily if they know that the agreement will be brought

to the Railroad Commission for approval; five other companies considered this an important reason, but listed "securing antitrust immunity" as the primary reason. One company contended that approval was required by the Railroad Commission rules, a clearly erroneous belief.

10. A few royalty interest owners could not be located. One working interest owner refused to sign, but waived any objections to the unit. Indeed, at the hearing, this company testified that the unit proposal was satisfactory. (Webb Holland, District Reservoir Engineer for Quintana Petroleum Corp., telephone interview with author, June 29, 1983).

11. Railroad Comm'n Order No. 2–71,907, in *Rules and Regulations: Texas Railroad Commission* (Austin, Tex., R. W. Byram, 1958—) sec. VII, p. B-3 [hereafter cited as *Byram's Rules*].

12. Ibid.

13. In its briefs in the *Port Acres* case, the commission argued that the 1949 Act did not authorize the agency to approve pooling agreements. The commission cited as support Section 101.014 of the 1949 Act, which states that "none of the provisions in this chapter shall be construed to require the approval of the commission of voluntary agreements for the joint development and operation of jointly owned properties." Motion for Rehearing of Railroad Comm'n in Halbouty v. Railroad Comm'n, Docket No. A–8200 pp. 4–7 (1961). The commission's conclusion is correct, but its reasoning seems faulty. Section 101.014 seems designed to allow joint operating agreements among cotenants to escape the burdens of the 1949 Act. Pooling agreements are for the joint development of *separately* owned properties. Therefore, they would fall under the 1949 Act if they involved secondary recovery operations. This would happen seldom. The Railroad Commission's stretch to interpret Section 101.014 so as to exclude pooling from the 1949 Act can be attributed to the peculiar position that the *Port Acres* case forced upon the commission. In defense of its one-third per well–two-thirds acreage prorationing orders, the commission was compelled to rebut any argument that public policy, as reflected in the 1949 Act, favored pooling. More than likely, pooling agreements were not mentioned in Section 101.014 of the 1949 Act because they were not ordinarily necessary for secondary recovery operations, not because public policy opposed voluntary pooling. Also, the risk of incurring antitrust liability through pooling into drilling units was probably slight.

14. Railroad Commission interviews and Company Questionnaires. All persons interviewed agreed that less than 10 percent of the unitization applications are contested. In fact, the commission's standard "fill-in-the-blanks" form for final orders approving unitization requests begins: "This is the unprotested application of______for unitization and secondary recovery. . . . " The lack of protest is not attributable to a lack of notice of the hearings. Notice is sent to all owners of interests in the proposed unit and to all offset operators and owners of unleased mineral interests adjacent to the proposed unit. Notice is also sent to all surface owners and to the county clerk of the county in which the proposed unit is located under the commission's new Rule 46 governing underground injection wells.

Rule 46 was promulgated to conform with the federal Safe Drinking Water Act. Texas Administrative Code tit. 16, sec. 3.46 (1982).

Generally three months elapse between an applicant's request for a unitization hearing and the hearing itself.

15. Peden, *Oil and Gas Hearing Aid,* pp. 72–73.

16. See, for example, R. W. Byram and Co., *Texas Oil and Gas Handbook* (Austin, Tex., R. W. Byram, 1982) p. 10.

17. One Railroad Commission staff member definitely would accept such an application; another would not, however (Railroad Commission interviews). In one instance, in 1971, the unitization hearing was recessed for lack of sufficient royalty interest consent, but the hearing for approval of the related secondary recovery operation continued. See *Oil Regulation Report,* vol. 38, Dec. 20, 1971 (Austin, Texas State House Reporter, Inc.) p. 8.

The State of Texas as a landowner will not sign a unitization agreement until it has been approved by the Railroad Commission. In these cases, the commission will approve unitization applications without 65 percent of the royalty interest owners signed up, but with the knowledge that the state of Texas will consent later and raise the percentage. See, for example, the N. Cowden unitization request, *Oil Regulation Report,* vol. 40, August 1, 1973, p. 10.

18. Railroad Commission interviews. The twenty questions appear in R. W. Byram, *Texas Oil and Gas Handbook,* p. 10; and Peden, *Oil and Gas Hearing Aid,* pp. 70–71.

19. Texas Natural Resources Code Ann. sec. 101.013(a)(2) (Vernon 1978).

20. Railroad Commission interviews; and Peden, *Oil and Gas Hearing Aid,* p. 69.

21. In hearings on experimental projects, the commission does request detailed information on the novel aspects of the program. See Peden, *Oil and Gas Hearing Aid,* p. 76.

22. Myers, *The Law of Pooling and Unitization,* vol. 1, sec. 7.02, pp. 249–250.

23. The SACROC orders approving the unitization agreement and the secondary recovery operation are quoted in full in Myers, *The Law of Pooling and Unitization,* vol. 2, at secs. 15.13 and 15.14. Even after a year of study, considerable controversy existed in 1954 over the amount of additional oil to be recovered. The opinions of expert witnesses ranged from 143 million barrels to 795 million barrels. Time has shown the latter estimate to be quite accurate.

24. The commission granted a two-stage incentive allowable to SACROC's carbon dioxide tertiary recovery project. Front-end capital costs equaled $45 million, but the predicted increase in recovery would not appear until many years later. When discounted to present value, the project was not economically viable. The commission granted an increased allowable of 209,000 barrels per day (versus the existing 174,000 barrels per day) effective on commencement of construction of the carbon dioxide facility and an allowable of 244,000 barrels per day effective

when carbon dioxide injection begins. See Paul Hull, "SACROC—An Engineering Conservation Triumph," *Interstate Oil Compact Commission Bulletin* vol. 12, no. 1 (1970), p. 33.

This is not to say that the Railroad Commission always approved inefficient projects by following this policy. In many instances, the economics of the unitized operation suffered because of the severe market demand prorationing restrictions placed on nonexempt projects by the exempt stripper wells, capacity water floods, and discovery allowables. Yet the unitized operation could produce oil at a lower cost than the exempt wells. Also, because the unit's total allowable was calculated under the existing field rules, which often included a large, per-well prorationing factor, the unit was penalized for not drilling the maximum number of permissible wells. For example, the original SACROC unit operated under a unit allowable equal to the number of wells times the 1947 yardstick plus 25 percent bonus allowable for saltwater injection. (This 25 percent was increased to 75 percent for the carbon dioxide project.) See Hull, "SACROC," pp. 30–38. There is nothing"efficient" about the base allowable which simply relies on the 1947 yardstick and the number of wells. Therefore, the agency's grant of additional bonus allowables does not necessarily induce inefficient projects.

25. In reviewing all the unitization applications reported in the *Oil Regulation Reports* for 1969, 1971, and 1976, only one example was found which seemed to involve this situation. In 1969 an independent operator sought to unitize a few hundred acres in the shallow Lehn–Apco oil field already being water flooded by twenty-eight injection wells on a lease basis. The applicant expressed the opinion that with unitization, the unit area could be produced to a "lower economic limit," thereby increasing ultimate recovery. See *Oil Regulation Report,* vol. 37, April 18, 1969, p. 12. The application was approved. See *Oil Regulation Report* vol. 37, May 9, 1969.

26. The commission has authorized the unitization of more than one horizon at a time in order to save operating costs. The agency approved the combination of the Goldsmith (5,600') and Clearfork fields into one unitized operation for confluent production. The unit applicant testified that 4.95 million barrels of oil could be recovered by water flooding the two reservoirs. Dual completions and confluent production would reduce investment costs by $660,000 and operating costs by $1,750,000, resulting in the recovery of an additional 650,000 barrels of oil. See *Oil Regulation Report,* vol. 37, April 18, 1969, pp. 1–3. The statewide rule allowing lease-allowable transfers also demonstrates the commission's concern with reducing operating inefficiencies. The commission's recent concern for preventing economic waste in the context of Rule 37 exception requests is discussed in the text of chapter 8 at notes 21–36.

27. For example, the West Tyler (Paluxy "B" Sand) field was unitized for pressure maintenance purposes in September 1976, only three years after its discovery. The proposed unit had 585 surface acres and only five producing oil wells at the time of the unitization hearing. The unit contained twenty-one different tracts and the field rule provided for 80-acre proration units. *Secondary Recovery Railroad Comm'n Summaries 1976,* p. 80.

28. See Peden, *Oil and Gas Hearing Aid*, p. 72. Incredible as it may seem, a few protests have been made by owners who did not want an increase in taxable income from the unit's operations and by adjacent nonsigned operators who did not want to install better lease equipment to handle their tract's increased production.

29. Railroad Commission interviews.

30. Ibid.; and C. Burton Bransletter, Sandra J. Gorka, and Robert Goldsmith, Jr., "Fieldwide Unitization in 1983," in *Advanced Oil, Gas and Mineral Law Course* (Austin, State Bar of Texas, 1983) pp. B-5–B-9.

31. According to Railroad Commission interviews, the commission does not require that notice of the hearing be sent to the royalty interest owners of the adjacent nonjoining tracts. The commission assumes that the lessees of these tracts will protect the interests of their own royalty owners against possible drainage to the unit or injury from the unit's operations. The same notice requirements apply to lease-allowable transfers. One commentator is somewhat critical of the Railroad Commission's failure to police the statewide lease-allowable system more closely to assure that the correlative rights of royalty interest owners are protected. Under the lease-allowable system's procedures, royalty owners are not given notice of an adjacent operator's exception request to place a well nearer to a lease line than the minimum distance of 233 feet. The agency relies on the royalty owner's lessee to protect the tract from possible drainage. An operator holding adjacent leases from different landowners may have an incentive to drain one lease at the expense of the other, or adjacent lessees may collude to this effect. The commentator proposes that the commission require more information on lease-allowable exception requests, particularly data on the ownership interest of tracts adjacent to the lease for which the exception is being sought. See Comment, "The Lease Allowable System: New Method of Regulating Oil Production in Texas," *Texas Law Review* vol. 47, no. 4 (1969) pp. 668–669. The possibility of harm to royalty interests in the voluntary unitization process is much less significant because all royalty interests within the proposed unit area receive notice of the unitization hearing, and because the 1949 Act requires that all royalty interests on tracts adjacent to the proposed unit be offered the opportunity to join. Also, the royalty interest owner still has the protection of the common law of implied covenants under which the lessee who drains his own lessor often will be held liable (see ch. 7 text at notes 78 to 124).

32. This phrase appeared in the order approving the SACROC unit in 1953 (see Myers, *The Law of Pooling and Unitization*, vol. 2, sec. 15.13), and in many subsequent unitization orders. This phrase may have originated from the Texas attorney general's interpretation of his approval authority under Section 21 of the 1935 Act. He interpreted this statute as granting him authority to approve voluntary unitization agreements in gas fields as long as the agreement contained nothing repugnant to the conservation laws, but not as authority to determine whether the agreement was fair or equitable. See Blakely M. Murphy, ed., *Conservation of Oil and Gas: A Legal History, 1948* (Chicago, Ill., American Bar Association, 1949), p. 470, n. 47. It is not clear why the attorney general held this

belief because one of the stated purposes of Section 21, in addition to conservation, was to ensure "the equitable distribution of royalty payments" and that "the market for natural gas may be more equitably distributed among the various landowners and operators" (see text of ch. 3 at note 162). It is far more difficult for the railroad commissioners (as opposed to the attorney general) to absolve themselves from the duty to protect correlative rights because their very function is to enforce the conservation laws, which often expressly require that such rights be protected, and because the language of the 1949 Act is much more express on this duty than the 1935 Act. In any event, the phrase no longer appears in the commission's orders.

33. Magnolia Petroleum Co. v. Railroad Comm'n, 141 Tex. 96, 170 S.W.2d 189 (Tex. 1943); Whelan v. Placid Oil Co., 274 S.W.2d 125 (Tex. Civ. App. 1955); American Petroleum Corp. v. Railroad Comm'n, 395 S.W.2d 403 (Tex. Civ. App. 1965); Lone Star Gas Co. v. Murchison, 353 S.W.2d 870 (Tex. Civ. App. 1962).

34. Myers, *The Law of Pooling and Unitization,* vol. 1, sec. 7.02, pp. 247–248.

35. Ibid.

36. A somewhat more defensible position in support of the commission's denial of authority to consider the fairness of the terms of the unitization agreement is that Section 101.013(3) requires only that the commission find that the rights of all owners are protected under the unit's "operations." The commission could argue that this authorizes the agency to inquire only into whether the physical operations of the unit (such as its effect on oil and gas flows between property boundaries) protects all the owners.

37. This is, of course, why it is very important that notice of the hearing be sent to the nonsigners. Any commission laxity in enforcing the notice requirements is unlikely to survive court review. In Railroad Comm'n v. Graford, 557 S.W.2d 946 at 953–54 (Tex. 1977), the court held that the due process clauses of the U.S. and Texas Constitutions required that the owners of unleased mineral lands, whose interests would be materially affected by the commission order, had a right to participate in the commission hearing on the proposed order. In this case, a group of unleased landowners had not been sent notice of the hearing on a proposed field consolidation. They nonetheless learned of the hearing, but the commission refused to allow them to participate, arguing that they were accorded due process because they could challenge the substantive merits of the order in court. The Texas Supreme Court did not consider judicial review to be an adequate substitute for the "opportunity to be heard at the real decision-making level; i.e., before the Commission."

38. At a recent institute on commission rules and procedures, one speaker noted that "[w]hile it is supposed by many operators that the Railroad Commission can compel unitization, it can not do so. . . . " See Skipper Lay, "Voluntary Pooling and Unitization," *Oil and Gas: Texas Railroad Commission Rules & Regulations* (Austin, State Bar of Texas, 1982) pp. C-1–C-26. Peden's guide to the

Railroad Commission's policies also notes that "most emphatically, the Commission does not have nor claim to have any authority to alter the terms of any lease, yet surprisingly often the Commission is requested to take action which would require that authority." See Peden, *Oil and Gas Hearing Aid*, p. 73. Perhaps, given the Railroad Commission's campaign against flaring and waste, it is understandable that some individuals are confused about the agency's unitization authority. If some landowners think that unitization is compulsory, then the commission's assumption that all agreements are voluntarily negotiated may be misplaced.

39. Myers, *The Law of Pooling and Unitization*, vol. 2, sec. 15.14.

40. Francis Oil & Gas Inc. v. Exxon Corp., 661 F.2d 873, 875 (10th Cir. 1981).

41. Francis Oil, an Oklahoma corporation, first filed suit in the federal district court in Oklahoma. It then also filed a class action suit in the Texas state court on behalf of all producers similarly situated. The defendants removed this class action to the federal district court in Texas. This court ruled in favor of Francis Oil that the unitization agreement did not effect a reallocation of the benefits of stripper-well pricing to the unit, but left these benefits to the producers of the stripper oil. The defendant, Exxon, appealed this ruling to the Temporary Emergency Court of Appeals (TECA), the special court established to hear federal oil- and gas-pricing regulatory cases. TECA reversed the lower court's decision. See Francis Oil & Gas, Inc. v. Exxon Corp., 687 F.2d 484 (Temp. Emer. Ct. App. 1982). Meanwhile, Francis Oil's original action in the Oklahoma federal court was dismissed when the defendant, Exxon, was granted a motion for judgment on the pleadings on the basis that Francis Oil had failed to join all indispensable parties. Francis Oil appealed this decision to the Tenth Circuit, which held that the judgment on the pleadings was improperly granted and remanded the case to the Oklahoma federal district court to determine additional facts regarding the necessary joinder of other parties. See Francis Oil & Gas Inc. v. Exxon, 661 F.2d 873 (10th Cir. 1981).

42. 661 F.2d at 874.

43. Ibid., at 874. The unit probably would have been feasible without the joinder of this 7.5 percent. However, the 1949 Act requires that all owners in the area reasonably defined by development be offered an opportunity to join the unit.

44. Michael Medina, attorney for Francis Oil & Gas, Inc., telephone interview with author, March 3, 1983 [hereafter cited as Medina interview].

45. This was the basis of TECA's decision against Francis Oil in Francis Oil & Gas Inc. v. Exxon, 687 F.2d 484 (Temp. Emer. Ct. App. 1982).

46. *Secondary Recovery Railroad Comm'n Summaries 1976*, pp. 178–181.

47. Ibid.; and see generally, Ronald D. Frank, "Impact of Current Multi-Tier Pricing Regulations of Crude Oil and Natural Gas on Exploration and Development Activity," *Exploration and Economics of the Petroleum Industry*, vol. 17 (Albany, N.Y., Matthew Bender, 1979) pp. 231–232.

48. In fact, by joining the unit, Francis's allocation was more than 100 barrels per day versus its pre-unitization total of 40 barrels per day. See Francis Oil & Gas Inc. v. Exxon Corp., 687 F.2d 484, 489 n. 7 (Temp. Emer. Ct. App. 1982).

Although all federal crude oil-pricing regulations were removed in 1981 the crude oil windfall profits tax uses many of the same classifications of crude oil. Stripper-well oil produced by independents is taxed at a lower rate than other types of crude. Thus, while Francis Oil originally sought only about $13,500 in damages against Exxon, the total amount of money at stake to Francis Oil and six other similar producers is close to $500,000 (Medina interview).

49. Medina interview. The importance of preserving tax advantages to stripper-well owners in order to induce them to unitize is discussed in Bransletter and coauthors, "Fieldwide Unitization," pp. B-19–B-28.

50. William Murray, former Texas railroad commissioner, telephone interview with author, February 4, 1983. Murray left the commission in 1963, long before the Yates field negotiations and hearings.

51. See William L. Horner, "Calculation of Just and Equitable Shares," in Jimmy M. Stoker, ed., *Institute on Mineral Law,* vol. 2 (Baton Rouge, Louisiana State University Press, 1954) pp. 71–73.

52. William Murray, former Texas railroad commissioner, telephone interview with author, February 4, 1983. The SACROC protestors had small edge tracts with thin sands. The field prorationing formula was 50 percent acreage–50 percent per well. The protesters were better off under the field formula than the unit formula. Indeed, if the unit could not prevent oil from migrating to the edges (the water flood was center-to-edge) the nonconsenters would also gain from net migration.

53. The Railroad Commission's order approving SACROC's secondary recovery program required elaborate transfer-allowable procedures. If any adjoining operators objected to a proposed transfer, the commission would have to call another hearing. Myers, *The Law of Pooling and Unitization,* vol. 2, sec. 15.14 (Rule 6).

54. Milton Fox, Texas state representative, telephone interview with author, January 27, 1983.

55. Francis Oil's failure to protest publicly was not a result of burdensome administrative costs, lack of opportunity, or bureaucratic roadblocks. The Railroad Commission staff is easily accessible. In fact, one of the major criticisms of the agency for many years was its easy accessibility to producers and its use of informal procedures, including frequent *ex parte* contacts involving cases in the hearing process. See David F. Prindle, *Petroleum Politics and the Texas Railroad Commission* (Austin, University of Texas Press, 1981) pp. 145–156; Kenneth Culp Davis and York Y. Willbern, "Administrative Control of Oil Production in Texas," *Texas Law Review* vol. 22, no. 2 (1944) pp. 149–193; and Joe Greenhill and Robert C. McGinnis, "Practice and Procedure in Oil and Gas Hearings in Texas," *Southwestern Law Journal* vol. 18, no. 3 (1964), pp. 406–436.

The Railroad Commission's adoption of formal procedural standards and practices in recent years has increased the agency's professionalism but without detracting from its basic accessibility and the informal atmosphere of its hearings. Producers and royalty owners may, and often do, represent themselves at the hearings.

56. In Oklahoma, unhappy royalty interest owners almost succeeded in repealing the 1945 compulsory unitization statute and did succeed in getting it amended in 1951. The care that many unit proponents take to secure consent is described in R. F. Bryant, Jr., "The Negotiation and Execution of a Voluntary Unit Agreement," *Institute on Oil and Gas Law and Taxation,* vol. 3 (Albany, N.Y., Matthew Bender, 1952).

57. Three consultants for the Poly Brooks Trust expressed some doubt, but no dissent, about the technical aspects of the pressure maintenance operation, and an attorney for the trust urged the agency not to forget the nonsigners and to monitor the field for their protection. See *Secondary Recovery Railroad Comm'n Summaries 1976,* pp. 181–182.

58. The unsigned working interests in Quintana's pseudo-secondary recovery proposal testified at the hearing that the unitization plan was fair. See text of this chapter at note 14. See also *Oil Regulation Report,* vol. 38, October 11, 1971, pp. 7–8 (Amoco, a nonsigner, supported the Reinecke unitization); ibid., June 15, 1971, pp. 6–7 (Forest Oil, a nonsigner supported the Ownby and Waples Platter unitization).

59. The commission's standard order approving unitization agreements reads: "The owners of interests not desiring to enter the unit on the yardstick basis may refuse to join the unitized operation and continue to participate in the production from the field on an independent basis. . . . " Further, "no proration order of the Commission will be promulgated pursuant to any terms of the unit agreement." See, for example, Railroad Comm'n Docket No. 7B–78,877 re: Application of North Ridge Corp. for Unitization and Waterflood Approval in the Gipe (Strawn, Lower) Field, Sept. 27, 1982.

60. See, for example, the SACROC order which specified exactly which wells could be used for injection and imposed limits on allowable transfers. See Myers, *The Law of Pooling and Unitization,* vol. 2, sec. 15.14.

61. Peden, *Oil and Gas Hearing Aid,* pp. 67 and 69. The commission also requires that separate production records be kept for any tract with an unsigned interest.

62. *Oil Regulation Report,* vol. 41, August 15, 1973, pp. 1–2. The 1949 Act requires a finding by the commission that all owners have been offered an opportunity to join the unit on the "same yardstick basis." See Texas Natural Resources Code Ann. sec. 101.013(a)(6) (Vernon 1978). The commission obviously interpreted this requirement to allow a choice of two yardsticks.

63. Railroad Comm'n Order No. 6–64,407, effective Jan. 1, 1975, in *Byram's Rules,* sec. VIII, p. 738. These total payments were measured by the greater of (a)

the quantities of oil and gas produced by primary recovery operations on the nonsigner's tract after the effective date of the unit; or (b) the quantities of recoverable primary oil and gas in place under the tract at the effective date of the unit. Thus, if production were reduced on a nonsigner's tract, he or she would be paid on the settlement basis, not according to the rule of capture. The Hawkins unitization involved 324 working interest owners and 2,088 royalty interest owners. With this formula, over 98 percent of the working interests and 89 percent of the royalty interests consented to the unitization. *Oil Regulation Report,* vol. 42, September 18, 1974, pp. 7–9. Also see ch. 8, note 127, describing the guaranteed royalty payments made by Exxon to landowners who refused to join a unitization agreement in the Katy field.

64. None of the 1976 unitization applications shown in table 6-1 were contested. In 1969 only 4 out of 49 unitization applications were protested. See note 69 in this chapter. One often finds, however, that fiercely contested hearings on other types of applications have preceded the peacefulness of the unitization hearing. Unit proponents can apply some pressure on noncooperative operators in a field to unitize by opposing the latters' requests for changes in, or exceptions to, field rules. For example, before the Hawkins field was unitized, a small operator with one lease in the field requested a net gas-oil ratio for one well. Exxon, Conoco, Texaco, and Amoco opposed the application, contending that the production of additional gas would cause waste. Exxon's geologist and engineer testified that substantial waste was already occurring in the Hawkins field. So convincing was the majors' testimony that the hearing examiner wondered whether he should recess the hearing until the entire matter could be thoroughly investigated. The applicant argued for a speedy decision on his one small well, rather than letting the "tail wag the dog." The examiner replied that "there apparently is a much larger ox in the ditch than the one we came here to treat." The examiner finally decided to continue the hearing and consider the fieldwide waste issue in the future. See *Oil Regulation Report,* vol. 41, May 11, 1973, pp. 1–2.

65. For example, TexLand Petroleum Inc.'s application was protested by a royalty interest owner who argued that the net-pay isopach maps used by the applicant should have shown more productive acreage for his tract. The royalty owner had no objections to the water flooding if no injection wells were placed on his tract. TexLand agreed. Railroad Comm'n Docket No. 8A-75,750, re: Application of TexLand Pet. Inc. (Sept. 15, 1980).

66. Peden, *Oil and Gas Hearing Aid,* pp. 72–73.

67. Railroad Commission of Texas, *Discussions of Law, Practice, and Procedure* (Austin, Tex., 1982) p. 56.

68. Railroad Commission interviews. In Railroad Comm'n Docket No. 8-78, 610, Phillips Petroleum Co. applied for approval of a unitization agreement for the South Fault Block of the Andector (Ellenburger) field. Amoco protested the application on the basis that Phillips refused to let Amoco join the unit, even though Amoco was the operator of a tract adjacent to the proposed unit area. Phillips asserted that it had given Amoco an opportunity to enter the unit on the

same yardstick basis as the owners of other interests in the unit, but that Amoco had rejected this offer by a letter dated July 8, 1981. Phillips contended that the 1949 Act only required that a one-time offer be extended and that the act required no further duty once the initial offer had been rejected. A lengthy hearing was held on this protested application on August 13 and 16, 1982, with Amoco seeking to force its way into the proposed unit. The hearing examiner did not issue a Memorandum of Decision in the case, and the unit was to expire by its own terms at the end of 1982 if not approved. As the termination date approached, Phillips entered a settlement agreement with Amoco. Amoco withdrew its protest, and the unitization application was approved by the commission on December 13, 1982, as an unprotested application. The approved unit did not include Amoco's tract, but Amoco was allowed to enter the unit a few months later as part of the settlement. Thus the issue of whether a unit proponent must keep open the offer to join until the unit is approved remains unresolved.

The background to this protest is not an uncommon one. After Phillips first made the offer to Amoco to join the unit, more detailed engineering data became available to Phillips showing that the offer to Amoco was inordinately generous. Thus Phillips was not unhappy that Amoco rejected it. Later, Amoco seemed to realize its mistake and wanted to join the unit on the original terms. When Phillips refused, Amoco brought its protest to the commission. Amoco also filed for a Rule 37 exception on its tract, which Phillips was prepared to protest. The settlement between the two rivals allowed Amoco to enter the unit a few months after its approval by the commission on terms less favorable than those of the original offer. Amoco also dismissed its request for a Rule 37 exception as part of the settlement (Elizabeth Harris, attorney for Phillips Petroleum Co., personal interview with author, May 21, 1984).

This episode demonstrates the same tactic of delay used by the commission in the Yates field to encourage all operators to cooperate (see ch. 5 text at note 70). The commission's refusal to issue an order on a controversial contested issue forces the opposing parties to bargain with each other to reach a result which is more mutually satisfactory than the stalemate created by the commission's inaction.

69. All of the issues of the *Oil Regulation Report* for 1969 were reviewed to ascertain the reasons for and disposition of protested unitization applications. Of the forty-nine reported unitization hearings in 1969, only four were protested as follows: In the July 15, 1969, Sivells Bend hearing, two protestors were unhappy with the participation formula. See *Oil Regulation Report* vol. 37, July 15, 1969, p. 8. In the April 9, 1969, hearing on the Trix Liz unit, a royalty owner's attorney requested a provision in the Railroad Commission order that the order not in any way limit any contractual or implied duty of the lessee. See *Oil Regulation Report* vol. 37, April 9, 1969, pp. 9–10. In the Thornton field hearing, an individual opposed the unit's application by letter. See *Oil Regulation Report* vol. 37, September 9, 1969, p. 1. In all three cases the Railroad Commission approved the unitization application without modification within a few weeks after the hearing. The fourth protest involved the issue of inadequate notice. The Cooke County regular field hearing started on December 12, 1968, but was recessed

when an operator testified that he had not received adequate notice of the hearing to prepare his protest to the unit. The hearing reopened on January 1, 1969. At this hearing, an operator stated that he was not offered an opportunity to join the unit on any basis. Finally on May 28, 1969, more than five months after the first hearing, the unitization application was approved, but one request for an injection well was denied. See *Oil Regulation Report* vol. 37, Jan. 31, 1969, pp. 2 and 6; and ibid., May 28, 1969. All the other hearings in 1969 were uneventful. The applicants' requests were fully granted, except in two cases. The Bellvue, S.E. field applicant was denied the use of two specific injection wells, and the Jordan field applicant was denied a unit allowable at the time, although the unit received transfer allowables. See *Oil Regulation Report* vol. 37, July 1, 1969, p. 2; ibid., vol. 37, April 18, 1969, p. 4; and ibid., April 30, 1969, p. 5. The commission probably assigned the Jordan field applicant a unit allowable after a substantial response to the water-flood injection was realized.

70. See, for example, Prutzman, Cage, Fletcher, Keith, Miller, and Winn, "Chronicle of Creating a Fieldwide Unit," *National Institute for Petroleum Landmen*, vol. 5 (Albany, N.Y., Matthew Bender, 1964) pp. 126–127.

71. See ch. 4 text at notes 65 to 66. McDonald argues that public control over unitized projects is not necessary and results in redundant and pernicious regulations and administrative costs. See Stephen L. McDonald, *Petroleum Conservation in the United States: An Economic Analysis* (Baltimore, Md., Johns Hopkins University Press for Resources for the Future, 1971) pp. 207–208.

72. Myers, *The Law of Pooling and Unitization*, vol. 1, pp. 244–246. After the Texas courts invalidated prorationing formulas with large per-well factors, the commission often promulgated prorationing orders using the same formula to allocate allowables as the participation factor that was used in the unitization agreement proposed for the field. See, for example, Pickens v. Railroad Comm'n, 387 S.W.2d 35 (Tex. 1965).

73. A typical order reads:

> The Railroad Commission of Texas retains all powers and duties with regard to conservation of oil and gas in this field, and no proration order of the Commission will be promulgated pursuant to any terms of the unit agreement; and the allocation formula for the wells included in the agreement shall remain and continue in full force and effect as if the agreement had not been approved.

See Peden, *Oil and Gas Hearing Aid*, p. 144. Under this system, a unitized field may still be overdrilled if the existing field prorationing formula encourages dense drilling.

74. See ch. 5 text at notes 72 to 79; and Rule 48, Texas Administrative Code tit. 16, sec. 3.48 (1982). The Yates field's grant of an outright doubling of its MER before a production increase was realized was unusual but appropriate, given the energy crisis of the 1970s and the complex negotiations required in the field.

75. Railroad Commission interviews. It was rumored that the Railroad Commission threatened to withhold the doubled MER from nonconsenters in the

Yates field, and this obviously influenced some of the operators in this field (see ch. 6 text between notes 47 and 48). Typically, however, any increased MER inured to the benefit of everyone in the field, whether in or out of the unit. See, for example, *Secondary Recovery Railroad Comm'n Summaries 1976*, pp. 15–16 (Tom O'Connor field received an increased MER from 15,000 to 18,000 barrels per day upon unitization; the nonjoiners shared in the rise). In this regard, the commission has been much less activist than the Arkansas Oil and Gas Commission, which issued an order in 1951 approving a voluntary unitization agreement and included provisions to reduce the nonsigners' share of the field's production to ensure that they did not receive more than their fair share. See Sullivan, *Legal History, 1958*, pp. 39–41. This order clearly violated the common law rule of capture which would prevail in the absence of compulsory unitization. The order was not attacked in court, probably because the Arkansas legislature passed a compulsory unitization law the same year. The Arkansas Supreme Court has been willing to replace the rule of capture with "equitable unitization" of nonsigners. See Dobson v. Arkansas Oil & Gas Comm'n, 218 Ark. 160, 235 S.W.2d 33 (1951).

76. See ch. 4 text at notes 203 to 204. The legality of a commission order revising an established per-well prorationing formula and replacing it with one more encouraging to unitization is fully discussed in chapter 8 text at notes 71 to 78.

Chapter 7

1. Texas Co. v. Daugherty, 107 Tex. 226, 176 S.W. 717 (1915). The opposing concept adopted in several other jurisdictions is the theory of "nonownership" or "qualified ownership," that is, that the landowner does not own the oil and gas in place in the ground, but only owns the exclusive right to explore for and produce the oil and gas. The minerals are owned only when reduced to possession. See Richard W. Hemingway, *The Law of Oil and Gas* (2 ed., Saint Paul, Minn., West Publishing, 1983) sec. 1.3.

2. Stephens County v. Mid-Kansas Oil & Gas Co., 113 Tex. 160 at 167, 254 S.W. 290 at 292 (1923). A. W. Walker, Jr., has commented as follows on Texas' concept of property rights in oil and gas: "There is no reason for giving an injured party a cause of action for the violation of some legal right resulting from a reasonable use of adjacent land if the aggrieved party's remedy of self-help is completely adequate for his proper protection." See A. W. Walker, Jr., "Property Rights in Oil and Gas and Their Effect Upon Police Regulation of Production," *Texas Law Review* vol. 16, no. 3 (1938) p. 374.

3. See, for example, Ohio Oil Co. v. Indiana, 177 U.S. 190 (1900); Bandini Co. v. Superior Court, 284 U.S. 8 (1931); and Champlin Refining Co. v. Corporation Comm'n, 286 U.S. 210 (1931). The Texas Supreme Court reduced the

power of the common law even further by holding that the rule of capture was inapplicable to drainage caused by the negligent acts of the adjoining landowner. See Elliff v. Texon Drilling Co., 146 Tex. 575, 210 S.W.2d 558 (1948). Furthermore, dictum in this case would limit application of the rule of capture only to legitimate drainage caused by an adjoining landowner operating within "the spirit and purpose of conservation statutes and orders of the Railroad Commission." 210 S.W.2d at 562. In Loeffler v. King, 228 S.W.2d 201 (Tex. Civ. App. 1950), *rev'd on other grounds,* 149 Tex. 626, 236 S.W.2d 772 (Tex. 1951), the court held that a mineral interest owner whose land was being drained by illegally spaced wells had a cause of action in damages against the owner of these wells for the drainage.

4. 155 Tex. 221, 285 S.W.2d 201 (1955), *cert. denied,* 351 U.S. 933 (1956).

5. Prior to this case, Ryan had sued to set aside the Railroad Commission order that had denied him a Rule 37 exception permit. The district court affirmed the Ryan denial, but also set aside the permit granted to Pickens on the basis that the Railroad Commission could not grant a well permit to the separate tracts until they had been pooled for joint development by the two lessees. Both parties appealed, and the court of civil appeals reversed the trial court and rendered judgment affirming the commission's grant of Pickens's permit and its denial of Ryan's. Pickens v. Ryan Consolidated Petroleum Corp., 219 S.W.2d 150 (Tex. Civ. App. 1949). The appellate court ruled that neither the Railroad Commission nor the trial court had the power to force a pooling agreement on unwilling owners. However, the decision did not address the issue of whether Ryan could share in the production from Pickens's well after it had been drilled.

6. Ryan Consolidated Petroleum Corp. v. Pickens, 24 Tex. Sup. Ct. Rep. 288 (1955), 4 O.&G.R. 701 (1955). This initial opinion reversed by a 5–4 margin the judgments of the lower courts which had denied Ryan relief on the basis that "(i)t is well established that there can be no compulsory unitization in Texas." Ryan Consolidated Petroleum Corp. v. Pickens, 266 S.W.2d 526 (Tex. Civ. App. 1954).

7. Ryan Consolidated Petroleum Corp. v. Pickens, 155 Tex. 221 at 226, 285 S.W.2d 201 at 207 (1955). The court also held that Ryan had notice of two other well-established rules: the voluntary subdivision rule and the commission's preference right rule. Under the latter, the person who first leases one of the voluntarily subdivided tracts that has been carved out of a larger tract will receive the well permit, unless geological conditions warrant placing the well on another section of the larger tract. Pickens had obtained his lease a few months prior to Ryan's lease. Because Ryan had or should have had knowledge of all of these rules, the fact that he purchased the lease on this voluntary subdivision manifested poor judgment, and equity need not provide relief from this. Despite these arguments, many commentators consider the majority opinion in *Ryan v. Pickens* to be patently unjust. See, for example, Comment, "*Ryan v. Pickens:* The Case for Compulsory Pooling in Texas," *Southwestern Law Journal* vol. 10, no. 2 (1956) pp. 182–197.

8. Hassie Hunt Trust v. Proctor, 215 Miss. 84, 60 So.2d 551 (1952); Griffith v. Gulf Refining Co., 215 Miss. 15, 60 So.2d 518 (1952). These and other equitable pooling cases are discussed in Norman B. Gillis, Jr., "Involuntary Equitable Pooling in Mississippi," *Mississippi Law Journal* vol. 27, no. 1 (1955) pp. 10–31; and Melvin W. Parse, Jr., "Equitable Pooling or Judicial Compulsion?" *Texas Law Review* vol. 34, no. 4 (1956) pp. 623–638. See also Rose v. Damm, 278 Mich. 388, 270 N.W. 722 (1936); and Quinn v. Pere Marquette Ry. Co., 256 Mich. 143, 239 N.W. 376 (1931).

9. 285 S.W.2d 201 at 210–215 (1955). That the railroad commissioners were uncomfortable with the denial of a permit to one of the two lessees is evidenced by their lengthy deliberations in the case. On November 14, 1945, Pickens applied for a well permit on his tract. Two days later Ryan applied for a permit on his tract. The commission denied both applications in December. On February 26, 1946, rehearings on both applications were granted. Rehearing was held on March 1, 1946. On April 3, the commission granted the Pickens application, but ordered that the well be placed on the boundary line between the two leases and that the two applicants share equally in this one well. Rehearings again were granted on both applications, and in July 1946, the commission granted Pickens the permit and denied Ryan's request.

In a subsequent case denying equitable pooling, the court recognized that its holding would allow the owner of a small tract to drain the appellants, who were adjacent landowners, but rationalized the injustice as follows: "We think the real reason appellants feel that they have been defrauded is the proration rule that exists in the Port Alto area which permits a person to secure a lease on one fifty-foot lot and produce oil or gas therefrom without pooling with the surrounding lots. So long as gas may be produced in this field upon such basis, producing companies will no doubt take advantage of such proration rule. It seems that in the future this kind of proration may *not* be permitted in view of the Normanna case...." Waters v. Bruner, 355 S.W.2d 230 at 235 (Tex. Civ. App. 1962). The injustice continues, however. See Brown v. Getty Reserve Oil, 626 S.W.2d 810 (Tex. Civ. App. 1982).

10. 276 S.W. 669 (Tex. Comm'n App. 1925, opinion adopted).

11. Ibid. at 672.

12. Ibid. The court also considered the injustice which might result to Japhet, who had paid $10,000 for the 10 acres, if he were forced to share a third of his production. This injustice was presumed without any evidence that Japhet would have paid less for the tract had he known royalties would have to be shared. The court simply assumed that the parties' silence on the issue of apportionment of royalties bespoke an intent not to share.

13. William O. Huie, "Apportionment of Oil and Gas Royalties," *Harvard Law Review* vol. 78, no. 6 (1965) p. 1534. Other jurisdictions have adopted equitable pooling to apportion royalties in the *Japhet v. McRae* situation. See Parse, "Equitable Pooling," pp. 624–627. Some scholars have commented that it is better to keep the nonapportionment rule because it provides certainty and

stability in the law by making it unnecessary for courts to attempt to divine the parties' intent when the contract is silent on this issue. Leases often contain entireties clauses, which would apportion royalties in a *Japhet v. McRae* situation.

14. Mueller v. Sutherland, 179 S.W.2d 801 (Tex. Civ. App. 1944). The dissent in *Mueller* distinguished *Japhet* on the basis that *Japhet* was decided in 1925 before the field was subject to well-spacing rules. See also, Nale v. Carroll, 266 S.W.2d 519 (Tex. Civ. App. 1954), *aff'd,* 289 S.W.2d 743 (Tex. 1956)(nonapportionment rule applied to two grantees of a subdivided tract even though the producing well had been drilled under a permit issued to the common grantor for the tract as a whole).

15. 144 S.W.2d 303 (Tex. Civ. App. 1940).

16. 144 S.W.2d 303 at 305.

17. Ibid.

18. In Southland Royalty Co. v. Humble Oil & Refining Co., 151 Tex. 324, 249 S.W.2d 914 (1952), Justice Calvert stated that it was "not entirely unappealing" to overrule the *Parker* holding. Ibid. at 916. The Fifth Circuit Court of Appeals has commented on the basic irreconcilability of the two lines of cases. Howell v. Union Producing Co., 392 F.2d 95, 96-101 (5th Cir. 1968).

19. 141 Tex. 425, 174 S.W.2d 43 (1943).

20. The court described the disadvantages which would be suffered by the lessee if forced to take a lease which Mrs. Lee had not accepted: the lessee's freedom in selecting well locations would be circumscribed by controversies and possible liability for drainage if a well were placed on only one of the two tracts. If wells were drilled on both tracts, separate measuring tanks and recordkeeping would be required in order to account to Mrs. Lee for her unpooled share of production. 174 S.W.2d at 47.

21. In Montgomery v. Rittersbacher, 424 S.W.2d 210 (Tex. 1968), a nonexecutive, royalty interest owner was allowed to ratify a lease containing a pooling and entirety clause more than seven years after a producing well had been drilled on adjacent land pooled with his tract. By this time, it was obvious that no well would be drilled on the tract in which the royalty owner had a one-sixteenth interest. Therefore, it was to his advantage to ratify the lease with its pooling clause and receive a diluted, but still substantial, share of the oil produced from the pooled unit. Also see May v. Cities Service Oil Co., 444 S.W.2d 822 (Tex. Civ. App. 1969), which allowed a calculating nonexecutive to ratify a pooling agreement despite his own delay and earlier refusal, once it became obvious that ratification was in his best interest. One justice in the *Montgomery v. Rittersbacher* case dissented as to the majority's choice of the effective date of ratification and then stated that if laches had been pleaded, the nonexecutive may have been barred from asserting the right to share in royalties because of his long delay in seeking ratification. This justice cited Nugent v. Freeman, 306 S.W.2d 167 (Tex. Civ. App. 1957) in which a nonexecutive royalty owner was denied the right to ratify and share in royalties produced from a well on adjoining, pooled acreage because of his three-year delay in seeking ratification.

22. For example, Louisiana permits the executive to pool nonexecutive interests. LeBlanc v. Haynesville Mercantile Co., 230 La. 299, 88 So. 2d 372 (1956).

23. Howard R. Williams and Charles J. Meyers, *Oil and Gas Law,* vol. 2 (New York, Matthew Bender, 1980) sec. 339.3; and Howard R. Williams, "Stare Decisis and the Pooling of Nonexecutive Interests in Oil and Gas," *Texas Law Review* vol. 46, no. 7 (1968) pp. 1013–1032. Were the law to allow the nonexecutive to be pooled without his consent, he would still be protected by the requirement that the executive exercise "utmost fair dealing" in protecting the nonexecutive's interests. This duty of care originated in the case of Schlittler v. Smith, 128 Tex. 628, 101 S.W.2d 543 (1937), and is analyzed in Lee Jones, Jr., "Non-Participating Royalty," *Texas Law Review* vol. 26, no. 5 (1948) pp. 569–606. Williams has stated that the uncertain nature of this duty of care has "undoubtedly had some deterrent effect upon the accomplishment of voluntary programs for cooperative development of producing formations." See Howard R. Williams, "Conservation of Oil and Gas," *Harvard Law Review* vol. 65, no. 7 (1952) p. 1158.

24. The Texas cases are thoroughly analyzed and criticized in Williams, "Stare Decisis."

25. 138 Tex. 341, 159 S.W.2d 472 (1942).

26. For example, a plaintiff-landowner who claimed that the pooling clause in her lease authorized allocation of royalty proceeds on an acreage basis only, sued to free her interest from the pooling agreement entered into by her lessee, which allocated production on a different basis. The case was dismissed for failure to join all other lessees and lessors in the pooled unit. See Leach v. Brown, 251 S.W.2d 553 (Tex. Civ. App. 1952). The plaintiff then filed an amended petition using the device of a class action to join many of the parties. The defendant-lessee again succeeded in having the case dismissed for failure to join necessary parties. The court of civil appeals also dismissed the plaintiff's appeal. See Leach v. Brown, 287 S.W.2d 304 (Tex. Civ. App. 1955). On further appeal, the Texas Supreme Court reversed and remanded the case for trial on the merits of the plaintiff's appeal. Leach v. Brown, 156 Tex. 66, 292 S.W.2d 329 (1956). The court of civil appeals then certified the suit as a proper one for a class action proceeding and remanded it to the trial court. See Leach v. Brown, 298 S.W.2d 185 (Tex. Civ. App. 1956). The trial court granted summary judgment for the lessee, holding that the unitization agreement was authorized by the pooling clause in the plaintiff's lease. On appeal, however, the court found the pooling to be unauthorized and invalid. See Leach v. Brown, 353 S.W.2d 920 (Tex. Civ. App. 1962). Thus, it took more than four years for the lessor to establish the right to sue the lessee, and another six years to litigate the issue on its merits.

Lessees who seek to litigate the validity of unitization agreements face a similar obstacle course. For example, when Francis Oil sued Marathon and Exxon for improperly accounting for the production proceeds of the Yates field unit that Francis Oil had joined (*see* discusson in ch. 6 text at notes 40 to 49), Francis first filed suit in the federal district court in Oklahoma. Francis Oil is an Oklahoma corporation, and this was a convenient forum. The trial court dismissed the suit on the pleadings for failure to join indispensable parties under Rule 19(b) of the

Federal Rules of Civil Procedure. If Francis were required to join all of the 194 working interest owners in the unit, some of the parties aligned as defendants were citizens of Oklahoma and this would destroy the diversity jurisdiction of the federal court. On appeal, the court reversed and remanded for further fact-finding in the trial court on the issue of whether joinder was required. Francis Oil & Gas Inc. v. Exxon Corp., 661 F.2d 873 (10th Cir. 1981). Francis Oil had also filed the same cause of action in the Texas state court which had jurisdiction over the Yates field. After research revealed the *Veal v. Thomason* joinder rule, Francis Oil filed a bilateral class-action suit in this Texas court. The defendants removed this class-action suit to federal court. The federal district court in Texas ruled in favor of Francis Oil on the merits of its claims alleging underpayment from unit proceeds. This decision was appealed to the Temporary Emergency Court of Appeals, which reversed and granted summary judgment in favor of defendants. See Francis Oil & Gas Inc. v. Exxon Corp., 687 F.2d 484 (Temp. Emer. Ct. App. 1982).

27. For this reason, one writer believes that the major oil companies, who are most often the operators of large unitization projects in Texas, are quite content with the *Veal v. Thomason* rule. It provides excellent title insurance for the unit. See Robert H. Dedman, "Indispensable Parties in Pooling Cases," *Southwestern Law Journal* vol. 9, no. 1 (1955) p. 79.

28. Maurice H. Merrill, "Recent Unitization Cases," *Oklahoma Law Review* vol. 6, no. 2 (1953) p. 171.

29. See Joseph J. French, Jr., and Frank W. Elliot, Jr., "The Legal Effect of Voluntary Pooling and Unitization: Theories and Party Practice," *Texas Law Review* vol. 35, no. 3 (1957) pp. 404–405; and Dedman, "Indispensable Parties in Pooling Cases," pp. 84–87. Also see, for example, Phillips Petroleum Co. v. Peterson, 218 F.2d 926 (10th Cir. 1954).

30. Williams and Meyers, *Oil and Gas Law,* vol. 6, secs. 928.5 and 929.2. Other writers have suggested that remedial legislation be enacted. Dedman proposed a statute which permits constructive service on every party to the unit agreement by serving the operator of the unit and the owners of the particular tract of land at issue. See Dedman, "Indispensable Parties in Pooling Cases," p. 81. Others have proposed liberalizing the procedural rules on joinder: see French and Elliott, "The Legal Effect of Voluntary Pooling and Unitization," pp. 422–423. The latter proposal has been effectuated. Rule 39 of the Texas Rules of Civil Procedure was revised in 1971 and now allows a court to determine "whether in equity and good conscience" the action should proceed among the parties before it, or should be dismissed because of the absence of indispensable parties. This change should lessen the number of parties classified as indispensable and allow lawsuits to proceed even in their absence, if equity so requires. See Roy W. McDonald, *Texas Civil Practice in District and County Courts,* vol. 1 (Wilmette, Ill., Callaghan, 1979) sec. 3.23; and Texas Oil & Gas Corporation v. Ostrom, 638 S.W.2d 231 (Tex. Civ. App. 1982). Rule 39 is modeled on Rule 19 of the Federal Rules of Civil Procedure. The Tenth Circuit Court of Appeal's interpretation of Rule 19 in Francis Oil's lawsuit against the Yates unit gave Francis Oil a fighting chance to

overcome the joinder problem. See Francis Oil & Gas Inc. v. Exxon Corp., 661 F.2d 873 (10th Cir.) discussed in note 26 in this chapter.

31. Williams and Meyers, *Oil and Gas Law*, vol. 6, sec. 928.

32. Phillips Petroleum Co. v. Peterson, 218 F.2d 926 (10th Cir. 1954).

33. Williams and Meyers, *Oil and Gas Law*, vol. 6, sec. 936; and French and Elliott, "The Legal Effect of Voluntary Pooling and Unitization," p. 410. See also Kenoyer v. Magnolia Petroleum Co., 173 Kan. 183, 245 P.2d 176 (1952) (pooling clause does not violate rule against perpetuities).

34. Williams and Meyers, *Oil and Gas Law*, vol. 4. sec. 669.9.

35. Hemingway, *The Law of Oil and Gas*, sec. 5.1. The barriers to development of oil and gas lands burdened by multiple ownership, both under the common law and under remedial legislation such as compulsory pooling and receivership acts, are analyzed in Ernest E. Smith, "Methods for Facilitating the Development of Oil and Gas Lands Burdened with Outstanding Mineral Interests," *Texas Law Review* vol. 43, no. 2 (1964) pp. 129–167.

36. See, for example, the cases involving such disputes in Hemingway, *The Law of Oil and Gas*, sec. 5.1(B). Some jurisdictions are more liberal than Texas in allowing the drilling cotenant to recover the costs of marginal wells or dry holes which form part of a larger development program that can be reasonably expected to benefit the nondrilling cotenants. In Texas, the developing cotenant cannot even recover an interest charge for the reasonable use of capital advanced in the operations. Cox. v. Davison, 397 S.W.2d 200 (Tex. 1966). The Texas rule has been criticized because it retards resource development. See Williams and Meyers, *Oil and Gas Law*, vol. 2, sec. 504.3.

37. Shaw & Estes v. Texas Consolidated Oils, 299 S.W.2d 307 (Tex. Civ. App. 1957).

38. For example, in West Virginia, the owner of a 1/768 undivided mineral interest enjoined development of the tract even though all the other cotenants had leased and though the holdout was demanding an exorbitant bonus and the defendant-lessee was willing to account to him for 1/768 of the proceeds without deducting any costs. Law v. Heck Oil Co., 106 W. Va. 296, 145 S.E. 601 (1928). Louisiana and Illinois also follow this minority rule. See Williams and Meyers, *Oil and Gas Law*, vol. 2, sec. 504.2.

39. Whelan v. Placid Oil Co., 274 S.W.2d 125 (Tex. Civ. App. 1954).

40. 398 S.W.2d 276 (Tex. 1966).

41. Superior Oil Co. v. Roberts, 390 S.W.2d 550 (Tex. Civ. App. 1965), *rev'd*, 398 S.W.2d 276 (Tex. 1966).

42. Roberts v. Superior Oil Co., 398 S.W.2d 276, 279 (Tex. 1966).

43. Texas Pacific Coal & Oil Co. v. Kirtley, 288 S.W. 619 (Tex. Civ. App. 1926). The difference in income between remaining a mineral interest owner or ratifying a lease and taking a royalty interest share can be substantial. For

example, assume X and Y are cotenants of Blackacre, each owning an undivided half interest. X leases the entire tract for a one-eighth royalty. Y may elect to ratify the lease and claim a one-sixteenth royalty (one-half of one-eighth) which is one-sixteenth of the oil and gas production cost-free. Or, Y can elect to take one-half the profits equal to one-half the production revenues minus one-half the costs. The profits share is not diluted by the fractional royalty interest in the lease, but it is cost-bearing. If the well is very profitable, the latter option is probably advantageous. If the well is marginal or has high costs, a cost-free royalty share may be preferred.

44. Williams and Meyers, *Oil and Gas Law,* vol. 2, sec. 505.2.

45. Montgomery v. Rittersbacher, 424 S.W.2d 210 (Tex. 1968), discussed above in note 21.

46. In Westbrook v. Atlantic Richfield Co., 502 S.W.2d 551 (Tex. 1973), the supreme court allowed landowners to ratify a fieldwide unit as mineral interest owners (see text of this chapter at notes 73 to 74). By ratifying, Roberts would no longer be able to claim a full one-half profits share from any future well drilled on the 1.5 acres. He would receive only a share of the profits based on his tract's contribution to the unit's proceeds as specified in the unit's allocation formula. Thus, if Roberts ratified, he would not be able to claim benefits from the unit and at the same time repudiate its unfavorable effects on his rights, a situation which the Texas Supreme Court quite correctly sought to prevent. See quotation in text of this chapter at note 42.

47. The nonexecutive can assert a breach of the executive's duty of good faith and fair dealing in appropriate circumstances. See Hemingway, *The Law of Oil and Gas,* sec. 2.2(D).

48. The Texas court did note that "there is nothing to show that Superior has prevented plaintiffs from developing the mineral resources of the town lots in any way." Superior Oil Co. v. Roberts, 398 S.W.2d at 277 (Tex. 1966). Also, the court noted that no unit wells were placed within 1,200 feet of Roberts's lots (the minimum well-spacing distance). This seems to be the full extent of the court's willingness to safeguard the rights of a cotenant who refuses to join a unitized lease and who is subsequently drained by unit wells on adjoining tracts.

49. In Bullard v. Broadwell, 588 S.W.2d 398 (Tex. Civ. App. 1979), Bullard purchased a nonexecutive, undivided one-third mineral interest, subject to his grantor's right to execute oil and gas leases. Broadwell owned the remaining two-thirds mineral interest and the right to lease the entire tract. Broadwell, without leasing, drilled a producing well on the tract and sought to pay Bullard one-third of the usual one-eighth royalty. Bullard sued for a one-third profits share. The appellate court held that in the absence of any valid lease on the tract, Bullard was not limited to a royalty interest share when his cotenant engaged in self-development of the tract. The court refused to follow the suggestion made in a scholarly authority that the parties in this situation probably intended that the nonexecutive would share in lease proceeds only. See Williams and Meyers, *Oil and Gas Law,* vol. 2, sec. 329.

50. Court decisions holding in favor of unit operators against nonjoiners generally express the importance of unitized operations to the development of additional oil and gas and to the state's economy. See, for example, California Co. v. Britt, 247 Miss. 718, 154 So.2d 144 (1963); Baumgartner v. Gulf Oil Corp., 184 Neb. 384, 168 N.W.2d 510 (1969), *cert. denied,* 397 U.S. 913 (1970); and Dobson v. Arkansas Oil & Gas Comm'n, 218 Ark. 160, 235 S.W.2d 33 (1951).

51. A cotenant has an absolute right to partition. This right cannot be denied on equitable grounds that it creates great hardship to the other cotenants. Equitable considerations may, however, dictate whether participation will be ordered in kind or by sale. Partition in kind is generally favored by the Texas courts. Henderson v. Chesley, 116 Tex. 355, 292 S.W. 156 (1927); and Moseley v. Hearrell, 141 Tex. 280, 171 S.W.2d 337 (1943). Partition in kind was granted in Humble Oil & Refining Co. v. Lasseter, 95 S.W.2d 730 (Tex. Civ. App. 1936), and a 1.5-acre tract was created which subsequently received a well permit, allowing the tract to drain a disproportionate share of oil from the adjoining tract. See Humble Oil & Refining Co. v. Lasseter, 120 S.W.2d 541 (Tex. Civ. App. 1938).

52. If the court orders partition by sale, either to prevent the creation of a separate tract which could thwart the unitized operations or because of the difficulty of partitioning in kind in a nonhomogeneous reservoir, then a plaintiff like Roberts will be forced in effect to sell his interest to the unit. This is the ultimate "squeeze play" in the minds of small-tract owners. In Zimmerman v. Texaco, Inc., 409 S.W.2d 607 (Tex. Civ. App. 1966), the owner of an undivided one-twelfth mineral interest brought suit against Texaco, the lessee of the remaining interests, alleging that "Texaco Inc. has practiced upon her the oldest, most time worn, and most despicable fraud known to the oil industry which is charging this plaintiff excessive costs in equipping and operating the lease and deliberately cutting down her production to 'squeeze her out' and make her sell her interest at a low price." 409 S.W.2d 607 at 611.

53. The Railroad Commission staff could not locate a docket that addressed the Altair field unitization, nor any other docket involving the same issue in another field. Letter to author from Susan Cory, Acting General Counsel of the Oil and Gas Division of the Railroad Commission, November 4, 1983.

54. Williams and Meyers, *Oil and Gas Law,* vol. 6, sec. 942. However, the California Supreme Court has upheld the constitutionality of a statute which does not allow an unleased mineral interest owner to share in profits. Hunter v. Justices' Court, 36 Cal. 2d 315, 223 P.2d 465 (1950).

55. Texas Natural Resources Code Ann. sec. 102.052 (Vernon 1978). See also Ernest E. Smith, "The Texas Compulsory Pooling Act," *Texas Law Review* vol. 44, no. 3 (1966) pp. 399–407.

56. Williams and Meyers, *Oil and Gas Law,* vol. 6, sec. 921.10. The API Model Form of Unit Agreement and the Rocky Mountain Operating Agreement Form so provide also (ibid., secs. 920.1 and 920.3).

57. The courts often use the terms *pooling* and *unitization* interchangeably. Most cases involving unitization clauses in leases actually concern pooling clauses.

58. If a lease does not contain a pooling clause, the lessee may still pool his working interest without pooling his lessor's interest. Knight v. Chicago Corp., 144 Tex. 98, 188 S.W.2d 564 (1945); Pinchback v. Gulf Oil Corp., 242 S.W.2d 242 (Tex. Civ. App. 1951). The lessee who pools only his working interest remains subject to all the lease covenants. He must account to the lessor for a full royalty on production from the leased tract; the lease will terminate without production from the tract at the end of the primary term; and the lessee must protect the tract against drainage and develop it as would a reasonably prudent operator.

59. 301 S.W.2d 185 (Tex. Civ. App. 1957).

60. Ibid. at 187–188. In another case, a Texas court upheld a pooled unit against the lessor's claim that the pooled land was insufficiently described to meet the conveyancing requirements of the Statute of Frauds. Kuklies v. Reinhert, 256 S.W.2d 435 (Tex. Civ. App. 1953)(Chief Justice McDonald dissenting). Prior to these cases, some scholars doubted the validity of the usual, blanket pooling provision and advised lessees to negotiate more definite and detailed agreements before executing the act of pooling. See Jones, "Non-Participating Royalty," p. 599.

61. 343 S.W.2d 726 (Tex. Civ. App. 1961).

62. Ibid. at 732.

63. 558 S.W.2d 509 (Tex. Civ. App. 1977).

64. Elliott v. Davis, 553 S.W.2d 223 (Tex. Civ. App. 1977); and Pritchett v. Forest Oil Corp., 535 S.W.2d 708 (Tex. Civ. App. 1976).

65. Smith v. Killough, 461 S.W.2d 510 (Tex. Civ. App. 1970); Banks v. Mecom, 410 S.W.2d 300 (Tex. Civ. App. 1966); Expando Prod. Co. v. Marshall, 407 S.W.2d 254 (Tex. Civ. App. 1966). See generally, Williams and Meyers, *Oil and Gas Law*, vol. 4, sec. 670.2.

66. Grimes v. LaGloria Corp., 251 S.W.2d 755 (Tex. Civ. App. 1952).

67. Leach v. Brown, 353 S.W.2d 920 (Tex. Civ. App. 1962).

68. Sauder v. Frye, 613 S.W.2d 63 (Tex. Civ. App. 1981).

69. 403 S.W.2d 325 (Tex. 1966).

70. Ibid. at 332–333.

71. Ibid. at 331.

72. Wallace F. Lovejoy and Paul T. Homan, *Economic Aspects of Oil Conservation Regulation* (Baltimore, Md., Johns Hopkins University Press for Resources for the Future, 1967) p. 81.

73. Westbrook v. Atlantic Richfield Co., 502 S.W.2d 551 (Tex. 1973).

74. The supreme court reversed the lower court's ruling which had held that the landowners' ratification of the Fairway Unit had revived the lease, and therefore the landowners were limited to royalty owners' shares of unit production. The supreme court's solicitude for preserving the landowners' right to ratify the unit as mineral interest owners in the *Westbrook* case contrasts strangely with its stance in Roberts v. Superior Oil Co., 398 S.W.2d 276 (Tex. 1966) as discussed in the text of this chapter at notes 40 to 56.

75. The winning landowners in *Jones v. Killingsworth* continued to litigate. Their second suit sought damages for an alleged wrongful conspiracy against them by their lessee and the assignees of the lease. Jones v. Hunt Oil Co., 456 S.W.2d 506 (Tex. Civ. App. 1970). The court in this case granted summary judgment for defendant-lessee, finding, *inter alia,* that the landowners had stipulated in the first lawsuit that the pooling had been done in good faith, and that the landowners had subsequently ratified the pooled unit and participated in it (presumably as mineral interest owners). Landowners in other fields also brought suit on claims of unauthorized pooling, inspired by the *Jones v. Killingsworth* precedent. See, for example, Banks v. Mecom, 410 S.W.2d 300 (Tex. Civ. App. 1966).

76. Leopard v. Stanolind Oil & Gas Co., 220 S.W.2d 259 (Tex. Civ. App. 1949). The majority opinion in this case actually veered from a strict constructionist policy and found that the lease authorized units exceeding 40 acres for gas, but not for oil. The court was undoubtedly influenced by the fact that pooling for gas in 1943 was necessary and in the public interest during World War II when materials were scarce. The dissent hewed to the strict approach and found the lease had terminated. The dissent viewed Stanolind's delivery of unit royalty proceeds to the landowners as an "artifice" intended to secure to Stanolind a right which they knew they did not possess.

Other cases finding ratification of otherwise unauthorized pooling include Yelderman v. McCarthy, 474 S.W.2d 781 (Tex. Civ. App. 1971); and Rainwater v. Mason, 283 S.W.2d 435 (Tex. Civ. App. 1955).

77. In the following cases the courts construed lease clauses or upheld pooling and unitization agreements in a manner evincing favorable regard rather than hostility towards the lessees' pooling acts: Phillips Petroleum Co. v. Bivins, 423 S.W.2d 340 (Tex. Civ. App. 1967); Herod v. Grapeland Joint Account, 366 S.W.2d 623 (Tex. Civ. App. 1963); Blocker v. Christie, Mitchell & Mitchell Co., 340 S.W.2d 320 (Tex. Civ. App. 1960); Miles v. Amerada Petroleum Corp., 241 S.W.2d 822 (Tex. Civ. App. 1950); Kuklies v. Reinert, 256 S.W.2d 435 (Tex. Civ. App. 1953). Regarding this basic rule of contract interpretation, see Freeport Sulphur Co. v. American Sulphur Royalty Co., 117 Tex. 439, 6 S.W.2d 1039 (1928); Texas & Pacific Coal & Oil Co. v. Barker, 117 Tex. 418, 6 S.W.2d 1031 (1928); Waggoner Estate v. Sigler, 118 Tex. 509, 19 S.W.2d 27 (1929); and Clifton v. Koontz, 160 Tex. 82, 325 S.W.2d 684 (1959).

78. Brewster v. Lanyon Zinc Co., 140 F. 801 at 814 (8th Cir. 1905). This is the landmark case embodying the principles of implied covenant law. For a discussion of the different judicial rationales used to imply covenants in oil and gas leases,

see Jacqueline Lang Weaver, "Implied Covenants in Oil and Gas Law Under Federal Energy Price Regulations," *Vanderbilt Law Review* vol. 34, no. 6 (1981) pp. 1485–1492.

79. See Hemingway, *The Law of Oil and Gas,* secs. 8.1–8.9.

80. See, for example, John F. Eberhardt, "Effect of Conservation Laws, Rules and Regulations on Rights of Lessors, Lessees, and Owners of Unleased Mineral Interests," *Institute on Oil and Gas Law and Taxation,* vol. 5 (Albany, N.Y., Matthew Bender, 1954) pp. 125–214; Patrick H. Martin, "A Modern Look at Implied Covenants to Explore, Develop, and Market Under Mineral Leases," *Institute on Oil and Gas Law and Taxation,* vol. 27 (Albany, N.Y., Matthew Bender, 1976) pp. 177–213; and Stephen F. Williams, "Implied Covenants for Development and Exploration in Oil and Gas Leases—The Determination of Profitability," *University of Kansas Law Review* vol. 27, no. 3 (1979) pp. 443–458.

81. Implied covenant law does not force lessees to drill wells that are unprofitable. The lessee is obligated to drill only those wells which a reasonably prudent operator would drill. Nonetheless, it may be less costly and more profitable for a lessee to space wells more widely or to pool or unitize than to drill a well on each of his lessors' separate tracts. His lessors do not have the same economic cost calculus and may prefer closer spacing and separate drilling to unitization.

82. The unitization agreement may provide that the royalty burden of a nonconsenting lessor be shared by all the members of the unit rather than absorbed solely by the lessee. See Williams and Meyers, *Oil and Gas Law,* vol. 6, sec. 937.3. The API Model Form of Unit Agreement has alternative provisions on this issue. Ibid., sec. 920.2, pp. 184 and 191 (Article 11.7 of the model form). See also, Kingwood Oil Co. v. Bell, 244 F.2d 115 (7th Cir. 1957) in which a lessee subject to a large overriding royalty in addition to his lessor's one-eighth royalty refused to join a unitization program for secondary recovery unless the other operators agreed to relieve him of some of his obligations; such an agreement was reached. Had it not been, and had the lessee refused to participate in the unit program, the lessor with the one-eighth royalty may have successfully sued the lessee for breach of an implied covenant, upon proof that the lessee could have profitably secured additional production by unitizing the tract, considering only the lessor's one-eighth royalty in the measurement of profitability. See Williams and Meyers, *Oil and Gas Law,* vol. 6, sec. 935.

83. But see Dobson v. Arkansas Oil and Gas Comm'n, 218 Ark. 160, 235 S.W.2d 33 (1951). In this case, the plaintiff was a nonconsenting lessor who sought to invalidate a unitization order issued by the commission which included her interest. The Arkansas Supreme Court held that, absent a compulsory unitization law, the commission had no power to include lessor's interest in the unit. The plaintiff's lessee had joined the unit and would have to account to the lessor on a non-unitized basis. The tract at issue was favorably situated and had a well on it, which produced 312 barrels per day after unitization versus its previous allowable of 250 barrels per day. The court held that the lessor was entitled to royalty

on only 250 barrels per day rather than the 312 barrels actually produced. Otherwise, the lessor would be unjustly enriched by the repressuring efforts of others. This holding obviously violates the rule of capture, but the court wanted to discourage holdouts so that voluntary unitization would be easier to achieve. The Arkansas legislature passed a compulsory unitization law the next year.

Other states have refused to so limit the nonconsenter's royalty. See Smith Petroleum Co. v. Van Mourik, 302 Mich. 131, 4 N.W.2d 495 (1942). While Texas has no case on point, it seems unlikely that the Texas courts would follow the Arkansas precedent. The strength of the rule of capture and lack of equitable pooling and unitization principles seem too firmly established in Texas jurisprudence.

84. The elements of a cause of action for breach of the development covenant are usually more complex than for the drainage covenant. Failure to reasonably develop sometimes results in the permanent loss of oil from the lessor's tract, just as in a drainage case, and the elements are then the same. For example, an inadequate number of wells, poorly placed to conserve reservoir pressure, may leave oil in the ground that could have been produced by additional drilling. Or, the failure to drill enough wells on an updip tract in a water-drive field may result in the permanent loss of oil. See General Crude Oil Co. v. Harris, 101 S.W.2d 1098 (Tex. Civ. App. 1937)(lessor recovered damages for breach of the development covenant from a lessee who had drilled only one well on 400 acres on the eastern edge of the East Texas field). Often, however, failure to develop results in delayed production but no permanent loss of oil. For example, a lessee with a tract of 80 acres in a field under 40-acre spacing and prorationing rules will often prefer to drill only one well on the 80 acres if this one well can drain the entire tract. The lessor may prefer that the operator drill two wells on 40-acre spacing so that the oil can be recovered in a shorter time period. By drilling only one well, the lessee saves drilling and operating costs at the expense of a slower rate of production. This slower rate causes the lessor to lose the use of the capital represented by the minerals remaining in the reservoir for a longer time. The lessor who receives a cost-free royalty does not experience the lessee's offsetting benefit of reduced costs. This type of development dispute is much more difficult to resolve because proof that additional drilling would be profitable to the lessee does not necessarily prove that a reasonably prudent operator would drill. The courts must strike a balance between drilling enough wells to recover the oil at a rate that is fair to the lessor and recognizing the lessee's interest in reducing costs and receiving a fair share of profits. Each case must be decided on its particular facts. See Williams and Meyers, *Oil and Gas Law*, vol. 5, secs. 815 and 832. The importance of well-spacing and prorationing orders by the Railroad Commission in the context of the development covenant is obvious. If a second well will not increase the tract's total allowable, because the first well's allowable will be reduced proportionately under an allowable system based on surface acreage, no breach of covenant will be found. See Clifton v. Koontz, 160 Tex. 82, 325 S.W.2d 684 (1959). The elimination of allocation formulas with large, per-well allowables following the Normanna field decision (see ch. 4 text at notes 196 to 218) considerably reduced the lessor's ability to succeed in development covenant cases. The

wider, statewide spacing rules adopted by the commission in 1965 had a similar effect. A lessee does not violate the prudent operator standard if his failure to drill an offset or development well is due to state prohibitions on this drilling. For this reason, the scene of many implied covenant battles has shifted from the courts to the regulatory commissions. See Williams and Meyers, *Oil and Gas Law,* vol. 5, secs. 832.2 and 866–867; and In re Champlin Refining Co., 296 P.2d 176 (Okla. 1956). State conservation regulation by no means has eliminated implied covenant law however. The prudent operator now has a duty to seek favorable administrative action, such as obtaining exceptions to well-spacing rules or to prorationing orders. See Weaver, "Implied Covenants in Oil and Gas Law," pp. 1531–1557.

85. Unitized operations may greatly increase ultimate recovery in the long run, but reduce short-term royalty payments. For example, if the gas from a field is no longer marketed but is used to repressure the field, the unit will experience a drop in income that may not be made up for several years. In Dobson v. Arkansas Oil and Gas Comm'n, 218 Ark. 160, 235 S.W.2d 33 (Ark. 1951), the unitized operation caused a landowner's royalty payments to fall from $9,600 to $2,800 in the first year.

86. See table 6-1. In large fields, some royalty interest owners simply cannot be found.

87. Rhoads Drilling Co. v. Allred Co., 123 Tex. 229, 70 S.W.2d 576 (1934) (breach of implied covenant for failure to install a pump on an oil well). In Waseco Chemical & Supply Co. v. Bayou State Oil Corp., 371 So.2d 305 (La. Ct. App. 1979), the court found the lessee breached the implied covenant to reasonably develop the leased tract by failing to employ fire flooding (*in-situ* combustion) to increase the tract's oil production. Fire flooding could recover 60 percent of the oil in place versus the 5 percent being recovered by the lessee through stripper-well operations. The case was decided under Louisiana law, but it uses the same principles of implied covenant law that apply in Texas. In this case, the lessee did not need to unitize in order to conduct the fire flood. The Oklahoma courts also recognize an implied covenant to perform secondary recovery. In Re Shailer's Estate, 266 P.2d 613 (Okla. 1954).

88. Sivert v. Continental Oil Co., 497 S.W.2d 482 (Tex. Civ. App. 1973). The case involves the construction of an express clause in a unitization agreement, but the same reasoning would apply in interpreting implied clauses.

89. 622 S.W.2d 563 at 568 (Tex. 1981).

90. The evidence showed that the commission already had granted twenty-two Rule 37 exception permits to Exxon and Amoco in this field (ibid. at 570). The duty to seek favorable administrative action is now firmly established in Texas' jurisprudence of implied covenant law.

91. The lower court had awarded an additional $1.9 million in exemplary damages upon a jury finding that Amoco had deliberately hastened the flooding of the Alexander's tracts through a plug-back program. Amoco's actions had resulted in a dramatic drop in the Alexander tract's production in 1973 from 9,000 barrels to 1,900 barrels per month, and the lower court held that this evidence

supported an award of exemplary damages for tortious conduct. Amoco Production Co. v. Alexander, 594 S.W.2d 467 at 479 (Tex. Civ. App. 1979). The supreme court reversed the award of punitive damages, holding that breach of an implied covenant was a cause of action in contract, not in tort. Amoco Production Co. v. Alexander, 622 S.W.2d 563 at 571 (Tex. 1981).

92. 622 S.W.2d 563 at 569. The court's holding does not seem limited to cases in which the lessee owes different royalty fractions to separate landowners, or to cases in which the lessee has actively sought to hasten the fieldwide drainage.

93. Ibid. at 570. Clearly the court would consider the probability that the Railroad Commission would not grant approval of a unitization agreement which was not signed by at least 65 percent of the royalty interest owners in the field in determining whether Amoco acted as a reasonably prudent operator. See ch. 6 at notes 16 and 17.

94. The supreme court found that Amoco knew that a 25 percent increase in the field's production rate, granted by the Railroad Commission at Amoco's request, would speed up the watering out of downdip tracts. 622 S.W.2d 563 at 570. Amoco itself argued that a change in the fieldwide prorationing formula was the only way to protect the correlative rights of all owners in the field. It is possible that a court could find breach of the implied covenant to protect against drainage due to a lessee's failure to produce more slowly.

95. See, for example, Waters v. Bruner, 355 S.W.2d 230 (Tex. Civ. App. 1962); Spurlock v. Hinton, 225 S.W.2d 203 (Tex. Civ. App. 1949). In Phillips Petroleum Co. v. Bivins, 423 S.W.2d 340 (Tex. Civ. App. 1967) and Pinchback v. Gulf Oil Corp., 242 S.W.2d 242 (Tex. Civ. App. 1951), the courts held that a lessee's implied covenant to develop was satisfied by a pooling arrangement, but the lessees had to account to the unpooled, nonconsenting lessors for a full one-eighth royalty, in addition to the proportionate share of royalty paid to the pooled lessors.

96. Byrd v. Shell Oil Co., 178 S.W.2d 573 (Tex. Civ. App. 1944); Miller v. Railroad Comm'n, 185 S.W.2d 223 (Tex. Civ. App. 1945); and Woolley v. Railroad Comm'n, 242 S.W.2d 811 (Tex. Civ. App. 1951). These cases are discussed in ch. 4 at note 103. See also Dailey v. Railroad Comm'n, 133 S.W.2d 219 (Tex. Civ. App. 1939).

97. The Alexanders might have recovered even greater damages for Amoco's failure to unitize than from its failure to drill replacement wells, because royalties from unitized production would continue even after the replacement wells had watered out. In Williams v. Humble Oil & Refining Co., 432 F.2d 165 (5th Cir. 1970), *cert. denied,* 402 U.S. 934 (1971), the federal court, in a case arising from Louisiana, concluded that a lessor could recover damages for the failure of his lessee to pool his tract into a drilling unit with adjacent tracts leased by the same lessee, thereby protecting the tract against drainage from wells on the adjacent land. The court measured damages as the amount that the plaintiff would have received from the pooled unit had it been formed. 432 F.2d at 173. Because pooling is almost always done on a surface acreage basis, actual damages would

be easy to measure in such a case compared to damages for the failure to unitize, where no standard participation formula applies.

98. A lessee could negotiate to pay "in lieu" royalties to the downdip landowners, that is, in lieu of drilling wells, these landowners would receive additional royalty from wells located off their tracts. The in lieu royalty in no way diminishes the royalty owed the updip landowners. The lessee who pays these double royalties will reach the break-even point in profitable operations sooner than if the field were unitized. The oil field will be abandoned with a greater percentage of oil left in the ground unless the lessee can renegotiate the royalty clauses in the lease. The payment of in lieu royalties may still be less costly than additional drilling however. For example, in Phillips Petroleum Co. v. Bivins, 423 S.W.2d 340 (Tex. Civ. App. 1967), the lessee, whose lease did not authorize pooling, nonetheless pooled tracts under lease from two different landowners and paid a full one-eighth royalty to the owner of the tract with the well located on it and an additional royalty out of its own working interest to the other landowner. This must have been a more efficient way of protecting the second landowner from drainage than drilling a second well on his tract. It is a very expensive way to pool however.

99. A lessee's duties to downdip lessors would not excuse the breach of any covenants to updip lessors. See quotation in the text of this chapter at note 92. See also U.S. Steel Corp. v. Whitley, 636 S.W.2d 465 at 472 (Tex. Civ. App. 1982).

100. 159 F.2d 174 (5th Cir. 1946), *cert. denied,* 331 U.S. 817 (1947).

101. Ibid. at 177.

102. 161 S.W.2d 571 (Tex. Civ. App. 1942).

103. Implied covenants may be negated by express clauses in the lease contract. See Gulf Production Co. v. Kishi, 129 Tex. 487, 103 S.W.2d 965 (1937).

104. 410 S.W.2d 187 (Tex. 1966). The facts are given in the lower court's opinion at 401 S.W.2d 623 (Tex. Civ. App. 1966).

105. 410 S.W.2d at 188 (Tex. 1966).

106. In Herod v. Grapeland Joint Account, 366 S.W.2d 623 (Tex. Civ. App. 1963), the plaintiff-landowner sought to free his tract from a unitized lease. The court held that the lease and the unit division orders executed by the plaintiff authorized unitization. The court also inferred from the plaintiff's earlier conduct that he had intended to unitize; he had sued prior lessees for damages for failing to unitize when nearby operators instituted competitive cycling operations in the field. 366 S.W.2d at 625, 627.

107. See Hemingway, *The Law of Oil and Gas,* sec. 8.7; and David K. Brooks, "Liability of an Oil and Gas Lessee for Causing Drainage: A Standard for Texas," *Texas Law Review* vol. 51, no. 3 (1973) pp. 546–577.

108. Amoco Production Co. v. Alexander, 622 S.W.2d 563 at 572 (Tex. 1981).

109. Tide Water Assoc. Oil Co. v. Stott, 159 F.2d 174 at 179 (5th Cir. 1946), *cert. denied,* 331 U.S. 817 (1947).

110. It is unlikely, but possible, that a court would go beyond the *Tide Water* decision and hold that whenever the lessee makes a fair offer to his lessor to unitize and the offer is refused, the lessee owes no further duties to the lessor and may proceed without risk of any liability under implied covenant law to the non-consenter. See the parallel discussion involving tort liability of the unit to non-consenters in the text of this chapter at notes 157 to 162.

111. "As the plan offered... was reasonable and fair in all aspects, the appellants amply fulfilled any duty of fair dealing which may have been imposed upon them by the lessor-lessee relationship." 159 F.2d 174 at 179.

112. Ibid.

113. Texas Natural Resources Code Ann. sec. 101.013 (Vernon 1978).

114. Railroad Commission interviews; and text of this chapter at notes 117 and 118.

115. See, for example, Railroad Comm'n v. Manziel, 361 S.W.2d 560 (Tex. 1962); Pickens v. Railroad Comm'n, 387 S.W.2d 35 (Tex. 1965); Imperial American Resources Fund, Inc. v. Railroad Comm'n, 557 S.W.2d 280 (Tex. 1977); and Railroad Comm'n v. Graford Oil Corp., 557 S.W.2d 946 (Tex. 1977).

116. The courts often grant equitable relief such as injunctions or specific performance to plaintiffs who prove that their common law remedy of damages is inadequate. See, for example, Waggoner Estate v. Sigler Oil Co., 118 Tex. 509, 19 S.W.2d 27 (1929).

117. 324 S.W.2d 279 (Tex. Civ. App. 1959).

118. Ibid. at 283–284. The commission's order recited that the McLachlan's controversy with the Stroubes was "a private legal matter."

119. Ibid. Because of this ratification, the McLachlans did not appeal the Travis district court's decision which upheld the commission order authorizing the well conversions.

120. Ibid. at 284–285.

121. In *McLachlan v. Stroube,* the McLachlans had no quarrel with the unitization agreement or the planned secondary recovery operation, which indeed they favored (ibid. at 283). Their quarrel was with their assignee's interpretation of the effect of unitization on the McLachlans' sliding-scale royalty in the contract between the McLachlans and Stroube. Warren Petroleum was the unit operator. For a parallel discussion of the lack of injunctive relief against tortious unit operations, see text of this chapter at notes 125 to 162.

122. In Carter Oil v. Dees, 340 Ill. App. 449, 92 N.E.2d 519 (1950), the lessee-plaintiff sought a declaratory judgment allowing him to convert one of four producing wells on a tract to an injection well to repressure the tract. The lessor refused to consent to this conversion, because it would push some oil off the leased tract. The court found that the oil drained away would be more than offset by increased production due to the repressuring, and rendered judgment in favor of the lessee. Usually in implied covenant cases to protect against drainage, proof

of substantial drainage, whether or not offset by migration of oil to the tract, is all that is required. But the Illinois court considered secondary recovery operations to be a valid exception to this rule. See Williams and Meyers, *Oil and Gas Law*, vol. 5, secs. 822.3 and 869. Texas would probably follow the Illinois precedent because of its sound reasoning and the public interest in encouraging secondary recovery. See Railroad Comm'n v. Manziel, quoted in the text of this chapter at note 136.

123. The lessee who is not a common lessee to several landowners in the field, some of whom have consented to unitization and some of whom have not, would not be able to avail himself of the *Tide Water v. Stott* holding (discussed in the text of this chapter at notes 100 to 112). Even if he were a common lessee, the *Tide Water* holding that the lessee did not breach the offset covenant was based on the fact that drilling additional wells on the lessor's tract would not be profitable. If such drilling were profitable, liability would exist. See also, Syverson v. North Dakota State Industrial Comm'n, 111 N.W.2d 128 at 133 (N.D. 1961). The lessee holding other leases in the unitized area from landowners who have consented to the unitization may elect to surrender the leasehold of the nonconsenting lessor. If the surrender occurs before drainage begins, the lessee will not be liable for drainage. The unit operator may still be liable in tort for the drainage however. See text of this chapter at notes 125 to 177.

124. The Railroad Commission would not permit allowables from the converted wells to be transferred off the tract on which the McLachlans had a royalty interest. 324 S.W.2d 279 at 286. This protection of the McLachlans' interests was more than is usually accorded to owners of unsigned interests. The commission typically requires only that the unit operator not convert the *last* producing well on the tract of an unsigned interest owner into an injection well. The added protection was probably a result of the McLachlans' vigorous protest of the unitization's approval. Thus, vigilant landowners may succeed in securing protection of their rights if they are active and alert in bringing the issue to the commission, even though the commission does not take the initiative in protecting the rights of signers and nonsigners to a unitization agreement.

125. Comanche Duke Oil Co. v. Texas Pacific Coal & Oil Co., 298 S.W. 554 (Tex. Civ. App. 1927). A later case held that no trespass is committed by mere vibrations from the explosion of dynamite on nearby land however. Kennedy v. General Geophysical Co., 213 S.W.2d 707 (Tex. Civ. App. 1948).

126. Elliff v. Texon Drilling Co., 146 Tex. 575, 210 S.W.2d 558 at 561 (1948).

127. In Hastings Oil Co. v. Texas Co., 149 Tex. 416, 234 S.W.2d 389 (1950), the court noted that injunctive relief against alleged oil and gas trespassers should be liberally granted because the trespass injury goes to the immediate and irreparable destruction of the minerals.

128. 162 Tex. 39, 344 S.W.2d 420 (1961).

129. 162 Tex. 26, 344 S.W.2d 411 (1961).

130. 344 S.W.2d 414–415.

131. No further litigation seems to have ensued after this pronouncement by the court, so it is still unclear whether sandfracing can be permanently enjoined.

132. The *Delhi–Taylor* cases were decided on February 22, 1961, rehearing denied on April 5, 1961; the *Normanna* decision dated from March 8, 1961.

133. 344 S.W.2d 411 at 417. The defendant cited as authority the case of Corzelius v. Railroad Comm'n, 182 S.W.2d 412 (Tex. Civ. App. 1944), in which the court upheld a commission order which allowed a third party to drill a directional well invading the defendant's mineral estate in order to control a fiery gas well, which was threatening the property and lives of others. The court in *Corzelius* refused to enjoin the directional drilling as a trespass because it was legally authorized and in the public interest. In the *Delhi–Taylor* case under discussion, the Texas Supreme Court refused to regard the *Corzelius* holding as a precedent authorizing the commission to license trespasses generally or to license deliberate actions of lessees which invaded adjoining mineral estates for the purpose of increasing production. 344 S.W.2d 411 at 417.

134. 344 S.W.2d 411 at 417.

135. 361 S.W.2d 560 (Tex. 1962).

136. 361 S.W.2d 560 at 568–569.

137. Raymond M. Myers, *The Law of Pooling and Unitization,* vol. 1 (Albany, N.Y., Banks and Co., 1967) sec. 2.06 (1). Even before the *Manziel* case had been decided, another authority stated that commission approval of a repressuring operation should shield a unit from tort liability to nonconsenters. See Leo J. Hoffman, *Voluntary Pooling and Unitization* (New York, Matthew Bender, 1954) p. 232.

138. 361 S.W.2d 560 at 566.

139. See, for example, Humble Oil & Refining Co. v. L & G Oil Co., 259 S.W.2d 933 (Tex. Civ. App. 1953), holding that Humble's asserted right to possession of the surface of a leased tract could not be adjudicated by the commission because the agency has no authority to authorize a trespass; Gregg v. Delhi-Taylor Oil Co., 162 Tex. 26, 344 S.W.2d 411 (1961), holding that the courts—not the commission—have the power to determine whether a subsurface trespass is occurring, unless the legislature, by a valid statute, has explicitly granted exclusive jurisdiction to the agency. (The Texas 1949 voluntary unitization act was passed to authorize the commission to grant limited antitrust immunity to a voluntarily formed unit, not to grant tort immunity to the unit. The commission expressly disavowed authority to adjudicate the trespass dispute in the *Delhi–Taylor* case); Lonestar Gas Co. v. Murchison, 353 S.W.2d 870 (Tex. Civ. App. 1962), holding that a suit by the gas company for conversion of its stored natural gas was not within the jurisdiction of the Railroad Commission, even though the commission had approved both Lonestar's use of the depleted reservoir for gas storage and Murchison's application to complete a producing well into the same reservoir. The court wrote:

> The Railroad Commission of Texas has no jurisdiction to adjudicate ownership of property. . . . The Railroad Commission is nothing but an administrative body

> appropriately clothed with authority to perform salutary functions of economic nature for the conservation of natural products. Its function is to administer conservation laws only and not to take the place of courts in their historical jurisdictional field.

353 S.W.2d 870 at 881. Also see the cases discussed in note 133 in this chapter.

140. Amerada Petroleum Corp. v. Railroad Comm'n, 395 S.W.2d 403 (Tex. Civ. App. 1965). In McLachlan v. Stroube, 324 S.W.2d 279 (Tex. Civ. App. 1959), the commission disavowed any authority to adjudicate the contract dispute between the two parties. See text of this chapter at note 118. As we have seen, for many years the commission's final orders approving unit agreements under the 1949 Act denied any authority to pass on the equity of the participation formula in the agreement, despite the fact that the 1949 Act requires a finding that the rights of all signers are protected. See ch. 6 text at notes 32 to 37. In Robinson v. Robbins Petroleum Corp., 501 S.W.2d 865 (Tex. 1963), the supreme court held that the commission order approving a unitized water-flooding operation did not purport to diminish the title or otherwise extend the burden upon the separately owned surface estate. Also see, Magnolia Petroleum v. Railroad Comm'n, 141 Tex. 96, 170 S.W.2d 189 (1943); Trapp v. Shell Oil Co., 145 Tex. 323, 198 S.W.2d 424 (1946); Pan American Production Co. v. Hollandsworth, 294 S.W.2d 205 (Tex. Civ. App. 1956); Whelan v. Placid Oil Co., 274 S.W.2d 125 (Tex. Civ. App. 1954); and Shell Oil v. Railroad Comm'n, 133 S.W.2d 791 (Tex. Civ. App. 1940).

141. Woods Exploration & Producing Co. v. Aluminum Co. of America, 382 S.W.2d 343 (Tex. Civ. App. 1965).

142. 186 Kan. 6, 348 P.2d 613 (1960).

143. These facts are given in Tidewater Oil Co. v. Jackson, 320 F.2d 157 at 160 (10th Cir. 1963), *cert. denied,* 375 U.S. 942 (1963).

144. Ibid.

145. Ibid. at 163.

146. Ibid.

147. Champlin Exploration, Inc. v. Railroad Comm'n, 627 S.W.2d 250 (Tex. Civ. App. 1982). In this case, Champlin had complained to the commission that a nearby operator, Humble Exploration Co., had been assigned excessive allowables in violation of the field rules. After a hearing, the commission agreed in part with Champlin and ordered a reduction in Humble's future allowables. Champlin appealed to the district court, alleging that the commission's order did not go far enough in ordering Humble to make up the previous overproduction and that the order erred in its conclusion that correlative rights had not been adversely affected by the violation of the field rules. Champlin feared that the commission's conlusion about correlative rights would bind it in a subsequent claim for damages against Humble and sought to strike this language from the order. The court denied Champlin's claim and held that the commission's determination concerning correlative rights would not be regarded as *res judicata* in any subsequent judicial proceeding brought by Champlin. Prior to the *Champlin* case, considerable

confusion existed on the subject of whether a commission finding or order could operate collaterally to estop a party from alleging a fact contrary to the order in a separate cause of action brought against a private defendant. Several cases held that collateral estoppel would exist in appropriate situations. See Bolton v. Coats, 533 S.W.2d 914 at 917 (Tex. 1975); and Exxon Corp. v. First National Bank of Midland, 529 S.W.2d 110 (Tex. Civ. App. 1975), and earlier cases cited therein. Other cases suggested that collateral estoppel did not exist. See, for example, Zimmerman v. Texaco, Inc., 413 S.W.2d 387 (Tex. 1967). Given the uncertainty in the prior law on this issue, it is possible that the Texas Supreme Court will overrule the *Champlin* decision on this point. The supreme court refused the writ of error in the *Champlin* case, but this fact is not dispositive of the court's acceptance of the reasoning in the lower court's opinion. Even if the Texas Supreme Court were to hold that collateral estoppel could apply to administrative determinations in appropriate circumstances, the *Tidewater Oil Co. v. Jackson* case is precedent for refusing to apply this principle when the administrative agency has no jurisdiction to decide the issue involved in the subsequent judicial proceeding.

148. 320 F.2d 157 at 161-162. The correctness of the federal court's interpretation of Kansas law is affirmed in Robert L. Driscoll, "Secondary Recovery of Oil and Gas: Significance of Agency Approval," *University of Kansas Law Review* vol. 13, no. 4 (1965) pp. 489–492.

149. Even if Kansas law did not impose strict liability, Tidewater probably would have been liable because the federal district court found that Tidewater had intentionally injected water at unreasonable rates. 320 F.2d 157 at 163 (1963). For further background on the dramatic record made in the district court concerning Tidewater's reckless and wanton disregard of the Jackson brother's rights, see Lee Jones, Jr., "Tort Liabilities in Secondary Recovery Operations, " *Rocky Mountain Mineral Law Institute,* vol. 6 (Albany, N.Y., Matthew Bender, 1961) pp. 652–661.

150. See the court's ruling in the *Manziel* case in the text of this chapter at note 136. The Texas Supreme Court has repudiated strict liability for surface pollution of an adjacent landowner's tract caused by an oil and gas operator's overflowing saltwater storage pits. See Turner v. Big Lake Oil Co., 128 Tex. 155, 96 S.W.2d 221 (1936). Such surface invasions of salt water have none of the social utility of subsurface invasions induced by secondary recovery operations, which increase the ultimate recovery of oil and gas. For a critical view of the *Turner* case and of the failure of Texas courts to hold oil and gas operators strictly liable for the many hazardous activities involved in drilling, see Leon Green, "Hazardous Oil and Gas Operations: Tort Liability, " *Texas Law Review* vol. 33, no. 5 (1955) pp. 574–587.

151. William H. Rodgers, Jr., *Environmental Law* (Saint Paul, Minn., West Publishing, 1977) sec. 2.13; and Page Keeton and Lee Jones, Jr., "Tort Liability and the Oil and Gas Industry II," *Texas Law Review* vol. 39, no. 3 (1961) pp. 256 and 262.

152. According to the Restatement of Torts, the intentional invasion of another's interest in land is unreasonable if "(a) the gravity of the harm outweighs the utility of the actor's conduct, or (b) the harm caused by the conduct is serious and

the financial burden of compensating for this and similar harm to others would not make the continuation of the conduct not feasible." American Law Institute, *Restatement of the Law (Second) Torts* (Saint Paul, Minn., American Law Institute Publishers, 1979) sec. 826. Many factors are considered in the process of weighing the harm against the utility of the invasion. See ibid., secs. 827–830.

153. The lack of injunctive relief against invasions of deleterious substances as a result of secondary recovery operations is just one more example of the generalized rule for private nuisances adopted in the *Restatement (Second) Torts* sec. 822, comment (d):

> For the purpose of determining liability for damages for private nuisance, an invasion may be regarded as unreasonable even though the utility of the conduct is great and the amount of harm is relatively small. But for the purpose of determining whether the conduct producing the invasion should be enjoined, additional factors must be considered. It may be reasonable to continue an important activity if payment is made for the harm it is causing....

See, for example, Boomer v. Atlantic Cement Co., 26 N.Y. 2d 219, 309 N.Y.S.2d 312, 257 N.E.2d 870 (1970) (injunction granted against operation of a large cement plant representing an investment of $45 million and 300 jobs, but the injunction to be vacated after defendant paid plaintiffs permanent damages to compensate them for the economic loss to their property caused by invasions of dirt and smoke).

154. In Robinson v. Robbins Petroleum Corp., 501 S.W.2d 865 (Tex. 1963), Robinson, the owner of the surface estate, sued Robbins, the oil and gas lessee of the underlying mineral estate (which was owned by Wagoner), for the wrongful taking of salt water from the plaintiff's tract to repressure the reservoir. The court held that the salt water did, indeed, belong to the surface-estate owner. The lessee had an implied easement to use as much water as was reasonably necessary to develop the minerals under Robinson's tract, but had no right to use the water to benefit other tracts. The court then wrote:

> The fact that the Railroad Commission entered orders approving the recovery units may be relevant to the propriety of the use of the water for production from lands of the Wagoner lease, but no statute or order purports to diminish the title or otherwise extend the burden upon Robinson's surface estate (p.868).

155. For example, in approving the protested SACROC unit, the Railroad Commission limited the unit's transfer of allowables in order to protect the correlative rights of the nonconsenters and to fulfill the required findings in the 1949 Act that the rights of nonsigners be protected. See note 53 in ch. 6. While one commentator's statement that this order would constitute a defense in a suit for damages (see Myers, *The Law of Pooling and Unitization,* vol. 1, sec. 14.05) seems wrong in light of the analysis presented in this chapter, the carefully drawn terms of the order would certainly be persuasive evidence of the reasonableness of the unit's injection operations.

156. Likewise, a commission order is not an impenetrable shield to implied covenant lawsuits brought by nonconsenting lessors. For example, the Railroad

Commission routinely prohibits the lessee of a tract with an unsigned royalty interest from converting the last producing well on the tract to an injection well. But the commission does not determine what the production allowable of this well should be. The unit operator is given a unit allowable to distribute among all the wells in the unit. Thus, the nonconsenting lessor could still allege a cause of action under implied covenant law if the production from this last well were so low that the tract was being drained by wells on other lands in the unit.

157. 361 S.W.2d 560 at 568 (Tex. 1962), quoting from Williams and Meyers, *Oil and Gas Law,* vol. 1, sec. 204.5.

158. In California Co. v. Britt, 247 Miss. 718, 154 So.2d 144 (1963), the court held that a unit had no trespass liability for injecting gas and sweeping oil from a nonconsenter's tract when the holdout had been offered a fair opportunity to join the unit. The court held that the rule of capture applied to the oil drained away from the nonconsenter's tract. Because no wells had been drilled on the holdout's tract, the issue of whether the unit would have been liable for injuring any wells owned by the nonconsenter did not arise, as it had in the *Tidewater Oil Co. v. Jackson* case.

159. The owners of structurally favored tracts to which the oil or gas migrated could still benefit from the positive rule of capture, however, and thus the motivation to hold out would remain in some cases. This fact prompted the Arkansas legislature to pass a compulsory unitization statute following Dobson v. Arkansas Oil & Gas Comm'n, 218 Ark. 160, 235 S.W.2d 33 (1951).

A unit formed by compulsory process is no more immune from tort liability than a voluntary unit. The benefit of compulsory process is that it eases the formation of fieldwide units so that there are no holdouts to complain of injury. If a compulsory unit is formed that only encompasses part of the field, tort liability to the nonjoined owners still exists, as attested by the Oklahoma cases cited in note 160 in this chapter. In Louisiana, a compulsory unitization order must be poolwide, and this explains the absence of cases on a unit's tort liability in Louisiana. See generally, George H. Bowen, "Secondary Operations—Their Values and Their Legal Problems," in *Institute on Oil and Gas Law and Taxation,* vol. 13 (Albany, N.Y., Matthew Bender, 1962) pp. 331–375.

160. See, for example, Young v. Ethyl Corp., 521 F.2d 771 (8th Cir. 1975), *aff'd in part, rev'd in part,* 581 F. 2d (8th Cir.), *cert. denied,* 439 U.S. 1089 (1979); Mowrer v. Ashland Oil and Refining Co., 518 F.2d 659 (7th Cir. 1975); Greyhound Leasing and Financial Corp. v. Joiner City Unit, 444 F.2d 439 (10th Cir. 1971); Gulf Oil v. Hughes, 371 P.2d 81 (Okla. 1962); West Edmond Hunton Lime Unit v. Lillard, 265 P.2d 730 (Okla. 1954); West Edmond Salt Water Disposal Ass'n v. Rosecrans, 204 Okla. 9, 226 P.2d 965 (1950), *appeal dismissed,* 340 U.S. 924 (1950); Boyce v. Dundee Healdton Sand Unit, 560 P.2d 234 (Okla. Ct. App. 1975). In Baumgartner v. Gulf Oil Corp., 184 Neb. 384, 168 N.W.2d 510 (1969), *cert. denied,* 397 U.S. 913 (1970), the plaintiff-lessee had refused a fair offer to unitize. The court refused to find tort liability against the unit, which subsequently flooded out the plaintiff's land, but nonetheless allowed recovery in damages (if proven) on the grounds that plaintiff's correlative rights required protection.

161. Sec, for example, California Public Resources Code sec. 3320.5 (West 1981) which states that "no working or royalty interest owner shall be liable for any loss or damage resulting from repressuring or other operations connected with the production of oil and gas which are conducted, without negligence, pursuant to and in accordance with a cooperative or unit agreement ordered or approved by the Supervisor pursuant to this Article." This immunity from tort liability applies only to units formed to halt land subsidence.

162. An early commentator opined that the state's exercise of its police power through approval of the unitization agreement should and probably would absolve the unit from having to pay compensation to injured property owners. Hoffman, *Voluntary Pooling and Unitization,* pp. 232–233. However, a later authority writes that "...as a water flood program may destroy legally acquired and existing economic values indirectly for a public purpose but which will directly serve to increase production and income to private producers, it would seem to be sufficiently different from the traditional exercise of police power as to require the payment of compensation." Hemingway, *The Law of Oil and Gas,* sec. 4.2. In *Tidewater Oil Co. v. Jackson,* the court of appeals assumed that Kansas law would proceed upon the basic and inescapable proposition that "(n)o man's property can be taken, directly or indirectly, without compensation." 320 F.2d 157 at 163 (10th Cir. 1963). Article I, section 17 of the Texas Constitution provides: "No person's property shall be taken, damaged, or destroyed for or applied to public use without adequate compensation being made, unless by the consent of such person. . . ."

163. See Carter Oil Co. v. Dees, 340 Ill. App. 449, 92 N.E.2d 519 (1950) discussed in note 122 above in this chapter; and Reed v. Texas Co., 22 Ill. App. 2d 131, 159 N.E.2d 641 (1959).

164. Railroad Comm'n v. Manziel, 361 S.W.2d 560 at 564 (Tex. 1962). The court in *Manziel* noted that millions of barrels of salt water had been reinjected into the East Texas field. Ibid. at 568. Were the Manziel theory of subsurface trespass accepted, the court feared that such saltwater reinjection would be halted. The fact that no trespass suits had arisen in East Texas belies this fear. Most producers had consented to the reinjection and proof of net drainage would often be impossible to establish.

165. 184 Neb. 384, 168 N.W.2d 510 (1969), *cert. denied,* 397 U.S. 913 (1970).

166. 247 Miss. 718, 154 So.2d 144 (1963). See also Syverson v. North Dakota State Industrial Comm'n, 111 N.W.2d 128 (N.D. 1961); and Reed v. Texas Co., 22 Ill. App. 2d 131, 159 N.E.2d 641 (1959) (both denying injunctions against unitized water flooding operations in actions brought by nonconsenters).

167. 398 S.W.2d 276 (Tex. 1966). This case is fully discussed in the text of this chapter at notes 40 to 56.

168. 168 N.W.2d 510 at 513. The Nebraska Commission refused to grant a well permit to the plaintiff who thus had to litigate his right to it in court before it was granted. This episode demonstrates how commission orders can create a virtual right of eminent domain to units which are authorized by the state, even when the state has no compulsory unitization powers.

169. Ibid. at 516–517. The court admitted that the *Manziel* fact situation was not analogous (because it involved an appeal of a commission order, not a private tort suit for damages), but proceeded nonetheless to quote from the Texas case.

170. This measure of damages based on lost profits is like that awarded against the unit for tortious conduct in *Tidewater Oil Co. v. Jackson*, 320 F.2d 157 at 164 (10th Cir. 1963). It should be noted that the plaintiff Baumgartner made no claim for the oil that naturally migrated off his tract toward unit wells to the north. The plaintiff admitted that the rule of capture applied to this natural migration. He sued for damages due to the additional migration resulting from the deliberate injection of water by the unit operator. By granting the plaintiff an opportunity to prove these damages, the Nebraska court accepted the plaintiff's premise that the rule of capture does not apply to secondary recovery operations. By contrast, the Mississippi court in the *Britt* case applied the rule of capture to secondary recovery migration and found no liability whatsoever for the unit's drainage of a holdout.

171. See Eugene Kuntz, "Correlative Rights of Parties Owning Interests in a Common Source of Supply of Oil or Gas," in *Institute on Oil and Gas Law and Taxation*, vol. 17 (Albany, N.Y., Matthew Bender, 1966) pp. 236–241.

172. See Brown v. Humble Oil & Refining Co., 126 Tex. 296, 83 S.W.2d 935 (1935), quoted in chapter 4 text at note 114.

173. Railroad Comm'n v. Manziel, 361 S.W.2d 560 (Tex. 1962), and Pickens v. Railroad Comm'n, 387 S.W.2d 35 (Tex. 1965), discussed in chapter 8 text at notes 102 to 105. These two cases define correlative rights in the context of judicial review of agency action and thus are arguably not precedents for the measure of damages to an injured party's correlative rights in common law tort cases. However, the judicial review cases can provide persuasive support for a limited measure of damages to holdouts in tort cases.

174. See Myers, *The Law of Pooling and Unitization*, vol. 2, sec. 15.14; and ch. 6 text at note 39.

175. Assume in a *Baumgartner*-type situation that the profits from joining the unit under its proposed participation formula based on productive acre-feet of sand equaled $12,000, and the profits from drilling on one's own under the field's one-third per well–two-thirds acreage formula equaled $27,000 (even without any in-migration of oil induced by the unit's adjacent repressuring operations). The nonjoined plaintiff injured by the unit's operations is entitled to recover $27,000 under the *Baumgartner* holding.

176. In one case, the Arkansas Supreme Court refused to follow the rule of capture, which would have allowed holdouts whose tract was structurally advantaged to benefit from the unit's increased recovery. In Dobson v. Arkansas Oil & Gas Comm'n, 218 Ark. 160, 235 S.W.2d 33 (1951), the court used the equitable principle of unjust enrichment to deny to nonconsenting royalty-interest owners the right to take their fractional share of the 312 barrels a day produced from the well on their drilling unit. The court limited the holdouts to their fractional share of 250 barrels a day, the amount which the well produced before the unit's

repressuring went into effect. It is extremely doubtful that the Texas Supreme Court would follow such a radical change in the common law. The Texas court has consistently refused to recognize the principle of equitable pooling or unitization. See text of this chapter at notes 1 to 14 and 40 to 42. The *Manziel* case which enunciates a strong pro-unitization stance by the Texas judiciary is of little assistance in a *Dobson*-type situation, which involves the positive—not the negative—rule of capture.

177. See ch. 6 text at notes 50 to 64. The use of time-split participation formulas, which allocate primary production under one set of factors and secondary recovery oil under a different set, is most revealing of the nonconsenters' ability to secure the economic rents from established prorationing rules.

178. Williams and Meyers, *Oil and Gas Law,* vol. 1, sec. 217.

179. Williams and Meyers, *Oil and Gas Law,* vol. 1, secs. 218–218.11.

180. Pooling and unitization provisions typically provide compensation to lessors whose surface lands bear the burden of unitized operations. Cecil N. Cook, "Rights and Remedies of the Lessor and Royalty Owner Under a Unit Operation," in *Institute on Oil and Gas Law and Taxation,* vol. 3 (Albany, N.Y., Matthew Bender, 1952) p. 118; and Williams and Meyers, *Oil and Gas Law,* vol. 6, sec. 921.14.

181. Both fresh and salt water belong to the surface estate owner in Texas. Robinson v. Robbins Petroleum Corp., 501 S.W.2d 865 (Tex. 1963).

182. 483 S.W.2d 808 (Tex. 1972).

183. Ibid. at 812. In a previous case, the court held that a lessee's surface pumps constituted an unreasonable use of the surface by interfering with a farmer's rolling sprinkler system. The use of cellar pumps or electric pumps were reasonable alternatives that would allow accommodation of both the surface and mineral estates. Getty Oil Co. v. Jones, 470 S.W.2d 618 (Tex. 1971). The court in *Whitaker* limited *Getty Oil* to cases in which reasonable alternatives existed on the leased premises.

184. The dissent in *Sun Oil v. Whitaker* pointed out that farm income in Texas totaled $3.4 billion in 1969, compared to the $4 billion value of oil production.

185. The result in *Sun Oil v. Whitaker* would probably be the same even if the oil and gas lease had been subsequent to, rather than prior to, the surface deed. In Texas, the surface deed will be treated as a severance of the mineral estate, and will be subject to its dominance. Hemingway, *The Law of Oil and Gas,* sec. 5.5; and Stanolind Oil & Gas Co. v. Wimberley, 181 S.W.2d 942 (Tex. Civ. App. 1944).

186. 309 S.W.2d 876 (Tex. Civ. App. 1958).

187. 501 S.W.2d 865 (Tex. 1973).

188. Ibid. at 868.

189. The dissent in *Sun Oil v. Whitaker* commented that "in this day of unitizations," it made no sense to restrict the mineral estate's duty to the surface to only

those reasonable alternatives available on the leased premises. 483 S.W.2d 808 at 820-821. The Oklahoma Supreme Court in Holt v. Southwest Antioch Sand Unit, Fifth Enlarged, 292 P.2d 988 (Okla. 1955) held that an oil and gas lessee could take salt water from one tract and inject it into wells on other premises without liability to the surface owner of the first tract. This result is directly contrary to the established common law as enunciated in *Robinson v. Robbins Petroleum Corp.*, and may reflect the Oklahoma court's effort to encourage unitized operations.

190. 321 S.W.2d 167 (Tex. Civ. App. 1959). See also, Humble Oil & Refining Co. v. L & G Oil Co., 259 S.W.2d 933 (Tex. Civ. App. 1953).

191. These events occurred in the Normanna field as the opening foray to the battle between the pooled owners and the unpooled small-tract owners, which culminated in the overthrow of the one-third–two-thirds prorationing formula and the eventual victory for compulsory pooling legislation. See ch. 4 text at notes 196 to 214.

192. Williams and Meyers, *Oil and Gas Law,* vol. 1, sec. 218.6.

Chapter 8

1. 49 S.W.2d 837 (Tex. Civ. App. 1932), *rev'd on other grounds and dismissed,* 122 Tex. 243, 56 S.W.2d 1075 (1933).

2. 49 S.W.2d 837 at 841.

3. Ibid. at 843. The Texas appellate court's willingness to uphold the commission's prorationing order despite the legislative prohibition against preventing "economic waste" contrasts starkly with the federal district court's reluctant approval of such an order even after the legislature had enacted market demand prorationing. See Danciger Oil & Refining Co. v. Smith, 4 F. Supp. 236 (N.D. Tex. 1933).

4. The justices did not require that the commission find that Danciger was producing wastefully on his leases in order to sustain the fieldwide order. It sufficed that Danciger's wells might be destroying the reservoir energy of the pool as a whole. 49 S.W.2d at 842. The justices clearly viewed the field as the relevant area to consider for the regulation and control of waste, a viewpoint which is consonant with unitization principles. See also Railroad Comm'n v. Fain, 161 S.W.2d 498 (Tex. Civ. App. 1942).

5. Danciger Oil & Refining Co. v. Railroad Comm'n, 122 Tex. 243, 56 S.W.2d 1075 (1933). Despite the supreme court's reversal, the court of civil appeals in Austin continued to use the *Danciger* opinion as precedent for the commission's broad authority to prevent waste. See, for example, Corzelius v. Railroad Comm'n, 182 S.W.2d 412 at 416 (Tex. Civ. App. 1944).

6. Railroad Comm'n v. Shell Oil Co., 146 Tex. 286, 206 S.W.2d 235 at 240 (1947).

7. Danciger Oil & Refining Co. v. Railroad Comm'n, 49 S.W.2d 837 at 841 (Tex. Civ. App. 1932), *rev'd on other grounds,* 122 Tex. 243, 56 S.W.2d 1075 (1933).

8. Railroad Comm'n v. Flour Bluff Oil Corp., 219 S.W.2d 506 at 509 (Tex. Civ. App. 1949).

9. Union Pacific Railroad Co. v. Oil and Gas Conservation Comm'n, 131 Colo. 528, 284 P.2d 242 at 243 (1955). The order excepted gas used on the lease premises and gas required to supply nearby domestic and municipal needs.

10. The relevant statutes are cited in the case at 284 P.2d 244–245. One section defined waste in oil fields as: "underground waste, inefficient, excessive, or improper use or dissipation of reservoir energy, including gas energy and water drive." The Colorado statutes specifically prohibited the flaring of casinghead gas in an "excessive or unreasonable amount."

11. Ibid. at 248.

12. Ibid. The same reasoning was used by a California court in denying relief to a unit operator whose cycling efforts in a condensate field were being destroyed by noncycling lessees on adjacent tracts. Relief was to be procured legislatively. Western Gulf Oil Co. v. Superior Oil Co., 92 Cal. App. 299, 206 P.2d 944 (1949).

13. Ultimately in 1957, five years after the commission issued its no-flare order, a unitization agreement for the Rangely field was signed by 99 percent of the working interests and 93 percent of the royalty interests, but only after the Colorado commissioners borrowed another idea from their Texas counterparts. The Colorado commission lowered the gas–oil ratio in the field and issued a show-cause order that indicated that the field allowable would be cut by 20 percent unless a voluntary unitization plan was presented to the agency for approval. Robert E. Sullivan, ed., *Conservation of Oil and Gas: A Legal History, 1948–1958* (Chicago, Ill., American Bar Association, 1960) p. 62. Six companies controlled all of the Rangely field except for 207 acres and five wells, making voluntary unitization fairly easy to achieve as compared with some of the fields subject to the Railroad Commission's orders that involved hundreds of different owners. Because the Colorado operators agreed to unitize, the 20 percent reduction in allowables was never instituted, so it is unknown if it would have survived judicial review.

14. 387 S.W.2d 35 (Tex. 1965).

15. Ibid. at 45.

16. Ibid. at 46.

17. Ibid. at 47.

18. Atlantic Refining Co. v. Railroad Comm'n, 162 Tex. 274, 346 S.W.2d 801 (1961); Halbouty v. Railroad Comm'n, 163 Tex. 417, 357 S.W.2d 364 (1962). These cases are discussed in chapter 4 text at notes 196 to 204.

19. 357 S.W.2d 364 at 376. The dissenting opinion in both cases cited Article 6014(g). The dissent framed the issue in each case as whether the small-tract owner could be compelled to unitize, and then stated that such a result could only be obtained legislatively.

20. See also Dailey v. Railroad Comm'n, 133 S.W.2d 219 (Tex. Civ. App. 1939), discussed in the text of this chapter at notes 113 to 117, which suggests that the commission would have the statutory authority to refuse Rule 37 exception permits to small-tract owners who had refused a fair offer to pool, if the commission followed proper rule-making procedures in promulgating this new approach to Rule 37 exception requests.

21. Texas Administrative Code tit. 16, sec. 3.37 (1982), commonly known as Rule 37.

22. Gulf Land Co. v. Atlantic Refining Co., 134 Tex. 59, 131 S.W.2d 73 at 80 (1939).

23. A Rule 37 exception based on confiscation, that is, on the protection of correlative rights, might still be obtained however.

24. Railroad Comm'n v. Shell Oil Co., 139 Tex. 66, 161 S.W.2d 1022 (Tex. 1942); Wrather v. Humble Oil & Refining Co., 147 Tex. 144, 214 S.W.2d 112 (1948).

25. 571 S.W.2d 497 (Tex. 1978).

26. Ibid. at 501–502.

27. Ivan D. Hafley, "Rule 37 Exceptions: Well Spacing and Density Requirements," *Oil and Gas: Texas Railroad Commission Rules & Regulations* (Austin, State Bar of Texas, 1982) pp. H-19–H-21.

28. Champlin Refining Co. v. Oklahoma Corp. Comm'n, 286 U.S. 210 (1932).

29. Ibid. at 232, 234. Had the only effect of market demand prorationing been found to raise prices, without any impact on the physical waste of oil, the prorationing orders would probably have been struck down, despite the fact that their professed purpose was to prevent waste. However, the Supreme Court found that market demand prorationing bore a reasonable relationship to the prevention of physical waste. The lower federal courts involved in the litigation over the East Texas prorationing orders had a much more difficult time closing their eyes to the real purpose of the orders. In People's Petroleum Producers v. Sterling, 60 F.2d 1041 at 1045, the court wrote:

> We have been, and are, greatly impressed with the manifold evidences of the desire, the dominant purpose, on the part of the oil industry, to get and keep crude prices up, and with, to say the least of it, the complaisant if not compliant attitude of the public officials toward that desire. . . .

However, because of the settled rule that the courts may not inquire into the purposes or motives behind legislative acts—except as this purpose is exhibited in the operation of the law—the district court refused to grant a temporary injunction against the East Texas order, pending a hearing on the merits of whether the order prevented physical waste.

30. See Texas Natural Resources Code Ann. secs. 85.046, 85.202, 86.012, and 86.042 (Vernon 1978). Section 85.046(6) of the code defines waste as "physical waste or loss incident to or resulting from drilling, equipping, locating, spacing, or operating a well or wells in a manner that reduces or tends to reduce the total ultimate recovery of oil or gas from any pool." This section clearly empowers the commission to pass well-spacing rules like Rule 37 on the basis that physical waste (such as the risk of fire hazards, blowouts, or excessive rates of production) is thereby prevented. Rule 37 has never been attacked on the basis that it illegally sought to prevent economic waste rather than physical waste. However, the rule has been honored more in the breach than in its observance. For decades, the protection of the correlative rights of small-tract owners was of greater concern to both the commission and the legislature than the prevention of the economic waste of drilling unnecessary wells. Texas' statutory definitions of correlative rights do not incorporate any right to be protected against having to drill unnecessary wells. Compare the IOCC's 1959 model form quoted in chapter 3 note 130. As we have seen in chapters 3 and 4, the drilling of unnecessary wells is considered a political virtue in Texas, not a sin, and therefore Texas' statutes do not expressly authorize the prevention of economic waste. See also Houston G. Williams and George M. Porter, "Practice Before the Wyoming Oil and Gas Conservation Commission," *Land and Water Law Review* vol. 10, no. 2 (1975) pp. 403–409, describing the Wyoming legislature's refusal to authorize its commission to prevent waste defined in terms of "the drilling of wells not reasonably necessary to effect an economic maximum ultimate recovery of oil and gas from a pool," even though such wells would unnecessarily increase production costs and the costs of products to the consumer.

31. Of course, many court decisions involving correlative rights and per-well factors in prorationing orders had discussed the economic profitability of small-tract drilling. See ch. 4 text at notes 111 and 117. In *Exxon Corp v. Railroad Comm'n*, BTA conceded that use of the existing well bore to recover additional oil would also drain large quantities of gas from the field, but the commission and the court found that drainage from Exxon would be no more harmful from the existing well bore than from a new well at a regular location. Therefore, Exxon's complaint about possible drainage from BTA's exception well was to be made by seeking review of the field's prorationing formula, not the Rule 37 permit.

32. Railroad Comm'n Docket No. 87,482, In re: Application of Hughes and Hughes to Plug Back Well No. 3. Final order issued October 5, 1981; Proposal for Decision issued June 23, 1981, and supplemented on September 30, 1981.

33. The Pettus, S.E. (Wilcox 9000′) field involved in this case is a continuous pressure-depletion gas reservoir, and one well is capable of draining the entire reservoir. The commission had determined at a prior hearing that Hughes and Hughes had 90.79 percent of the net acre-feet in the field and Texas Oil and Gas had 8.23 percent. Hughes and Hughes had two existing gas wells and Texas Oil and Gas had one. The field rules provided for 320-acre spacing and a prorationing

formula based on 100 percent net acre-feet, which should have accorded Hughes and Hughes about 90 percent of the allowables. However, the market demand for gas from this field exceeded the delivery capacity of Hughes and Hughes' two wells. The two wells could not fully produce their assigned allowables, and the remainder was assigned to the Texas Oil and Gas well. As a result, this well began producing as much as 34.7 percent of the field allowable. Hughes and Hughes sought to increase the delivery capacity of their acreage by operating a third well so that no further allowables from their leases would be reassigned to Texas Oil and Gas. In essence, this case involved the age-old problem of the large-tract owner being drained by an adjacent small-tract operator. Even though the proration formula was based entirely on acre-feet, the production of gas from a field like this one, which is at a 100 percent market-demand factor, is essentially on a per-well basis, with each well limited only by its delivery capacity. No issue of waste arises to justify curtailing production below market demand, and under the gas prorationing statutes, the commission must fix gas allowables at the lawful market demand for the gas or at the volume that can be produced without waste, whichever is less. Texas Natural Resources Code Ann. sec. 86.086 (Vernon 1978). Thus, the gas prorationing *system*, not the prorationing formula, allows the owner of a well on a small tract to drain a larger tract. For further discussion of the problems that arise under the gas prorationing system, see text of this chapter at notes 50 to 67.

34. On cross-examination at the commission hearing, Mr. Foster, the engineering expert for Hughes and Hughes, admitted that the cost of drilling a new well would be recovered in about 104 days, even with its $950,000 price tag. Railroad Comm'n Docket No. 87,482, In re: Application of Hughes and Hughes to Plug Back Well No. 3, Exceptions of Texas Oil and Gas Corp. to the Examiner's Proposal for Decision and Recommended Order (July 22, 1981).

35. Technically, the permit was granted to prevent confiscation, but, in fact, the Rule 37 exception was required only to save drilling costs. See also, Railroad Comm'n v. Lone Star Gas Co., 587 S.W.2d 110 (Tex. 1979) (upholding a commission order granting applicant the use of an existing bore to recover his fair share of native gas in an underground gas storage reservoir).

36. Hughes and Hughes did not attempt to support their Rule 37 application with any arguments relating physical waste to economic waste. Such arguments are more difficult to make in the context of depletion gas fields rather than oil fields. Hughes and Hughes might have argued that the capital saved in one operation could be used to explore and develop other oil and gas reserves, resulting in greater ultimate recovery of oil or gas. However, there is no guarantee that cost savings will be invested in additional drilling in Texas, or for that matter in any oil and gas activity. Nonetheless, from the perspective of economic efficiency, the commission's decision to allow Hughes and Hughes to save $900,000 in drilling costs is a good one.

37. 361 S.W.2d 560 (Tex. 1962).

38. Article 6029(4) now appears in Texas Natural Resources Code Ann. sec. 85.202(a)(4) (Vernon 1978).

39. 361 S.W.2d 560 at 570 (Tex. 1962).

40. Also see, Railroad Comm'n v. Fain, 161 S.W.2d 498 (Tex. Civ. App. 1942) (upholding a Commission prorationing order which would injure four wells in a field because the order was necessary to prevent the physical waste of oil in the field as a whole).

41. 143 Tex. 509, 186 S.W.2d 961 (1945).

42. Corzelius v. Harrell, 179 S.W.2d 419 (Tex. Civ. App. 1944), *rev'd on other grounds*, 143 Tex. 509, 186 S.W.2d 961 (1945). On the other hand, Harrell's cycling operation was gradually replacing wet gas with dry gas under Corzelius's tract, and this did not allow Corzelius to produce his fair share of the condensate. Thus, Corzelius and Harrell were draining different elements of the gas field from each other, and the commission order was not fair to either party.

43. This statute now appears as Texas Natural Resources Code Ann. sec. 86.081 (Vernon 1978). The legislative history of this section of the 1935 gas conservation act appears in the text of chapter 3 at notes 148 to 173.

44. Corzelius v. Harrell, 143 Tex. 509 at 515, 186 S.W.2d 961 at 968 (1945). By the time the supreme court heard the appeal, the commission had amended the contested order which the lower courts had enjoined. Because the trial court's opinion enjoining the commission from issuing orders similar to the one appealed was so broad and could handicap the commission in writing future orders, the supreme court reversed the lower court's opinion for mootness and dismissed the case. The U.S. Supreme Court had intimated, in Thompson v. Consolidated Gas Utilities Corp., 300 U.S. 55 (1937), that Section 10 of Article 6008 was a valid exercise of the state's power to legislate in order to protect correlative rights. See also, Railroad Comm'n v. Permian Basin Pipeline Co., 302 S.W.2d 238 (Tex. Civ. App. 1957), *cert denied*, 358 U.S. 37 (1958) (upholding the commission's power to prorate gas and require ratable purchasing by the pipeline in order to protect correlative rights).

45. A. W. Walker, Jr., "Developments in Oil and Gas Law During World War II," *Texas Law Review* vol. 25, no. 1 (1946) pp. 9–10; H. P. Pressler, Jr., "Legal Problems Involved in Cycling Gas in Gas Fields," *Texas Law Review* vol. 24, no. 1 (1945) p. 33; Howard R. Williams, "Problems in the Conservation of Gas," *Rocky Mountain Mineral Law Institute*, vol. 2 (Albany, N.Y., Matthew Bender, 1956) p. 300. The commission escaped the difficult task of protecting the correlative rights of both Harrell and Corzelius to the gas and condensate in the Bammel field because Corzelius's well caught fire and cratered in August 1943. The spectacular blowout caused nearby water wells to become geysers and rapidly spread to other gas wells; crops were destroyed; and sixty or seventy homes were threatened by surface eruptions of seeping gas. The commission issued an order authorizing Harrell to directionally drill a well from his land to be bottomed on Corzelius's tract in an effort to stem the flow from the cratered well. Corzelius protested the order on the basis that the commission had no authority to empower Harrell to commit a trespass. This litigation resulted in another strong judicial opinion upholding the commission's broad authority to issue orders preventing

waste. Corzelius v. Railroad Comm'n, 182 S.W.2d 412 (Tex. Civ. App. 1944). Before the blowout occurred, the commission had issued a second order in the Bammel field which required that all gas produced from the reservoir be processed to extract efficiently all condensate, and that the residue gas either be returned to the reservoir or used for light and fuel, provided that not more than 20 million cubic feet per day could be withdrawn. This order differed from the first one only by seeming to require that Corzelius process his gas through Harrell's plant before selling it. It is still not clear how this second order protected Harrell's right to receive a fair share of the gas. Also, Harrell's cycling operation was displacing wet gas with dry gas from Corzelius's tract, and Corzelius might have alleged a trespass cause of action against Harrell or attacked the commission order as unfair to his right to recover condensate.

46. 152 Tex. 439, 259 S.W.2d 173 (1953).

47. 259 S.W.2d 173 at 177.

48. Ibid. at 177.

49. 405 S.W.2d 338 (Tex. 1966). In this case, the court upheld a commission order requiring an oil operator who had overproduced to compensate by underproducing in the future. The majority opinion was directly contrary to a 1933 opinion by the Texas attorney general which denied the commission any such power. See ch. 3 at note 129. A dissenting opinion argued that the commission had exceeded its statutory authority because the required underproduction was imposed as a penalty for filing false reports and because the statutory penalties for this fraud were exclusive and did not include adjusting future allowables. The gas prorationing laws, unlike the oil statutes, expressly allow balancing of over- and underproduction. See Texas Natural Resources Code Ann. sec. 86.090 (Vernon 1978).

50. 405 S.W.2d 313 (Tex. 1966), *cert. denied,* 385 U.S. 991 (1966).

51. Benz-Stoddard v. Aluminum Co. of America, 368 S.W.2d 94 (Tex. 1963).

52. Railroad Comm'n v. Aluminum Co. of America, 380 S.W.2d 599 (Tex. 1964).

53. Railroad Comm'n v. Woods Exploration & Producing Co., 405 S.W.2d 313 (Tex. 1966), *cert. denied,* 385 U.S. 991 (1966).

54. Woods Exploration & Producing Co. v. Aluminum Co. of America, 284 F. Supp. 582 (S.D. Tex. 1968); 304 F. Supp. 845 (S.D. Tex. 1969), *rev'd in part and remanded,* 438 F.2d 1286 (5th Cir. 1971), *cert. denied,* 404 U.S. 1047; 382 S.W.2d 343 (Tex. Civ. App. 1964).

55. In 405 S.W.2d 313 at 316, the court gave the following illustration of the gross unfairness of the system:

> One of the reservoirs in the Appling Field is designated as Segment 5, 7500′, and 18 wells were completed therein. Seventeen of these wells were located on 7.132 acres, while the remaining well was situated on a 328-acre unit. Since the 17 small tract wells have a large aggregate delivery capacity, their operators were able to

make correspondingly large forecasts for the month of July, 1964. The total delivery capacity of all 18 wells was approximately 898,000 MCF per month, and the production forecasts resulted in the determination that reasonable market demand, and hence the reservoir allowable, for the month was approximately 880,000 MCF. Under the basic 1/3–2/3 allocation formula, the large tract well was entitled to an allowable of 591,000 MCF, but the productive capacity of the best well completed in the reservoir was approximately 87,000 MCF per month. The productive capacity of the single large tract well was only about 34,000 MCF, and in accordance with the usual Commission procedures some 557,000 MCF were allocated to the town lot wells for the one month in addition to their shares of the reservoir allowable as determined by application of the 1/3–2/3 formula.

One reason that the owners of small-tract wells could file forecasts far in excess of the true delivery capacity of their wells was that their wells were multiple completions into several strata. The Middle Kopnicky stratum was the deepest and most productive, and the multiple completions resulted in drainage from the Middle Kopnicky to the thinner, upper-zone sands through well-bore communication. This drainage distorted the producing capability of the upper-zone well completions. Ibid. at 315–316.

56. Sections 1 and 22 of Article 6008 now appear as Texas Natural Resources Code Ann. secs. 86.001 and 86.041 (Vernon 1978).

57. Texas Natural Resources Code Ann. sec. 86.086 (Vernon 1978).

58. Reliance on pipeline purchasers' forecasts would not always solve the problem, especially if the pipeline owner colluded with certain producers in the field to discriminate against others. See, for example, Railroad Comm'n v. Permian Basin Pipeline Co., 302 S.W.2d 238 (Tex. Civ. App. 1957), *cert. denied,* 358 U.S. 37 (1958).

59. 405 S.W.2d 313 at 320–330.

60. 405 S.W.2d 313 at 329. Also see ch. 3 text at notes 148 to 176.

61. 405 S.W.2d 313 at 318, n. 4.

62. 157 S.W.2d 695 (Tex. Civ. App. 1941).

63. The court construed Article 6049d(6), now appearing at Texas Natural Resources Code Ann. sec. 85.054 (Vernon 1978).

64. See ch. 3 text at notes 162 to 176.

65. The majority opinion dispensed with the *Spraberry* and *Corzelius* cases by noting that "none of them throws any light on the question now before us." 405 S.W.2d 313 at 319.

66. In an earlier case, the supreme court struck down a simple and effective method of estimating expected market demand for each reservoir. Beginning with the Henze field in 1955, the commission did not project future market demand for the field; instead it used the total actual production from the reservoir as a base for calculating the individual well allowables. Each well received a

percentage factor representing its fair share of total production. After the actual monthly production total for the reservoir was reported, each well's percentage factor was applied to the total to obtain the well's monthly allowable. If a well had produced more than its monthly allowable, it was cut back in the next months until the overproduction was balanced by underproduction. The Henze method treated all producers fairly and avoided relying on the forecasts and nominations of interested parties who were not always eager to cooperate. But in Rudman v. Railroad Comm'n, 162 Tex. 579, 349 S.W.2d 717 (1961), the supreme court invalidated this system of gas prorationing because the relevant statute required that the market demand for each reservoir and the allowables from each well be determined in advance. The *Woods* case in the Appling field reflects the same narrow interpretation of the commission's authority to operate the mechanics of the gas-prorationing system as the *Rudman* case. Also see, Gage v. Railroad Comm'n, 582 S.W.2d 410 (Tex. 1979) and Railroad Comm'n v. Graford, 557 S.W.2d 946 (Tex. 1977), discussed in the text of chapter 10 at notes 46 to 55, for another example of a narrow judicial interpretation of the commission's gas prorationing authority.

67. Additional wells may be unnecessary to the ultimate recovery of gas, but still be efficient if society values current production more than production in the future. For example, assume a gas field with a market demand of 1 million MCF per day. Three gas wells meet this demand by producing at their maximum capacity of 333,333 MCF per day each. Two wells are owned by X with 90 percent of the field's reserves, and one well is owned by Y with 10 percent of the reserves. X receives a Rule 37 exception permit to protect his correlative rights and drills a fourth well, which also has a capacity of 333,333 MCF per day. If market demand remains at 1 million MCF per day, the field's four wells can now produce more than is demanded. With a prorationing formula based on reserves, X's wells will be prorated to 90 percent of their capacity and will produce 900,000 MCF per day. Y's well will be prorated to produce the remaining 10 percent of market demand. Correlative rights are protected, but the fourth well is completely unnecessary from a conservation standpoint. If, on the other hand, the demand for gas is so strong that the market will purchase 1,333,333 MCF per day, the fourth well is not an inefficient well. At this market demand, all four wells in the field will continue to produce at capacity. X will now produce 75 percent of the field's allowable and Y will produce 25 percent. Correlative rights are still not adequately protected, however. X could seek another Rule 37 exception well.

68. 583 S.W.2d 307 (Tex. 1979).

69. Ibid. at 311.

70. See, for example, Henderson Co. v. Thompson, 300 U.S. 258 (1937) (upholding a Texas statute which prohibited use of sweet gas in the manufacture of carbon black against the complaint that it would allow drainage from the owners of sweet gas to the owners of sour gas); Murray v. R&M Well Servicing and Drilling Co., 297 S.W.2d 225 (Tex. Civ. App. 1956) (upholding lengthy shutdown of a gas-cap well in order to prevent waste even though the oil wells in the same horizon continued to produce); Railroad Comm'n v. Fain, 161

S.W.2d 498 (Tex. Civ. App. 1942) (upholding prorationing order, even though it would cause waste to occur in four wells, because greater waste to thirty-six other wells was thereby prevented.).

71. See chapters five and six. By contrast, in Canada, the Alberta Conservation Commission sets the allowables so low on tracts whose owners refuse to unitize that no incentive exists to go it alone. Howard R. Williams and Charles J. Meyers, *Oil and Gas Law,* vol. 6 (New York, Matthew Bender, 1980) sec. 933.1. The Arkansas commission has also adjusted allowables to prevent holdouts from unduly benefitting from the unit's repressuring. See Dobson v. Arkansas Oil & Gas Comm'n, 218 Ark. 160, 235 S.W.2d 33 (1951). The Mississippi commission has manipulated the assignment of acreage to prorationing units in order to deter holdouts. Corley v. Mississippi State Oil and Gas Board, 105 So.2d 633 (Miss. 1958).

72. Atlantic Refining Co. v. Railroad Comm'n, 162 Tex. 274, 346 S.W.2d 801, 811 (1961).

73. Railroad Comm'n v. Aluminum Co. of America, 380 S.W.2d 599 (Tex. 1964).

74. Ibid. at 604. The court also felt that application of the Normanna rule to all fields would cast an intolerable administrative burden on the commission. Ibid. at 601.

75. Ibid. at 602.

76. The lower courts had found substantial evidence of changed conditions between 1956, the date of the original prorationing order, and 1961, when the Aluminum Co. attacked the order. None of the changed conditions related to physical waste, however. See Railroad Comm'n v. Aluminum Co. of America, 368 S.W.2d 818, 821-826 (Tex. Civ. App. 1963).

77. The statute requires that the commission prorate oil on a "reasonable basis." Texas Natural Resources Code Ann. sec. 85.053 (Vernon 1978). Preventing holdouts from being unjustly enriched seems reasonable. The Spraberry field case (Railroad Comm'n v. Rowan Oil Co.), 152 Tex. 439, 259 S.W.2d 173 (1953) and Railroad Comm'n v. Sample, 405 S.W.2d 338 (Tex. 1966) accord the commission broad discretion in administering oil allowables to protect correlative rights.

78. Staff of the Oil and Gas Division of the Railroad Commission, personal interviews with author, January 13–14, 1983. It was rumored in the oil industry grapevine that the commission would not grant increased MER allowables to operators who refused to join the unitization plan for the Yates field in 1976. This rumor influenced at least one small operator to sign the unit agreement. See ch.6 text between notes 47 and 48. A much earlier lawsuit seeking to force the commission to change the prorationing formula in the Yates field had failed, largely because the formula had been acquiesced in for many years and no changed conditions were found to justify upsetting established property rights. Standard Oil Co. of Texas v. Railroad Comm'n, 215 S.W.2d 633 (Tex. Civ. App. 1948).

79. In addition to the cases discussed in the article, see Gage v. Railroad Comm'n, 582 S.W.2d 410 (Tex. 1979) and Railroad Comm'n v. Graford Oil Co., 557 S.W.2d 946 (Tex. 1977) (both holding that the commission lacks authority to prorate gas from commingled production). These cases are discussed fully in the text of chapter 10 at notes 46 to 55; Harrington v. Railroad Comm'n, 375 S.W.2d 892 (Tex. 1964) (the commission lacks authority to penalize the operator of deviated wells by denying permission to redrill and straighten the wells; statutory penalties are exclusive); Rudman v. Railroad Comm'n, 162 Tex. 579, 349 S.W.2d 717 (1961) (the commission lacks authority to prorate gas by Henze method). Several cases have interpreted the Mineral Interest Pooling Act to deny the commission the authority to issue a forced pooling order: Railroad Comm'n v. Coleman, 460 S.W.2d 404 (Tex. 1970) (the commission lacks authority to force pool upon application of a royalty interest owner); Broussard v. Texaco, Inc. 479 S.W.2d 270 (Tex. 1972) (the commission lacks authority to force pool interests of lessee of two adjoining tracts, both tracts having productive acreage in excess of standard proration unit); Northwest Oil Co. v. Railroad Comm'n, 462 S.W.2d 371 (Tex. Civ. App. 1971) (the commission lacks authority to force pool upon the application of mineral interest owners who did not propose to drill a well); Windsor Gas Corp. v. Railroad Comm'n, 529 S.W.2d 834 (Tex. Civ. App. 1975) (the commission lacks jurisdiction to force pool tracts of land when pooling applicant has not made a fair and reasonable offer to pool voluntarily). The court's narrow construction of the Mineral Interest Pooling Act seems justified by the act's legislative history; the commission was granted the right to issue compulsory orders only on a sparing and limited basis. See ch. 4 text at notes 214 to 218.

80. Railroad Comm'n v. Shell Oil Co., 139 Tex. 66, 161 S.W.2d 1022 (1942). This standard of review comported with that used in the federal courts at the time. See John J. Watkins and Debora S. Beck, "Judicial Review of Rulemaking Under the Texas Administrative Procedure and Texas Register Act," *Baylor Law Review* vol. 34, no. 1 (1982) p. 19.

81. See, for example, Thomas M. Reavley, "Substantial Evidence and Insubstantial Review in Texas," *Southwestern Law Journal* vol. 23, no. 2 (1969) pp. 239–255; and A. W. Walker, Jr., "The Application of the Substantial Evidence Rule in Appeals From Orders of the Railroad Commission," *Texas Law Review* vol. 32, no. 6 (1954) pp. 640–659. Because Texas legislators often felt that the substantial evidence rule placed too much power in an agency's hands, statutes sometimes provide that judicial review of agency orders be by trial *de novo*, not by the substantial evidence rule. Trial *de novo* nullifies, rather than suspends, the order of the agency, and the issue is retried in the court as a regular civil suit. The court ignores the agency proceedings, and the district court's judgment substitutes for the agency order. The statutes authorizing review of Railroad Commission orders (Texas Natural Resources Code Ann. secs. 85.241-.243 (Vernon 1978)) are silent regarding the type of judicial review, and so the substantial evidence rule is applied. Trial *de novo* still exists under APTRA when a statute specifies its use. Texas Revised Civil Statutes Ann. art. 6252-13(a), sec. 19 (Vernon Supp. 1984). See also, Robert W. Hamilton and J. J. Jewett III, "The Administrative Procedure and Texas Register Act: Contested Cases and Judicial Review," *Texas Law Review* vol. 54, no. 2 (1976) pp. 293–307.

82. Texas Revised Civil Statutes Ann. art. 6252-13(a), sec. 19(d)(3)(Vernon Supp. 1984).

83. Texas Revised Civil Statutes Ann. art. 6252-13(a), sec. 19(e)(Vernon Supp. 1984).

84. Ibid. APTRA also requires that the court reverse or remand agency decisions which violate constitutional or statutory provisions, which exceed the agency's statutory authority, which are made upon unlawful procedure, or which are affected by other error of law. APTRA defines the scope of judicial review only for contested cases, that is, cases subject to adjudicatory hearings. However, the substantial evidence standard may also apply to commission rules issued under general notice-and-comment procedures. See Watkins and Beck, "Judicial Review of Rulemaking," pp. 18–30.

85. For example, in Imperial American Resources Fund, Inc. v. Railroad Comm'n, 557 S.W.2d 280 (Tex. 1977), the court upheld the commission findings and order as reasonably supported by substantial evidence, stating that "as in most cases of this nature, the evidence offered by the opposing parties is conflicting, so much so that contrary findings might have been supportable by substantial evidence." Ibid. at 286.

86. 142 Tex. 293, 177 S.W.2d 941 (1944).

87. 193 S.W.2d 824 (Tex. Civ. App. 1946), *aff'd,* 331 U.S. 791 (1947), *aff'd on rehearing,* 332 U.S. 786 (1948).

88. 193 S.W.2d 824 at 834. The validity of the prorationing order in the *Danciger* case, discussed in the text of this chapter at notes 1 to 5, was decided at the Texas civil appeals court level. The federal courts seldom reviewed Railroad Commission orders after 1943. In Burford v. Sun Oil Co., 319 U.S. 315 (1943), *reh'g denied,* 320 U.S. 214 (1943), the Supreme Court held that the federal courts should refuse to review decisions of the commission relating to regulation of the oil and gas industry until after they had been considered by the state courts. Because such questions involved difficult and complex issues of local policy in which the state had a very large interest, concentrating judicial review in the state courts would allow these courts, like the commission, "to acquire a specialized knowledge which is useful in shaping the policy of regulation of the ever-changing demands of this field." Ibid. at 327. This would further the Texas courts' role as "working partners with the Railroad Commission in the business of creating a regulatory system for the oil industry." Ibid. at 326.

89. The district court had declared invalid the commission order. The court of civil appeals, using the substantial evidence rule, reversed the judgment of the district court and affirmed the commission order. Railroad Comm'n v. Marrs, 161 S.W.2d 1037 (Tex. Civ. App. 1942). The supreme court then reversed the appellate court and invalidated the order.

90. Trapp v. Shell Oil Co., 145 Tex. 323, 198 S.W.2d 424 (1946). The supreme court thereby overruled statements in the *Marrs* case which conflicted with the substantial evidence rule. 198 S.W.2d 424 at 433. When the *Trapp* case was first tried, the district court used the substantial evidence rule and upheld the commission order. The appellate court reversed the district court's ruling and remanded

the case to be tried in accordance with the *Marrs* decision, which had not used the substantial evidence rule. The trial court, in the second trial, invalidated the commission order. Ibid. at 441. This sequence shows that the scope of judicial review can significantly affect substantive rights. The second opinion of the district court was upheld on appeal. 189 S.W.2d 26 (Tex. Civ. App. 1945). Ultimately, on appeal to the supreme court, the lower court judgments were reversed, and the supreme court upheld the commission order under the substantial evidence rule, without a further remand. 198 S.W.2d at 441. Chief Justice Alexander's dissent provides a good view of why the substantial evidence rule was, and still is, so controversial. 198 S.W.2d at 442–450.

91. Railroad Comm'n v. Humble Oil & Refining Co., 193 S.W.2d 824 at 832 (Tex. Civ. App. 1946).

92. Ibid. at 832–833. The same concern that the large-tract owners' reduced drilling and operating costs put small-tract producers at a disadvantage is evident in F. A. Gillespie & Sons Co. v. Railroad Comm'n, 161 S.W.2d 159 (Tex. Civ. App. 1942). The commission had established 10-acre spacing and prorationing units in an oil field with allowables based completely on surface acreage. Atlantic Refining owned many 5-acre tracts and assigned leases on twenty-seven of them to Gillespie. Gillespie secured pooling agreements and combined these twenty-seven tracts into nine tracts of 15-acres each and drilled nine wells. He procured allowables on the basis of 15 acres per well. Shasta, an adjoining lessee, protested Gillespie's grant of allowables based on 15-acre units because the field rules still provided for a 10-acre spacing pattern. The court held that the commission had no right to violate its own field rules and grant Gillespie allowables based on 15-acre spacing because this discriminated against other operators in the field who incurred greater costs by drilling on a 10-acre pattern. The holding in this case is obviously anti-pooling and anti-efficiency, although the court probably would have upheld the 15-acre spacing and allowable units if the commission had promulgated a field rule authorizing such units.

93. 193 S.W.2d 824 at 828. Humble Oil had no quarrel with the concept of a per-well allowable factor; the company simply considered a 50 percent factor excessive, and suggested 40 percent would be fairer to the large tracts.

94. The substantial evidence rule helped to sustain even more irrational prorationing formulas than that in the Hawkins field. For example, in the New Refugio oil field, the allowable formula was 100 percent per well with no acreage or well potential factor. The formula was upheld against Mackhank's challenge that it was confiscatory. In particular, the court noted that Mackhank could have used self-help to drill more densely and protect itself against drainage, at least until World War II intervened and imposed drilling restrictions on producers. Railroad Comm'n v. Mackhank Petroleum Co., 186 S.W.2d 351 at 356 (Tex. Civ. App. 1945). A second challenge sought to invalidate the Commission's exemption of two wells, owned by W. R. R. Oil Co., from shutdown days on the basis that these wells would then drain oil from Mackhank, whose wells were shut in for seven days per month under market demand prorationing. The appellate court had ruled that the exemption was unreasonable because the commission

could shut in W. R. R.'s two wells and transfer their allowables to W. R. R.'s third well, thus avoiding all possibility of inequitable discrimination. 186 S.W.2d 351 at 358–359. On appeal, the Texas Supreme Court held that substantial evidence existed to support the commission's exemption order on the basis of waste prevention, and that "the Court cannot strike down an administrative order on the ground that the evidence heard by the Court indicated that a more equitable one could be entered." Railroad Comm'n v. Mackhank Petroleum Co., 144 Tex. 393, 190 S.W.2d 802 at 804 (1945). See also, Standard Oil Co. of Texas v. Railroad Comm'n, 215 S.W.2d 633 (Tex. Civ. App. 1948) (upholding the prorationing order in the Yates field based on average daily well potentials).

95. Railroad Comm'n v. Aluminum Co. of America, 380 S.W.2d 599 (Tex. 1964) (holding that laches barred large-tract owners from seeking to invalidate the one-third–two-thirds gas-prorationing formula). See also, Midas Oil Co. v. Stanolind Oil & Gas Co., 142 Tex. 417, 179 S.W.2d 243 (1944) (holding that the statutory right of an adjoining lessee to contest the granting of a Rule 37 exemption may be lost by laches or waiver if the contestant unreasonably delays in instituting suit until after the permittee has made expenditures in reliance on his permit).

96. 126 Tex. 296, 83 S.W.2d 935 (1935), *on motion for rehearing,* 126 Tex. 296, 87 S.W.2d 1069 (1935).

97. 83 S.W.2d 935 at 944.

98. 87 S.W.2d 1069 at 1069-1070.

99. Ibid. at 1070.

100. Railroad Comm'n v. Humble Oil & Refining Co., 193 S.W.2d 824 at 834 (Tex. Civ. App. 1946), *aff'd,* 331 U.S. 791 (1947), *aff'd on rehearing,* 332 U.S. 786 (1948).

101. 193 S.W.2d 824 at 833.

102. 387 S.W.2d 35 (Tex. 1965).

103. Pickens's other attack on the prorationing order, as an invalid attempt by the commission to force unitization, is discussed in the text of this chapter at notes 14 to 19.

104. 361 S.W.2d 560 (Tex. 1962).

105. Ibid. at 573. The Whelans owned 41 percent to 46 percent of the total productive acre-feet in the reservoir. They had recovered only 35 percent of the oil during primary production. The irregular injection well would allow them to recover 51 percent of the oil during secondary recovery. Without this injection well, the Whelans would recover only 43 percent of the oil during secondary recovery. Even though this 43 percent figure equaled the Whelans' original ownership of total reserves, the court upheld the commission order giving the Whelans 51 percent of the secondary recovery oil.

106. See, for example, Brown v. Humble Oil & Refining Co., 126 Tex. 296, 83 S.W.2d 935 (1935); Gulf Land Co. v. Atlantic Refining Co., 134 Tex. 59, 131

S.W.2d 73 (1939). In such cases, even the substantial evidence rule could not sustain the commission's orders granting permits. In *Gulf Land,* a 6.88-acre tract in East Texas had been voluntarily subdivided into six smaller tracts after oil was discovered. All six tracts had acquired Rule 37 exception wells. Clearly, five of these were illegal, but no owner or operator had bothered to attack the commission permits for any of these six wells. Only when the lessee of one of the six small tracts sought and received a permit for a second well (for a total of seven wells on the 6.88 acres) did the operator adjacent to the 6.88 acres bring suit against the commission order granting the well.

107. Gulf Land Co. v. Atlantic Refining Co., 134 Tex. 59, 131 S.W.2d 73 (1939); Railroad Comm'n v. Shell Oil Co., 139 Tex. 66, 161 S.W.2d 1022 (1942), the Trem Carr case. The courts were a little slow in rejecting the "more wells, more oil" theory. Before 1940, the Texas courts found that Rule 37 exception wells were required to prevent waste on tracts of less than three acres in the East Texas field. See, for example, Humble Oil & Refining Co. v. Railroad Comm'n, 112 S.W.2d 222 (Tex. Civ. App. 1937); Humble Oil & Refining Co. v. Turnbow, 133 S.W.2d 191 (Tex. Civ. App. 1939).

108. Wrather v. Humble Oil & Refining Co., 147 Tex. 144, 214 S.W.2d 112 (1948); Hawkins v. Texas Co., 209 S.W.2d 338 (Tex. 1948).

109. Byrd v. Shell Oil Co., 178 S.W.2d 573 (Tex. Civ. App. 1944). See also, Miller v. Railroad Comm'n, 185 S.W.2d 223 (Tex. Civ. App. 1945); Kraker v. Railroad Comm'n, 188 S.W.2d 912 (Tex. Civ. App. 1945); Woolley v. Railroad Comm'n, 242 S.W.2d 811 (Tex. Civ. App. 1951). But see F. A. Gillespie & Sons Co. v. Railroad Comm'n, 161 S.W.2d 159 (Tex. Civ. App. 1942), discussed in this chapter at note 92.

110. Railroad Comm'n v. Williams, 163 Tex. 370, 356 S.W.2d 131 (1961). The courts still needed to stay the commission's hand at times to change its habitual practice of granting Rule 37 exception permits. See Coloma Oil & Gas Corp. v. Railroad Comm'n, 163 Tex. 483, 358 S.W.2d 566 (1962); Railroad Comm'n v. Humble Oil & Refining Co., 424 S.W.2d 474 (Tex. Civ. App. 1968). Despite the crackdown on Rule 37 grants, the commission was still generous to slant-well operators caught up in the East Texas dragnet in the early 1960s. Many operators who had illegally drilled slant wells applied for Rule 37 exception permits for the bottom hole locations of their deviated wells, and received them. The court approved the commission's policy of granting these exceptions. Stewart v. Humble Oil & Refining Co., 377 S.W.2d 830 (Tex. 1964).

111. Railroad Comm'n v. Williams, 163 Tex. 370, 356 S.W.2d 131 (1961).

112. Brown v. Humble Oil & Refining Co., 126 Tex. 296, 83 S.W.2d 935 at 944 (1935), *on motion for rehearing,* 126 Tex. 296, 87 S.W.2d 1069 (1935).

113. 133 S.W.2d 219 (Tex. Civ. App. 1939).

114. Ibid. at 222. See also, F. A. Gillespie & Sons Co. v. Railroad Comm'n, 161 S.W.2d 159 (Tex. Civ. App. 1942), discussed at note 92 of this chapter.

115. The commission issued such an order for the North Houston field on May 9, 1939. Robert E. Hardwicke, "Oil-Well Spacing Regulations and Protection of Property Rights in Texas," *Texas Law Review* vol. 31, no. 2 (1952) p. 120, n. 42.

116. Ibid. at 121–122, n. 43; Robert E. Hardwicke and M. K. Woodward, "Fair Share and the Small Tract in Texas," *Texas Law Review* vol. 41, no. 1 (1962) pp. 93–101. The most serious problems involve the legal rights of lessors. If the lessee of Tract A were granted a Rule 37 exception permit only on condition that he share the production from the well with the lessee and lessor of Tract B, the lessor of Tract A might nonetheless demand a full royalty from the well on his own tract. The lessor of Tract B might demand a well of his own and sue his lessee for breach of an implied covenant. In Colorado Interstate Gas Co. v. Sears, 362 S.W.2d 396 (Tex. Civ. App. 1962), the court affirmed the commission's grant of a Rule 37 permit to Sears, the lessee of a 7.1-acre tract, even though Sears had made no attempt to pool into one of the adjoining 640-acre units owned by Colorado Interstate. Colorado Interstate testified that it would have granted Sears' request to pool, but the court held that because Colorado's leases did not contain pooling clauses, it was uncertain that Colorado could have pooled its lessors with the Sears' tract. Therefore, the trial court had properly excluded evidence about Sears' failure to request pooling.

117. Halbouty v. Railroad Comm'n, 163 Tex. 417, 357 S.W.2d 364 at 376 (1962), *cert. denied,* 371 U.S. 888 (1962).

118. Robert E. Hardwicke, *Antitrust Laws v. Unit Operation of Oil and Gas Pools* (rev. ed., Dallas, Tex., Society of Petroleum Engineers of the American Institute of Mining and Metallurgical Engineers, 1961) p. 146. Recycling would cause oil production to drop from 17,400 barrels to 14,700 barrels per day in the short term, however. Under these conditions, voluntary unitization is very difficult to achieve.

119. Railroad Commission of Texas, *Secondary and Enhanced Recovery Operations in Texas to 1980: Bulletin 80* (Austin, Tex., 1980) p. 20; and Blakely M. Murphy, ed., *Conservation of Oil and Gas: A Legal History, 1948* (Chicago, Ill., American Bar Association, 1949) p. 470.

120. Railroad Comm'n v. Shell Oil Co., 146 Tex. 286, 206 S.W.2d 235 at 242 (1947).

121. Railroad Comm'n v. Sterling Oil & Refining Co., 147 Tex. 547, 218 S.W.2d 415 (1949).

122. 218 S.W.2d 415 at 427.

123. Railroad Comm'n v. Flour Bluff Oil Co., 219 S.W.2d 506 (Tex. Civ. App. 1949).

124. Ibid. at 508. It is interesting to note that the court completely misses the holdout problem that frustrates private efforts to avoid waste under the rule of capture and calls for the compulsion of law.

125. Railroad Comm'n v. Rowan Oil Co., 152 Tex. 439, 259 S.W.2d 173 (1953).

126. 259 S.W.2d 173 at 177.

127. See ch. 6 text at notes 28 to 61. Another good example of the propensity to settle appears in the background facts of Schlipf v. Exxon Corp., 626 S.W.2d 74 (Tex. Civ. App. 1981), *aff'd,* 644 S.W.2d 453 (Tex. 1982). Exxon, the operator of a large unit in the Katy gas field, settled a lawsuit brought by a group of unsigned landowners. In exchange for dismissing the lawsuit against the unit and for executing the unitization agreement, Exxon promised to pay the Schlipfs a guaranteed royalty payment of $9,350 per royalty acre from January 1, 1969 to January 1, 1999. If the Schlipfs' share of unitized royalty proceeds did not amount to this sum, Exxon would guarantee the remainder. If the Schlipfs' share of unitized proceeds exceeded this sum, the settlement agreement provided that Exxon would pay the excess as a lump sum at the end of the thirty years. When gas prices zoomed upward in the 1970s, the Schlipfs demanded that the excess royalties be paid immediately, rather than in 1999, and this demand precipitated a lawsuit. The court held the parties to the plain terms of their settlement agreement, and the Schlipfs' demand was denied.

One unitization agreement is extraordinary for its litigiousness. In Susanoil, Inc. v. Continental Oil Co., 516 S.W.2d 260 (Tex. Civ. App. 1973), and Susanoil, Inc. v. Continental Oil Co., 519 S.W.2d 230 (Tex. Civ. App. 1975), the plaintiff-lessees sought damages against Continental for fraud in obtaining a unitization agreement for water flooding. The plaintiffs alleged that during the negotiations which led to their executing the unitization agreement, Continental assured them that all owners of existing equipment in the field would be equally treated, and all would assign their leases and equipment to the unit operator rent-free. Continental then made an undisclosed agreement with another group of lessees in the field allowing this group to remove some of their well equipment and realize its salvage value. After Susanoil's managers learned of this agreement, they demanded that Continental pay them the reasonable rental value of their leasehold equipment used by the unit. Both of the cases involve procedural issues raised in the course of litigation; there is no record of the outcome on the substantive issue of fraud. The plaintiffs did not seek to attack the commission order granting approval of Continental's unitization application. The plaintiffs probably could have invoked the provision in the 1949 Act which requires that all owners in the field be given an opportunity to enter the unit on the same yardstick basis in order to invalidate the commission's approval. In Sivert v. Continental Oil Co., 497 S.W.2d 482 (Tex. Civ. App. 1973), the same plaintiffs alleged that Continental, the unit operator, had violated the terms of the unitization agreement by ceasing the water-flood operation for more than ninety consecutive days. The court rendered judgment for Continental. The dissension in this unit operation is the exception, not the rule, at least in the extent to which it surfaced into lawsuits.

128. 358 S.W.2d 706 (Tex. Civ. App. 1962).

129. Ibid. at 711.

130. Comment, "Prospects for Compulsory Fieldwide Unitization in Texas," *Texas Law Review* vol. 44, no. 3 (1966) pp. 528–529.

131. Texas Natural Resources Code Ann. sec. 101.004 (Vernon 1978). Sections 85.002 and 86.004 of this code also assert the superiority of the antitrust laws to any of the conservation laws and orders administered by the Railroad Commission.

132. See ch. 3 text at note 78. For many years, one of the forbidden types of conduct was a trust, defined as "a combination of capital, skill, or acts by two or more persons, firms, corporations, or associations of persons," formed to "regulate, fix, or limit the output of any article or commodity which may be manufactured, mined, produced or sold." Texas Revised Civil Statutes Ann. art. 7797 (1911). A monopoly was defined as "a combination of two or more corporations effected by bringing the direction of their affairs under common management or control" to create or tending to create a trust. Texas Business & Commerce Code Ann. sec. 15.01 (Vernon 1968). Voluntary unitization agreements were clearly susceptible to antitrust charges under these laws, and for this reason, the 1949 voluntary unitization act was important to producers, especially the majors.

133. The Texas Free Enterprise and Antitrust Act of 1983, now appearing in the Texas Business & Commerce Code Ann. secs. 15.01 to 15.40 (Vernon Supp. 1984). Reform of the state's antitrust law was desirable because the prior law, almost unchanged since 1889, was poorly drafted to meet the demands of a more complex and changing economic system. This prior law consisted of a list of prohibited types of conduct. If a business practice fell into one of the listed categories, it was declared illegal, even though it might have been a practice that actually promoted competition. Conversely, some practices that may have lessened competition were exonerated from antitrust liability. See Steve Baron, "A New Era in Texas Antitrust: The Texas Free Enterprise and Antitrust Act of 1983," *Trial Lawyers Forum* (October–December 1983) p. 16. Reform had been urged for decades prior to 1983. See M. S. Breckenridge, "Some Phases of the Texas Anti-trust Law," *Texas Law Review* vol. 3, no. 4 (1925) pp. 335–362; and *Texas Law Review* vol. 4 (1926) pp. 129–153; and Charles B. Nutting, "The Texas Anti-trust Law: A Post Mortem," *Texas Law Review* vol. 14, no. 3 (1936) pp. 293–304.

134. Texas Business & Commerce Code Ann. sec. 15.04 (Vernon Supp. 1984).

135. Ibid. at secs. 15.05(a), 15.05(b).

136. Ibid. at sec. 15.05(g). See also sec. 15.05(d)(4).

137. Operators with cooperative projects that do not meet the requirements of the 1949 Act can nonetheless receive commission approval of transfer allowable rules, well locations, injection rates, and the like for their project. Arguably, under the 1983 Act, agency approval of these items would immunize the operators from charges alleging that allowables, injection rates, or well locations were being used anticompetitively in violation of the antitrust laws. However, the commission's approval of these items would not amount to specific approval of

the cooperative nature of the joint undertaking, such as approval under the 1949 Act provides, and thus might not immunize the operators from antitrust liability for combining or contracting to restrain trade.

138. In Pickens v. Railroad Comm'n, 387 S.W.2d 35 (Tex. 1965), the court sustained the commission's prorationing order in the Fairway field, even though the plaintiffs argued that the order was designed to force operators to unitize. The court wrote that "the fact that persons or corporations owning 88 or 92 percent of the production had been able to agree to the 50–50 formula which they recommended to the Commission, does not prove a relinquishment of authority by the Commission to anyone," but in a footnote, the court appended the statement that "no question regarding the antitrust laws is here raised or passed upon." Ibid. at 45. This seems a muted warning to the operators who controlled 90 percent of the field that the antitrust laws were available if evidence of coercion appeared. Under the 1983 Texas antitrust act, the operators subject to the prorationing order would seem immune from allegations of antitrust violations arising from the prorationing order. See also, Zimmerman v. Texaco, Inc., 409 S.W.2d 607 (Tex. Civ. App. 1966) (cotenant of a one-twelfth interest alleged a squeeze play and conspiracy against her by the lessee of the other eleven-twelfths interest, but this count was dropped in amended petition); Danaho Refining Co. v. Pan American Petroleum Corp., 383 S.W.2d 941 (Tex. Civ. App. 1964) (antitrust action by a small refiner against operators who had unitized their oil and gas leases and constructed their own recycling plant; instructed verdict for the defendants affirmed on the basis that no evidence existed to support the claims of antitrust violations); Railroad Comm'n v. Permian Basin Pipeline Co., 302 S.W.2d 238 (Tex. Civ. App. 1957), *cert. denied,* 358 U.S. 37 (1958) (commission order requiring ratable takes by a pipeline upheld in light of evidence that pipeline company and the field's largest producer were cooperating to exclude other producers from using gathering and plant facilities and that fictitious nominations and forecasts of demand for gas were causing nonratable takes); Pabst v. Roxana Petroleum Corp., 125 Tex. 52, 80 S.W.2d 956 (1935) (plaintiff-lessors who refused to sign an agreement pooling their small tract with adjacent lands for sulfur production held to have no cause of action against lessee for conspiracy or antitrust violations when the lessee proceeded to produce sulfur from the adjacent tracts and allegedly destroyed the value of plaintiff's land).

139. Woods Exploration and Producing Co. v. Aluminum Co. of America, 382 S.W.2d 343 (Tex. Civ. App. 1964) (holding that the Railroad Commission had no jurisdiction over antitrust claims. The district court of Calhoun County had dismissed the case on the grounds that venue lay only in the district court of Travis County which has exclusive jurisdiction for review of Railroad Commission orders. The appellate court reversed and remanded the case for trial in the Calhoun County court on the antitrust claims arising under state law). The federal cases are discussed below at notes 140 to 168.

140. In one of the several antitrust suits against the major oil companies spawned by the energy crisis in the 1970s, the City of Long Beach filed an action alleging that seven majors who had unitized a tract of land leased from the city

had violated the Sherman Act by combining and conspiring to establish crude oil prices at unreasonably low and artificial levels, resulting in revenue losses to the city. In re Petroleum Products Antitrust Litigation, 419 F. Supp. 712 (J.P.M.D.L. 1976). This case has not yet been decided on the merits.

While monopolists generally strive to raise prices, the use of market power to reduce prices in order to drive out competitors is often perceived as a threat by independent producers. See ch. 3 text at notes 12 to 15. In 1950 the federal government filed an antitrust complaint against seven major oil companies and the Conservation Committee of California Oil Producers for conspiring to monopolize and restrain trade in the Pacific States. The complaint alleged that the seven majors purchased 90 percent of the crude oil produced by independent producers in these states and that they privately prorated crude oil production so as to eliminate the independents' profit margin. See Comment, "Conservation and Price Fixing in the California Petroleum Industry," *Southern California Law Review* vol. 29, no. 4 (1956) pp. 472–473. The case was terminated by consent judgments entered into by all the defendants except the Conservation Committee and Texaco, as to each of which the case was dismissed. See United States v. Standard Oil Co., 1958 Trade Cases para. 69,212 (S.D. Cal. 1958); United States v. Standard Oil Co., 1959 Trade Cases para. 69,240 (S.D. Cal. 1958); and United States v. Standard Oil Co., 1959 Trade Cases para. 69,399 (S.D. Cal. 1959). The consent decree forbade the six majors from holding membership in any group of crude oil producers which sponsored, recommended, or carried out programs to control the production of crude oil in the Pacific States area with the objective of fixing the price of crude oil. However, the majors were not prevented from entering into agreements providing for joint operation of lands such as unit plans, ratable taking plans, joint operating agreements, poolwide MER agreements, well spacing agreements, and secondary recovery and pressure maintenance programs. 1959 Trade Cases para. 69,399 at p. 75,529.

141. The complaint and related documents in the *Cotton Valley* case appear in Hardwicke, *Antitrust Laws,* pp. 211–228.

142. United States v. Cotton Valley Operators Committee, 77 F. Supp. 409 (W.D. La. 1978), *aff'd mem.*, 339 U.S. 940, *reh'g denied,* 339 U.S. 972 (1950). In its opinion requiring a more definite statement of complaint, the court wrote in 77 F. Supp. 409 at 414:

> Undoubtedly, if persons, such as are the defendants, according to the complaint, without compulsion for conservation, voluntarily entered into an agreement to pool 85 per cent of production in a field such as this, and to exclude from participation all others..., it could hardly be contended, that such a course was not within the purview of the anti-trust laws. If this were permissible as to one field, no reason can be seen why it could not be extended to others, and thus bring about a condition similar to that which precipitated dissolution of the original Standard Oil Company years ago.

143. 438 F.2d 1286 (5th Cir. 1971), *cert. denied,* 404 U.S. 1047 (1972); 509 F.2d 784 (5th Cir. 1975), *cert denied,* 423 U.S. 833 (1976).

144. See Woods Exploration and Producing Co. v. Aluminum Co. of America, 382 S.W.2d 343 (Tex. Civ. App. 1964); and Woods Exploration and Producing Co. v. Aluminum Co. of America, 284 F. Supp. 582 (S.D. Tex 1968); 304 F. Supp. 845 (S.D. Tex 1969). The state court action was dropped when the small-tract owners won much of the antitrust battle in the federal courts. The federal courts can enjoin parties from litigating claims in the state courts that have already been decided at the federal level. See Woods Exploration, 438 F.2d 1286 at 1311–1316 (5th Cir. 1971).

145. 438 F.2d 1286 at 1288 (5th Cir. 1971).

146. Section 2 of the Sherman Act makes it unlawful to "monopolize, or attempt to monopolize, or combine or conspire with any other person or persons to monopolize any part of the trade or commerce among the several states." 15 U.S.C. sec. 2 (1974).

147. 438 F.2d 1286 at 1309–1310.

148. 509 F.2d 784 at 789–793 (5th Cir. 1975).

149. 438 F.2d 1286 at 1308.

150. Texas Natural Resources Code Ann. sec. 101.013(a)(6) (Vernon 1978), discussed in ch. 6 text at notes 30 to 31.

151. Woods Exploration & Producing Co. v. Aluminum Co. of America, 438 F.2d 1286 at 1304–1307 (5th Cir. 1971). The offense of monopoly under Section 2 of the Sherman Act has two elements: (1) possession of monopoly power in the relevant market, and (2) the willful acquisition or maintenance of that power. The related offenses of attempting or conspiring to monopolize do not require actual possession of monopoly power, but do require proof of a specific intent to destroy competition or build a monopoly. Other cases in which the rule of capture was used to thwart or destroy competition in a field include Hague v. Wheeler, 157 Pa. 324, 27 A. 714 (1893); and Louisville Gas Co. v. Kentucky Heating Co., 117 Ky. 71, 77 S.W. 368 (1903). These cases were brought under the common law of malicious waste, not under any antitrust laws, but the facts in each case show why the courts would consider a single field to be the relevant market area for defining monopoly power.

152. 438 F.2d 1286 at 1292–1299.

153. 284 F. Supp. 582 at 585–594.

154. Parker v. Brown, 317 U.S. 341 (1942).

155. 304 F. Supp. 845 (S.D. Tex. 1969).

156. Woods Exploration & Producing Co. v. Aluminum Co. of America, 438 F.2d 1286 at 1292–1298 (5th Cir. 1971).

157. 509 F.2d 784 at 786.

158. See Railroad Comm'n v. Woods Exploration & Producing Co., 405 S.W.2d 313 at 316 (Tex. 1966), *cert. denied,* 385 U.S. 991 (1966).

159. Interestingly, the Texas court of civil appeals had addressed the antitrust issue in the collateral case brought by the large-tract owners to invalidate the one-third–two-thirds prorationing formula in the Appling field. The small-tract owners argued in defense of the one-third–two thirds order that it prevented the furtherance of an antitrust conspiracy to eliminate them as competitors, and that the Texas conservation laws expressly preserved the effectiveness of the antitrust laws. The district court had sustained the large-tract owners' exception to this pleading. The Texas appellate court affirmed this exception, based on the rule established in Eastern R.R. President's Conference v. Noerr Motor Freight, Inc., 365 U.S. 127 (1960), that a combination to influence government action does not violate the antitrust laws. The court then held that the Texas antitrust laws had no bearing on the validity of the prorationing order. Railroad Comm'n v. Aluminum Co. of America, 368 S.W.2d 818 at 826 (Tex. Civ. App. 1963). The Texas Supreme Court's opinion reversing the lower court's invalidation of the one-third–two-thirds formula does not address the antitrust issue. 380 S.W.2d 599 (Tex. 1964). However, the court did note that there was evidence showing that the large-tract owners had refused some pooling and unitization offers from the small-tract owners. Ibid. at 603. One cannot but speculate that the supreme court's affirmance of the one-third–two-thirds formula, which so greatly favored the small-tract producers, was influenced by the rumored antitrust violations in the field.

Arguably, the 1983 antitrust act reduces the vitality of the antitrust laws in the oil and gas industry because private actions required or affirmatively approved by the commission are no longer subject to the state antitrust laws. In this event, the penalty for filing false gas nominations is limited to that imposed by the conservation statutes or by the state penal code. See, for example, Texas Natural Resources Code Ann. secs. 85.381–.388 and 91.143 (Vernon 1978). In addition, a producer who has been drained because of these false filings might have a cause of action against the operator who made the false statements for damages equal to the value of any oil and gas illegally drained. Phillips Petroleum Co. v. American Trading and Production Corp., 361 S.W.2d 942 (Tex. Civ. App. 1962); American Trading and Production Corp. v. Phillips Petroleum Co., 449 S.W.2d 794 (Tex. Civ. App. 1969). However, the act of filing false nominations is certainly not required or approved by the state, and an antitrust action may still lie. See Pan American Petroleum Corp. v. Hardy, 370 S.W.2d 904 (Tex. Civ. App. 1963) (holding that defendant-lessee who made false representations to the Railroad Commission causing the agency to make an erroneous classification, cannot rely upon and benefit from this classification and his own wrongdoing; parties injured by such wrongful conduct may recover actual and exemplary damages).

160. 317 U.S. 341 at 350–351.

161. For example, in Cantor v. Detroit Edison Co., 428 U.S. 579 (1976), six justices found the *Parker* doctrine inapplicable, but no majority of justices concurred in a single opinion; two justices wrote concurring opinions, and three

justices expressed a vigorous dissent. In *Cantor,* four justices stated that the *Parker* holding was limited to action taken by state officials. (The plaintiff in *Parker* had sued only the state prorationing committee, not any private raisin producers.) Thus, if an oil and gas producer sues only other producers in the unit for antitrust violations and does not call into question the legality of any state law or act of a state agency or official, four justices would not shield the unit from the federal antitrust laws. 428 U.S. 579 at 585–592 (1976). However, in Southern Motor Carriers Rate Conference, Inc. v. United States, 53 U.S.L.W. 4422 (March 26, 1985), seven justices held that *Parker* immunity was not limited to activities of public officials, because then any plaintiff could frustrate a state regulatory program merely by filing suit against the regulated private parties rather than against the state officials implementing the plan. This murkiness in the law of *Parker v. Brown* has provided fertile ground for discussions of the state action doctrine by many scholars, most of them critical of the doctrine. See, for example, Roland W. Donnem, "Federal Antitrust Law Versus Anticompetitive State Regulation," *American Bar Association Antitrust Law Journal* vol. 39, no. 4 (1970) pp. 950–967; S. Paul Posner, "The Proper Relationship Between State Regulation and the Federal Antitrust Laws," *New York University Law Review* vol. 49, no. 5 (1974) pp. 693–739; Roger C. Simmons and John R. Fornaciari, "State Regulation as an Antitrust Defense: An Analysis of the *Parker v. Brown* Doctrine," *University of Cincinnati Law Review* vol. 43, no. 1 (1974) pp. 61–99; Paul E. Slater, "Antitrust and Government Action: A Formula for Narrowing *Parker v. Brown,*" *Northwestern University Law Review* vol. 69, no. 1 (1974) pp. 71–109; Paul R. Verkuil, "State Action, Due Process and Antitrust: Reflections on *Parker v. Brown,*" *Columbia Law Review* vol. 75, no. 2 (1975) pp. 328–358. For the difficulties of reconciling the Court's original decision in *Parker v. Brown* with subsequent cases, see Lawrence A. Sullivan, *Handbook of the Law of Antitrust* (Saint Paul, Minn., West Publishing Co., 1977) pp. 731–739; and Phillip Marcus, *Antitrust Law and Practice* (Saint Paul, Minn., West Publishing, 1980) at sec. 349.

162. Goldfarb v. Virginia State Bar, 421 U.S. 773 (1975), *reh'g denied,* 423 U.S. 886 (1976).

163. 53 U.S.L.W. 4422 (March 26, 1985).

164. Ibid. The two-pronged test had been used earlier in California Retail Liquor Dealers v. Midcal Aluminum, Inc., 445 U.S. 97 (1980).

165. See ch. 4 text at notes 22 to 66.

166. The Supreme Court in *Parker* stressed the consonance of the state plan with federal agriculture laws. 317 U.S. 341 at 352–359. It can be argued that Congress has expressly consented to the states' prorationing and oil and gas conservation legislation by approving the Interstate Oil Compact Commission.

167. See ch. 2 text at notes 25 to 38. Also see, John Cirace, "An Economic Analysis of the 'State-Municipal Action' Antitrust Cases," *Texas Law Review* vol. 61, no. 3 (1982) pp. 481–515 (arguing that states should be able to displace competition and be exempted from federal antitrust laws where the purpose of the displacement is to correct substantial market failures, imperfections, or in-

stability); and Milton Handler, "The Current Attack on the Parker v. Brown State Action Doctrine," *Columbia Law Review* vol. 76, no. 1 (1976) pp. 1–20 (arguing that the criticisms of the state action doctrine are unjustified and that the Sherman Act was never intended to outlaw surrogate methods of social and economic control in favor of enforced competition and a free market philosophy; the *Parker* doctrine is integral to our concept of federalism which endows the states with broad latitude to experiment with novel systems of social and economic controls).

168. Arthur T. Smith, "Antitrust Aspects of Joint Operations," *Rocky Mountain Mineral Law Institute,* vol. 16 (Albany, N.Y., Matthew Bender, 1971) pp. 338–341. Very few unitization agreements provide for joint marketing, so they are not vulnerable to the type of complaint issued against the Cotton Valley unit.

Chapter 9

1. All secondary recovery operations in Texas produced 585,420,000 barrels of oil in 1979. Railroad Commission of Texas, *Secondary and Enhanced Recovery Operations in Texas to 1980: Bulletin 80* (Austin, Tex., 1980) p. V [hereafter cited as RRC, *Bulletin 80*]. Texas produced 978,544,000 barrels of oil in 1979. Railroad Commission of Texas, *1981 Annual Report of the Oil and Gas Division* (Austin, Tex., 1981) p. 19 [hereafter cited as *RRC 1981 Annual Report*].

2. RRC, *Bulletin 80,* p. II. The 821 unitized projects listed in *Bulletin 80* produced an average of 577,000 barrels of oil each in 1979. By contrast, the remaining 2,477 nonunitized, secondary recovery projects produced a total of 111,756,000 barrels in 1979, for an average of 45,117 barrels per project.

3. Calculated from RRC, *Bulletin 80,* p. 1–311, by summing Item 45 for all active unitized projects and dividing by 1979 total crude oil production in Texas of 978,544,000 barrels.

4. Henrietta M. Larson and Kenneth W. Porter, *History of Humble Oil and Refining Company: A Study in Industrial Growth* (Salem, N.H., Ayer, 1959) p. 405.

5. Data from the Oklahoma Corporation Commission. This commission defines secondary recovery much as the Railroad Commission does: the introduction of energy or fluid into a common source of supply for the purpose of increasing the ultimate recovery of oil according to a plan approved by the Oklahoma commission. This definition includes pressure maintenance and *in-situ* combustion as well as secondary or tertiary recovery. About 57 million barrels of oil were produced from all secondary recovery projects in 1982 out of Oklahoma's total annual production of 148 million barrels.

6. Oklahoma also has some large tracts, such as Indian reservations, which can be operated efficiently without unitization, in which case the 39 percent figure underestimates Oklahoma's percentage of oil from unitized projects, just as the 48 percent figure for Texas is an underestimate.

7. Oklahoma experienced a boom in oil production in the late 1920s and early 1930s as did Texas. Consequently, it also has many old, large fields. The Texas governor borrowed the idea of using martial law to control the oil fields in the early 1930s from Oklahoma's governor. For other parallels, see W. P. Z. German, "Legal History of Conservation of Oil and Gas in Oklahoma," in *Legal History of Conservation of Oil and Gas* (Chicago, Ill., American Bar Association, 1939) pp. 110–213; and Blakely M. Murphy, ed., *Conservation of Oil and Gas: A Legal History, 1948* (Chicago, Ill., American Bar Association, 1949) pp. 369–422.

8. La. Act 157, La. Revised Statutes Ann. sec. 30:5(B). In practice, the commissioner does not usually assert his power to order mandatory cycling unless a substantial number of the working interests in the field voluntarily consent thereto. Thomas M. Winfiele, "New Legislation Relating to the Conservation Department," in Carlos E. Lazarus, ed., *Institute on Mineral Law* vol. 8 (Baton Rouge, Louisiana State University Press, 1961) p. 17.

9. La. Revised Statutes Ann. sec. 30:10. Louisiana's 1940 Act also provided that no rule or order of the commissioner could require a producer to drill and operate any unnecessary well in order to obtain his fair share of production from the pool. La. Revised Statutes Ann. sec. 30:9. Louisiana's strong interest in preventing economic waste is strikingly different from the attitude of Texas.

10. La. Revised Statutes Ann. sec. 30:5(C). The 1960 compulsory unitization law expressly acknowledges the prevention of unnecessary drilling as a purpose of unitization. In 1960 the Louisiana legislature also passed Act 442, La. Revised Statutes Ann. sec. 30:9(C). Under this act, the commissioner may permit operators to drill wells anywhere in a drilling unit, rather than at its center, if necessary to achieve efficient drainage. Thus, operators in a water-drive field can place their wells farther from the water level to avoid drilling a second unit well to replace the original well when it waters out. See Winfiele, "New Legislation Relating to the Conservation Department," pp. 19–21. Texas operators in this position would probably drill the center well and then drill Rule 37 exception wells later as replacements, using the lease-allowable transfer rules to shut down the original wells with high water-oil ratios. See Amoco Production Co. v. Alexander, 622 S.W.2d 563 (Tex. 1981). This hardly is as efficient as drilling just one optimally placed well. In all respects, Louisiana's statutes are in marked contrast to those of Texas in their prevention of wasteful drilling, their promotion of pooling and unitization, and their express protection of correlative rights by minimizing net drainage from tracts.

11. Also, 84 percent of the condensate produced in Louisiana in 1982 came from unitized fields. Data from J. W. Hecker, chief engineer, Office of Conservation, Department of Natural Resources, Baton Rouge, Louisiana. Louisiana's early and consistent devotion to pooling and unitization also is evidenced by the fact that in 1964 the commissioner held 501 public hearings, almost all involving some type of unitization. Austin W. Lewis, "Rules of Regulatory Bodies: Louisiana," *National Institute for Petroleum Landmen,* vol. 7 (Albany, N.Y., Matthew

Bender, 1966) p. 124. Louisiana's commissioners have also used their discretionary administrative powers to encourage unitization. For example, before the 1960 Act was passed, an innovative approach to poolwide unitization was implemented by the commissioner in two oil fields. He established individual drilling units with acre-feet participation for each pool. When 100 percent of the interested parties in an individual drilling unit executed a unitization agreement, the drilling unit was combined with similar units into a single entity called a "proration unit," with transfer-allowable privileges. This sequential unitization ultimately resulted in the entire field's operation as a unit. See Winfiele, "New Legislation Relating to the Conservation Department," p. 12. The Louisiana commissioner actively sponsored the 1960 unitization law (ibid., p. 19). In fact, one commissioner criticized attorneys practicing before him for being too "timid" and failing to recommend changes in Louisiana's conservation statutes to improve efficiency in the oil fields. George C. Gibson, "Practice Before the Office of Conservation," in Patrick H. Martin, ed., *Institute on Mineral Law* vol. 26 (Baton Rouge, Louisiana State University Law Center, 1979) pp. 52–53.

12. Texas data from *RRC 1981 Annual Report*, p. 19; Louisiana data from telephone interview with the Louisiana Office of Conservation, July 22, 1983; and Oklahoma data from telephone interview with the Oklahoma Corporation Commission, July 22, 1983. The use of average daily production per well as a measure of efficiency is admittedly impressionistic. Geological conditions in Oklahoma may require closer well spacing and lower producing rates to recover oil and gas efficiently. The variance in this statistic among these three states is so large, however, that it is unlikely that geological conditions alone account for the total difference. The three states are neighbors and share many of the same geological characteristics and access to markets; they even share some of the same fields.

13. Governor's Special Study Committee of the Interstate Oil Compact Commission, *A Study of Conservation of Oil and Gas in the United States 1964* (Oklahoma City, Okla., Interstate Oil Compact Commission, Dec. 11, 1964) p. 53. Railroad Commission of Texas, *Secondary and Enhanced Recovery Operations in Texas*, Bulletins 72, 74, 76, 78 and 80 for the years 1972 through 1980, respectively.

14. *Rules and Regulations: Texas Railroad Commission* (Austin, Tex., R. W. Byram & Co., 1958—) sec. VII pp. B-1 to B-7; sec. IV, pp. B-1 to B-4; sec. VIII, pp. B-1 to B-8; sec. IX, pp. B-1 to B-15; sec. V, pp. B-1 to B-20; sec. VI, pp. B-1 to B-17 [hereafter cited as *Byram's Rules*].

15. Ibid. Unit applicants must submit an estimate of the additional oil expected to be recovered and the Railroad Commission does examine this statistic for accuracy.

16. Ibid.

17. Calculated from RRC, *Bulletin 80*, items 20 and 21 of pp. 14–35, and 104–159.

18. When the Big Wells (San Miguel) field was unitized for a pressure maintenance operation only four years after its discovery in April 1969, the Superior Oil

Company testified in support of the application that it had never heard of a field of this size (18,500 acres in the unit) being unitized so quickly and so early in its life. *Oil Regulation Report* vol. 41, May 23, 1973 (Austin, Tex., Texas State House Reporter) pp. 1–2.

19. RRC, *Bulletin 80,* p. II.

20. W. J. McAnelly, Jr., "A Review of Poolwide Unitization Under Act 441 of 1960," in George W. Hardy III, ed., *Institute on Mineral Law,* vol. 15 (Baton Rouge, Louisiana State University Press, 1968) p. 15.

21. For example, in District 3, the six fields that took more than the average time of fifteen years to unitize were all fields discovered between 1928 and 1948 when large per-well factors were especially prevalent. The other sixteen fields in District 3 took less than 9.4 years to unitize on the average. All but three of these sixteen were discovered between 1948 and 1970.

22. Steven N. Wiggins and Gary D. Libecap, "Oil Field Unitization: Contractual Failure in the Presence of Imperfect Information" (draft, College Station, Texas A&M University, Department of Economics, March 1984) tab. A.1. The authors of this paper hypothesize that imperfect information about tract values is a root cause of the failure to reach mutually satisfactory unitization agreements, independent of the problem of holdouts who obstruct negotiations in order to gain a larger participation factor. Four corollaries follow from this theory: (1) Lessees with the most productive leases will be less likely to join the unit in its early stages because there is greater uncertainty regarding the value of the lease over the longer time frame; (2) as fields age, unitization is easier to achieve because more data are available to establish fair formulas; (3) lessees with large tracts have the most to gain from forming the unit early and so will be more flexible in accepting a range of possible participation formulas; and (4) lessees with large, productive tracts that can be operated viably alone will often withdraw from negotiations for a fieldwide unit, and form separate, smaller units. The authors tested these corollaries with data from company files on the unit negotiation process in seven Texas oil fields and found significant support for their hypotheses.

23. Little primary oil was left in the giant Slaughter and Wasson fields and this explains why they were easier to unitize than Yates and Conroe, where the Railroad Commission had to apply considerable pressure on the operators. The latter two fields involved pressure maintenance operations; the former only typical secondary recovery water flooding. William Murray, former Texas railroad commissioner, telephone interview with author, July 19, 1983. Still, both the Slaughter and Wasson fields are comprised of partial, not fieldwide units. See text at note 31 in this chapter.

24. Data calculated from RRC, *Bulletin 80,* items 15 and 16 pp. 14–35 and 104–159.

25. Ibid.

26. Ibid., items 17, 18, and 25.

27. Ibid. District 3 is obviously the most highly unitized of the three districts studied. The wells in District 3 are the most productive in Texas, averaging 28 barrels per day in 1982, more than double the Texas average daily well production ("Producing Oil Wells Reach Record Level," World Oil, February 15, 1984, p. 104.

28. District 8 has five times as many unitized projects as Districts 3 and 4 combined. Thus its lower percentage of fully unitized fields is more representative of the typical unitization project.

29. Gary D. Libecap and Steven N. Wiggins, "The Influence of Private Contractual Failure on Regulation: The Case of Oil Field Unitization" (draft, College Station, Texas A&M University, Department of Economics, February 1984) tab. 1. Table 1 also lists the comparable figure for Oklahoma as 38 percent and for Wyoming as 82 percent. Libecap and Wiggins attribute the large differences among the three states in the percentage of production from fieldwide units to differences in regulatory policies. Texas has no compulsory unitization statute. Oklahoma has such a statute, but it requires the consent of 63 percent of the working interest owners first, and it only applies to secondary recovery operations. Wyoming is composed of large tracts of federal land, and the federal government enforces exploratory units on these lands. Because tracts are unitized in Wyoming before any information uncertainties arise from drilling data, fieldwide unitization agreements are easily negotiated (usually on a surface-acreage basis). Libecap and Wiggins estimate that regulatory policies in Oklahoma resulted in 14 percent more production from fieldwide units in that state in 1975 than in Texas, and that federal policies in Wyoming resulted in 59 percent more production from fieldwide units than in Texas (ibid., p. 16).

30. Some fields are so large that several separately unitized areas are formed under different operators because no one company could easily manage the entire field. For example, the Wasson field is divided into six separately unitized areas, averaging 11,486 acres each and operated by Amoco, Arco, Cornell, Shell, Sun, and Texaco, respectively. RRC, *Bulletin 80,* pp. 182–183. This field is probably operated efficiently.

31. RRC, *Bulletin 80,* pp. 176–182. The 89,000-acre Levelland field follows the Slaughter pattern. See RRC, *Bulletin 80,* pp. 166–170.

32. Libecap and Wiggins, "The Influence of Private Contractual Failure," p. 5.

33. Data calculated from RRC, *Bulletin 80,* item 16, pp. 14–35 and 104–159.

34. Data calculated from RRC, *Bulletin 80.*

35. Ibid.

36. Ibid.

37. Data calculated from *Byram's Rules*. Amoco's total includes unit applications by Pan American Petroleum Corp. and Stanolind, two companies now merged with it. Exxon's total includes unit applications by Humble Oil and Refining, its former name. Sun's total includes unit applications by Texas Pacific

Oil Co., Sunray Midcontinent, and Forest Oil, three companies now merged with Sun. Arco's total includes Sinclair Oil & Gas Co., now merged with it. Mobil's total includes Magnolia Petroleum Co., now merged with Mobil. Gulf's total includes Kewanee Oil Co., now merged with Gulf. Chevron's total includes Standard Oil of Texas, its former name. Telephone interview with the Texas Railroad Commission, July 20, 1983.

38. Data calculated from *Byram's Rules.*

39. J. R. Murray, "Crude Oil Producing Ability by Railroad Commission District," Railroad Commission of Texas Report No. 18 (Austin, Tex. 1977) at Summary (unpaginated).

40. Ibid. The remaining 22 percent of the oil was produced from 8,434 fields that averaged about 75 barrels of oil production per day. Texas had 8,641 producing oil fields in 1976.

41. RRC, *Bulletin 80,* pp. 2–311.

42. Ibid.

Chapter 10

1. The commission undertook internal reviews during the no-flare campaign in the 1950s, and also after the Arab oil embargo in 1973. The commission's investigation of the Conroe field led to the discovery of waste and the field's subsequent unitization in 1977. Staff of the Oil and Gas Division of the Railroad Commission, personal interviews with author, January 13–14, 1983 [hereafter cited as Railroad Commission interviews].

2. The commission's lack of information on fields that are potential candidates for unitization has been strongly criticized. See Wallace F. Lovejoy and Paul T. Homan, *Economic Aspects of Oil Conservation Regulation* (Baltimore, Md., Johns Hopkins University Press for Resources for the Future, 1967) pp. 80–81.

3. Responses by the sixteen largest oil producers in Texas to the author's questionnaires, 1983 (copies in author's file) [hereafter cited as Company Questionnaires].

4. See ch. 4 text at note 177. These 2 billion barrels were in addition to the 9 billion barrels that could be recovered because of the use of secondary recovery technologies.

5. Milton Fox, Texas state representative, telephone interview with author, January 27, 1983.

6. William Murray, former Texas railroad commissioner, telephone interview with author, February 4, 1983.

7. Some caution must be exercised in using this opinion evidence. Oil operators might not admit publicly that they know of fields that would benefit from unitization for fear of incurring lawsuits from royalty interest owners alleging breach of an implied covenant to increase the recovery from the field. One major oil company mentioned that the Hastings field, a fairly large oil field, required unitization. Amoco is currently seeking to unitize this field. The Hastings field is the same one involved in the litigation in Amoco Production Co. v. Alexander, 622 S.W.2d 563 (Tex. 1981) discussed in the text of chapter 8 at notes 89 to 108.

8. Company Questionnaires. It could be that all seven of these majors had the Bryan (Woodbine) field in mind, because of the widespread publicity about this field, as discussed in this chapter at notes 9 to 18.

9. Railroad Commission interviews. Of course, without a systematic survey of fields potentially requiring unitization, the staff cannot be fully knowledgeable on this issue.

10. Compulsory pooling has curtailed some overdrilling, but by no means all. Applications to pool into wells in the Bryan (Woodbine) are often strongly protested. See, for example, *Oil Regulation Report,* vol. 50, March 16, 1983, pp. 4–5.

11. John Makeig, "Bryan Faces Revenue Loss from Order on Oil Wells," *Houston Chronicle,* March 11, 1983.

12. *Oil Regulation Report,* vol. 50, March 9, 1983, p. 1.

13. *Houston Chronicle,* March 11, 1983. Pipeline companies had laid pipes to a dozen or more wells in Bryan, but city restrictions, heavy rainfalls, and other problems had slowed the companies' efforts to handle the large quantity of gas produced by the wells.

14. Railroad Comm'n Docket No. 3-79,887, Final Order dated July 11, 1983 (copy in author's files). The final order allowed individual wells to be granted an exception to the no-flare order for ninety days from completion to allow time for testing and pipeline connections. After ninety days, temporary administrative exceptions could be obtained, but only upon proof that (1) the costs of gathering by pipeline or truck would exceed the revenues from sale of the oil and gas, or (2) the volume of gas was so minimal that it did not justify connection, or (3) unusual circumstances existed. Interim orders prohibiting flaring had been issued between March 21, 1983 and July 11, 1983, the date of the final order.

15. Vernetta Mickey, "Operators Return to Economical Chalk," *Drill Bit,* September 1983, pp. 21–22.

16. Ibid.

17. Ibid. In October 1983, the commission reduced the field's gas–oil ratio again, from 830–1 to 600–1. This represents an 85 percent cut in allowables from the basic statewide gas–oil ratio of 2000–1. From April 1983 to February 1984, total field production from the Bryan (Woodbine) declined from 835,000 barrels to 189,000 barrels of oil. The commission has called hearings every six months to monitor the progress of the unitization effort. Still, progress has been slow

because the operators cannot all agree on a participation formula. One group of about 80 percent of the working interest owners have formed the Bryan (Woodbine) Operators Committee (BWOC) and have signed a unitization agreement with a formula based 80 percent on original oil in place, 10 percent on net acre-feet, and 10 percent on productive acreage. However, Getty Oil and others oppose this formula and have proposed that the commission cut the field's top allowable by another 84 percent, from 190 barrels per day to 30 barrels per day. The BWOC views this proposal as an attempt to place "severe economic coercion" on its members. The BWOC has asked the commission to replace the temporary, field prorationing rule, which is based only on productive acreage, with a permanent field rule based on the same 80–10–10 formula used in the BWOC's unitization agreement. According to the BWOC, this would encourage further unitization efforts and would remove the incentive that nonparticipants have under the temporary rules to elect to hold out and drain others. In requesting the agency to take this action, this group of independents reminded the commission of the many instances in the past when the commission adopted rules which "greatly encouraged" unitization. *Oil Regulation Report,* vol. 37, May 8, 1984.

18. When drilling in the Woodbine formation was at its peak in 1981, a well cost as much as $1.2 million to complete. In late 1982 and 1983, the cost fell to about a half-million dollars because of reduced rig rates and service and supply fees brought on by the recession and a declining price of crude oil. Vernetta Mickey, "Bryan—City of Rigs and Regs," *Drill Bit,* September 1983, p. 27.

19. *Oil Regulation Report,* vol. 39, Jan. 29, 1971, pp. 2–3.

20. Letter from Fred Young to author, February 8, 1982.

21. *Oil Regulation Report,* Aug. 1, 2, 3, 8, 14, 15, 16, 17, 22, 24, 28, 29, 1973.

22. *Oil Regulation Report,* vol. 41, Aug. 1, 1973, pp. 1, 8–9.

23. *Oil Regulation Report,* vol. 41, Aug. 1, 1973, pp. 1, 8–9; and August 29, 1973, pp. 6–9.

24. *Oil Regulation Report,* vol. 41, Aug. 28, 1973, pp. 6–8. See also testimony of Wilbert Lasater on behalf of thirty-one independent producers in *Oil Regulation Report* vol. 41, August 1, 1973, pp. 7–8; and *Oil Regulation Report,* vol. 41, August 3, 1973, pp. 1–2.

25. *Oil Regulation Report,* vol. 41, Aug. 1, 1973, pp. 3–4. The East Texas field had about 15,000 producing wells in 1973. An estimated 1,500 wells could drain the field efficiently. See Lovejoy and Homan, *Economic Aspects,* p. 121.

26. *Oil Regulation Report,* vol. 41, August 1, 1973, pp. 4–6; *Oil Regulation Report,* vol. 41, August 22, 1973, pp. 1–3.

27. Jack Keever, "East Texas Oilmen Rail Against Suggestion to End 'Bonus Rule'," *Houston Chronicle,* Aug. 27, 1978.

28. Ibid. The commission's concern about the impact on local communities of changes in the East Texas regulatory system reportedly defeated other proposals for reform in 1962 and 1965. See Lovejoy and Homan, *Economic Aspects,* p. 121.

The operation of fewer wells in some oil and gas fields would undoubtedly have a significant impact on local communities. During the serious recession in the oil industry in 1982 and 1983, Lee County lost most of its oil well service business and the county's tax base declined by $32 million. Many school districts that depended heavily on oil and gas taxes as a revenue base had to raise local property taxes and eliminate teaching positions. John Toth, "Oil Dollar Drop Pinches Area Budgets," *Houston Chronicle,* Oct. 30, 1983. Even large cities like Houston have suffered from the slump in oil prices and drilling. Over 100,000 jobs in Houston's oil-dominated economy have been lost from mid-1982 to the start of 1984. Thomas Petzinger, Jr., "In Houston, Oil Employees Suffer," *Wall Street Journal,* Nov. 29, 1984.

29. William J. Murray, Jr., "Engineering Aspects of Unit Operations," *Institute on Oil and Gas Law and Taxation,* vol. 3 (Albany, N.Y., Matthew Bender, 1952) p. 17.

30. See also, Imperial American Resources Fund, Inc. v. Railroad Comm'n, 557 S.W.2d 280 (Tex. 1977), illustrating the administrative burden on the agency to resolve Rule 37 disputes regarding optimal well locations in non-unitized gas fields.

31. 622 S.W.2d 563 (Tex. 1981), discussed in ch. 7 text at notes 89 to 98.

32. For example, the Kettleman Hills (North Dome) field in California was only partially unitized in 1931, with the result that the non-unitized owners of less than 5 percent of the acreage seriously compromised the success of the unit. See A. Allen King, "Pooling and Unitization of Oil and Gas Leases," *Michigan Law Review* vol. 46, no. 3 (1948) pp. 327–328.

33. The postembargo, federal crude oil-pricing and -allocation statutes charged the federal regulatory agencies with the goal of administering the new laws to preserve and foster an economically sound and competitive petroleum industry. Independent producers were allowed to retain the favorable, percentage-depletion tax allowance, which was repealed in 1975 with respect to the major integrated firms. The Crude Oil Windfall Profit Tax Act of 1980 exempts the independent producer's first 1,000 barrels per day of production from taxation. For many years, the Federal Power Commission allowed gas producers selling less than 10 million cubic feet of gas per year to charge 130 percent of the maximum price permitted large producers. See Jacqueline Lang Weaver, "Implied Covenants in Oil and Gas Law Under Federal Energy Price Regulations," *Vanderbilt Law Review* vol. 34, no. 6 (1981) pp. 1555–1557. The Independent Petroleum Association of America (IPAA) and the National Stripper Well Association are well-organized and well-financed lobbying groups at the federal level, and their testimony was often solicited at the numerous congressional hearings on energy policy. See, for example, the statement of C. John Miller, president, Independent Petroleum Association of America in *Profitability of Domestic Energy Company Operations, 1974,* Hearings on Excess Profits Tax Legislation Before the Senate Finance Committee, 93rd Cong., 2d sess. (1974) pp. 3, 16.

34. Fewer than half the states with compulsory unitization laws authorize the regulatory agency to initiate unitization on its own motion. See David W. Eckman, "Statutory Fieldwide Oil and Gas Units: A Review for Future Agreements," *Natural Resources Lawyer* vol. 6, no. 3 (1973) p. 384, n. 1. Mississippi passed such a weak compulsory unitization statute in 1964 that only one field was ever unitized under its provisions. The statute was finally amended in 1972. Robert G. Rogers and E. Spivey Gault, "Mississippi Compulsory Fieldwide Unitization," *Mississippi Law Journal* vol. 44, no. 1 (1973) p. 187.

35. Eckman, "Statutory Fieldwide Oil and Gas Units," p. 385. Wyoming's compulsory unitization law requires voluntary agreement by 80 percent of the owners first, but the Wyoming commission will not necessarily approve a compulsory order unless an even higher percentage has agreed to the unit. Houston G. Williams and George M. Porter, "Practice Before the Wyoming Oil and Gas Conservation Commission," *Land and Water Law Review* vol. 10, no. 2 (1975) pp. 353–409. Even in Louisiana, with its strong conservation legislation, producers have complained that the required joinder of 75 percent of royalty interest owners is too high and impedes unitization efforts. U.S. Congress Office of Technology Assessment, *Enhanced Oil Recovery Potential in the United States* (Washington, D.C., Government Printing Office, 1977) app. C, p. 212.

36. Wyoming's 1971 compulsory unitization statute prohibits the creation of units in which production is allocated to the separate tracts solely on an acreage basis. This feature effectively precludes the use of the statute to form exploratory units, even though such units are common on federal lands in Wyoming. Morris G. Gray and Oscar E. Swan, "Fieldwide Unitization in Wyoming," *Land and Water Law Review* vol. 7, no. 2 (1972) p. 449. Ohio's 1965 compulsory law does not allow unitization to save drilling expenses. The statute requires proof that a *substantial* increase in the ultimate recovery of oil and gas will result. Charles J. Meyers and Howard R. Williams, "Petroleum Conservation in Ohio," *Ohio State Law Journal* vol. 26, no. 4 (1965) p. 614.

37. For example, despite passage of a new, stronger compulsory unitization law in 1974 in Florida (amending a prior 1945 statute), a dissident minority of owners in the Jay field blocked unitization for almost a year. Comment, "Compulsory Unitization in Florida: A New Emphasis in the Energy Crisis?," *University of Florida Law Review* vol. 27, no. 1 (1974) p. 209. Much litigation has ensued in Oklahoma over the nuisance liability of a unit to owners of tracts outside the unit but within the same reservoir, despite the availability, since 1945, of compulsory process to aid formation of fieldwide units. See, for example, West Edmund Hunton Lime Unit v. Lillard, 265 P.2d 730 (Okla. 1954); Greyhound Leasing & Financial Corp. v. Joiner City Unit, 444 F.2d 439 (10th Cir. 1971); Boyce v. Dundee Healdton Sand Unit, 560 P.2d 234 (Okla. Ct. App. 1975); West Edmund Salt Water Disposal Ass'n v. Rosecrans, 204 Okla. 9, 226 P.2d 965 (1950), *appeal dismissed,* 340 U.S. 924 (1950).

38. H.B. 311 is the Texas bill that almost passed in 1973. This bill did not provide strong mechanisms to unitize fields. It applied only to secondary recovery operations; it required many detailed findings, which could spark long delays

in the unitization process; the allocation formula for primary reserves preserved the small-tract owners' vested regulatory advantage in old-style proration formulas with large per-well formulas; and the commission had no authority to require unitization on its own motion—at least 75 percent of the working-interest owners had to have agreed to the unitization plan first. The only type of bill which might have passed in 1973 would have had to expressly exclude the largest oil fields in Texas that still needed unitization: East Texas, Yates, and Hawkins. See ch. 4 at notes 155 to 185.

39. Mr. Doherty's plea for federal, compulsory unitization legislation sprang from the urgent need for supplies of oil and gas during World War I and the enormous waste of these resources during the first decades of the twentieth century (see ch. 1 text, at note 1). Subsequently, state prorationing, pooling, and well-spacing legislation remedied many of the faults and much of the waste exposed by Doherty in the production and regulatory techniques used before 1930. Federal compulsory unitization laws no longer seemed necessary, especially when vast overcapacity in domestic oil fields served to assure our national defense needs. The petroleum industry ably performed the prodigious task of fueling the war effort during World War II. So successful was its response to the call to arms that Eugene Rostow's postwar proposal for federal legislation to force unitization of oil and gas fields went largely unheeded. See Eugene V. Rostow, *A National Policy for the Oil Industry* (Northford, Conn., Elliot's Books, 1948). Rostow's proposal was premised on antitrust grounds: that market demand prorationing amounted to price-fixing and that unitization should occur after divestiture splintered the vertically integrated majors into smaller companies. The industry that had performed so well during the war and that was then serving the postwar consumer boom did not seem to need the radical reforms proposed by Rostow. In the 1970s antitrust became a *cause célèbre* as the American public vented its outrage against the major oil companies over severe supply shortages and staggering price increases. The legislative response was direct and immediate: price controls on crude oil and natural gas, windfall profits taxes, serious debates over divestiture, and several massive antitrust lawsuits. Yet through all the debate and action, no legislator, professor, or oil industry iconoclast rose to call for federal legislation that would force Texas to operate its fields more efficiently in the public interest through use of compulsory unitization process.

One cannot help but wonder, however, whether Texas would have passed a compulsory unitization law had the Arab oil embargo of October 1973 occurred earlier in May when the legislature had a proposed compulsory unitization bill, H.B. 311, under debate (see ch. 4). After all, the 1967 embargo had impelled a statewide lease-allowable system. However, not all embargoes have inspired greater efficiency in the statewide prorationing system. The 1956 Suez crisis caused demand for Texas crude to soar, especially for Gulf Coast crude which was easily transportable to other parts of the nation. The commission refused to increase the allowables of Gulf Coast fields above that of fields in West Texas, which were operating at the maximum capacity of their pipeline connections. The commission interpreted its mandate to allocate statewide production "on a fair and reasonable basis" to require that all fields receive the same proportionate increase in the market demand factor, even though this meant that some demand would go unmet. See John Vafai, "Market Demand, Prorationing and Waste—A Statutory Confusion," *Ecology Law Quarterly* vol. 2, no. 1 (1972) p. 139. The railroad commissioners truly viewed themselves as managers of the entire Texas economy. The federal government did not unseat the commissioners from their positions of power, even when these state officials rendered the national interest subservient to that of Texas.

40. In 1971 the President's Council of Economic Advisors issued a report that characterized market demand prorationing as a shield for high-cost, high-priced domestic oil rather than as a conservation measure. A member of the President's Council of Economic Advisors reportedly told the press that if the commission reduced Texas' total allowable in March 1971, the federal government would suspend the Connally Act, which prohibited the interstate shipment of "hot oil" produced in violation of the state's prorationing orders. The report provoked much animosity from the railroad commissioners. Byron Tunnell, chairman of the Railroad Commission, blasted the report as "a deliberate effort by some in Washington to malign the State of Texas and intimidate the Commission." See *Oil Regulation Report*, vol. 38, March 8, 1971, pp. 1–3.

41. Congress, recognizing that state control of production rates might not always accord with the national interest, enacted a statute in 1975 authorizing the president of the United States to require oil and gas production from both federal and private lands at the maximum efficient rate (MER) or, in times of severe disruptions in energy supplies, at the "temporary emergency production rate" (TEPR). Also in 1975, the Federal Power Commission (FPC) asserted authority to require gas producers to accelerate the development of gas reservoirs and to impose minimum production rate obligations on gas lessees through the FPC's certification procedures. See Dan A. Bruce, " 'Maximum Efficient Rate'—Its Use and Misuse in Production Regulation," *Natural Resources Lawyer* vol. 9, no. 3 (1976) pp. 442 and 451; and Note, "The Prudent Operator Standard and FERC Authority," *Texas Law Review* vol. 57, no. 4 (1979) pp. 661–674. Such federal intervention into the states' traditional preserve of regulating oil- and gas-producing rates and into private leases and sales contracts sets a precedent for federal legislation requiring efficient conservation practices such as unitization.

42. Office of Technology Assessment, *Enhanced Oil Recovery Potential in the United States*, pp. 211–212. The Office of Technology Assessment (OTA) report acknowledged that only a field-by-field study of ownership patterns could determine the extent of this obstacle to unitization. The OTA report concluded that if joinder problems were found to be serious constraints on enhanced oil recovery, the federal government could recommend that states adopt a statute allowing compulsory unitization when only 60 percent of the owners in a field consented to a unit plan. Alternatively, Congress could require states to have a compulsory unitization law in order to qualify for federal support or to avoid having a federal agency assume the state's responsibility for unitization. See ibid. pp. 12 and 85–88. (The energy crisis inspired Congress to pass legislation that conditioned a state's receipt of federal highway trust fund revenues on the state's enacting a 55

miles-per-hour speed limit in place of the existing 70-mile-per-hour limit. Emergency Highway Energy Conservation Act, 23 U.S.C. sec. 154 (Supp. 1983). Texas lawmakers hurriedly enacted a 55-mile-per-hour speed limit bill in 1973. See ch. 4 text at note 191.

In 1978 a congressional committee held hearings on ways to increase incentives to produce secondary and tertiary oil. When asked for its position on the OTA's legislative proposals, the Carter administration responded that federal efforts should be limited to encouraging states to enact strong compulsory unitization bills, but the benefit of such legislation would be "questionable." Further, "(f)ederal legislation mandating unitization would probably be opposed by the states and may have a detrimental effect on future unitization efforts." See statement of R. Dobie Langenkamp, deputy assistant secretary, Oil, Natural Gas, and Shale Resources, Department of Energy, in *Incentives for Tertiary Enhanced Recovery Techniques, Crude Oil Production,* Hearings on S. 2623 and S. 2999 before the Committee on Energy and Natural Resources, 95th Cong., 2d sess. (1967) p. 109. When Senator Bentsen of Texas was asked to comment on the OTA report recommending that a compulsory unitization statute be passed in Texas, he replied: "Well, you are getting into a complex subject for a fellow that's not a petroleum engineer" and deferred on the point to someone more knowledgeable (ibid. p. 32). Senator Hansen joined the colloquy by noting that in his four years as governor of Wyoming, he had observed that the states had "rather ingenious" ways of handling the problem of unitizing fields and, therefore, such matters should be left to state authority. This ended the discussion.

43. The federal government has the constitutional power to enact a strong compulsory unitization statute that would preempt all the weaker state laws. The federal government is a recognized leader in promoting unitization on federal lands and has considerable expertise in this area. See Leroy H. Hines, *Unitization of Federal Lands* (Denver, Colo., F. H. Gower, 1953). Such a radical degree of intervention into the states' traditional preserve has never been entertained and is most unlikely. If such a federal law were passed and states refused to enforce it, the federal energy bureaucracy would have to expand considerably to fill this function. If the federal government attempted to coerce states into enforcing the federal law, litigation would undoubtedly ensue. Such litigation has plagued the enforcement of the federal Clean Air Act, with most circuit courts of appeals holding that neither the Clean Air Act nor the Constitution allows a federal agency (in this case the Environmental Protection Agency) to order states to perform federal functions. See David Currie, *Air Pollution: Federal Law and Analysis* (Wilmette, Ill., Callaghan, 1981) sec. 4.29. The fact that Congress has not even passed milder legislation, such as granting the benefits of federal price and tax incentives only to states which have enacted compulsory unitization laws, bespeaks considerable satisfaction with the states' current conservation practices.

44. David F. Prindle, *Petroleum Politics and the Texas Railroad Commission* (Austin, University of Texas Press, 1981) pp. 139–140.

45. Ibid., p. 140.

46. This is Statewide Rule 10, Texas Administrative Code tit. 16, sec. 3.10 (1982).

47. Gage v. Railroad Comm'n, 582 S.W.2d 410 (Tex. 1979).

48. Railroad Comm'n v. Graford Oil Corp., 557 S.W.2d 946 (Tex. 1977). The court's strict interpretation of the commission's statutory authority over natural gas prorationing in the *Gage* and *Graford* cases follows an established pattern of narrowly construing the agency's powers in this one area, as contrasted to the court's expansive view of the commission's waste-prevention powers. See ch. 8 at notes 50 to 61.

49. Frank Douglass and H. Philip Whitworth, "Practice Before the Oil and Gas Division of the Railroad Commission of Texas," *St. Mary's Law Journal* vol. 13, no. 4 (1982) pp. 737–738. Commingling of production existed in at least 2,448 fields in the state. See C. C. Small, "Downhole Commingling, Multi-Zone Proration, Net GOR, and Special Allowables," *Oil and Gas: Texas Railroad Commission Rules & Regulations* (Austin, State Bar of Texas, 1982) p. D-5.

50. Small, "Downhole Commingling," pp. D-7–D-8.

51. In Mote Resources Inc. v. Railroad Comm'n, 618 S.W.2d 877 (Tex. Civ. App. 1981), the small-tract producers requested and received an injunction against enforcement of the commission's prorationing order pending disposition of the commission's appeal. The appellate court ruled that, absent an injunction, the small-tract owners would be deprived of an opportunity to produce 143 million cubic feet of gas per month during pendency of the appeal. In Railroad Comm'n v. Mote Resources, 645 S.W.2d 639 (Tex. Civ. App. 1983), the appellate court affirmed the holding of the district court that the commission's order of October 20, 1983, was invalid, because the commission had no authority to prorate commingled production at that time.

52. Texas Natural Resources Code Ann. secs. 85.053, 85.055, and 86.081 (Vernon Supp. 1982).

53. Texas Natural Resources Code Ann. sec. 85.053(b)(iii)(Vernon Supp. 1982). Another provision denies the commission any power to extend the vertical or areal limits of fields discovered between January 1, 1940, and June 1, 1945. Ibid. at sec. 85.053(b)(i). While other reservoirs may be affected by this limitation, the dates were chosen to apply primarily to the Boonsville field. See Douglass and Whitworth, "Practice Before the Oil and Gas Division of the Railroad Commission," p. 740, n. 104.

54. Small, "Downhole Commingling," p. D-9.

55. For example, the Oklahoma Supreme Court supported the Corporation Commission's finding that twenty-one stringer sands constituted a single common source of supply subject to the unitization statutes because downhole commingling had created effective communication between all the stringers, which had originally been separate. See Jones v. Continental Oil Co., 420 P.2d 905 (Okla. 1966). See also, Jones Oil Co. v. Corporation Comm'n, 382 P.2d 751 (Okla. 1963), *cert. denied,* 375 U.S. 931 (1963).

The Texas Supreme Court had some room to maneuver the definition of a "common reservoir" into meaning an area which appears to be underlaid by an "accumulation of oil or gas" rather than a "common accumulation of oil or gas," but did not choose to so maneuver. See Railroad Comm'n v. Graford Oil Corp., 557 S.W.2d 946 at 950 (Tex. 1977).

56. Michael T. Halbouty, chairman of the board, chief executive officer, Michael T. Halbouty Energy Company, telephone interview with author, July 7, 1983.

57. Richard B. Stewart, "The Reformation of American Administrative Law," *Harvard Law Review* vol. 88, no. 8 (1975) p. 1736. The traditional model of administrative law accords judicial review the role of channeling administrative discretion within the boundaries set by the legislature. This model has been eroded because the legislature has delegated to administrative agencies increasingly broad and vague statutory mandates, and the courts have largely upheld these statutes against the challenge that they invalidly delegated legislative power to the administrative branch. These vague statutes, however, threaten democratic values by allowing major questions of socioeconomic policy to be determined by unelected officials. This has caused an uncomfortable void in the legitimacy of administrative policy. Stewart is critical of the judiciary's attempt to reform administrative law by developing a new model of "interest representation" to check agency discretion. By liberalizing the rules of standing and expanding the rights of interested members of the public to participate in agency proceedings, the courts have sought to assure that agencies act as "mini-legislatures," that is, that all interests affected by the exercise of the legislative powers delegated to the agencies have been fairly represented in the agencies' decision making. Stewart criticizes the pluralistic theory of legitimacy as unworkable; it causes considerable delay in agency decision making as more and more parties are entitled to formal trial-type procedures, and it makes compromise and harmony among conflicting interests more difficult to obtain. Stewart suggests, as an alternative, the regular popular election of agency members. This political model would legitimate the delegation of broad discretion to agency officials. Stewart admits, however, that the intelligent appraisal of and voting for agency members by the general public are unlikely, and specialized interest groups will often control who is elected. In fact, he uses the historical experience of the states which provided for the election of railroad commission members in the nineteenth century as examples of the failure of this political model. Incompetent commissioners were often elected and, in some cases, the railroads acquired even more domination of the agency. Most states abandoned popular election for appointed commissions. That the Texas Railroad Commission bucked this historical trend to appointed agencies is a tribute to its increasing expertise and its ability to serve the public interest (although the Texas legislature did seriously debate whether the commission should be appointive in 1931 (see ch. 3, note 42)). Professor Stewart suggests a "nominalist theory" of administrative law as a possible alternative to the traditional and the "interest representation" models. The nominalist model would appraise agencies on a case-by-case basis and then match specific remedies (such

as deregulation, greater specification of goals by Congress, payment of fees to groups to represent the public interest before the agency) to perceived agency deficiencies (ibid. at 1805–1810). Stewart's suggestion invites the close study of individual agencies' institutional contexts and characteristics. This study of the Texas Railroad Commission illustrates the successful use of popularly elected agency officials to legitimate agency discretion.

58. The Texas legislature passed the Underground Natural Gas Storage and Conservation Act in 1977 to promote the conservation and orderly withdrawal of gas in Texas. Texas Natural Resources Code Ann. secs. 91.171–91.184 (Vernon Supp. 1982). This act grants to gas utilities the right of eminent domain to acquire storage facilities for natural gas in depleted reservoirs. The right is conditioned on Railroad Commission approval of the reservoir formation to be condemned, and on the storer's obtaining, by negotiated means, at least two-thirds of the ownership of the mineral and royalty interests in the subsurface stratum. The commission also is empowered to resolve disputes between the condemnor and any purchaser of native gas remaining in the reservoir, a function akin to resolving correlative rights issues in the allocation of a field's reserves. The analogy between the gas storage act and compulsory unitization is obvious: without the power of eminent domain, multiple ownership of a depleted gas reservoir poses serious problems to efficient gas storage. A few holdouts can block an underground storage project by refusing to lease, except at an exorbitant price, reflecting their monopolistic position. The gas storer who injects gas into the formation, without the permission of all the owners of the stratum, may be liable in trespass if the gas migrates into the holdout's unleased part of the reservoir. The Texas courts have greatly aided the gas storer by refusing to apply the rule of capture to injected gas (see Lone Star Gas Co. v. Murchison, 353 S.W.2d 870 (Tex. Civ. App. 1962)), but the courts cannot eliminate all the legal and economic problems facing gas storers in multiply owned reservoirs. The Texas legislature passed the Underground Natural Gas Storage Act to serve the public interest in the efficient storage, withdrawal, and conservation of natural gas. No ideological scruples against government intrusion into private property rights prevented the passage of this condemnation statute. However, the independent producers of Texas have no vested regulatory advantage in depleted oil and gas reservoirs, as they do in producing ones.

59. In his book, *Antitrust and the Oil Monopoly: The Standard Oil Cases, 1890–1911* (Westport, Conn., Greenwood Press, 1979), Bruce Bringhurst documents his thesis that antitrust lawsuits are often used as political tools. It is popular for state governors, attorneys general, and legislators to run on a platform denouncing major oil companies as monopolists. In fact, however, these politicians welcome the new technology, efficiency, and economies which the large companies bring to market. Therefore, most state antitrust lawsuits against Standard Oil of New Jersey and its affiliates in the years from 1890 to 1911 did nothing to stop the advance of Standard Oil into new markets and new states because the officials bringing the lawsuits did not actually intend to effect any substantive changes in industrial structure. Similarly, Texas legislators have been able to maintain their

political popularity and vote against compulsory unitization bills because unitization is secured in Texas through other means.

60. The rule of reason is used by federal courts to interpret the Sherman Anti-Trust Act, upon which Texas antitrust law is now modeled. See, for example, Standard Oil Co. of New Jersey v. United States, 221 U.S. 1 (1911). At one time, immunity from Texas' antitrust laws was critical for voluntary unitization agreements, because the Texas antitrust laws differed substantially from the rule of reason approach, and voluntary unitization agreements were extremely susceptible to state antitrust charges.

61. This paragraph is similar to the model act drafted in 1947 by the legal committee of the Interstate Oil Compact Commission, and to other states' voluntary unitization laws. American Petroleum Institute, *Secondary Recovery of Oil in the United States* (2 ed., Dallas, Tex., American Petroleum Institute, 1950) pp. 46–68.

62. Texas actually had such a statute approving voluntary unitization agreements in gas fields from 1935 to 1949. See ch. 3 text at notes 148 to 176. The intractable legal problems of prorationing gas in nonunitized fields are discussed in the text of chapter 8 at notes 50 to 67. The authority to approve unitization agreements based on the protection of correlative rights would also encourage unitization in some oil fields. See ch. 7 text at notes 89 to 112.

63. Texas Natural Resources Code Ann. sec. 101.011 is fully cited in appendix II, and is critically analyzed in the text of chapter 4 at notes 31 to 51.

64. Considerable confusion exists on this issue in the legal literature. See ch. 6 text at notes 1 to 3 and at note 9. Section 101.013 is cited in full in appendix II, and is critically analyzed in the text of chapter 4 at notes 52 to 66.

65. Section 101.017 also prohibits the cooperative marketing of crude petroleum, condensate, and distillate. Because liquids are so much easier to transport and market, this prohibition should be retained. The legal committee of the Interstate Oil Compact Commission recommended that states allow cooperative marketing of gas, but not of crude oil. See American Petroleum Institute, *Secondary Recovery of Oil*, p. 65.

66. In Town of Chino Valley v. State Land Dep't, 119 Ariz. 243, 580 P.2d 704 (1978), the court upheld a statute that precluded injunctive relief against state-approved transfers of water which injured another's groundwater supply, as long as the statute provided for the award of just and reasonable damages. Such a statute essentially grants the power of eminent domain to private persons operating under state-approved permits.

67. Office of Technology Assessment, *Enhanced Oil Recovery Potential in the United States*, pp. 12 and 88.

68. See "Barrels of Waste," *Houston Business Journal*, Jan. 16, 1984 (describing the proposal of Professors Libecap and Wiggins of Texas A&M University). Such an approach is not new. In 1958 an economist suggested that the tax benefits of

percentage depletion should be granted only to royalties on oil produced from unitized fields. See Lovejoy and Homan, *Economic Aspects*, p. 77.

69. Letter from Mack Wallace, chairman of the Railroad Commission of Texas to President Reagan, October 20, 1983 (copy in author's possession). The letter was widely publicized in the Texas press. See, for example, John Toth, "Oil Dollar Drop Pinches Area Budgets," *Houston Chronicle*, Oct. 21, 1983.

Table of Cases

(References are to pages in text or to footnote numbers in which case is cited)

Amazon Petroleum Corp. v. Railroad Commission, ch. 3, 145*n*

Amerada Petroleum Corp. v. Railroad Commission, ch. 6, 33*n*; ch. 7, 140*n*

American Petroleum Corp. v. Hardy, ch. 8, 159*n*

American Trading and Production Co. v. Phillips Petroleum Co., ch. 8, 159*n*

Amoco Production Co. v. Alexander, 222–225, 228, 256, 257, 335; ch. 1, 9*n*; ch. 9, 10*n*

Amoco Production Co. v. Underwood, 216

Atlantic Refining Co. v. Bright & Schiff, 252–253; ch. 4, 201*n*

Atlantic Refining Co. v. Railroad Commission, 127, 128, 131, 197, 236, 267–268, 283, 284, 290–291; ch. 4, 196*n*

Bandini Co. v. Superior Court, ch. 7, 3*n*

Banks v. Mecom, ch. 7, 65*n*, 75*n*

Barnard v. Monongahela Natural Gas Co., ch. 2, 25*n*

Baumgartner v. Gulf Oil Corp., 244, 245, 246, 247; ch. 4, 148*n*; ch. 7, 50*n*, 160*n*

Benz-Stoddard v. Aluminum Co. of America, ch. 8, 51*n*

Biskamp v. General Crude Oil Co., ch. 3, 175*n*

Blocker v. Christie, Mitchell & Mitchell Co., ch. 7, 77*n*

Bolton v. Coats, ch. 7, 147*n*

Boomer v. Atlantic Cement Co., ch. 7, 153*n*

Boyce v. Dundee Healdton Sand Unit, ch. 7, 160*n*; ch. 10, 37*n*

Brewster v. Lanyon Zinc Co., ch. 7, 78*n*

Broussard v. Texaco Inc., ch. 8, 79*n*

Brown v. Getty Reserve Oil, ch. 7, 9*n*

Brown v. Humble Oil & Refining Co., 291–292; ch. 4, 114*n*; ch. 7, 172*n*; ch. 8, 106*n*, 112*n*

Brown v. Smith, 206–207, 208, 209

Bullard v. Broadwell, ch. 7, 49*n*

Burford v. Sun Oil Co., ch. 8, 88*n*

Byrd v. Shell Oil Co., ch. 4, 103*n*; ch. 7, 96*n*; ch. 8, 109*n*

California Co. v. Britt, 244; ch. 7, 50*n*, 158*n*, 170*n*
California Retail Liquor Dealers Association v. Midcal Aluminum, Inc., ch. 8, 164*n*
Canadian River Gas Co. v. Terrell, ch. 3, 156*n*
Cantor v. Detroit Edison Co., ch. 8, 161*n*
Carter Oil Co. v. Dees, ch. 7, 122*n*, 163*n*
Champlin Exploration v. Railroad Commission, ch. 7, 3*n*, 147*n*
Champlin Refining Co. v. Oklahoma Corp. Commission, ch. 3, 121*n*; ch. 7, 84*n*; ch. 8, 28*n*
Cities Service Gas Co. v. Peerless Oil & Gas Co., ch. 5, 86*n*
Clifton v. Koontz, ch. 7, 77*n*, 84*n*
Clymore Production Co. v. Thompson, ch. 5, 6*n*
Coloma Oil & Gas Corp. v. Railroad Commission, ch. 8, 110*n*
Colorado Interstate Gas Co. v. Sears, ch. 8, 116*n*
Comanche Duke Oil Co. v. Texas Pacific Coal & Oil Co., ch. 7, 125*n*
Commonwealth Edison Co. v. Montana, ch. 4, 145*n*
Consolidated Gas Utilities Co. v. Thompson, ch. 3, 169*n*
Constantin v. Smith, ch. 3, 113*n*
Corley v. Mississippi State Oil and Gas Board, ch. 8, 71*n*
Corzelius v. Harrell, 273–274, 277, 280, 287–288
Corzelius v. Railroad Commission, ch. 3, 173*n*; ch. 7, 133*n*
Cox v. Davison, ch. 7, 36*n*; ch. 8, 5*n*

Dailey v. Railroad Commission, 296, 297; ch. 7, 96*n*; ch. 8, 20*n*
Danaho Refining Co. v. Pan American Petroleum Co., ch. 8, 138*n*
Danciger Oil & Refining Co. v. Railroad Commission, 262–264; ch. 3, 29*n*, 121*n*
Danciger Oil & Refining Co. v. Smith, ch. 8, 3*n*
Delhi-Taylor Oil Corp. v. Holmes, 235
Dobson v. Arkansas Oil & Gas Commission, ch. 4, 148*n*; ch. 5, 86*n*; ch. 6, 75*n*; ch. 7, 50*n*, 83*n*, 85*n*, 159*n*; ch. 8, 71*n*

Eastern R. R. President's Conference v. Noerr Motor Freight, Inc., ch. 8, 159*n*
Elliff v. Texon Drilling Co., ch. 2, 25*n*; ch. 7, 3*n*, 126*n*
Elliott v. Davis, ch. 7, 64*n*
Expando Production Co. v. Marshall, ch. 7, 65*n*
Exxon Corp. v. First National Bank of Midland, ch. 7, 147*n*
Exxon Corp. v. Railroad Commission, 268–271

Francis Oil & Gas Co. v. Exxon Corp. (10th Cir.), ch. 6, 40*n*, 41*n*, 42*n*, 43*n*; ch. 7, 26*n*, 30*n*
Francis Oil & Gas Co. v. Exxon Corp. (Temp. Emer. Ct. App.), ch. 6, 41*n*, 45*n*, 48*n*; ch. 7, 26*n*
Freeport Sulphur Co. v. American Sulphur Royalty Co., ch. 7, 77*n*

Gage v. Railroad Commission, ch. 8, 66*n*, 79*n*; ch. 10, 47*n*
General Crude Oil Co. v. Harris, ch. 7, 84*n*
Getty Oil Co. v. Jones, ch. 7, 183*n*
F. A. Gillespie & Sons Co. v. Railroad Commission, ch. 8, 92*n*, 109*n*, 114*n*
Goldfarb v. Virginia State Bar, ch. 8, 162*n*

Goldsmith & Powell v. State, ch. 5, 80*n*
Gregg v. Delhi-Taylor Oil Corp., 236–237; ch. 7, 139*n*
Greyhound Leasing & Financial Corp. v. Joiner City Unit, ch. 7, 160*n*; ch. 10, 37*n*
Griffith v. Gulf Refining Co., ch. 7, 8*n*
Grimes v. La Gloria Corp., ch. 7, 66*n*
Gulf Land Co. v. Atlantic Refining Co. (Tex. Sup. Ct.), ch. 8, 22*n*
Gulf Land Co. v. Atlantic Refining Co. (5th Cir.), ch. 8, 106*n*, 107*n*
Gulf Oil Corp. v. Hughes, ch. 7, 160*n*
Gulf Production Co. v. Kishi, ch. 7, 103*n*

Haby v. Stanolind Oil & Gas Co., ch. 5, 40*n*
Hague v. Wheeler, ch. 8, 151*n*
Halbouty v. Darsey, ch. 4, 205*n*
Halbouty v. Railroad Commission, 127, 131, 162–164, 197, 267, 283, 284, 290, 292, 297; ch. 4, 197*n*, 202*n*; ch. 5, 94*n*
Harrington v. Railroad Commission, ch. 8, 79*n*
Hassie Hunt Trust v. Proctor, ch. 7, 8*n*
Hastings Oil Co. v. Texas Co., ch. 7, 8*n*, 127*n*
Hawkins v. Texas Co., ch. 8, 108*n*
Henderson v. Chesley, ch. 7, 51*n*
Henderson v. Thompson, ch. 8, 70*n*
Herod v. Grapeland Joint Account, ch. 7, 77*n*, 106*n*
Holt v. Southwest Antioch Sand Unit Fifth Enlarged, ch. 7, 189*n*
Howell v. Union Producing Co., ch. 7, 18*n*
Humble Oil & Refining Co. v. L. & G. Oil Co., ch. 7, 139*n*, 190*n*
Humble Oil & Refining Co. v. Lasseter (Tex. Civ. App.-Texarkana), ch. 7, 51*n*
Humble Oil & Refining Co. v. Lasseter (Tex. Civ. App.-Austin), ch. 7, 51*n*
Humble Oil & Refining Co. v. Railroad Commission, ch. 4, 114*n*; ch. 8, 107*n*
Humble Oil & Refining Co. v. Turnbow, ch. 8, 107*n*
Hunter v. Justices' Court, ch. 7, 54*n*
Hunter Co. v. McHugh, ch. 4, 195*n*
Hutchins v. Humble Oil & Refining Co., 226–228

Imperial American Resources Fund Inc. v. Railroad Commission, ch. 7, 115*n*; ch. 8, 85*n*

Jackson v. State Corporation Commission, 239–241
Japhet v. McRae, 204–205; ch. 4, 194*n*
Jones v. Continental Oil Co., ch. 10, 55*n*
Jones Oil Co. v. Corporation Commission, ch. 10, 55*n*
Jones v. Hunt Oil Co., ch. 7, 75*n*
Jones v. Killingsworth, 217–218; ch. 7, 75*n*

Kansas-Nebraska Natural Gas Co. v. State Corporation Commission, ch. 5, 86*n*
Kennedy v. General Geophysical Co., ch. 7, 125*n*
Kenoyer v. Magnolia Petroleum Co., ch. 7, 33*n*
Kingwood Oil Co. v. Bell, ch. 7, 82*n*
Knight v. Chicago Corp., ch. 7, 58*n*
Kraker v. Railroad Commission, ch. 4, 103*n*, 120*n*; ch. 8, 109*n*
Kuklies v. Reinhert, ch. 7, 60*n*, 77*n*

Law v. Heck Oil Co., ch. 7, 38*n*
Leach v. Brown, ch. 7, 26*n*, 67*n*
Le Blanc v. Haynesville Mercantile Co., ch. 7, 22*n*
Leopard v. Stanolind Oil & Gas Co., ch. 7, 76*n*
Loeffler v. King, ch. 7, 3*n*
Lone Star Gas Co. v. Murchison, ch. 6, 33*n*; ch. 7, 139*n*; ch. 10, 58*n*
Louisville Gas Co. v. Kentucky Heating Co., ch. 8, 151*n*

MacMillan v. Railroad Commission, 43, 60; ch. 3, 9*n*
McLachlan v. Stroube, 232–234, 257–258; ch. 7, 140*n*
Magnolia Petroleum Co. v. Railroad Commission, ch. 6, 33*n*; ch. 7, 140*n*
Marrs v. Railroad Commission, 289–290; ch. 4, 115*n*, 122*n*
May v. Cities Service Oil Co., ch. 7, 21*n*
Midas Oil Co. v. Stanolind Oil & Gas Co., ch. 8, 95*n*
Miles v. Amerada Petroleum Corp., ch. 7, 77*n*
Miller v. Crown Central Petroleum Corp., 251
Miller v. Railroad Commission, ch. 4, 103*n*; ch. 7, 96*n*; ch. 8, 109*n*
Montgomery v. Rittersbacher, ch. 7, 21*n*, 45*n*
Moseley v. Hearrell, ch. 7, 51*n*
Mote Resources Inc. v. Railroad Commission, ch. 10, 51*n*
Mowrer v. Ashland Oil & Refining Co., ch. 7, 160*n*
Mueller v. Sutherland, ch. 7, 14*n*
Murray v. R & M Well Servicing & Drilling Co., ch. 8, 70*n*

Nale v. Carroll, ch. 7, 14*n*
Normanna Case. *See* Atlantic Refining Co. v. Railroad Commission
Northwest Oil Co. v. Railroad Commission, ch. 8, 79*n*
Nugent v. Freeman, ch. 7, 21*n*

Ohio Oil Co. v. Indiana, ch. 7, 3*n*

Pabst v. Roxana Petroleum Corp., ch. 8, 138*n*
Palmer Petroleum Corp. v. Phillips Petroleum Co., 89
Panama Refining Co. v. Ryan, ch. 3, 145*n*
Pan American Petroleum Corp. v. Hardy, ch. 8, 159*n*
Pan American Production Co. v. Hollandsworth, ch. 7, 140*n*
Parker v. Brown, 307–310; ch. 4, 11*n*
Parker v. Parker, 205
Pattie v. Oil & Gas Conservation Commission, ch. 2, 29*n*, 128*n*
Pennsylvania Coal Co. v. Mahon, ch. 2, 29*n*
People's Petroleum Producers v. Smith, ch. 3, 113*n*
People's Petroleum Producers v. Sterling, ch. 5, 101*n*; ch. 8, 29*n*
In re Petroleum Products Antitrust Litigation, ch. 8, 140*n*
Phillips Petroleum Co. v. American Trading and Production Corp., ch. 8, 159*n*
Phillips Petroleum Co. v. Bivins, ch. 7, 77*n*, 95*n*, 98*n*
Phillips Petroleum Co. v. Peterson, ch. 7, 32*n*
Pickens v. Railroad Commission, 266–267, 268, 292, 293, 294, 302; ch. 6, 72*n*; ch. 7, 115*n*; ch. 8, 138*n*
Pickens v. Ryan Consolidated Petroleum Corp., ch. 7, 5*n*
Pinchback v. Gulf Oil Corp., ch. 7, 58*n*, 95*n*
Port Acres Case. *See* Halbouty v. Railroad Commission
Pritchett v. Forest Oil Co., ch. 7, 64*n*

Quinn v. Pere Marquette Railway Co., ch. 7, 8*n*

Railroad Commission v. Aluminum Co. of America (Tex. Civ. App.-Austin), ch. 8, 76*n*, 159*n*
Railroad Commission v. Aluminum Co. of America (Tex. Sup. Ct.), 127; ch. 8, 52*n*, 73*n*, 95*n*
Railroad Commission v. Continental Oil Co., 278–279
Railroad Commission v. Bass, ch. 4, 97*n*
Railroad Commission v. Coleman, ch. 4, 217*n*; ch. 8, 79*n*
Railroad Commission v. Fain, ch. 8, 4*n*, 40*n*, 70*n*
Railroad Commission v. Flour Bluff Oil Corp., ch. 5, 32*n*; ch. 8, 123*n*
Railroad Commission v. Graford Oil Co., ch. 7, 115*n*; ch. 8, 66*n*, 79*n*; ch. 10, 48*n*, 55*n*
Railroad Commission v. Humble Oil & Refining Co. (Tex. Civ. App.-Austin, 1946), 106–107, 289–290, 293; ch. 4, 211*n*; ch. 10, 103*n*
Railroad Commission v. Humble Oil & Refining Co. (Tex. Civ. App.-Austin, 1968), ch. 8, 110*n*
Railroad Commission v. Humble Oil & Refining Co. (U.S. Sup. Ct.), ch. 4, 110*n*
Railroad Commission v. Lone Star Gas Co., ch. 8, 35*n*
Railroad Commission v. Mackhank Petroleum Co., ch. 8, 94*n*
Railroad Commission v. Manziel, 237–238, 241, 242, 244, 245, 246, 248, 256–257, 258, 271, 293–294, 300, 302, 328; ch. 7, 115*n*, 122*n*
Railroad Commission v. Marrs, ch. 4, 122*n*
Railroad Commission v. Mote Resources, ch. 10, 51*n*
Railroad Commission v. Permian Basin Pipeline Co., ch. 8, 44*n*, 58*n*, 138*n*
Railroad Commission v. Rowan Oil Co., 105, 108, 274–275, 279–280, 300; ch. 5, 36*n*, 37*n*; ch. 8, 77*n*
Railroad Commission v. Rowan & Nichols Oil Co., 105, 108, 291; ch. 3, 145*n*
Railroad Commission v. Sample, 275; ch. 8, 77*n*
Railroad Commission v. Shell Oil Co. (Tex. Sup. Ct., 1942), ch. 8, 24*n*, 80*n*, 107*n*
Railroad Commission v. Shell Oil Co. (Tex. Sup. Ct., 1947), ch. 5, 23*n*, 38*n*; ch. 8, 6*n*, 120*n*
Railroad Commission v. Shell Oil Co. (Tex. Sup. Ct., 1964), ch. 4, 204*n*
Railroad Commission v. Sterling Oil & Refining Co., ch. 5, 29*n*; ch. 8, 121*n*
Railroad Commission v. Williams, ch. 8, 110*n*, 111*n*
Railroad Commission v. Woods Exploration & Producing Co., 276, 279–280, 281; ch. 8, 158*n*
Rainwater v. Mason, ch. 7, 76*n*
Reed v. Texas Co., ch. 7, 163*n*, 166*n*
Rhoads Drilling Co. v. Allred Co., ch. 7, 87*n*
Roberts v. Superior Oil Co., 248; ch. 7, 42*n*, 74*n*
Robinson v. Robbins Petroleum Corp., 251–252; ch. 7, 140*n*, 158*n*, 181*n*
Rose v. Damm, ch. 7, 8*n*
Rowan & Nichols Oil Co. v. Railroad Commission, 105, 108, 291; ch. 4, 101*n*
Rudman v. Railroad Commission, ch. 8, 66*n*, 79*n*
Ryan Consolidated Petroleum Corp. v. Pickens, 203–204, 213, 256; ch. 4, 194*n*, 213*n*; ch. 5, 89*n*

Sauder v. Frye, ch. 7, 68*n*
Schlipf v. Exxon Corp., ch. 8, 127*n*
Schlittler v. Smith, ch. 7, 23*n*
In re Shailer's Estate, ch. 7, 87*n*
Shaw & Estes v. Texas Consolidated Oils, ch. 7, 37*n*
Shell Oil v. Railroad Commission, ch. 7, 140*n*
Shell Oil Co. v. Stansbury, 227, 228
Sivert v. Continental Oil Co., ch. 7, 88*n*; ch. 8, 127*n*
Smith v. Killough, ch. 7, 65*n*
Smith Petroleum Co. v. Van Mourik, ch. 7, 83*n*
Sneed v. Phillips Petroleum Co., ch. 3, 158*n*
Southern Louisiana Area Rate Cases v. Federal Power Commission, ch. 4, 223*n*
Southern Motor Carriers Rate Conference, Inc. v. United States, 309; ch. 8, 161*n*
Southland Royalty Co. v. Humble Oil & Refining Co., ch. 7, 18*n*
Spurlock v. Hinton, ch. 7, 95*n*
Standard Oil Co. v. Railroad Commission, 126; ch. 4, 122*n*; ch. 5, 69*n*
Standard Oil Co. of New Jersey v. United States, ch. 10, 60*n*
Standard Oil Co. of Texas v. Railroad Commission, ch. 8, 78*n*, 94*n*
Stanolind Oil & Gas Co. v. Wimberly, ch. 7, 185*n*
Staples v. Railroad Commission, 300
State v. Humble Oil & Refining Co., ch. 3, 82*n*
Stephens County v. Mid-Kansas Oil & Gas Co., ch. 7, 2*n*
Stewart v. Humble Oil & Refining Co., ch. 8, 110*n*
Sun Oil Co. v. Whitaker, 249–251; ch. 4, 67*n*
Superior Oil Co. v. Roberts, 210–215, 245, 256
Susanoil, Inc. v. Continental Oil Co. (Tex. Civ. App.-San Antonio, 1973), ch. 8, 127*n*
Susanoil, Inc. v. Continental Oil Co. (Tex. Civ. App.-San Antonio, 1975), ch. 8, 127*n*
Syverson v. North Dakota State Industrial Commission, ch. 7, 123*n*, 166*n*

Texaco v. Letterman, 216
Texaco Inc. v. Railroad Commission, 282–283, 286; ch. 5, 83*n*
Texas & Pacific Coal & Oil Co. v. Barker, ch. 7, 77*n*
Texas Pacific Coal & Oil Co. v. Kirtley, ch. 7, 43*n*
Texas Co. v. Daugherty, ch. 7, 1*n*
Texas Oil & Gas Corp. v. Ostrom, ch. 7, 30*n*
Texas Panhandle Gas Co. v. Thompson, ch. 3, 169*n*
Texoma Natural Gas Co. v. Railroad Commission, ch. 3, 153*n*
Texoma Natural Gas Co. v. Terrell, ch. 3, 153*n*
Thompson v. Consolidated Gas Utilities Corp., ch. 2, 29*n*; ch. 3, 171*n*; ch. 8, 44*n*
Tidewater Associated Oil Co. v. Stott, 226, 227, 228–230, 257; ch. 7, 123*n*
Tidewater Oil Co. v. Jackson, 239–241; ch. 7, 162*n*, 170*n*
Tiller v. Fields, 215
Town of China Valley v. State Land Dept., ch. 10, 66*n*
Trapp v. Shell Oil Co., ch. 4, 121*n*; ch. 7, 140*n*; ch. 8, 90*n*
Turner v. Big Lake Oil Co., ch. 7, 150*n*

Union Pacific Railroad Co. v. Oil & Gas Conservation Commission, ch. 5, 86*n*

United States v. Cotton Valley Operators Committee, ch. 8, 142*n*
United States v. Exxon Corp., ch. 1, 3*n*
United States v. Socony-Vacuum Oil Co., ch. 3, 109*n*
United States v. Standard Oil Company, ch. 8, 140*n*
United States Steel Corp. v. Whitley, ch. 7, 99*n*

Veal v. Thomason, 207–209

Waggoner Estate v. Sigler, ch. 7, 77*n*, 116*n*
Waseco Chemical & Supply Co. v. Bayou State Oil Co., ch. 7, 87*n*
Waters v. Bruner, ch. 7, 9*n*, 95*n*
West Edmond Hunton Lime Unit v. Lillard, ch. 7, 160*n*; ch. 10, 37*n*
West Edmond Salt Water Disposal Association v. Rosencrans, ch. 10, 37*n*
Westbrook v. Atlantic Richfield Co., ch. 7, 46*n*, 73*n*
Western Gulf Oil Co. v. Superior Oil Co., ch. 8, 12*n*
Whelan v. Placid Oil Co., ch. 6, 33*n*; ch. 7, 39*n*, 140*n*
Williams v. Humble Oil & Refining Co., ch. 7, 97*n*
Windsor Gas Corp. v. Railroad Commission, ch. 8, 79*n*
Woods Exploration and Producing Co. v. Aluminum Co. of America (S. D. Tex.), ch. 4, 224*n*; ch. 8, 144*n*
Woods Exploration and Producing Co. v. Aluminum Co. of America (Tex. Civ. App.), ch. 8, 139*n*, 144*n*
Woods Exploration and Producing Co. v. Aluminum Co. of America (5th Cir.), 305, 307, 310; ch. 7, 140*n*; ch. 8, 54*n*
Woolley v. Railroad Commission, ch. 4, 103*n*; ch. 7, 96*n*; ch. 8, 109*n*
Wrather v. Humble Oil & Refining Co., ch. 8, 24*n*, 108*n*

Yelderman v. McCarthy, ch. 7, 76*n*
Young v. Ethyl Corp., ch. 7, 160*n*

Zimmerman v. Texaco, Inc., ch. 7, 51*n*, 147*n*; ch. 8, 138*n*

Index

Adam, C. K., 328
Administrative law, 346–347
Administrative Procedure Act of *1946,* 143
Administrative Procedure and Texas Register Act of *1976* (APTRA), 288
  Substantial evidence rule and, 289, 301
Agua Dulce condensate field
  No-flare order, 139–140, 141, 143
  Pressure maintenance order, 159, 163
Allowables, field
  False nominations, 307
  "Living," 105
  Per-well, 31, 98, 103, 106, 285
  See also Bonus allowables; Lease allowable system; Special allowable field rules; Transfer allowables
Altair field, unitization agreement, 213
Amerada-Hess, 321
American Bar Association, 37
American Institute of Mining and Metallurgical Engineers, 37
American Petroleum Institute (API), 37
  Definition of secondary recovery operations, 85–86, 175
  And unitization, 32, 48, 101
  And worldwide prorationing, 55
Amoco Production Company, protest against Phillips Petroleum unitization agreement, ch. 6, 68*n*
Andector (Ellenburger) field, South Fault Block, ch. 6, 68*n*
Anti-Market Demand Prorationing Act, *1931*
  Article 6014(g) and, 160
  Committee hearings: on equitable prorationing, 45; on Railroad Commission incompetence, 43–45; on threat of oil monopoly, 40–43, on unitization, 45–49
  Oil price stabilization and, 50
  Railroad Commission waste prevention authority under, 39, 40, 48, 49, 50
  Unitization and, 38
Antitrust laws
  Federal, 302; Appling field owners violation of, 305–306; Cotton Valley agreements violation of, 79, 304; state action exemption from, 308
  State: concerns over unitization violation of, 32; exemptions from, 303
  Texas, 51–52, 302; cooperative agreements violation of, 79; major oil company East Texas violation of, 42; Market Demand Act to reinforce, 61–62; revision *1983,* 303, 304, 352 (ch. 4, 72*n*); voluntary unitization act immunity from, 77, 96, 174, 303, 304
Anton-Irish field, unitization, 148, 158
API. *See* American Petroleum Institute
Appling field
  Conflict between large- and small-tract gas producers, 276, 277
  Gas prorationing formula, 127
  Surface-acreage formula, 127
  Violation of Sherman Antitrust Act, 305
  See also Table of Cases, Railroad Commission v. Woods Exploration and Producing Co.

APTRA. *See* Administrative Procedure and Texas Register Act of *1976*
Arab oil embargo, 123, 338 (ch. 1, 2*n*)
Arkansas Oil and Gas Commission, ch. 6, 75*n*
Article 6014(g)
  Applied to gas cycling, 141–142
  Doctrine of equal coercion and, 197
  Judicial review of, 264–265, 287, 312
  Proposed removal of anti-unitization provision from, 350–351
  Railroad Commission interpretation of, 159–164, 197
  Waste provision, 137, 162, 359–360
Article 6029 (4), 272

Baker, Hines, ch. 5, 34*n*
Bammel gas field, 273, 280, 287. *See also Table of Cases,* Corzelius v. Harrell
Bentsen, Lloyd, ch. 10, 42*n*
Bonus allowables, 153
  East Texas, 155–156, 157, 158, 332
  Request for supervision of, 282–283
Boonesville (Bend Conglomerate Gas) field, suspension of prorationing in, 343–346
Briscoe, Governor Dolph, 123
Bryan (Woodbine) field
  Cost of development, 326
  Effect on oil and gas demand, 347
  Emergency no-flare order, 325
  Need for unitization, 324
  Reduction of gas-oil ratio, 326
Bryan (Woodbine) Operators Committee (BWOC), ch. 10, 17*n*
BTA Oil Producers, 269, 270
Burns, L. T. Estate, 321
Business and Commerce Code, Texas, 95
BWOC. *See* Bryan (Woodbine) Operators Committee

California
  Defeat of compulsory unitization bill, ch. 4, 182*n*
  Limited compulsory unitization law, ch. 1, 6*n*; ch. 2, 1*n*; agreements under, ch. 4, 12*n*
Carbon dioxide miscible flooding, EOR, 18
Casinghead gas
  No-flare order for, 78, 142–148, 196
  Prohibited marketing of, 196–197
  Recycling, 78
  Waste prevention of, 287
Cayuga field, 140
Central Prorationing Committee (CPC), 43, 44
Chemical flooding, EOR, 18
Clearfork field, ch. 6, 26*n*
Cochran, T. B., III, ch. 6, 9*n*
Cole Committee, 80
Collateral estoppel, 239, 240
Colorado Oil and Gas Conservation Commission, no-flare order of, 265–266
"Commission called" hearing, ch. 5, 63*n*
Common law doctrines
  Approach to unitization, 4, 201–209, 310
  Compulsory unitization to remove litigation problems under, 258
  Conservation practices and, 261
  See also Cotenancy; Cross-conveyancing theory; Implied covenants; Rule of capture; Surface estate liability; Tort liability
Common Purchaser Act, 69, 73
  Enforcement, 41
  Provisions, 56, 59–60 (ch. 3, 67*n*)
Community lease law, 205
Condensate fields
  Exemption from market demand prorationing, 165
  Gas cycling operations for, 19, 24, 72, 78 (ch. 4, 44*n*)
  Texas Gulf Coast, 33
  Voluntary unitization act on processing, 85
  Waste prevention in recovery operations: legislation on, 139; Railroad Commission order on, 139–142
Connally Act, ch. 3, 145*n*
Conoco-Driscoll field, 278
Conroe field
  Pressure maintenance, 151
  Unitization, 158, 339 (ch. 5, 45*n*; ch. 10, 1*n*)
  Waste from oil migration, 285
Conservation
  Compulsory pooling laws and, 22–23

Correlative rights and, 21, 70–71, 108, 282, 313
Defined, 6
Economic efficiency and, 6, 7
Texas: framework of law for, 134, 135; legislation, 132, 343; Railroad Commission policy for, 139–151, 159, 177, 348
During war years, 142, 158, 161
Cooperative projects, for development, 72
Act authorizing gas processing, ch. 4, 77*n*
Foundation of unitization policy, 77
Gas Conservation Act and, 78
Unitization versus, ch. 4, 6*n*
Violation of antitrust laws, 79
See also Voluntary unitization act of *1949*
Correlative rights
Conservation and, 70–71, 108, 282, 313
Defined, 21 (ch. 8, 30*n*)
Depletion gas fields, 281
Need for stability in, 284–285
Prorationing formula to protect, 282–284, 286
Protection of nonconsenters', 246–248, 311
State protection of, 22, 124 (ch. 3, 130*n*)
Unitization and, 27, 333–334
Voluntary unitization act and, 92–93, 183–194
See also Rule of capture
Cotenancy, 205
Effect on oil and gas development, 209–210
Unitization and, 210–213
Cotton Valley field
Antitrust action against, 79, 304
Cycling operations, 304 (ch. 4, 56*n*)
Council of Economic Advisers, report on market demand prorationing, ch. 10, 40*n*
CPC. *See* Central Prorationing Committee
Cross-conveyancing theory, 207–209
Effect on unitization, 255
Crude Oil Windfall Profit Tax of *1980*, ch. 10, 33*n*
Culberson, Commissioner Olin, 141 (ch. 5, 86*n*, 93*n*)
Damages, tort action
Measure of trespass, 243–244, 245
Danciger, Joe, ch. 3, 22*n*
Declaration of intent. *See* Article 6014(g)
Defense Department, requested oil reserve capacity, ch. 4, 145*n*
Depletion allowance, 337 (ch. 4, 90*n*; ch. 10, 33*n*)
Dissolved gas drive well, 10, 12, 14
Doctrine of equal coercion, 150, 187
Replacement of Article 6014(g) with, 197
Doherty, Henry L., 1, 48
On unitization, 37 (ch. 10, 39*n*, 53*n*)
Downhole commingling, 343–345
Drainage
Gravity, 12
Implied covenant to protect against, 222–225, 228; offset wells to prevent, 49; toward small tracts, 106 (ch. 4, 103*n*)
Drilling
Employment in, 112, 113
Excessive: compulsory pooling to reduce costs of, 124; economic waste of, 82, 270; to qualify as secondary recovery operation, 308
Rule 37 exception to save costs of, ch. 8, 35*n*
Dry gas fields, 19, 24
Benefits from unitizing, 26–27

East Texas field
Advocates of unitization in, 46–49
Bias toward independent producers in: Article 6014(g), 104; exceptions to well-spacing rule, 102; judicial approval of Rule 37 exceptions, 104–109; prorationing, 103
Costs of administering, 329–333
Distribution of oil production, *1931*, 49
House Bill 311 and, 122
Humble Oil purchases in, 41, 56–57
Lease-allowable system, 170, 331–332
Legal history of, ch. 3, 69*n*
Market demand factor, 329, 330, 332
Martial law, 60
Number of wells, *1938*, 67 (ch. 3, 143*n*)
Oil production, 39, 322
Political influence, 133, 198

East Texas field (*Cont.*)
Prorationing system, 39, 42, 133, 291
Rule 37 applied to, 65–66, 326
Saltwater pollution, 154–155, 157
Special allowable rules for, 154–158
Unnecessary drilling, 67
East Texas Saltwater Disposal Company, 156
Economic efficiency
Conservation and, 6, 7
Federal energy policy and, 337
Unitization and, 27–29
Ellenburger field. *See* Andector (Ellenburger) field
Eminent domain laws, ch. 2, 38*n*
Employment, in unnecessary oil operations, 112, 113
Energy crisis, *1970*s, 1, 2, 124
Energy Policy and Conservation Act of *1976,* ch. 4, 193*n*
Enhanced oil recovery (EOR), 18, 30, 33
Excess capacity
Benefits, 8, 116
National security and, ch. 4, 145*n*
Prorationing to reduce, 98, 117
State role in encouraging, 115–116
Exploratory units, 54, 57, 338
Exxon Company, U.S.A., 177, 188, 193, 316, 331. *See also* Humble Oil and Refining Company

Fairway field, 158, 255, 348
Problems in unitizing, 151, 218, 267, 312–313
Prorationing order, 217, 292, 293, 302
See also Table of Cases, Pickens v. Railroad Commission
False nomination forecasts, 307
Farish, William S., 46, 47, 48 (ch. 3, 17*n*, 47*n*, 96*n*)
Attitude toward unitization, 53, 54
Pricing policy, ch. 3, 123*n*
On prorationing, 55
Federal government
Energy crisis and, 124
Pricing rule for stripper-well oil, 188
Role in oil industry cost-price squeeze, 115
Role in Texas energy policies, 337–340
Federal Oil Conservation Board, 37
Federal Power Commissson, 117, 142, 146, 307 (ch. 4, 223*n*; ch. 10, 41*n*)
Fig Ridge field, 282, 286, 287, 313. *See also Table of Cases,* Texaco Inc. v. Railroad Commission
Flour Bluff field, 264
No-flare order, 145, 146, 299
Foran, E. V., 44 (ch. 3, 45*n*)
Fort Chadbourne field, 148, 158
Fox, Milton, 324 (ch. 4, 188*n*, 190*n*)
On compulsory unitization, 191–192, 324
Francis Oil & Gas, Inc., 187–191 (ch. 7, 26*n*)
Frankfurter, Felix, 340 (ch. 4, 113*n*)
Free riders, defined, ch. 2, 38*n*

Gas-cap wells, 10–12, 13
Limit on production, 24, 86
Recovery efficiency, 14, 23–24
Unitization and, ch. 2, 37*n*
Voluntary unitization act restrictions on, 86–87, 91
Gas Conservation Act of 1935
Purpose, 277–278
Sections 10 and 11: application to oil fields, 275; authorization of gas prorationing, 274, 280
Section 12, monthly allowable reservoir of gas, 277
Section 21: cooperative agreements approved under, 78; limitations, 144; protection of correlative rights, 70–71, 277–278, 282; repeal, 72; and voluntary unitization, 70–71, 72, 81, 140 (ch. 7, 32*n*)
Gas cycling, 24, 87, 91
Denial of special allowables for, 91
Described, 19 (ch. 4, 44*n*)
For maximum recovery of all liquid hydrocarbons, 72, 78, 148
PAW order on, 141
Technology improvement for, 140
Tort liability from, 234–235
Gas flaring, 113, 196
Prohibition order, 78–79, 139, 158, 159; judicial review of, 263–264, 297–300

Waste prevention from prohibiting casinghead, 78, 142–148
Gas-oil ratio rule, ch. 5, 77*n*, 83*n*
Gas prorationing laws, 21
Effectiveness, 22–23
Fairness, 26
For protection of correlative rights, 26–27, 270–271
Gas recovery
From gas-cap wells, 23–24
Methods, 19
Research on secondary, 20
Unconventional, 20, 26
See also Recovery, oil and gas
Gas recycling, 78, 91
Defined, ch. 4, 44*n*
Fort Chadbourne field, 149
Panhandle, 53
For wet gas production, 273

Gas repressuring, 53, 87, 198
Defined, ch. 4, 44*n*
East Texas field, 91
Railroad Commission and, 199
SACROC field, 150
Tort liability from, 234
Gas reservoirs
Condensate, 19–20, 33
"Tight sands," 20, 26
Unitization for efficient operation of, 2–3, 26–27
Water drive, 19
General American, 329
Getty Oil Company, 326
Goldsmith field, ch. 6, 26*n*
Good-faith standard, in pooling, 216
Governor's Energy Advisory Council, postembargo policy on unitization, 123–124
Gulf-McElroy field, 289
Gulf Oil
East Texas acreage, 41
Prorationing and, ch. 3, 120*n*

Halbouty, Michel T., 163, 346
Hardwicke, Robert E., 127, 278
Opposition to compulsory unitization, 80, 89
Support for prorationing, ch. 3, 69*n*
Harrison, Dan, ch. 3, 12*n*, 96*n*
Hastings field, 222 (ch. 10, 7*n*)
Hawkins field, 122, 293
Production, 322
Surface-acreage formula, 106, 134 (ch. 4, 103*n*)
Unitization, 193, 339
See also Table of Cases, Railroad Commission v. Humble Oil and Refining Company
Henze field, ch. 8, 66*n*
Hepburn Act, ch. 3, 71*n*
Heyser field, ch. 5, 45*n*
No-flare order, 145, 298
Hill, Attorney General, ch. 4, 186*n*
Holmes, R. C., ch. 3, 26*n*, 27*n*, 47*n*
House Bill 25. *See* Anti-Market Demand Prorationing Act
House Bill 67, *1947,* on unitization agreements, 81–82, 87–88, 96, 97, (ch. 4, 61*n*)
House Bill 266
Prohibition of waste, 139
Protection of correlative rights in gas fields, 147
House Bill 311, *1973,* 324, 338
Correlative rights and, 282
Defeat, 122–123
Unitization provisions, 118–121
House Bill 388, *1929,* 54
House Bill 782, *1935,* 139
Hughes and Hughes, 270, 281
Humble Oil and Refining Company, 50
Anti-Market Demand Prorationing Act and, 61
Block-leasing policy, 53, 350
Early history, 51–52
East Texas acreage, 41, 56–57
On economic waste, 40
Efforts at antitrust immunity, 53–54
Efforts to correct common pool problems, 58–59
Experiments with pressure maintenance, 53
Gas repressuring operations, 54, 149
Lease-allowable system and, 169
On no-flare orders, 145–146
Pipeline operations, 52, 57, 58
Price cuts, 56, 59 (ch. 4, 110*n*)
Support for East Texas unitization, 46–48, 54, 57–58
Support for prorationing, 55, 56, 59, 60

Ideology, and compulsory unitization opposition, 100
Implied covenants doctrine
Actions causing breach of, 232–234 (ch. 7, 84*n*)
Cycling operations damage and, 226–227, 229–230
Drainage and, 222–225, 228
Drilling and, 220–221
Effect on pooling, 221
Effect on unitization, 221–222, 253, 254
In oil and gas leases, 219–220
Imputed stripper-well exemption regulation, 190
Independent oil and gas producers
Access to pipelines, 59–60
Compulsory pooling bill and, 129, 130
East Texas benefits to, 102, 103, 104–109
Exploration and development expenditures, 116
Exploratory drilling by, 84, 112
Financial contribution to Railroad Commission election, ch. 4, 133*n*
Humble Oil competition and, 58
Mistrust of major oil companies, 31–32, 110
Opposition to unitization, 49, 109; Self-interested reasons for, 112–113
Political power, 59, 60, 67, 112–113, 345
Prorationing and, 41–42, 49, 78
Rule of capture and, 49–50, 63
Safeguards under House Bill 311, 119, 121
Support for unitization bill, 116
Independent Petroleum Association, 55
Indispensable parties rule, 207–208
Injection wells, 175–176, 193
Rule 46 on, ch. 6, 14*n*
In situ combustion, 88. *See also* Recovery, oil and gas, tertiary
Interior Department, 83
Interstate Oil Compact Commission (IOCC)
Definition of secondary recovery operations, 175
Model conservation act, ch. 3, 130*n*
Model voluntary unitization proposal, 81, 87, 88, 96
Report on unitization agreements under antitrust law, ch. 4, 10*n*
Study of state conservation laws, ch. 4, 91*n*

Jester, Beauford, ch. 5, 22*n*, 91*n*
Joinder rule
Compulsory unitization to solve problem of, 209
Cross-conveyancing theory and, 207–208
Justice Department, 79

Kansas, court decision involving tort liability, 239–241
Katy gas field, unitization, 141
Kelly-Snyder (SACROC) field, 158 (ch. 2, 46*n*)
Carbon dioxide tertiary recovery project, 180
Described, 149–150
Pressure maintenance program, 165, 166
Production, 317, 322
Railroad Commission orders for, 187, 246 (ch. 7, 155*n*)
Special allowables for, 151–152, 153
Voluntary unitization agreement, 180, 182, 191, 195
Kettleman Hills (North Dome) field, ch. 4, 12*n*
Partial unitization, ch. 10, 32*n*

Lake Creek condensate field, pressure maintenance, 141, 159 (ch. 5, 86*n*)
Langdon, Commissioner, 128, 171 (ch. 4, 193*n*; ch. 5, 126*n*)
Large-tract owners
Conflict with small-tract owners over gas prorationing, 276–277
Oil drainage toward small-tract owners, 104–105, 106
Yardstick for joining unitization agreement, 306
See also Major oil companies
Lease-allowable system, 169–170, 198 (ch. 6, 26*n*)

Royalty owners and, ch. 6, 31*n*
Transfers, 193, 276–277, 331–332
Leases, oil and gas, 112
Implied covenants doctrine applied to, 219–220
Pooling clauses, 215–219
Libecap, Gary D., 319 (ch. 9, 22*n*)
Liquid hydrocarbons
Importance during World War II, 143, 161
Separation from gas, 19
Voluntary unitization act on processing, 85
Waste in extracting, 69
Long Lake condensate gas field, 226
Louisiana, compulsory unitization law, 116
Applicable to condensate gas fields, *1940,* 37, 91 (ch. 4, 12*n*)
Applicable to oil and gas, *1960,* 91–92
Oil produced under, 316
Poolwide, ch. 7, 159*n*
Louisiana Conservation Commission, 304

McKnight, Peyton, 122
Major oil companies
Campaign for worldwide prorationing, 55
Efforts to purchase East Texas acreage, 41
Exploration and development expenditures, 116
Monopoly of Texas oil, 40–41
Pipeline monopoly, 41
Rule of capture and, 49
Unitization and, 48; compulsory, 101, 323–324; number of units under, 320–322; Railroad Commission approval for, 174–175
See also Large-tract owners
Mandatory Oil Import Program (MOIP), ch. 4, 90*n*
Marathon Oil Company, 188, 320 (ch. 5, 70*n*)
Marginal Well Act of *1931*
Amendment, *1933,* 66, 67
Constitutionality, ch. 4, 124*n*
Effects, 132
Judicial approval of, 105
Prorationing system, 103; exemption from, 103–104, 132; restrictions on non-exempt wells, 152
Provisions, 64–65
Marginal wells
Barrels of oil produced by, 66
Defined, 64 (ch. 4, 108*n*)
Fraudulent classification of, ch. 4, 104*n*
Prorationing, 103; exemptions from, 103–104, 132
See also Stripper wells
Market demand factor (MDF), 141
East Texas field, 329, 330, 332
Market Demand Prorationing Act, *1932*
Prorationing policy, 61, 63; for gas, 69
Reinforcement of antitrust policy, 61–62 (ch. 4, 71*n*)
As waste prevention measure, 62
Marketing, cooperative oil and gas, 84–85
Maximum efficient rate of production (MER)
Applications for reduced, 164–165
Described, 14
East Texas field, 329
Law authorizing production at, ch. 4, 193*n*; ch. 10, 41*n*
Prorationing, 22, 75, 348; and economic efficiency, ch. 2, 34*n*; yardstick for well-spacing versus, ch. 5, 111*n*
Railroad Commission estimates of, 141
Wartime petroleum demands and, 78
Yates field, 151, 188, 189–190
MDF. *See* Market demand factor
Medina, Michael, 189, 190 (ch. 7, 49*n*)
MER. *See* Maximum efficient rate of production
Merrill, Maurice, 208, 209
Meyers, Charles J., 208
Midcontinent Oil and Gas Association, 37
Mineral Interest Pooling Act, *1965,* 128, 132, 169, 199, 214, 270
Effects, 130–131
Provisions, 129–130
State lands excluded from, ch. 4, 215*n*
Mineral Leasing Act, 83
Miscible displacement, 88, 175
Mississippi, compulsory unitization law, 116
Mobil Oil, 327–328

MOIP. *See* Mandatory Oil Import Program
Monopoly
 Defined, ch. 2, 38*n*
 Texas crude oil, 40–43
 See also Antitrust laws
Montana, coal industry boom, ch. 4, 145*n*
Murray, William, 143, 151, 159, 165, 168
 On compulsory unitization, 113–115, 181
 On conflict between waste prevention and correlative rights protection, 333
 "Doctrine of equal coercion," 187
 "Permissive unitization," 150, 180
 On SACROC agreement, 191

National Guard, 60
National Industrial Recovery Act (NIRA), ch. 3, 109*n*, 145*n*
Natural gas. *See* listings under Gas
Nebraska
 Compulsory unitization law, 247
 Court decisions involving tort liability, 244, 245–246
Neff, Commissioner Pat M., 42 (ch. 3, 32*n*, 39*n*)
NIRA. *See* National Industrial Recovery Act
No-flare order. *See* Gas flaring
Nonapportionment rule, 204–205, 254–255
No-pit rule, 167, 170–172
Normanna field
 Drainage from neighboring tracts, 125
 Prorationing formula, 292
 See also Table of Cases, Atlantic Refining v. Railroad Commission
North Texas oil, Humble Oil purchases, 56
Nuisance
 Liability for, 241–242
 Water flood project as, 240

O'Connor, Tom, field, ch. 5, 71*n*
Office of Technology Assessment, 339
Ogallala water, ch. 4, 67*n*
Oil and gas companies. *See* Independent oil and gas producers; Large-tract owners; Major oil companies; Small-tract owners
Oil prices, ch. 1, 8*n*
 Excess production capacity and, 99
 Increase, *1971–73,* 117
 Post–World War II changes in, 98
 Proposed government support for, 7
 Rule of capture and, 27–28
Oil recovery
 Efficiency in: conservation and, 20; controlled rate for, 13–14, 15; pressure maintenance for, 14, 16; well location and, 14
 Enhanced, 18, 30, 33
 Excess capacity, 98, 99; unitization and, 109–110
 Government policies to encourage, 7
 Number of Texas fields for, *1985,* 322
 Operations, 9–10
 Stripper operations for, 16
 See also Maximum efficient rate of production; Recovery, oil and gas; Secondary oil recovery
Oil reservoirs, 7
 Dissolved gas drive, 10, 14
 Gas-cap drive, 10–12, 14, 23–24, 86–87, 91
 Range of error in value estimates of, ch. 2, 40*n*
 Structural advantages in, 23, 101
 Unitization for efficient operation of, 25, 26
 Water drive, 12–13, 14, 23
Oklahoma
 Compulsory unitization law, 37, 388 (ch. 4, 12*n*; ch. 9, 29*n*); attempt to repeal, ch. 6, 56*n*; consideration of structural advantage, ch. 4, 96*n*; effect on Texas law, 90; exemption of old reservoirs from, 116; litigation over, 89–90; restrictions on exploratory activities, 83
 Oil production under secondary recovery projects, 316
Oklahoma Corporation Commission, 316
Old Ocean field
 Pooling permit, 296
 Unitization, 148
Olney field, pressure maintenance, 53

Panhandle gas field, 71
  Described, 68
  Gas Conservation Act and, 69–70
  Law of *1899* restricting production, 68
  Physical waste in, 263
Parker, R. D., ch. 3, 17*n*, 135*n*
Patent Club, 42
PAW. *See* Petroleum Administration for War
Peden, Brooks, 165 (ch. 5, 76*n*, 93*n*; ch. 6, 38*n*)
Pegasus field, unitization, 148, 158
Penn, Robert R., ch. 3, 57*n*
Permissive Bill, *1929*, ch. 3, 58*n*
Petroleum Administration for War (PAW), 78, 141, 143, 161, 337
Phillips Petroleum Company, and Andector field unitization agreement, ch. 6, 68*n*
Pipeline cases, ch. 3, 71*n*
Pipelines
  Gas, 140; Panhandle, 68–69, 71; prorationing, 73
  Oil: Humble Oil Company, 52, 57, 58, 59; independents' efforts to insure access to, 59–60; monopoly, 41, 50
Political influence
  East Texas, 133, 170, 198
  Independent producers, 59, 60, 67, 112, 133, 170
Pollution
  Restrictions on oil industry, 99
  Saltwater, 154–155, 234–235, 251–252, 340
Pooling
  Community lease and, 205
  Compulsory, 21–22; effectiveness, 22, 23, 24; objective, 124; Texas, 8
  Defined, 7
  Nonexecutive interests and, 206–207, 310
  Oil and gas lease clauses on, 215; court's strict construction policy toward, 218–219; good-faith restraint in, 216; limitation on production, 218; limitation on size of unit, 217
  Rule 37 and, 104, 131, 204, 296–297
  Unitization versus, 7–8
  Voluntary unitization act and, 88–89, 95
  See also Mineral Interest Pooling Act
Port Acres field
  Denial of no-waste order for, 162–166
  Drainage, 125
  Drilling, 126
  See also Table of Cases Halbouty v. Railroad Commission
Pressure maintenance operation
  Described, 14, 16
  Difficulty in negotiating for, 318–319
  Field orders on, 159–164
  Humble Oil experiments on, 53
  Negative rule of capture to encourage, 242–243
  Risk, 30
  Secondary recovery operations versus, 16
  Special allowable rules to induce, 154, 156–158
  Texas fields, 33–34
  Voluntary unitization act on, 85
Prindle, David E., 341
Profitable obstructionism
  To compulsory unitization, 101, 122; East Texas incentive for, 102, 103, 104–109
  Defined, 30
  To voluntary unitization, 29–30
Prorationing system
  Conservation through, 21, 22
  Defined, ch. 2, 26*n*
  Gas allocation formula for: judicial review on, 125, 262–263, 267, 274, 276–280; Panhandle field, 69, 72–74; problems arising from, ch. 8, 33*n*; to protect correlative rights, 274, 334
  Market demand, ch. 3, 9*n*; advocates of, 42, 49; criticism of, 41–42, 44, 49 (ch. 10, 40*n*); defined, 40, 75 (ch. 2, 26*n*); efforts at statewide and worldwide, 55, 56; prices under, 98, 110; problems in controlling, 43–45; results of, 98; unitization versus, 45–49, 134; as waste prevention measure, 62
  Maximum efficient rate of production, 22, 75, 348 (ch. 2, 34*n*; ch. 5, 111*n*)
  Oil allocation formulas: judicial review of, 125, 262–263, 266–268, 269; per-well basis for, 31, 64, 98, 103, 106, 285, 301; to protect correlative

Oil allocation formulas (*Cont.*)
rights, 275, 282–286; special allowable under, 127–128, 198; surface-acreage basis for, 103–104, 106, 127, 268, 289; well potential for, 65–67; yardstick allowable, *1947*, ch. 4, 81*n*
Raisin, 308
Rule of capture versus, 22, 28
See also Market Demand Prorationing Act

Quintana Petroleum Corporation, 321
Pseudo-secondary recovery project, 176–177, 180, 181, 335 (ch. 6, 58*n*)

Railroad Commission
Administration of oil and gas conservation laws, 3
Administration of Rule 37 exception requests, 102, 103, 104, 131, 204
Administrative discretion over field rules, 137–138, 158–166, 196–197
Authority over gas use, 68–69, 71
Authority over tort liability, 238–239
Authority under Anti-Market Demand Prorationing Act, 38–39, 40, 48–49, 50, 117; judicial review of, 262–263
Administrative discretion over orders on unitization, 138, 166; allowable yardsticks, 167–169; effectiveness, 172: on lease-allowable system, 169–170; on no-pit rule, 171; on temporary spacing, 167
Authority under voluntary unitization act, 86, 87, 88, 90, 304; for control over agreements, 93–94, 96, 97; to insure protection of private rights, 92; judicial review of, 300–301; to prevent waste, 92
Formation, 52
Gas prorationing in Panhandle field, 72–74
Hearings on regulatory problems, ch. 5, 63*n*
Interpretation of Article 6014(g), 159–164
Interpretation of Marginal Wells Act, 103
Investigation of condensate gas field, 140–141
No-flare orders, 78–79, 139, 158, 159; judicial review of, 263–264
Nonpolicy toward compulsory unitization, 341–342
As political entity, 346
Pooling authority, 128
Position on small-tract drilling, 102, 103
Protection of correlative rights, 214–215; judicial interpretation of, 272–273, 274–275; prorationing formula for, 283–286; waste prevention and, 273–274
Regulation of freshwater use, 94
Requirements for agreement approvals, 363–364
Voluntary unitization act and: final orders for approval, 195; hearings on, 177–179; number of agreements approved, 317, 369; promotion of agreements, 4–5, 173, 199, 202, 315; secondary recovery agreements, 175–177
See also Waste, Railroad Commission authority over
Raisin-prorationing system, 308
Rangely field, Colorado, 266
Ranger oil field, ch. 2, 46*n*
Recovery, oil and gas
Controls, 21–22
Secondary: negative rule of capture to encourage, 242–243; "pseudo," 176, 177, 180, 181, 335 (ch. 6, 58*n*); public benefits of, 243, 272; Railroad Commission interpretation of, 175–176; Voluntary unitization act and, 85–86, 88, 175
Technology, 179–180
Tertiary, 3, 88, 180
Unitization effect on, 2–3, 35, 99
See also Gas recovery; Maximum efficient rate of production; Oil recovery; Secondary oil recovery
Repressuring. *See* Gas repressuring
Reservoirs. *See* Gas reservoirs; Oil reservoirs
Res judicata. *See* Collateral estoppel
Restatement of Torts, ch. 7, 152*n*

Rostow, Eugene, 79
Royalty owners
  Compulsory pooling bill and, 130
  Lease-allowable system and, ch. 6, 31*n*
  Opposition to unitization, 30, 31; self-interested reasons for, 111–112
  Safeguards under House Bill 311, 121
  Unitization applications and, 177, 178–179
  Voluntary unitization act and, 95, 96
Rule against perpetuities, 209
Rule of capture
  Defined, 21
  Effects of, 27, 201
  Hostility to unitization, 201, 253–254
  Incompatibility with conservation, 23, 34–35
  Judicial review of, 203–205
  Justification for, 203
  Limits on economic efficiency, 27–28
  Negative, 242–243
  Proposed ownership concept to replace, 49
  Well-potential prorationing formula and, 65
  See also Correlative rights
Rule of indispensable parties, 207–208
Rule of nonapportionment, 204–205
  Effect on unitization efforts, 254–255
Rule 8 "no-pit" order, 167, 170–172
Rule 37
  Applications under, 367–368
  Applied to East Texas field, 65–66, 326
  Exception requests: court role in policing, 127, 295, 296–297; to protect correlative rights, 286; Railroad Commission response to, 204, 294–295; for saltwater injection wells location, 293; for well-spacing, 102, 103; wells drilled under, 122, 125, 224–225, 313
  Pooling and, 104, 131, 296–297
  Recourse against drainage, ch. 4, 103*n*
  Temporary spacing rule, 167
  Waste prevention and, 268–272, 281
Rule 38, on temporary well-spacing, 167
Rule 46, on injection wells, ch. 6, 14*n*
Rule 49, gas-oil ratio, ch. 5, 77*n*, 83*n*
Rule 49(b), limits on gas-cap well production, 86–87 (ch. 2, 37*n*)
SACROC field. *See* Kelly-Snyder (SACROC) field
Sadler, Jerry, ch. 4, 110*n*
Saltwater
  Pollution: liability to surface estate for, 251–252; tort liability for, 234–235, 340
  Storage pits, 170
Saltwater injection systems
  Bonus allowables for, 155–157
  Ownership, 154–155
  Purpose, 157
Sandfracing, 235–236
Secondary oil recovery
  Described, 16–17
  Drawbacks, 17–18
  Oil produced from, 315–316
  Profitability, 3
  Property rights and, 24–25
  Risk, 30
  Texas fields, 33–34
  Trespass and, 311
  Unnecessary drilling to qualify for, 308
Seeligson field, ch. 5, 45*n*
  No-flare order for, 78, 142, 143–144, 298
Senate Bill 257, 344–345
Shafter Lake Devonian field, pressure maintenance program, 162
Shell Oil, East Texas acreage, 41
Sherman Antitrust Act, 79, 304, 305–306, 308
"Show cause" hearing, 150
Sinclair, H. F., ch. 3, 47*n*
Slaughter field, 322
  Partial unitization, 319–320
Small-tract owners
  Gas prorationing for greater allowables, 276–277
  "Living allowables" for, 105
  Prorationing formula for, 102–103, 105–106, 273
  Rule 37 exception for, 104–108, 305
  Special allowable field rules for, 127–128
  Voluntary subdivisions, 102
  Yardstick for joining unitization agreements, 306
Special allowable field rules, 151–152, 195
  Applied to small-tract owners, 127–128
  Bonus, 153, 155–156, 157, 158

Special allowable field rules (*Cont.*)
  Capacity allowables, 152–153
  East Texas field, 154–158
Spraberry field, ch. 5, 45*n*
  Casinghead gas flaring, 146
  Prorationing of production, 348
  Size of unitized projects, 320
  See also Table of Cases, Railroad Commission v. Rowan Oil Co.
Standard Oil Company of New Jersey, 48, 51, 55 (ch. 4, 110*n*; ch. 10, 59*n*)
State action doctrine, 307–310, 311
States
  Conservation measures, 21–24
  Interpretation of "police power" to regulate, ch. 2, 29*n*
  Regulation of oil and gas companies, 112–113, 115
  Regulation of unitization agreements, 309
Statewide orders on unitization, Texas, 138
  Allowable yardsticks, *1965* and *1966,* 167–169
  Effectiveness, 172
  Lease-allowable system, 169–170 (ch. 6, 26*n*)
  Rule 38 no-pit order, 170–172
  Spacing rules, *1962,* 167
Sterling, Ross, ch. 3, 75*n*
  On market demand prorationing, 43
  Martial law in East Texas, 60
  Receipt of advance royalties from Humble Oil, 44
  Request for conservation act, 39
Stewart, Richard B., 346
Stripper wells, 16
  Cost of shutting down, ch. 3, 144*n*
  Exemption from market demand prorationing, 132
  Imputed stripper well exemption regulation for, 190
  Number of, ch. 2, 12*n*
  Price of oil from, 188
Structural advantage, reservoir, 23, 101
Substantial evidence rule
  Anti-waste orders under, 147, 164
  Judicial review of Railroad Commission decisions under, 107, 135–136, 288–293, 301
Suez Canal crisis, 197, 338 (ch. 4, 89*n*)
Sugarland field, 46
  Pressure maintenance experiments, 53
Surface estate liability, 248
  Mineral estate owner's dominance in use of, 249–253

Tax shelter, oil industry, 99, 337
Teagle, Walter, 54
Technology, new recovery, 78, 179–180
Tennessee, compulsory unitization law, ch. 1, 5*n*
Terrell, C. V., ch. 3, 28*n*, 39*n*
Texas
  Antitrust law, *1889,* 42, 51–52, 79, 302; revisions, 303, 304, 352
  Compulsory pooling act, 8, 38
  Compulsory unitization and: benefits of, 349; failure to enact, 98, 100; need for, 33–34; potential for, 349; statutes against, 54
  Conservation laws: and unitization, 39; and rule of capture, 70
  Courts' promotion of voluntary unitization, 4–5, 312–314, 315
  Energy policy, ch. 4, 192*n*; federal involvement in, 337–340
  Oil production, 316
  Voluntary unitization, 3, 4, 38; *see also* Voluntary unitization act
Texas Attorney General opinions
  On Anti-Market Demand Prorationing Act, ch. 3, 64*n*
  On House Bill 67, 81–82, 96, 97 (ch. 4, 61*n*)
  On House Bill 311, ch. 4, 186*n*
  On Marginal Wells Act, 65 (ch. 4, 108*n*)
  No. 2932 on oil production levels, ch. 3, 129*n*; ch. 8, 49*n*
  On Seeligson field cooperative development, 144–145
Texas Common Purchaser Act, 307
Texas Free Enterprise and Antitrust Act, 303, 304, 352
Texas Governor's Energy Advisory Council, 34
Texas Independent Producers and Royalty Owners Association (TIPRO)
  On compulsory unitization, 100, 198 (ch. 4, 171*n*)
  Draft of compulsory pooling bill, ch. 4, 214*n*
  Proposed reduction in East Texas field MDF, 330, 332

Support for House Bill 311, 118, 122
"Watchdog groups," 118
Texas Natural Resources Code
See also Voluntary unitization act
Section 85.046. *See* Article 6014(g)
Section 85.056, on Railroad Commission power to protect public, ch. 4, 223*n*
Section 86.181, on gas uses, ch. 5, 4*n*
Sections 101.001–101.018, 361–365 (ch. 4, 71*n*)
Section 101.013, on Railroad Commission approval of agreements, 363–364 (ch. 7, 113*n*)
Sections 101.051–101.052, ch. 4, 76*n*
Sections 103.041–103.046, ch. 4, 77*n*
Statewide orders, ch. 5, 3*n*
Texas Oil and Gas Association, 61, 271
Texas Supreme Court
Decisions encouraging unitization, 221–223
Decisions on prorationing, 105–107, 125, 344
Injunction against sandfracing, 235
Interpretation of Railroad Commission authority: on gas proration, 274; on no-flare orders, 297–300; under substantial evidence rule, 107, 135–136, 288–292, 298; on waste prevention, 287, 297
Texas 2000 Commission Report, 34 (ch. 4, 192*n*)
Texas Water Code, 94
Thermal processes, EOR, *in situ,* 18
Thompson, Ernest O., 135, 139 (ch. 5, 30*n*, 91*n*)
On oil imports, ch. 4, 144*n*
On prorationing, 61, 63
Protection of independents, 113
TIPRO. *See* Texas Independent Producers and Royalty Owners Association
Tort liability
Effect of compulsory versus voluntary unitization on, ch. 7, 159*n*
Measure of trespass damages, 243–244, 245
Nonconsenters, 239–241, 244, 246–248, 306
Unitized operations leading to, 234–238, 253
Trans-Alaskan pipeline, 117
Transfer allowables, 151–152, 195
Controls over, 193
Gas prorationing and, 276–277
Off-lease, 331–332
Trespass, 25
Measure of damages for, 243–244, 245
Saltwater invasion liability under, 235, 237–238
Sandfracing as, 235–236
Secondary oil recovery and, 311
Unitization to avoid, 26

Unconventional gas recovery (UGR), 20, 26
Underground Natural Gas Storage and Conservation Act, ch. 10, 58*n*
Unitization
Common law approach to, 4, 201–209
Compulsory: concerns over income decline from, 111; potential benefits to Texas from, 349; drilling cost savings from, 329; elimination of implied covenant and tort liability problems under, 258; ideological campaign against, 100; legal deterrents to, 111; major oil companies on, 323–324; mistrust of, 110; pride of ownership versus, 110; profitable obstructionism to, 101–110; Texas failure to enact, 98, 100. *See also* House Bill 311
Defined, 1–2
East Texas field: advocates of, 46–49; independents' opposition to, 49–50
Economic efficiency from, 27–29
Elimination of conservation/correlative rights conflict with, 26–27
Formulas, 25 (ch. 4, 96*n*)
Oil produced under, 315–316
Partial, 319–320; deficiencies in, 326–327, 335–336
Permissive, 150, 180
Phased, ch. 4, 127*n*
Pooling versus, 7–8
Results of Texas approach to, 326–327, 334–335
Voluntary: antitrust laws and, 32, 302, 303, 308; cost of operations under, 33; court's role in encouraging 4–5, 312–314, 315; Gas Conservation Act on, 70–71; mistrust of, 31–32; multiple party ownership and, 32;

Voluntary (*Cont.*)
number of agreements approved and denied, 317, 369; pride of ownership versus, 31; problems in negotiating agreements, 318–319; profitability, 3; profitable obstructionism to, 29–30; prorationing system versus, 30–31; protection for nonsigners of agreements, 193; recommended legislation to encourage, 355–357; royalty owners and, 30, 31; special field orders for, 158–166; timeliness, 318–319; tort liability relating to, 234–238, 253. *See also* House Bill 67; Voluntary Unitization Act, *1949*

Van field, exploratory unitization agreement, 54, 57
Voluntary unitization act, *1949,* 77–78, 158
Antitrust immunity from, 132, 174, 303, 309
Factors leading to, 79–80
Hearings on, 177–179
Judicial review of, 300–301
Louisiana unitization law versus, 91–92
Proposed replacement of, 352–353
Role in Yates field unitization, 342
Section 101.002, right to enter into unitization agreements, 96; Railroad Commission interpretation of, 174–175
Section 101.003, reaffirmation of *1932* antitrust policy, ch. 4, 71*n*
Section 101.004, antitrust immunity, 95–96, 195
Section 101.011: approval for needed cooperative facilities, 176; authorization of joint gas processing agreements, 85; authorization of secondary recovery, 85–86, 88, 175–176; limits on gas-cap drive, 86–87; limits on pooling agreements, 88–89; proposed revisions in, 354, 355; restrictions on exploratory units, 82–84; restrictions on *in situ* combustion or tertiary recovery, 88; restrictions on marketing, 84–85
Section 101.012, refutation of coercive power, 95
Section 101.013: effect on East Texas field, 91; limits on areas covered by agreements, 181–183; limits on cost of recovery operations, 176–177, 179; limits on pooling agreements, 88–89, 174; proposed revisions in, 354–355; protection of correlative rights, 92–93, 183–184; provisions copied from Oklahoma statute, 90; required findings prior to unitization, 177–183, 214
Section 101.014: approval for joint operating agreements, 96–97; pooling agreements, ch. 6, 13*n*
Section 101.015, Railroad Commission control over unitization agreements, 93–94, 194
Section 101.016, permissable provisions, 94
Section 101.017: authorization of joint gas processing agreements, 85; Railroad Commission control over unitization agreements, 194
Section 101.018, protection of those refusing to enter agreements, 95
Shortcomings, 351–352
See also House Bill 67; Unitization, voluntary

Walker, A. W., 160
Wallace, Mack, 357–358
Wasson field, production, 322
Waste
Article 6014(g) on, 162, 359–360
Defined, 270
Economic, 40, 49; from casinghead gas flaring, ch. 5, 33*n*; from overdrilling, 82; public concern over, 272; Railroad Commission and, 262–263, 269; of raisins, 308
Field orders preventing: in casinghead gas fields, 78, 142–148; in condensate gas fields, 138–142; by improved recovery practices,

148–151; by special allowable rules, 151–158
Physical, 40; compulsory unitization to prevent, 338; public concern over, 272; Railroad Commission and, 50, 262
Railroad Commission authority over, 117, 137, 273–274, 312, 348; Anti-Market Demand Prorationing Act on, 38–39, 48–49, 50, 58; House Bill 388 on, 54; judicial review of, 262–263, 265, 269, 287; Voluntary unitization act on, 158
Substitutes for unitization in preventing, 134
Water, fresh
Disputes over use, 94
Secondary recovery operations use of, 249–251
See also Saltwater
Water drive reservoirs, ch. 4, 50*n*
East Texas, 33
Gas, 19
Oil, 12–13, 14, 23
Water flooding, ch. 4, 50*n*
Capacity allowables for, 152–153, 197–198
Railroad Commission approval for, 175, 180
Tort liability from saltwater invasion in, 234–235, 239, 251–252 (ch. 7, 162*n*)
Use of fresh water, 250–251
Waters-Pierce Company, 52
Webster field
Prorationing formulas, 193
Unitization, 339
Wells. *See* Drilling; Injection wells; Marginal wells; Stripper wells; Well-spacing laws
Well-spacing laws, 21
Allowable yardsticks for, 167–169
Effectiveness, 23
"Eight times area," rules for, ch. 4, 103*n*
Judicial review of, 105–107
Temporary, 167
Wartime, 78
See also Rule 37
West Texas, Humble Oil pipelines system, 47
West Virginia, compulsory unitization law, ch. 1, 5*n*
Wiggins, Steven N., 319 (ch. 9, 22*n*)
Williams, Howard R., 208
Working interest owners, 30, 177
Number required for unitization application, 178–179
Rights under House Bill 311, 119, 120, 121
World War II
Acceptance of unitization concept, 78
Allocation of scarce materials, 337
Oil and gas conservation, 142
Oil and gas prices, 78, 161
Pooling, ch. 7, 76*n*
Regulations to prevent waste, 158, 161
Wyoming, compulsory unitization law, ch. 1, 5*n*; ch. 10, 29*n*

Yardsticks
Elimination of incentives to overdrill from, 168
For joining unitization agreements, 306
In the voluntary unitization act, 184–185
Well-spacing, 167–169
Yates field, 57, 122 (ch. 3, 97*n*; ch. 4, 122*n*)
Efficiency in operations, 46–47
Production, 322; by Francis Oil & Gas, Inc., 187, 189, 191, (ch. 7, 26*n*)
Request for increase in MER allowables, 151, 188, 189–190, 199 (ch. 6, 75*n*)
Unitization, 151, 191, 339, 342 (ch. 8, 78*n*)
See also Table of Cases, Francis Oil & Gas Co. v. Exxon Corp.
Young, Fred, 328 (ch. 4, 143*n*; ch. 5, 48*n*)

For Product Safety Concerns and Information please contact our EU representative GPSR@taylorandfrancis.com
Taylor & Francis Verlag GmbH, Kaufingerstraße 24, 80331 München, Germany

www.ingramcontent.com/pod-product-compliance
Ingram Content Group UK Ltd.
Pitfield, Milton Keynes, MK11 3LW, UK
UKHW021836240425
457818UK00006B/217

* 9 7 8 1 6 1 7 2 6 0 2 4 7 *